	% Vol	% wt
O_2	20.95	23.08
N_2	78.09	75.58
Inert gas	0.94	1.28
CO_2	0.03	0.03

Jessey Mayer

55P

THE ENGINEERS' MANUAL

THE ENGINEERS' MANUAL

By

RALPH G. HUDSON, S. B.,

Emeritus Professor of Electrical Engineering

Massachusetts Institute of Technology

SECOND EDITION

Twenty-Second Printing

JOHN WILEY & SONS, Inc.

New York London

SECOND EDITION

Twenty-Second Printing, January, 1963

Printed in U. S. A.

PREFACE TO SECOND EDITION.

THIS work originated from the conception that the practicing engineer or engineering student would welcome a consolidation of the formulas and constants for which he is accustomed to search through several volumes and that the application of each formula might be explained more concisely than in texts devoted exclusively to the process of derivation. With this end in view those engineering formulas, mathematical operations and tables of constants which appear to be most useful are presented in systematic order and in a size of book designed to fit the pocket.

Each formula is preceded by a statement in which its application, the symbology of the involved physical quantities and definite units of measurement are indicated. It is believed that this method of presentation increases the speed of selection and understanding of a desired formula and insures greater accuracy of substitution since data units of any kind may be converted into specified units by reference to the table of conversion factors. The sequence of the formulas is based generally upon their order of derivation so that the understanding of a formula may be enlarged by inspection of the formulas which precede it. All catchwords, symbols and formulas are printed in full face type and each formula or group of formulas is numbered to facilitate reference to the text or cross reference between formulas.

For the practicing engineer the aim throughout has been to enable him to obtain results quickly and accurately even in a branch of engineering to which he can give little attention. For instructional purposes the object has been to present a summary of the important relations which may be derived from fundamental principles so that the student may give his undivided attention to the sources of engineering knowledge, the evolution of engineering formulas and their applications. It is suggested that class room exercises devoted to the derivation of the stated formulas be given to increase the student's comprehension of the origin of his working formulas and of the mathematical operations which intervene as well as to create discrimination between those relations which are fundamental, derived and empirical. In the solution of problems original data may be given in terms of units not specified in the formulas and for conditions not definitely prescribed in the text.

Many changes and additions were made in each printing of the first edition. After several printings substantial improvements were made in all parts of the book. In this (the second) edition the entire chapter on Heat and a large part of the chapter on Electricity have been rewritten and brought up to date. The author wishes to express his obligations to Professor C. L. Svenson and Mr. A. E. Fitzgerald for their co-operation in this work.

The second edition also contains revisions and extensions of all tables of physical constants, new steam tables, recomputations of all conversion factors affected by the latest definition of the British Thermal Unit, an enlarged table of conversion factors, and many additions throughout the book.

The author wishes to express again his appreciation for the assistance rendered by Professor H. B. Luther, Professor Dean Peabody, and the late Doctor Joseph Lipka in the preparation of the first edition.

RALPH G. HUDSON.

CAMBRIDGE, MASS.
January, 1939

MATHEMATICS

ALGEBRA

1 Powers and Roots

$a^n = a \cdot a \cdot a \cdot \ldots$ to n factors. $\qquad a^{-n} = \frac{1}{a^n}$.

$a^m \cdot a^n = a^{m+n}$; $\frac{a^m}{a^n} = a^{m-n}$. $\qquad (ab)^n = a^n b^n$; $\left(\frac{a}{b}\right)^n = \frac{a^n}{b^n}$.

$(a^m)^n = (a^n)^m = a^{mn}$. $\qquad (\sqrt[n]{a})^n = a$.

$a^{\frac{1}{n}} = \sqrt[n]{a}$; $\qquad a^{\frac{m}{n}} = \sqrt[n]{a^m}$. $\qquad \sqrt[n]{ab} = \sqrt[n]{a}\sqrt[n]{b}$; $\sqrt[n]{\frac{a}{b}} = \frac{\sqrt[n]{a}}{\sqrt[n]{b}}$.

$\sqrt[n]{\sqrt[m]{a}} = \sqrt[mn]{a}$.

2 Operations with Zero and Infinity

$a \cdot 0 = 0$; $a \cdot \infty = \infty$; $0 \cdot \infty$ is indeterminate, see page 37.

$\frac{0}{a} = 0$; $\frac{a}{0} = \infty$; $\qquad \frac{0}{0}$ " " " " 37.

$\frac{\infty}{a} = \infty$; $\frac{a}{\infty} = 0$; $\qquad \frac{\infty}{\infty}$ " " " " 37.

$a^0 = 1$; $\quad 0^a = 0$; $\qquad 0^0$ " " " " 37.

$\infty^a = \infty$; $\qquad \infty^0$ " " " " 37.

$a^\infty = \infty$, if $a^2 > 1$; $a^\infty = 0$, if $a^2 < 1$; $a^\infty = 1$, if $a^2 = 1$, see also page 37.

$a^{-\infty} = 0$, if $a^2 > 1$; $a^{-\infty} = \infty$, if $a^2 < 1$; $a^{-\infty} = 1$, if $a^2 = 1$, see also page 37.

$a - a = 0$; $\infty - a = \infty$; $\infty - \infty$ is indeterminate, page 37.

3 Binomial Expansions

$(a \pm b)^2 = a^2 \pm 2\,ab + b^2$.

$(a \pm b)^3 = a^3 \pm 3\,a^2b + 3\,ab^2 \pm b^3$.

$(a \pm b)^4 = a^4 \pm 4\,a^3b + 6\,a^2b^2 \pm 4\,ab^3 + b^4$.

$$(a \pm b)^n = a^n \pm \frac{n}{1}a^{n-1}b + \frac{n(n-1)}{1 \cdot 2}a^{n-2}b^2 \pm \frac{n(n-1)(n-2)}{1 \cdot 2 \cdot 3}a^{n-3}b^3 + \ldots$$

NOTE. n may be positive or negative, integral or fractional. When n is a positive integer, the series has $(n + 1)$ terms; otherwise the number of terms is infinite.

4 Polynomial Expansions

$(a+b+c+d+\dots)^2 = a^2+b^2+c^2+d^2+\dots+2a(b+c+d+\dots) + 2b(c+d+\dots)+2c(d+\dots)+\dots$

= sum of the squares of each term and twice the product of each term by the sum of the terms that follow it.

$(a+b+c)^3 = [(a+b)+c]^3 = (a+b)^3 + 3(a+b)^2c + 3(a+b)c^2 + c^3.$

5 Factors

$a^2 - b^2 = (a+b)(a-b).$

$a^2 + b^2 = (a+b\sqrt{-1})(a-b\sqrt{-1}).$

$a^3 - b^3 = (a-b)(a^2+ab+b^2).$

$a^3 + b^3 = (a+b)(a^2-ab+b^2).$

$a^4 + b^4 = (a^2+ab\sqrt{2}+b^2)(a^2-ab\sqrt{2}+b^2).$

$a^{2n} - b^{2n} = (a^n+b^n)(a^n-b^n).$

$a^n - b^n = (a-b)(a^{n-1}+a^{n-2}b+a^{n-3}b^2+\dots+b^{n-1}).$

$a^n - b^n = (a+b)(a^{n-1}-a^{n-2}b+a^{n-3}b^2-\dots-b^{n-1})$ if n is even.

$a^n + b^n = (a+b)(a^{n-1}-a^{n-2}b+a^{n-3}b^2-\dots+b^{n-1})$ if n is odd.

6 Ratio and Proportion

If $a:b = c:d$, or $\frac{a}{b}=\frac{c}{d}$, or $ad = bc$, then

$$\frac{b}{a}=\frac{d}{c}; \qquad \frac{a}{c}=\frac{b}{d}.$$

$$\frac{a\pm b}{c\pm d}=\frac{a}{c}=\frac{b}{d}; \qquad \frac{a\pm c}{b\pm d}=\frac{a}{b}=\frac{c}{d}.$$

$$\frac{a+b}{a-b}=\frac{c+d}{c-d}; \qquad \frac{a+c}{a-c}=\frac{b+d}{b-d}.$$

$$\frac{ma}{mb}=\frac{nc}{nd}; \qquad \frac{ma}{nb}=\frac{mc}{nd}.$$

$$\frac{a^n}{b^n}=\frac{c^n}{d^n}; \qquad \frac{\sqrt[n]{a}}{\sqrt[n]{b}}=\frac{\sqrt[n]{c}}{\sqrt[n]{d}}; \qquad \frac{a^{\frac{m}{n}}}{b^{\frac{m}{n}}}=\frac{c^{\frac{m}{n}}}{d^{\frac{m}{n}}}.$$

If $\frac{a}{b}=\frac{c}{d}=\frac{e}{f}=\dots$, then

$$\frac{a}{b}=\frac{c}{d}=\frac{e}{f}=\dots=\frac{a+c+e+\dots}{b+d+f+\dots}=\frac{pa+qc+re+\dots}{pb+qd+rf+\dots}.$$

If $\frac{a}{b}=\frac{c}{d}$ and $\frac{e}{f}=\frac{g}{h}$, then $\frac{ae}{bf}=\frac{cg}{dh}$.

7 Constant Factor of Proportionality, *k*

If $y = kx$, y varies as x, or y is proportional to x.

If $y = \frac{k}{x}$, y varies inversely as x, or y is inversely proportional to x.

If $y = kxz$, y varies jointly as x and z.

If $y = k\frac{x}{z}$, y varies directly as x and inversely as z.

8 Logarithms

(a) Definition. If b is a finite positive number, other than 1, and $b^x = N$, then x is the logarithm of N to the base b, or $\log_b N = x$. If $\log_b N = x$, then $b^x = N$.

(b) Properties of logarithms.

$$\log_b b = 1;\ \log_b 1 = 0;\ \log_b 0 = \begin{cases} +\infty, & \text{when } b \text{ lies between 0 and 1} \\ -\infty, & \text{when } b \text{ lies between 1 and } \infty \end{cases}$$

$$\log_b M \cdot N = \log_b M + \log_b N. \qquad \log_b \frac{M}{N} = \log_b M - \log_b N.$$

$$\log_b N^p = p \log_b N. \qquad \log_b \sqrt[r]{N^p} = \frac{p}{r} \log_b N.$$

$$\log_b N = \frac{\log_a N}{\log_a b}. \qquad \log_b b^N = N;\ b^{\log_b N} = N.$$

(c) Systems of logarithms.

Common (Briggsian) — base 10.

Natural (Napierian or hyperbolic) — base 2.7183 –, (designated by e or ϵ).

NOTE. The abbreviation of "common logarithm" is "log" and the abbreviation of "natural logarithm is "ln."

(d) Characteristic or integral part (c) of the common logarithm of a number (N).

If N is not less than one, c equals the number of integral figures in N, minus one.

If N is less than one, c equals 9 minus the number of zeros between the decimal point and the first significant figure, minus 10 (the -10 being written after the mantissa).

(e) Mantissa or decimal part (m) of the common logarithm of a number N.

If N has not more than three figures, find mantissa directly in table, page 250.

If N has four figures, $m = m_1 + \frac{f}{10}\left(m_2 - m_1\right)$, where m_1 is the mantissa corresponding to the first three figures of N, m_2 is the next larger mantissa in the table and f is the fourth figure of N.

(f) Number (N) corresponding to a common logarithm which has a characteristic (c) and a mantissa (m).

If N is desired to three figures, find the mantissa nearest to m in the table, page 250, and the corresponding number is N.

If N is desired to four figures, find the next smaller mantissa, m_1, and the next larger mantissa, m_2, in the table. The first three figures of N correspond to m_1 and the fourth figure equals the nearest whole number to $10\left(\frac{m - m_1}{m_2 - m_1}\right)$.

NOTE. If c is positive, the number of integral figures in N equals c plus one. If c is negative (for example, $9 - 10$ or -1), write numeric c minus one zeros between the decimal point and the first significant figure of N.

(g) Natural logarithm (ln) of a number (N).

Any number, N, can be written $N = N_1 \times 10^{\pm p}$, where N_1 lies between 1 and 1000. Then $\ln N = \ln N_1 \pm p \ln 10$.

If N_1 has not more than three figures, find $\ln N_1$ directly in table, page 252.

If N_1 has four figures, N_2 is the number composed of the first three figures of N_1, and f is the fourth figure of N_1, then

$$\ln N_1 = \ln N_2 + \frac{f}{10}[\ln (N_2 + 1) - \ln N_2].$$

(h) Number (N) corresponding to a natural logarithm, ln N.

Any logarithm, $\ln N$, can be written $\ln N = \ln N_1 \pm p \ln 10$, where $\ln N_1$ lies between $4.6052 = \ln 100$ and $6.9078 = \ln 1000$. Then $N = N_1 \times 10^{\pm p}$.

The first three figures of N_1 correspond to the next smaller logarithm, $\ln N_2$, in the table, and the fourth figure, f, of N_1 equals the nearest whole number to $10\left(\frac{\ln N_1 - \ln N_2}{\ln (N_2 + 1) - \ln N_2}\right)$.

9 The Solution of Algebraic Equations

(a) The quadratic equation.

If
$$ax^2 + bx + c = 0,$$
then
$$x = \frac{-b \pm \sqrt{b^2 - 4ac}}{2a} = \frac{2c}{-b \mp \sqrt{b^2 - 4ac}}$$

If $b^2 - 4ac \gtreqless 0$ $\begin{cases} \text{the roots are real and unequal,} \\ \text{the roots are real and equal,} \\ \text{the roots are imaginary.} \end{cases}$

The second equation serves best when the two values of x are nearly equal.

(b) The cubic equation.

Any cubic equation, $y^3 + py^2 + qy + r = 0$ may be reduced to the form $x^3 + ax + b = 0$ by substituting for y the value $\left(x - \frac{p}{3}\right)$. Here $a = \frac{1}{3}(3q - p^2)$, $b = \frac{1}{27}(2p^3 - 9pq + 27r)$.

Algebraic Solution of $x^3 + ax + b = 0$.

Let

$$A = \sqrt[3]{-\frac{b}{2} + \sqrt{\frac{b^2}{4} + \frac{a^3}{27}}}, \quad B = \sqrt[3]{-\frac{b}{2} - \sqrt{\frac{b^2}{4} + \frac{a^3}{27}}},$$

then

$$x = A + B, \quad -\frac{A+B}{2} + \frac{A-B}{2}\sqrt{-3}, \quad -\frac{A+B}{2} - \frac{A-B}{2}\sqrt{-3}.$$

If $\frac{b^2}{4} + \frac{a^3}{27} \gtreqless 0$ $\begin{cases} \text{1 real root, 2 conjugate imaginary roots,} \\ \text{3 real roots of which 2 are equal,} \\ \text{3 real and unequal roots.} \end{cases}$

Trigonometric Solution of $x^3 + ax + b = 0$.

In the case where $\frac{b^2}{4} + \frac{a^3}{27} < 0$, the above formulas give the roots in a form impractical for numerical computation. In this case, a is negative. Compute the value of the angle ϕ from $\cos \phi = \sqrt{\frac{b^2}{4} \div \left(-\frac{a^3}{27}\right)}$ (see page 260), then

$$x = \mp 2\sqrt{-\frac{a}{3}}\cos\frac{\phi}{3}, \mp 2\sqrt{-\frac{a}{3}}\cos\left(\frac{\phi}{3} + 120°\right), \mp 2\sqrt{-\frac{a}{3}}\cos\left(\frac{\phi}{3} + 240°\right),$$

where the upper or lower signs are to be used according as b is positive or negative.

In the case where $\frac{b^2}{4} + \frac{a^3}{27} > 0$, compute the values of the angles ψ and ϕ from $\cot 2\psi = \sqrt{\frac{b^2}{4} \div \frac{a^3}{27}}$, $\tan \phi = \sqrt[3]{\tan \psi}$; then the real root of the equation is

$$x = \pm 2 \sqrt{\frac{a}{3}} \cot 2\phi,$$

where the upper or lower sign is to be used according as **b** is positive or negative.

In the case where $\frac{b^2}{4} + \frac{a^3}{27} = 0$, the roots are

$$x = \mp 2 \sqrt{-\frac{a}{3}}, \quad \pm \sqrt{-\frac{a}{3}}, \quad \pm \sqrt{-\frac{a}{3}},$$

where the upper or lower signs are to be used according as **b** is positive or negative.

(c) **The biquadratic equation.**

Any biquadratic equation such as

$$y^4 + py^3 + qy^2 + ry + s = 0$$

may be reduced to the form

$$x^4 + ax^2 + bx + c = 0$$

by substituting for *y* the value $\left(x - \frac{p}{4}\right)$.

If $x^4 + ax^2 + bx + c = 0$, form first the cubic equation

$$t^3 + \left(\frac{a}{2}\right) t^2 + \left(\frac{a^2 - 4c}{16}\right) t - \frac{b^2}{64} = 0$$

and solve as indicated in 9 (b).

If the roots of the above cubic equation are **l**, **m**, and **n**, then the roots of the biquadratic equation are:

if **b** is *positive*,

$$x = -\sqrt{l} - \sqrt{m} - \sqrt{n}, \quad -\sqrt{l} + \sqrt{m} + \sqrt{n},$$
$$\sqrt{l} - \sqrt{m} + \sqrt{n}, \quad \sqrt{l} + \sqrt{m} - \sqrt{n};$$

if **b** is *negative*,

$$x = \sqrt{l} + \sqrt{m} + \sqrt{n}, \quad \sqrt{l} - \sqrt{m} - \sqrt{n},$$
$$-\sqrt{l} + \sqrt{m} - \sqrt{n}, \quad -\sqrt{l} - \sqrt{m} + \sqrt{n}.$$

(d) **Graphical solution of the cubic and biquadratic equations.**

To find the real roots of the cubic equation

$$x^3 + ax + b = 0,$$

draw the parabola (page 22) $y^2 = 2x$, and the circle (page 21), the coördinates of whose center are $x = \frac{4 - a}{4}$, $y = -\frac{b}{8}$, and which passes through the vertex of the parabola. Measure the ordinates of the points of intersection; these give the real roots of the equation.

To find the real roots of the biquadratic equation

$$x^4 + ax^2 + bx + c = 0,$$

draw the parabola $y^2 = 2x$, and the circle the coördinates of whose center are $x = \frac{4-a}{4}$, $y = -\frac{b}{8}$ and whose radius is $\sqrt{\left(\frac{4-a}{4}\right)^2 + \left(-\frac{b}{8}\right)^2 - \frac{c}{4}}$. Measure the ordinates of the points of intersection; these give the real roots of the equation.

NOTE. The one parabola $y^2 = 2x$ drawn on a large scale suffices for the solution of all cubic and biquadratic equations.

(e) **The binomial equation.**

If $x^n = a$, the n roots of this equation are:

if **a** is *positive*,

$$x = \sqrt[n]{a}\left(\cos\frac{2k\pi}{n} + \sqrt{-1}\sin\frac{2k\pi}{n}\right)$$

if **a** is *negative*,

$$x = \sqrt[n]{-a}\left(\cos\frac{(2k+1)\pi}{n} + \sqrt{-1}\sin\frac{(2k+1)\pi}{n}\right),$$

where k takes in succession the values $0, 1, 2, 3, \ldots, n-1$.

(f) **The general quadratic equation.**

If

$$ax^{2n} + bx^n + c = 0,$$

then

$$x^n = \frac{-b \pm \sqrt{b^2 - 4ac}}{2a},$$

and x is found as in 9 (e).

(g) **The general equation of the nth degree.**

$$P = p_0x^n + p_1x^{n-1} + p_2x^{n-2} + \ldots + p_{n-1}x + p_n = 0.$$

There are no formulas which give the roots of this general equation if $n > 4$. If $n > 4$, use one of the following methods. These are advantageous even when $n = 3$ or $n = 4$.

Method I. Roots by factors.

Find a number, r, by trial or guess such that $x = r$ satisfies the equation, that is, such that

$$p_0r^n + p_1r^{n-1} + p_2r^{n-2} + \ldots + p_{n-1}r + p_n = 0.$$

(Integer roots must be divisors of p_n.) Then $x - r$ is a factor of the left member of the equation. Divide out this factor, leaving an equation of degree one less than that of the original equation. Proceed in the same manner with the reduced equation.

Method II. Roots by approximation. (The "pinch" method.)

If for $x = a$ and $x = b$, the left member, P, of the equation has opposite signs, then a root of the equation lies between a and b. By this method the real roots may be obtained to any desired degree of accuracy. For example, let P have the signs given in the following tables:

x	. . . −2	−1	0	1	2 . . .
P	−	+	+	+	−

; roots lie between −2 and −1, between 1 and 2, ...

x	1 . . .	1.3	1.4	1.5	. . . 2
P	+	+	+	−	−

; a root lies between 1.4 and 1.5.

x	1.4 . . .	1.46	1.47 . . .	1.5
P	+	+	−	−

; a root lies between 1.46 and 1.47.

x	1.46	1.465	1.47
P	+	+	−

; a root lies between 1.465 and 1.47.

Therefore one root is $x = 1.47$ to the nearest second decimal.

10 Progressions

(a) Arithmetic progression.

$a, a + d, a + 2d, a + 3d, \ldots$, where d = common difference.

The nth term, $t_n = a + (n - 1)d$.

The sum of n terms, $S_n = \frac{n}{2}[2a + (n - 1)d] = \frac{n}{2}(a + t_n)$.

The arithmetic mean of a and b $= \frac{a + b}{2}$.

(b) Geometric progression.

$a, ar, ar^2, ar^3, \ldots$, where r = common ratio.

The nth term, $t_n = ar^{n-1}$.

The sum of n terms, $S_n = a\left(\frac{r^n - 1}{r - 1}\right) = \frac{rt_n - a}{r - 1}$.

If $r^2 < 1$, S_n approaches a definite limit as n increases indefinitely, and

$$S_\infty = \frac{a}{1 - r}.$$

The geometric mean of a and b $= \sqrt{ab}$.

Interest, Annuities, Sinking Funds

11 Amount (A_n) of a sum of money or principal (P) placed at a rate of interest (r)* for n years.

At simple interest: $A_n = P(1 + nr)$.

At interest compounded annually: $A_n = P(1 + r)^n$.

At interest compounded q times a year: $A_n = P\left(1 + \frac{r}{q}\right)^{nq}$.

12 Present value (P) of an amount (A_n) due in n years at a rate of interest (r).*

At simple interest: $P = \frac{A_n}{1 + nr}$.

At interest compounded annually: $P = \frac{A_n}{(1 + r)^n}$.

At interest compounded q times a year: $P = \frac{A_n}{\left(1 + \frac{r}{q}\right)^{nq}}$.

NOTE. The present value of an amount due in n years is the sum of money which placed at interest for n years will produce the given amount.

13 True discount (D) or the difference between the amount (A_n) due at the end of n years and its present value (P).

$$D = A_n - P.$$

14 Annuity (N) that a principal (P), drawing interest at the rate r,* will give for a period of n years.

* Expressed as a decimal.

Interest compounded annually: $N = P\dfrac{r(1+r)^n}{(1+r)^n - 1}$.

NOTE. An annuity is a fixed sum paid at regular intervals.

15 Present value (P) of an annuity (N) to be paid out for **n** consecutive years, the interest rate being **r**.*

Interest compounded annually: $P = N\dfrac{(1+r)^n - 1}{r(1+r)^n}$.

16 Amount of a sinking fund (S) created by a fixed (end of the year) investment (N) placed annually at compound interest (r)* for a term of n years.

$$S = N\frac{(1+r)^n - 1}{r}.$$

17 Fixed investment (N) placed annually at compound interest (r)* for a term of **n** years to create a sinking fund (S).

$$N = S\frac{r}{(1+r)^n - 1}.$$

TRIGONOMETRY

Definition of Angle

An angle is the amount of rotation (in a fixed plane) by which a straight line may be changed from one direction to any other direction. If the rotation is counter-clockwise the angle is said to be positive, if clockwise, negative.

Measure of Angle

A **degree** is $\frac{1}{360}$ of the plane angle about a point.

A **radian** is the angle subtended at the center of a circle by an arc equal in length to the radius.

18 Trigonometric Functions of an Angle

sine (sin) $\alpha = \dfrac{y}{r}$.

cosine (cos) $\alpha = \dfrac{x}{r}$.

tangent (tan) $\alpha = \dfrac{y}{x}$.

cotangent (cot) $\alpha = \dfrac{x}{y}$.

secant (sec) $\alpha = \dfrac{r}{x}$.

cosecant (csc) $\alpha = \dfrac{r}{y}$.

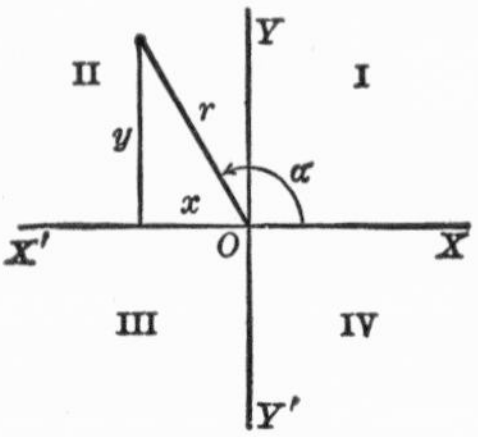

FIG. 18.

exsecant (exsec) $\alpha = \sec\alpha - 1$.

versine (vers) $\alpha = 1 - \cos\alpha$. coversine (covers) $\alpha = 1 - \sin\alpha$.

NOTE. **x** is positive when measured along **OX**, and negative, along **OX'** **y** is positive when measured parallel to **OY**, and negative, parallel to **OY'**.

* Expressed as a decimal

19 Signs of the Functions

Quadrant	sin	cos	tan	cot	sec	csc
I.	+	+	+	+	+	+
II.	+	−	−	−	−	+
III.	−	−	+	+	−	−
IV.	−	+	−	−	+	−

20 Functions of 0°, 30°, 45°, 60°, 90°, 180°, 270°, 360°

	0°	30°	45°	60°	90°	180°	270°	360°
sin.	0	$\frac{1}{2}$	$\frac{\sqrt{2}}{2}$	$\frac{\sqrt{3}}{2}$	1	0	−1	0
cos.	1	$\frac{\sqrt{3}}{2}$	$\frac{\sqrt{2}}{2}$	$\frac{1}{2}$	0	−1	0	1
tan.	0	$\frac{\sqrt{3}}{3}$	1	$\sqrt{3}$	∞	0	∞	0
cot.	∞	$\sqrt{3}$	1	$\frac{\sqrt{3}}{3}$	0	∞	0	∞
sec.	1	$\frac{2\sqrt{3}}{3}$	$\sqrt{2}$	2	∞	−1	∞	1
csc.	∞	2	$\sqrt{2}$	$\frac{2\sqrt{3}}{3}$	1	∞	−1	∞

21 Function of Angles in any Quadrant in Terms of Angles in First Quadrant

	$-\alpha$	$90° \pm \alpha$	$180° \pm \alpha$	$270° \pm \alpha$	$n(360)° \pm \alpha$
sin.	$-\sin\alpha$	$+\cos\alpha$	$\mp\sin\alpha$	$-\cos\alpha$	$\pm\sin\alpha$
cos.	$+\cos\alpha$	$\mp\sin\alpha$	$-\cos\alpha$	$\pm\sin\alpha$	$+\cos\alpha$
tan.	$-\tan\alpha$	$\mp\cot\alpha$	$\pm\tan\alpha$	$\mp\cot\alpha$	$\pm\tan\alpha$
cot.	$-\cot\alpha$	$\mp\tan\alpha$	$\pm\cot\alpha$	$\mp\tan\alpha$	$\pm\cot\alpha$
sec.	$+\sec\alpha$	$\mp\csc\alpha$	$-\sec\alpha$	$\pm\csc\alpha$	$+\sec\alpha$
csc.	$-\csc\alpha$	$+\sec\alpha$	$\mp\csc\alpha$	$-\sec\alpha$	$\pm\csc\alpha$

Note. In the last column, n = any integer.

22 Fundamental Relations Among the Functions

$\sin\alpha = \frac{1}{\csc\alpha}$; $\cos\alpha = \frac{1}{\sec\alpha}$; $\tan\alpha = \frac{1}{\cot\alpha} = \frac{\sin\alpha}{\cos\alpha}$.

$\csc\alpha = \frac{1}{\sin\alpha}$; $\sec\alpha = \frac{1}{\cos\alpha}$; $\cot\alpha = \frac{1}{\tan\alpha} = \frac{\cos\alpha}{\sin\alpha}$.

$\sin^2\alpha + \cos^2\alpha = 1$; $\sec^2\alpha - \tan^2\alpha = 1$; $\csc^2\alpha - \cot^2\alpha = 1$.

23 Functions of Multiple Angles

$\sin 2\alpha = 2\sin\alpha\cos\alpha$;
$\cos 2\alpha = 2\cos^2\alpha - 1 = 1 - 2\sin^2\alpha = \cos^2\alpha - \sin^2\alpha$.
$\sin 3\alpha = 3\sin\alpha - 4\sin^3\alpha$;
$\cos 3\alpha = 4\cos^3\alpha - 3\cos\alpha$.
$\sin 4\alpha = 4\sin\alpha\cos\alpha - 8\sin^3\alpha\cos\alpha$;
$\cos 4\alpha = 8\cos^4\alpha - 8\cos^2\alpha + 1$.
$\sin n\alpha = 2\sin(n-1)\alpha\cos\alpha - \sin(n-2)\alpha$,
$\cos n\alpha = 2\cos(n-1)\alpha\cos\alpha - \cos(n-2)\alpha$.

24 Functions of Half Angles

$\sin\frac{\alpha}{2} = \sqrt{\frac{1-\cos\alpha}{2}}$; $\cos\frac{1}{2}\alpha = \sqrt{\frac{1+\cos\alpha}{2}}$.

$\tan\frac{1}{2}\alpha = \frac{1-\cos\alpha}{\sin\alpha} = \frac{\sin\alpha}{1+\cos\alpha} = \sqrt{\frac{1-\cos\alpha}{1+\cos\alpha}}$.

25 Powers of Functions

$\sin^2\alpha = \frac{1}{2}(1 - \cos 2\alpha)$; $\cos^2\alpha = \frac{1}{2}(1 + \cos 2\alpha)$.
$\sin^3\alpha = \frac{1}{4}(3\sin\alpha - \sin 3\alpha)$; $\cos^3\alpha = \frac{1}{4}(\cos 3\alpha + 3\cos\alpha)$.
$\sin^4\alpha = \frac{1}{8}(\cos 4\alpha - 4\cos 2\alpha + 3)$; $\cos^4\alpha = \frac{1}{8}(\cos 4\alpha + 4\cos 2\alpha + 3$

$\sin^n\alpha = \frac{1}{(2\sqrt{-1})^n}\left(y - \frac{1}{y}\right)^n$; $\cos^n\alpha = \frac{1}{(2)^n}\left(y + \frac{1}{y}\right)^n$.

In the last two formulas, expand $\left(y \pm \frac{1}{y}\right)^n$ by 3 and then write $\left(y^k + \frac{1}{y^k}\right)$ $= 2\cos kx$ and $\left(y^k - \frac{1}{y^k}\right) = 2\sqrt{-1}\sin kx$.

26 Functions of Sum or Difference of Two Angles

$\sin(\alpha \pm \beta) = \sin\alpha\cos\beta \pm \cos\alpha\sin\beta$.
$\cos(\alpha \pm \beta) = \cos\alpha\cos\beta \mp \sin\alpha\sin\beta$.
$\tan(\alpha \pm \beta) = \frac{\tan\alpha \pm \tan\beta}{1 \mp \tan\alpha\tan\beta}$.

27 Sums, Differences and Products of Two Functions

$\sin\alpha \pm \sin\beta = 2\sin\frac{1}{2}(\alpha \pm \beta)\cos\frac{1}{2}(\alpha \mp \beta)$.
$\cos\alpha + \cos\beta = 2\cos\frac{1}{2}(\alpha + \beta)\cos\frac{1}{2}(\alpha - \beta)$.
$\cos\alpha - \cos\beta = -2\sin\frac{1}{2}(\alpha + \beta)\sin\frac{1}{2}(\alpha - \beta)$.

$$\tan\alpha \pm \tan\beta = \frac{\sin(\alpha \pm \beta)}{\cos\alpha\cos\beta}.$$

$$\sin^2\alpha - \sin^2\beta = \sin(\alpha + \beta)\sin(\alpha - \beta).$$

$$\cos^2\alpha - \cos^2\beta = -\sin(\alpha + \beta)\sin(\alpha - \beta).$$

$$\cos^2\alpha - \sin^2\beta = \cos(\alpha + \beta)\cos(\alpha - \beta).$$

$$\sin\alpha\sin\beta = \tfrac{1}{2}\cos(\alpha - \beta) - \tfrac{1}{2}\cos(\alpha + \beta).$$

$$\cos\alpha\cos\beta = \tfrac{1}{2}\cos(\alpha - \beta) + \tfrac{1}{2}\cos(\alpha + \beta).$$

$$\sin\alpha\cos\beta = \tfrac{1}{2}\sin(\alpha + \beta) + \tfrac{1}{2}\sin(\alpha - \beta).$$

28 Equivalent Expressions for sin α, cos α, and tan α

$$\sin\alpha = \sqrt{1 - \cos^2\alpha} = \frac{\tan\alpha}{\sqrt{1 + \tan^2\alpha}} = \frac{1}{\sqrt{1 + \cot^2\alpha}} = \frac{\sqrt{\sec^2\alpha - 1}}{\sec\alpha} = \frac{1}{\csc\alpha}$$

$$= \cos\alpha\tan\alpha = \frac{\cos\alpha}{\cot\alpha} = \frac{\tan\alpha}{\sec\alpha} = \frac{\sin 2\alpha}{2\cos\alpha} = \sqrt{\tfrac{1}{2}(1 - \cos 2\alpha)}$$

$$= 2\sin\frac{\alpha}{2}\cos\frac{\alpha}{2}.$$

$$\cos\alpha = \sqrt{1 - \sin^2\alpha} = \frac{1}{\sqrt{1 + \tan^2\alpha}} = \frac{\cot\alpha}{\sqrt{1 + \cot^2\alpha}} = \frac{1}{\sec\alpha} = \frac{\sqrt{\sec^2\alpha - 1}}{\csc\alpha}$$

$$= \sin\alpha\cot\alpha = \frac{\sin\alpha}{\tan\alpha} = \frac{\cot\alpha}{\csc\alpha} = \frac{\sin 2\alpha}{2\sin\alpha} = \sqrt{\tfrac{1}{2}(1 + \cos 2\alpha)}$$

$$= \cos^2\frac{\alpha}{2} - \sin^2\frac{\alpha}{2} = 1 - 2\sin^2\frac{\alpha}{2} = 2\cos^2\frac{\alpha}{2} - 1.$$

$$\tan\alpha = \frac{\sin\alpha}{\sqrt{1 - \sin^2\alpha}} = \frac{\sqrt{1 - \cos^2\alpha}}{\cos\alpha} = \frac{1}{\cot\alpha} = \sqrt{\sec^2\alpha - 1} = \frac{1}{\sqrt{\csc^2\alpha - 1}}$$

$$= \frac{\sin\alpha}{\cos\alpha} = \frac{\sec\alpha}{\csc\alpha} = \frac{\sin 2\alpha}{1 + \cos 2\alpha} = \frac{1 - \cos 2\alpha}{\sin 2\alpha} = \frac{2\tan\frac{\alpha}{2}}{1 - \tan^2\frac{\alpha}{2}}.$$

29 Definitions of Inverse or Anti-functions

$\mathrm{Sin}^{-1}\,a$ is defined as the angle whose sine is a. $\mathrm{Sin}^{-1}\,a$ has an infinite number of values. If α is the value of $\sin^{-1} a$ which lies between $-90°$ and $+90°\left(-\frac{\pi}{2}\text{ and }+\frac{\pi}{2}\text{ radians}\right)$, and if n is any integer,

$$\sin^{-1} a = (-1)^n\alpha + n\cdot 180° = (-1)^n\alpha + n\pi. \quad [\text{similarly for } \csc^{-1} a]$$

$\mathrm{Cos}^{-1}\,a$ is defined as the angle whose cosine is a. $\mathrm{Cos}^{-1}\,a$ has an infinite number of values. If α is the value of $\cos^{-1} a$ which lies between 0° and 180° (0 and π radians), and if n is any integer,

$$\cos^{-1} a = \pm\alpha + n\cdot 360° = \pm\alpha + 2n\pi. \quad [\text{similarly for } \sec^{-1} a]$$

$\mathrm{Tan}^{-1}\,a$ is defined as the angle whose tangent is a. $\mathrm{Tan}^{-1}\,a$ has an infinite number of values. If α is the value of $\tan^{-1} a$ which lies between 0° and 180° (0 and π radians), and if n is any integer,

$$\tan^{-1} a = \alpha + n\cdot 180° = \alpha + n\pi. \quad [\text{similarly for } \cot^{-1} a]$$

30 Some Relations Among Inverse Functions

$$\sin^{-1} a = \cos^{-1}\sqrt{1-a^2} = \tan^{-1}\frac{a}{\sqrt{1-a^2}} = \cot^{-1}\frac{\sqrt{1-a^2}}{a}$$

$$= \sec^{-1}\frac{1}{\sqrt{1-a^2}} = \csc^{-1}\frac{1}{a}.$$

$$\cos^{-1} a = \sin^{-1}\sqrt{1-a^2} = \tan^{-1}\frac{\sqrt{1-a^2}}{a} = \cot^{-1}\frac{a}{\sqrt{1-a^2}}$$

$$= \sec^{-1}\frac{1}{a} = \csc^{-1}\frac{1}{\sqrt{1-a^2}}.$$

$$\tan^{-1} a = \sin^{-1}\frac{a}{\sqrt{1+a^2}} = \cos^{-1}\frac{1}{\sqrt{1+a^2}} = \cot^{-1}\frac{1}{a} = \sec^{-1}\sqrt{1+a^2}$$

$$= \csc^{-1}\frac{\sqrt{1+a^2}}{a}.$$

$$\cot^{-1} a = \tan^{-1}\frac{1}{a};\quad \sec^{-1} a = \cos^{-1}\frac{1}{a};\quad \csc^{-1} a = \sin^{-1}\frac{1}{a}.$$

$$\text{vers}^{-1} a = \cos^{-1}(1-a);\ \text{covers}^{-1} a = \sin^{-1}(1-a);\ \text{exsec}^{-1} a = \sec^{-1}(1+a).$$

$$\sin^{-1} a \pm \sin^{-1} b = \sin^{-1}(a\sqrt{1-b^2} \pm b\sqrt{1-a^2}).$$

$$\cos^{-1} a \pm \cos^{-1} b = \cos^{-1}(ab \mp \sqrt{1-a^2}\sqrt{1-b^2}).$$

$$\tan^{-1} a \pm \tan^{-1} b = \tan^{-1}\frac{a \pm b}{1 \mp ab}.$$

$\sin^{-1} a + \cos^{-1} a = 90°$; $\tan^{-1} a + \cot^{-1} a = 90°$; $\sec^{-1} a + \csc^{-1} a = 90°$,
if $\sin^{-1} a$, $\tan^{-1} a$, $\csc^{-1} a$ lie between $-90°$ and $+90°$
and $\cos^{-1} a$, $\cot^{-1} a$, $\sec^{-1} a$ lie between $0°$ and $180°$.

31 Solution of Trigonometric Equations

By means of the relations expressed in 18 to 30 inclusive, reduce the given equation to an equation containing only a single function of a single angle. Solve the resulting equation by algebraic methods, 9, for the remaining function, and from this find the values of the angle, by 29 and table, page 278. Test all these values in the original equation and discard those which do not satisfy it.

Solution of Some Special Equations.

If $\sin\alpha = \sin\beta$, then $\alpha = (-1)^n\beta + n\cdot 180°$. (n = any integer)
If $\cos\alpha = \cos\beta$, then $\alpha = \pm\beta + 2n\cdot 180°$.
If $\tan\alpha = \tan\beta$, then $\alpha = \beta + n\cdot 180°$.
If $\cos\alpha = \sin\beta$, then $\alpha = \pm\beta \mp 90° + 2n\cdot 180°$.
If $\tan\alpha = \cot\beta$, then $\alpha = -\beta + 90° + n\cdot 180°$.
If $a\cos\alpha + b\sin\alpha = c$, and a, b, c are any numbers, and $c^2 \leqq a^2 + b^2$

then
$$\alpha = \tan^{-1}\frac{b}{a} + \cos^{-1}\frac{c}{\sqrt{a^2+b^2}}.$$

32 Properties of Plane Triangles

Notation. α, β, γ = angles; a, b, c = sides.
A = area; h_b = altitude on b; $s = \frac{1}{2}(a + b + c)$.
r = radius of inscribed circle; R = radius of circumscribed circle.

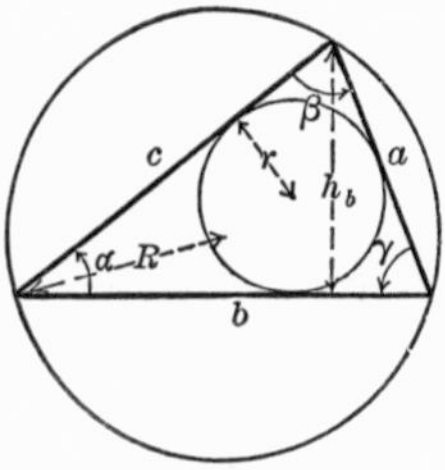

FIG. 32.

$\alpha + \beta + \gamma = 180° = \pi$ radians

$$\frac{a}{\sin\alpha} = \frac{b}{\sin\beta} = \frac{c}{\sin\gamma}.$$

$$\frac{a+b}{a-b} = \frac{\tan\frac{1}{2}(\alpha+\beta)}{\tan\frac{1}{2}(\alpha-\beta)}.^*$$

$$a^2 = b^2 + c^2 - 2\,bc\cos\alpha,^* \quad a = b\cos\gamma + c\cos\beta.^*$$

$$\cos\alpha = \frac{b^2 + c^2 - a^2}{2\,bc},^* \quad \sin\alpha = \frac{2}{bc}\sqrt{s(s-a)(s-b)(s-c)}.^*$$

$$\sin\frac{\alpha}{2} = \sqrt{\frac{(s-b)(s-c)}{bc}},^* \quad \cos\frac{\alpha}{2} = \sqrt{\frac{s(s-a)}{bc}},^*$$

$$\tan\frac{\alpha}{2} = \sqrt{\frac{(s-b)(s-c)}{s(s-a)}} = \frac{r}{s-a}.^*$$

$$h_b = c\sin\alpha^* = a\sin\gamma^* = \frac{2}{b}\sqrt{s(s-a)(s-b)(s-c)}.^*$$

$$r = \sqrt{\frac{(s-a)(s-b)(s-c)}{s}} = (s-a)\tan\frac{\alpha}{2}.^*$$

$$R = \frac{a}{2\sin\alpha}^* = \frac{abc}{4A}.$$

$$A = \tfrac{1}{2}bh_b^* = \tfrac{1}{2}ab\sin\gamma^* = \frac{a^2\sin\beta\sin\gamma}{2\sin\alpha}^* = \sqrt{s(s-a)(s-b)(s-c)} = rs.$$

33 Solution of the Right Triangle

Given any two sides, or one side and any acute angle, α, to find the remaining parts.

$$\sin\alpha = \frac{a}{c}, \quad \cos\alpha = \frac{b}{c}, \quad \tan\alpha = \frac{a}{b}, \quad \beta = 90° - \alpha.$$

$$a = \sqrt{(c+b)(c-b)} = c\sin\alpha = b\tan\alpha.$$

$$b = \sqrt{(c+a)(c-a)} = c\cos\alpha = \frac{a}{\tan\alpha}.$$

$$c = \frac{a}{\sin\alpha} = \frac{b}{\cos\alpha} = \sqrt{a^2 + b^2}.$$

$$A = \tfrac{1}{2}ab = \frac{a^2}{2\tan\alpha} = \frac{b^2\tan\alpha}{2} = \frac{c^2\sin 2\alpha}{4}.$$

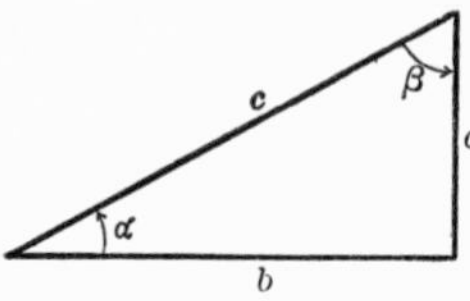

FIG. 33.

* Two more formulas may be obtained by replacing a by b, b by c, c by a, α by β, β by γ, γ by α.

34. Solution of Oblique Triangles. (For numerical work, use tables on page 278.)

I. Given any two angles α and β, and any side c.

$$\gamma = 180° - (\alpha + \beta); \quad a = \frac{c \sin \alpha}{\sin \gamma}; \quad b = \frac{c \sin \beta}{\sin \gamma}.$$

II. Given any two sides a and c, and an angle opposite one of these, say α.

$$\sin \gamma = \frac{c \sin \alpha}{a}, \quad \beta = 180° - (\alpha + \gamma), \quad b = \frac{a \sin \beta}{\sin \alpha}.$$

NOTE. γ may have two values, $\gamma_1 < 90°$ and $\gamma_2 = 180° - \gamma_1 > 90°$. If $\alpha + \gamma_2 > 180°$, use only γ_1.

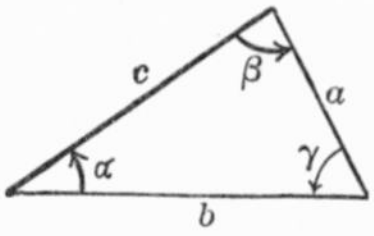

FIG. 34 (I).

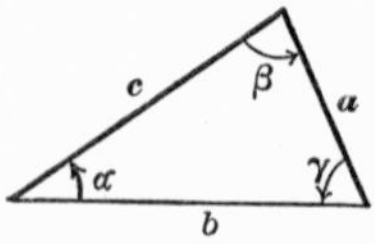

FIG. 34 (II).

III. Given any two sides b and c and their included angle α. Use any one of the following sets of formulas:

(1) $\frac{1}{2}(\beta + \gamma) = 90° - \frac{1}{2}\alpha; \quad \tan \frac{1}{2}(\beta - \gamma) = \frac{b - c}{b + c} \tan \frac{1}{2}(\beta + \gamma);$

$$\beta = \tfrac{1}{2}(\beta + \gamma) + \tfrac{1}{2}(\beta - \gamma); \quad \gamma = \tfrac{1}{2}(\beta + \gamma) - \tfrac{1}{2}(\beta - \gamma); \quad a = \frac{b \sin \alpha}{\sin \beta}.$$

(2) $a = \sqrt{b^2 + c^2 - 2\,bc \cos \alpha}; \quad \sin \beta = \frac{b \sin \alpha}{a}; \quad \gamma = 180° - (\alpha + \beta).$

(3) $\tan \gamma = \frac{c \sin \alpha}{b - c \cos \alpha}; \quad \beta = 180° - (\alpha + \gamma); \quad a = \frac{c \sin \alpha}{\sin \gamma}.$

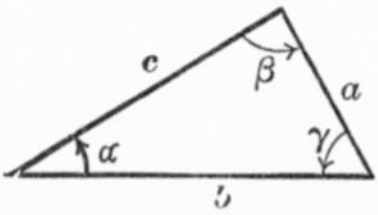

FIG. 34 (III).

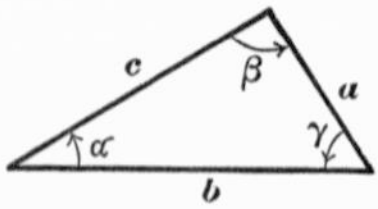

FIG. 34 (IV).

IV. Given the three sides a, b, and c.
Use either of the following sets of formulas.

(1) $s = \frac{1}{2}(a + b + c), \quad r = \sqrt{\frac{(s - a)(s - b)(s - c)}{s}}.$

$$\tan \tfrac{1}{2}\alpha = \frac{r}{s - a}, \quad \tan \tfrac{1}{2}\beta = \frac{r}{s - b}, \quad \tan \tfrac{1}{2}\gamma = \frac{r}{s - c}.$$

(2) $\cos \alpha = \frac{b^2 + c^2 - a^2}{2\,bc}, \quad \cos \beta = \frac{c^2 + a^2 - b^2}{2\,ca}, \quad \gamma = 180° - (\alpha + \beta).$

MENSURATION: LENGTHS, AREAS, VOLUMES

Notation: **a, b, c, d, s** denote lengths, **A** denotes area, **V** denotes volume.

35 Right Triangle

$A = \frac{1}{2} ab$. (For other formulas, see **33**)

$c = \sqrt{a^2 + b^2}$, $a = \sqrt{c^2 - b^2}$, $b = \sqrt{c^2 - a^2}$.

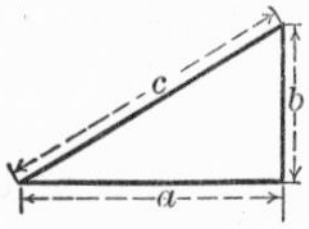

Fig. 35.

36 Oblique Triangle

$A = \frac{1}{2} bh$. (For other formulas, see **32, 34**)

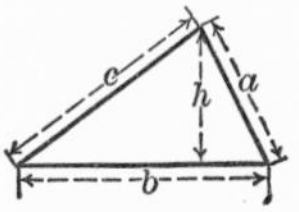

Fig. 36.

37 Equilateral Triangle

$A = \frac{1}{2} ah = \frac{1}{4} a^2\sqrt{3}$.

$h = \frac{1}{2} a \sqrt{3}$.

$r_1 = \frac{a}{2\sqrt{3}}$

$r_2 = \frac{a}{\sqrt{3}}$

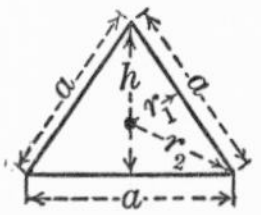

Fig. 37.

38 Square

$A = a^2$; $d = a\sqrt{2}$.

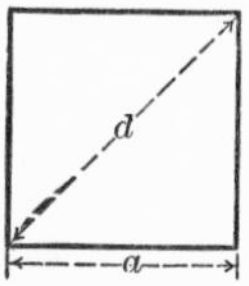

Fig. 38.

39 Rectangle

$A = ab$; $d = \sqrt{a^2 + b^2}$.

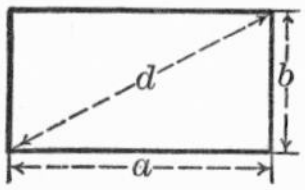

Fig. 39.

40 Parallelogram (opposite sides parallel)

$A = ah = ab \sin \alpha$.

$d_1 = \sqrt{a^2 + b^2 - 2\,ab \cos \alpha}$;

$d_2 = \sqrt{a^2 + b^2 + 2\,ab \cos \alpha}$.

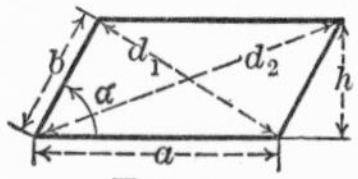

Fig. 40.

41 Trapezoid (one pair of opposite sides parallel)

$A = \frac{1}{2} h (a + b)$.

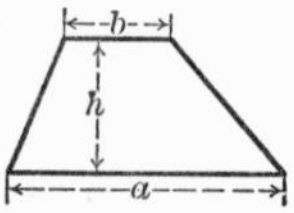

Fig. 41.

42 Isosceles Trapezoid (non-parallel sides equal)

$A = \frac{1}{2} h (a + b) = \frac{1}{2} c \sin \alpha (a + b)$
$= c \sin \alpha (a - c \cos \alpha) = c \sin \alpha (b + c \cos \alpha)$.

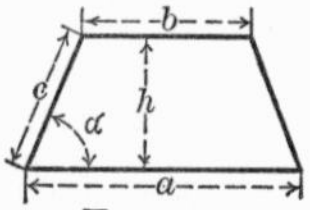

Fig. 42.

43 Trapezium (no sides parallel)

$A = \frac{1}{2}(ah_1 + bh_2)$ = sum of areas of 2 triangles.

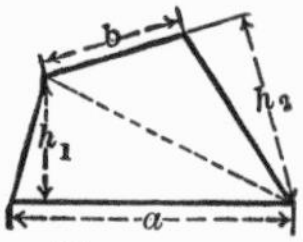

FIG. 43.

44 Regular Polygon of n Sides {all sides equal, all angles equal}

$$\beta = \frac{n-2}{n}\,180^\circ = \frac{n-2}{n}\,\pi \text{ radians.}$$

$$\alpha = \frac{360^\circ}{n} = \frac{2\pi}{n} \text{ radians.}$$

FIG. 44.

n	a	r	R	A
3	$2r\sqrt{3} = R\sqrt{3}$	$\frac{1}{6}a\sqrt{3}$	$\frac{1}{3}a\sqrt{3}$	$\frac{1}{4}a^2\sqrt{3} = 3r^2\sqrt{3} = \frac{3}{4}R^2\sqrt{3}$
4	$2r = R\sqrt{2}$	$\frac{1}{2}a$	$\frac{1}{2}a\sqrt{2}$	$a^2 = 4r^2 = 2R^2$
6	$\frac{2}{3}r\sqrt{3} = R$	$\frac{1}{2}a\sqrt{3}$	a	$\frac{3}{2}a^2\sqrt{3} = 2r^2\sqrt{3} = \frac{3}{2}R^2\sqrt{3}$
8	$2r(\sqrt{2}-1) = R\sqrt{2-\sqrt{2}}$	$\frac{1}{2}a(\sqrt{2}+1)$	$\frac{1}{2}a\sqrt{4+2\sqrt{2}}$	$2a^2(\sqrt{2}+1) = 8r^2(\sqrt{2}-1) = 2R^2\sqrt{2}$
n	$2r\tan\frac{\alpha}{2} = 2R\sin\frac{\alpha}{2}$	$\frac{a}{2}\cot\frac{\alpha}{2}$	$\frac{a}{2}\csc\frac{\alpha}{2}$	$\frac{na^2}{4}\cot\frac{\alpha}{2} = nr^2\tan\frac{\alpha}{2} = \frac{nR^2}{2}\sin\alpha$

45 Circle {C = circumference, α = central angle in radians}

$$C = \pi D = 2\pi R.$$

$$c = R\alpha = \frac{1}{2}D\alpha = D\cos^{-1}\frac{d}{R} = D\tan^{-1}\frac{l}{2d}.$$

$$l = 2\sqrt{R^2 - d^2} = 2R\sin\frac{\alpha}{2} = 2d\tan\frac{\alpha}{2} = 2d\tan\frac{c}{D}.$$

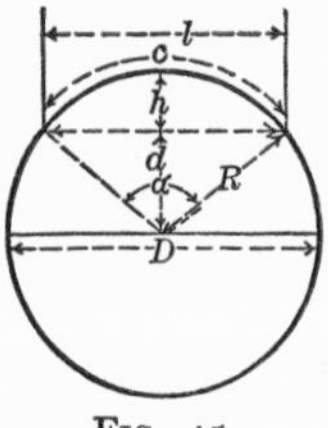

FIG. 45.

$$d = \frac{1}{2}\sqrt{4R^2 - l^2} = \frac{1}{2}\sqrt{D^2 - l^2} = R\cos\frac{\alpha}{2} = \frac{1}{2}l\cot\frac{\alpha}{2} = \frac{1}{2}l\cot\frac{c}{D}.$$

$$h = R - d.$$

$$\alpha = \frac{c}{R} = \frac{2c}{D} = 2\cos^{-1}\frac{d}{R} = 2\tan^{-1}\frac{l}{2d} = 2\sin^{-1}\frac{l}{D}.$$

$$A_{(circle)} = \pi R^2 = \frac{1}{4}\pi D^2 = \frac{1}{2}RC = \frac{1}{4}DC.$$

$$A_{(sector)} = \frac{1}{2}Rc = \frac{1}{2}R^2\alpha = \frac{1}{8}D^2\alpha.$$

$$A_{(\text{segment})} = A(\text{sector}) - A(\text{triangle}) = \tfrac{1}{2} R^2 (\alpha - \sin \alpha) = \tfrac{1}{2} R\left(c - R \sin \frac{c}{R}\right)$$

$$= R^2 \sin^{-1} \frac{l}{2R} - \tfrac{1}{4} l \sqrt{4R^2 - l^2} = R^2 \cos^{-1} \frac{d}{R} - d\sqrt{R^2 - d^2}$$

$$= R^2 \cos^{-1} \frac{R-h}{R} - (R-h)\sqrt{2Rh - h^2}.$$

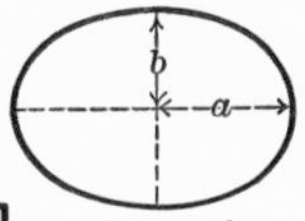

FIG. 46.

46 Ellipse * $A = \pi ab$.

Perimeter (s) =

$$\pi(a+b)\left[1 + \tfrac{1}{4}\left(\frac{a-b}{a+b}\right)^2 + \frac{1}{64}\left(\frac{a-b}{a+b}\right)^4 + \frac{1}{256}\left(\frac{a-b}{a+b}\right)^6 + \cdots\right].$$

$$\approx \pi \frac{a+b}{4}\left[3(1+\lambda) + \frac{1}{1-\lambda}\right] \qquad \lambda = \left[\frac{a-b}{2(a+b)}\right]^2$$

47 Parabola * $A = \tfrac{2}{3} ld$.

Length of arc (s) $= \tfrac{1}{2}\sqrt{16d^2 + l^2} + \dfrac{l^2}{8d} \ln\left(\dfrac{4d + \sqrt{16d^2 + l^2}}{l}\right)$

$$= l\left[1 + \tfrac{2}{3}\left(\frac{2d}{l}\right)^2 - \tfrac{2}{5}\left(\frac{2d}{l}\right)^4 + \cdots\right].$$

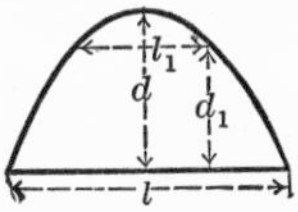

FIG. 47.

Height of segment (d_1) $= \dfrac{d}{l^2}(l^2 - l_1^2)$.

Width of segment (l_1) $= l\sqrt{\dfrac{d - d_1}{d}}$.

48 Cycloid * (r = radius of generating circle)

$A = 3\pi r^2$.

Length of arc (s) = 8 r.

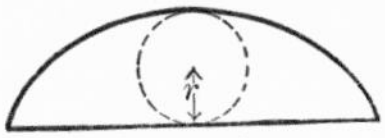

FIG. 48.

49 Catenary *

Length of arc (s) $= l\left[1 + \tfrac{2}{3}\left(\dfrac{2d}{l}\right)^2\right]$ approximately,

if **d** is small in comparison with **l**.

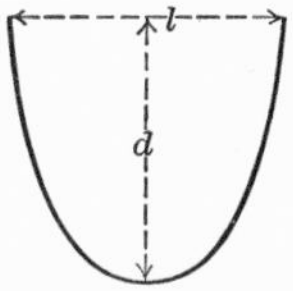

FIG. 49.

50 Area by Approximation

Let $y_0, y_1, y_2, \ldots, y_n$ be the measured lengths of a series of equidistant parallel chords, and let **h** be their distance apart, then the area enclosed by any boundary is given approximately by one of the following rules.

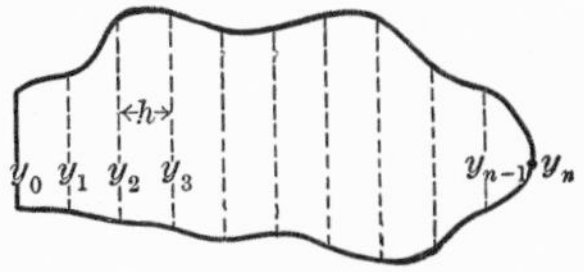

FIG. 50.

* For definition and equation, see Analytic Geometry, pp. 22–27.

$$A_T = h\left[\tfrac{1}{2}(y_0 + y_n) + y_1 + y_2 + \ldots + y_{n-1}\right]$$

(Trapezoidal Rule)

$$A_D = h\left[0.4(y_0 + y_n) + 1.1(y_1 + y_{n-1}) + y_2 + y_3 + \ldots + y_{n-2}\right]$$

(Durand's Rule)

$$A_S = \tfrac{1}{3}h\left[(y_0 + y_n) + 4(y_1 + y_3 + \ldots + y_{n-1}) + 2(y_2 + y_4 + \ldots + y_{n-2})\right]$$

(**Simpson's Rule,** where **n** is **even**).

The larger the value of **n**, the greater is the accuracy of approximation. In general, for the same number of chords, A_S gives the most accurate, A_T, the least accurate approximation.

51 Cube

$V = a^3$; $d = a\sqrt{3}$.
Total surface $= 6\,a^2$.

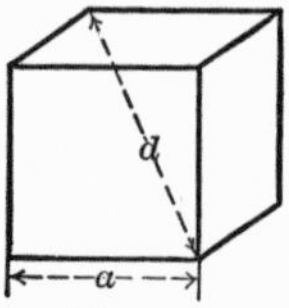

FIG. 51.

52 Rectangular Parallelopiped

$V = abc$; $d = \sqrt{a^2 + b^2 + c^2}$.
Total surface $= 2\,(ab + bc + ca)$.

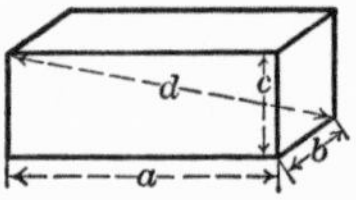

FIG. 52.

53 Prism or Cylinder

V = (area of base) × (altitude).
Lateral area = (perimeter of right section) × (lateral edge).

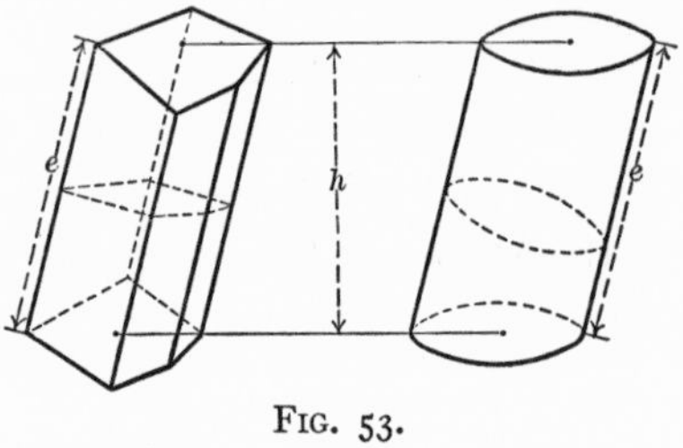

FIG. 53.

54 Pyramid or Cone

V = $\tfrac{1}{3}$ (area of base) × (altitude).
Lateral area of regular figure = $\tfrac{1}{2}$ (perimeter of base) × (slant height).

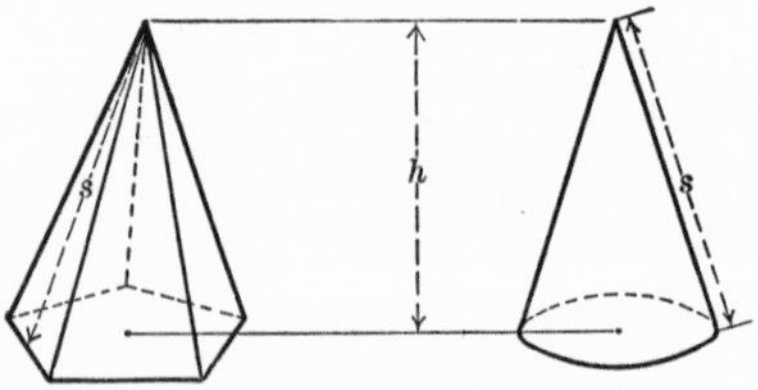

FIG. 54.

55 Frustum of Pyramid or Cone

$V = \tfrac{1}{3}(A_1 + A_2 + \sqrt{A_1 \times A_2})\,h$, where A_1 and A_2 are areas of bases, and **h** is altitude.

Lateral area of regular figure = $\tfrac{1}{2}$ (sum of perimeters of bases) × (slant height).

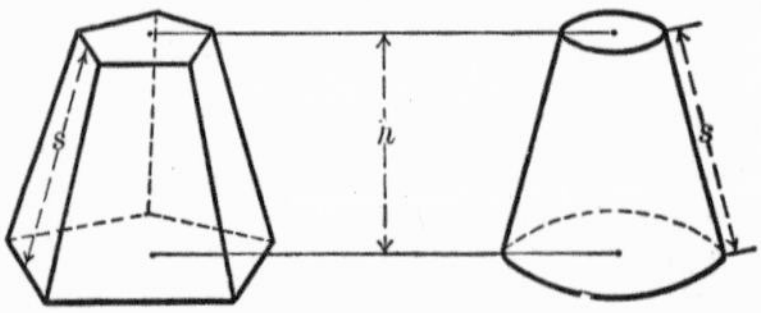

FIG. 55.

56 Prismatoid (bases are in parallel planes, lateral faces are triangles or trapezoids)

$$V = \frac{1}{6}(A_1 + A_2 + 4\,A_m)\,h,$$

where A_1, A_2 are areas of bases, A_m is area of mid-section, and h is altitude.

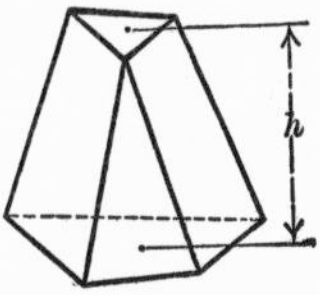

FIG. 56.

57 Sphere

$A_{(sphere)} = 4\,\pi R^2 = \pi D^2.$

$A_{(zone)} = 2\,\pi Rh = \pi Dh.$

$V_{(sphere)} = \frac{4}{3}\,\pi R^3 = \frac{1}{6}\,\pi D^3.$

$V_{(spherical\ sector)} = \frac{2}{3}\pi R^2 h = \frac{1}{6}\pi D^2 h.$

$V_{(spherical\ segment\ of\ one\ base)}$

$= \frac{1}{6}\,\pi h_1\,(3\,r_1^2 + h_1^2) = \frac{1}{3}\,\pi h_1^2\,(3\,R - h_1).$

$V_{(spherical\ segment\ of\ two\ bases)}$

$= \frac{1}{6}\,\pi h\,(3\,r_1^2 + 3\,r_2^2 + h^2).$

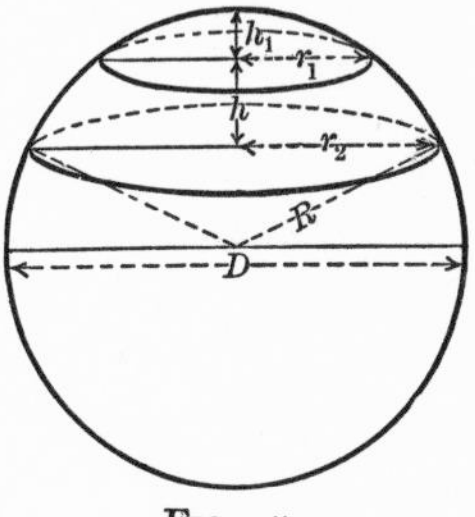

FIG. 57.

58 Solid Angle (ψ), at any point (P) subtended by any surface (S), is equal to the portion (A) of the surface of a sphere of unit radius which is cut out by a conical surface with vertex at P and the perimeter of S for base.

The unit solid angle (ψ) is called a **steradian**.

The total solid angle about a point $= 4\,\pi$ steradians.

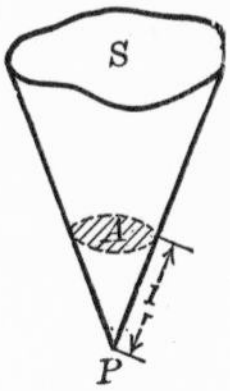

FIG. 58.

59 Ellipsoid

$$V = \frac{4}{3}\,\pi abc.$$

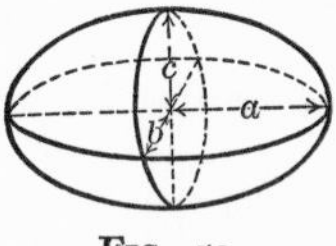

FIG. 59.

60 Paraboloidal segment

$V_{(segment\ of\ one\ base)} = \frac{1}{2}\,\pi r_1^2 h.$

$V_{(segment\ of\ two\ bases)} = \frac{1}{2}\,\pi d\,(r_1^2 + r_2^2).$

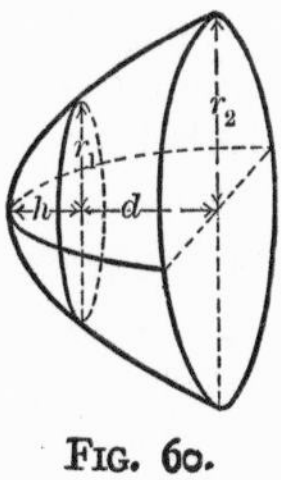

FIG. 60.

61 Torus

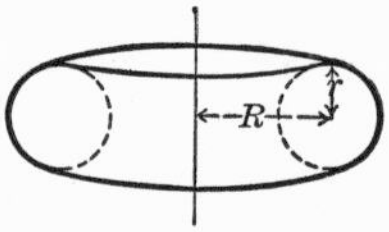

FIG. 61.

$V = 2\pi^2 Rr^2$.
Surface (S) $= 4\pi^2 Rr$.

62 Solid (V) or Surface (S) of Revolution,

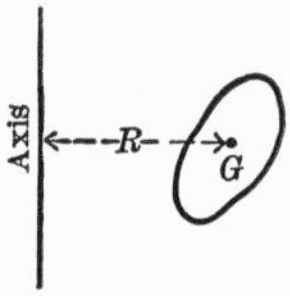

FIG. 62.

generated by revolving any plane area (A) or arc (s) about an axis in its plane, and not crossing the area or arc.

$V = 2\pi RA$; $S = 2\pi Rs$,

where R = distance of center of gravity (G) of area or arc from axis.

ANALYTIC GEOMETRY

I. Plane

63 Rectangular Coördinates. (Fig. 63)

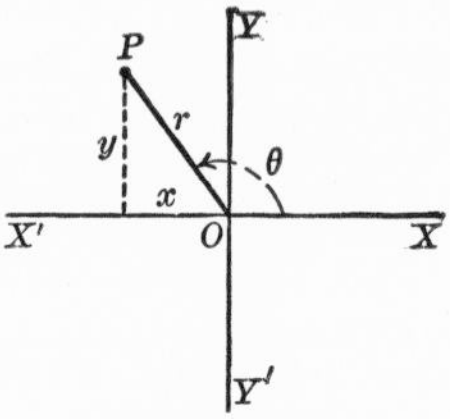

FIG. 63, 64.

Let two perpendicular lines, X'X (x-axis) and Y'Y (y-axis) meet in a point O (origin). The position of any point P (x, y) is fixed by the distances x (abscissa) and y (ordinate) from Y'Y and X'X, respectively, to P.

NOTE. x is + to the right and − to the left of Y'Y, y is + above and − below X'X.

64 Polar Coördinates. (Fig. 64)

Let O (origin or pole) be a point in the plane and OX (initial line) be any line through O. The position of any point P (r, θ) is fixed by the distance r (radius vector) from O to the point and the angle θ (vectorial angle) measured from OX to OP.

NOTE. r is + measured along terminal side of θ, r is − measured along terminal side of θ produced; θ is + measured counter-clockwise, θ is − measured clockwise.

65 Relations connecting Rectangular and Polar Coördinates

$x = r\cos\theta$, $y = r\sin\theta$.

$$r = \sqrt{x^2+y^2},\ \theta = \tan^{-1}\frac{y}{x},\ \sin\theta = \frac{y}{\sqrt{x^2+y^2}},\ \cos\theta = \frac{x}{\sqrt{x^2+y^2}},\ \tan\theta = \frac{y}{x}.$$

66 Points and Slopes. (Fig. 66)

Let P_1 (x_1, y_1) and P_2 (x_2, y_2) be any two points, and let α be the angle from **OX** to P_1P_2, measured counter-clockwise.

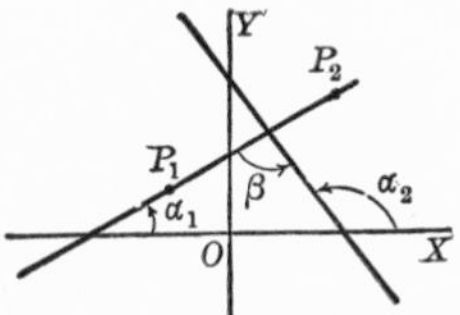

FIG. 66.

$P_1P_2 = d = \sqrt{(x_2 - x_1)^2 + (y_2 - y_1)^2}$.

Mid-point of P_1P_2 is $\left(\frac{x_1 + x_2}{2}, \frac{y_1 + y_2}{2}\right)$.

Point which divides P_1P_2 in the ratio $m_1 : m_2$ is

$$\left(\frac{m_1x_2 + m_2x_1}{m_1 + m_2}, \frac{m_1y_2 + m_2y_1}{m_1 + m_2}\right).$$

Slope of $P_1P_2 = \tan\alpha = m = \frac{y_2 - y_1}{x_2 - x_1}$.

Angle between two lines of slopes m_1 and m_2 is $\beta = \tan^{-1}\frac{m_2 - m_1}{1 + m_1m_2}$.

Two lines of slopes m_1 and m_2 are perpendicular if $m_2 = -\frac{1}{m_1}$.

67 Locus and Equation

The collection of all points which satisfy a given condition is called the **locus** of that condition; the condition expressed by means of the variable coördinates of any point on the locus is called the **equation** of the locus.

The locus may be represented by equations of three kinds:

Rectangular equation involves the rectangular coördinates (x, y).

Polar equation involves the polar coördinates (r, θ).

Parametric equations express x and y or r and θ in terms of a third independent variable called a parameter.

The following equations are given in the system in which they are most simply expressed; sometimes several forms of the equation in one or more systems are given.

68 Straight Line. (Fig. 68)

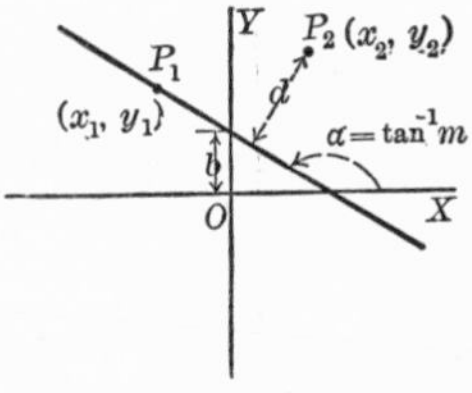

FIG. 68.

$Ax + By + C = 0$ [$-A \div B$ = slope]

$y = mx + b$. [m = slope, b = intercept on **OY**]

$y - y_1 = m(x - x_1)$. [m = slope, P_1 (x_1, y_1) is a known point on line]

$d = \frac{Ax_2 + By_2 + C}{\pm\sqrt{A^2 + B^2}}$. [$d$ = distance from a point P_2 (x_2, y_2) to the line $Ax + By + C = 0$]

69 Circle.

Locus of a point at a constant distance (radius) from a fixed point **C** (center). [For mensuration of circle, see 45]

(1) $\begin{cases} (x - h)^2 + (y - k)^2 = a^2. & C\ (h, k),\ \text{rad.} = a. \\ r^2 + b^2 - 2\,br\cos(\theta - \beta) = a^2. & C\ (b, \beta),\ \text{rad.} = a. \end{cases}$

(2) $\begin{cases} x^2 + y^2 = 2\,ax. & C\ (a, 0),\ \text{rad.} = a. \\ r = 2\,a\cos\theta. & C\ (a, 0),\ \text{rad.} = a. \end{cases}$

(3) $\begin{cases} x^2 + y^2 = 2\,ay. \\ r = 2\,a \sin\theta. \end{cases}$ C (o, a), rad. = a. $C\left(a, \frac{\pi}{2}\right)$, rad. = a.

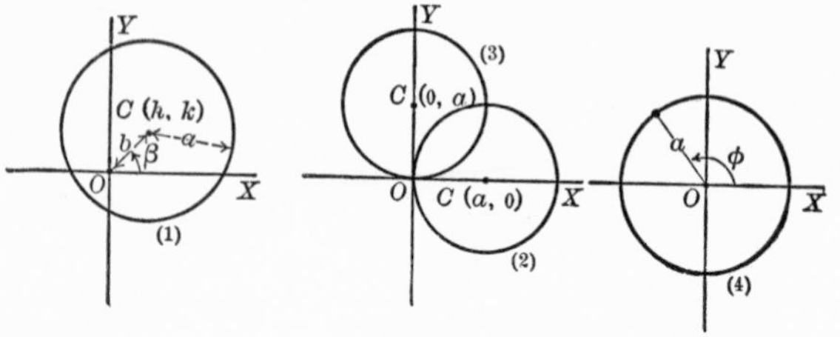

FIG. 69.

(4) $\begin{cases} x^2 + y^2 = a^2. \\ r = a. \\ x = a\cos\phi,\ y = a\sin\phi. \end{cases}$

C (o, o), rad. = a.
C (o, o), rad. = a.
C (o, o), rad. = a, ϕ = angle from OX to radius.

70 Conic. Locus of a point whose distance from a fixed point (focus) is in a constant ratio, e (called eccentricity), to its distance from a fixed straight line (directrix). [Fig. 70]

$\begin{cases} x^2 + y^2 = e^2\,(d + x)^2. \\ r = \dfrac{de}{1 - e\cos\theta}. \end{cases}$ [d = distance from focus to directrix]

The **conic** is called a **parabola** when e = 1, an **ellipse** when e < 1, a **hyperbola** when e > 1.

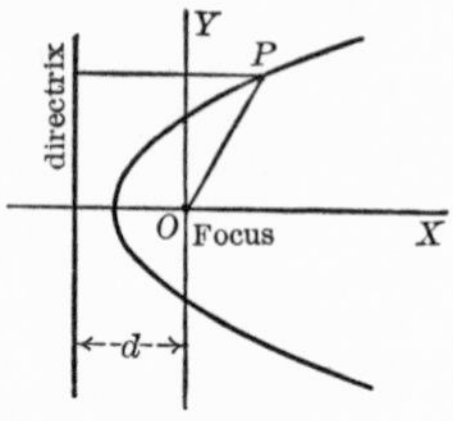

FIG. 70.

71 Parabola. Conic where e = 1. [For mensuration of parabola, see **47**]

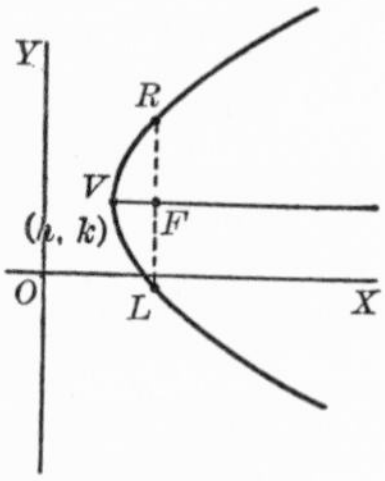

FIG. 71 (1).

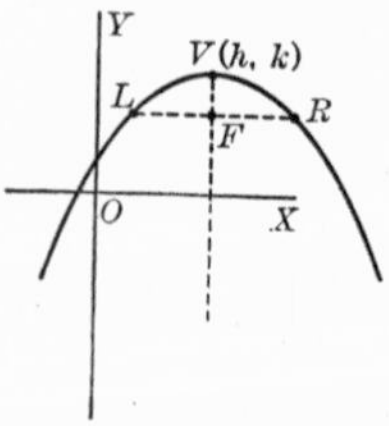

FIG. 71 (2).

(1) $\begin{cases} (y - k)^2 = a\,(x - h). \\ y^2 = ax. \end{cases}$ Vertex (h, k), axis ∥ OX. [Fig. 71 (1)]
Vertex (o, o), axis along OX.

(2) $\begin{cases} (x - h)^2 = a\,(y - k). \\ x^2 = ay. \end{cases}$ Vertex (h, k), axis ∥ OY. [Fig. 71 (2)]
Vertex (o, o), axis along OY.

Distance from vertex to focus = VF = $\frac{1}{4}$ a.
Latus rectum = LR = a.

72 Ellipse. Conic where $e < 1$. [For mensuration of ellipse, see 46]

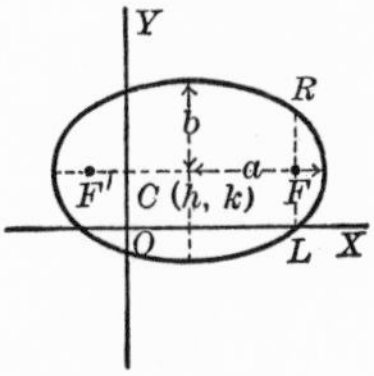

FIG. 72 (1).

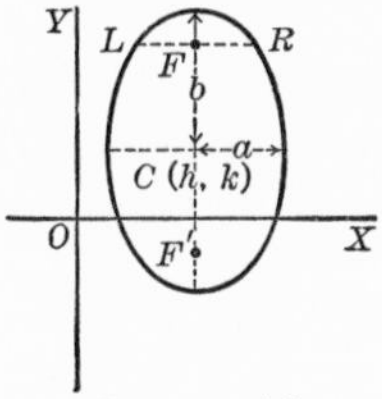

FIG. 72 (2).

$$\begin{cases} \dfrac{(x-h)^2}{a^2} + \dfrac{(y-k)^2}{b^2} = 1. & \text{Center } (h, k), \text{ axes } \parallel OX, OY. \\ \dfrac{x^2}{a^2} + \dfrac{y^2}{b^2} = 1. & \text{Center } (0, 0), \text{ axes along } OX, OY. \end{cases}$$

	a > b, Fig. 72 (1)	b > a, Fig. 72 (2)
Major axis	$2a$	$2b$
Minor axis	$2b$	$2a$
Distance from center to either focus	$\sqrt{a^2 - b^2}$	$\sqrt{b^2 - a^2}$
Latus rectum	$\dfrac{2b^2}{a}$	$\dfrac{2a^2}{b}$
Eccentricity, e	$\dfrac{\sqrt{a^2 - b^2}}{a}$	$\dfrac{\sqrt{b^2 - a^2}}{b}$
Sum of distances of any point from the foci, PF′ + PF	$2a$	$2b$

73 Hyperbola. Conic where $e > 1$.

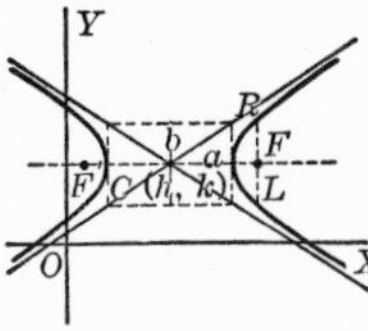

FIG. 73 (1).

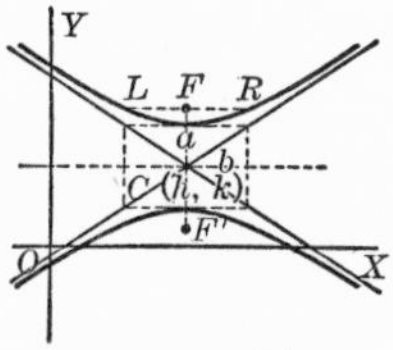

FIG. 73 (2).

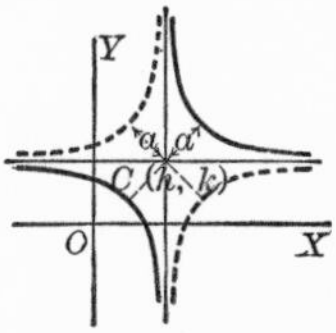

FIG. 73 (3).

$$(1)\begin{cases} \dfrac{(x-h)^2}{a^2} - \dfrac{(y-k)^2}{b^2} = 1. & C\ (h, k), \text{ transverse axis } \parallel OX. \quad \text{[Fig. 73 (1)]} \\ \dfrac{x^2}{a^2} - \dfrac{y^2}{b^2} = 1. & C\ (0, 0), \text{ transverse axis along } OX. \end{cases}$$

$$(2)\begin{cases} \dfrac{(y-k)^2}{a^2} - \dfrac{(x-h)^2}{b^2} = 1. & C\ (h, k), \text{ transverse axis } \parallel OY. \quad \text{[Fig. 73 (2)]} \\ \dfrac{y^2}{a^2} - \dfrac{x^2}{b^2} = 1. & C\ (0, 0), \text{ transverse along } OY. \end{cases}$$

Transverse axis $= 2a$; **conjugate axis** $= 2b$.

Distance from center to either focus $= \sqrt{a^2 + b^2}$.

Latus rectum $= \frac{2\,b^2}{a}$.

Eccentricity, $e = \frac{\sqrt{a^2 + b^2}}{a}$.

Difference of distances of any point from the foci $= 2\,a$.

Asymptotes are two lines through the center to which the branches of the hyperbola approach indefinitely near; their slopes are $\pm \frac{b}{a}$ [Fig. 73 (1)] or $\pm \frac{a}{b}$ [Fig. 73 (2)].

Rectangular (equilateral) hyperbola, $b = a$. The asymptotes are perpendicular.

$$(3)\begin{cases}(x - h)(y - k) = \pm \frac{a^2}{2}. & \text{Center } (h, k), \text{ asymptotes } \parallel OX, OY.\\ xy = \pm \frac{a^2}{2}. & \text{Center } (0, 0), \text{ asymptotes along } OX, OY.\end{cases}$$

Where the + sign gives the smooth curve in Fig. 73 (3).
Where the − sign gives the dotted curve in Fig. 73 (3).

74 Cubical [Fig. 74 (1)] and Semicubical [Fig. 74 (2)] Parabolas

75 Witch. [Fig. 75]

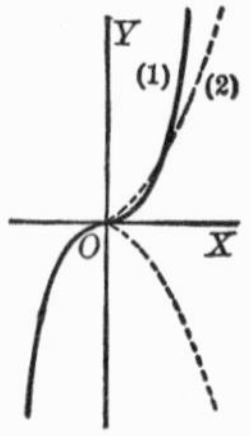

FIG. 74.

(1) $y = ax^3$.
(2) $y^2 = ax^3$.

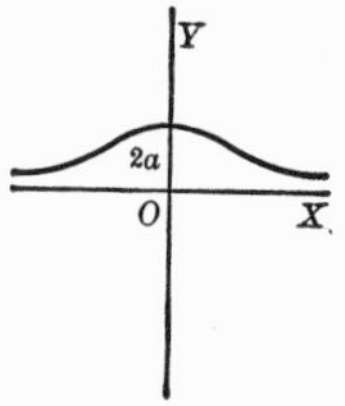

FIG. 75.

$$y = \frac{8\,a^3}{x^2 + 4\,a^2}.$$

76 Cissoid. [Fig. 76]

77 Strophoid. [Fig. 77]

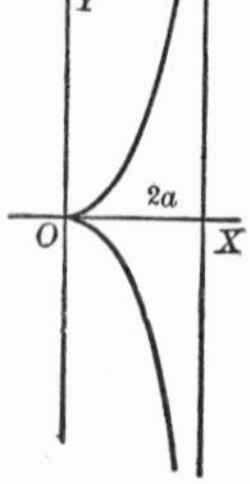

FIG. 76.

$$y^2 = \frac{x^3}{2\,a - x}.$$

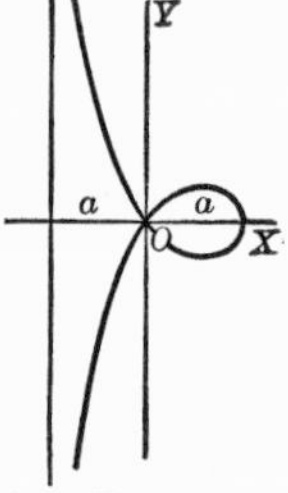

FIG. 77.

$$y^2 = x^2\left(\frac{a - x}{a + x}\right).$$

78 Sine Wave. [Fig. 78]

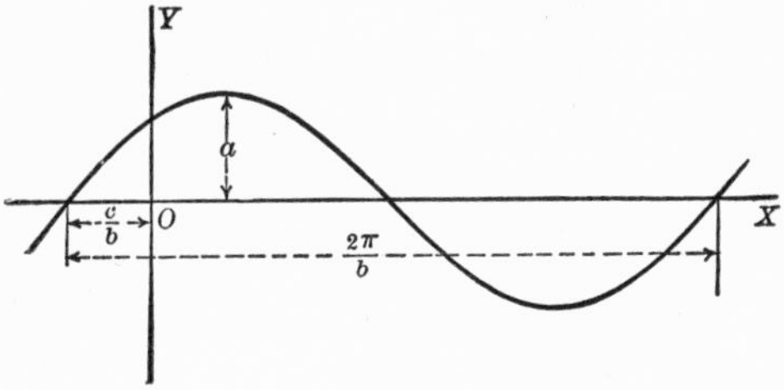

FIG. 78.

$$\begin{cases} y = a \sin (bx + c). \\ y = a \cos (bx + c') = a \sin (bx + c), \text{ where } c = c' + \dfrac{\pi}{2}. \\ y = m \sin bx + n \cos bx = a \sin (bx + c), \text{ where } a = \sqrt{m^2 + n^2},\ c = \tan^{-1} \dfrac{n}{m}. \end{cases}$$

The curve consists of a succession of waves, where

a = **amplitude** = maximum height of wave.

$\frac{2\pi}{b}$ = **wave length** = distance from any point on wave to the corresponding point on the next wave.

$x = -\frac{c}{b}$ (called the **phase**) marks a point on **OX** from which the positive half of the wave starts.

79 Tangent [Fig. 79 (1)] and Cotangent [Fig. 79 (2)] Curves

80 Secant [Fig. 80 (1)] and Cosecant [Fig. 80 (2)] Curves

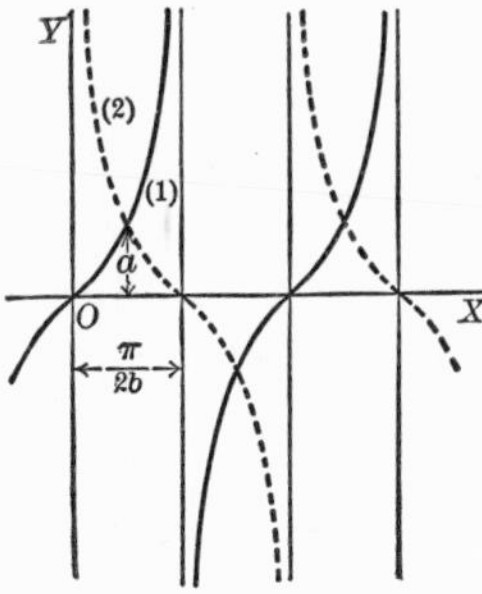

FIG. 79.

(1) $y = a \tan bx$.

(2) $y = a \cot bx$.

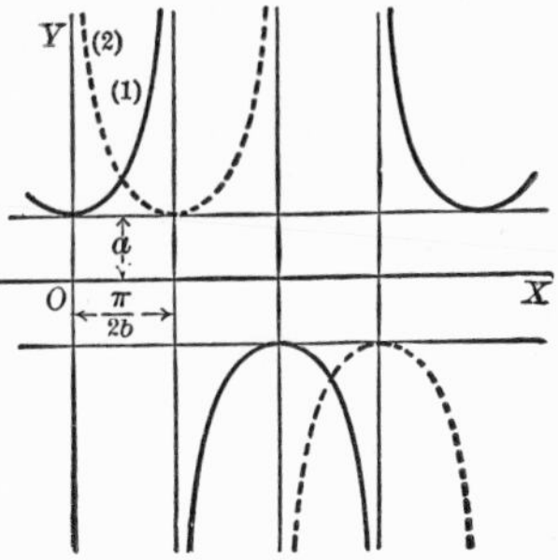

FIG. 80.

(1) $y = a \sec bx$.

(2) $y = a \csc bx$.

xponential or Logarithmic Curves. [Fig. 81]

(1) $y = ab^x$ or $x = \log_b \frac{y}{a}$.

(2) $y = ab^{-x}$ or $x = -\log_b \frac{y}{a}$.

(3) $x = ab^y$ or $y = \log_b \frac{x}{a}$.

(4) $x = ab^{-y}$ or $y = -\log_b \frac{x}{a}$.

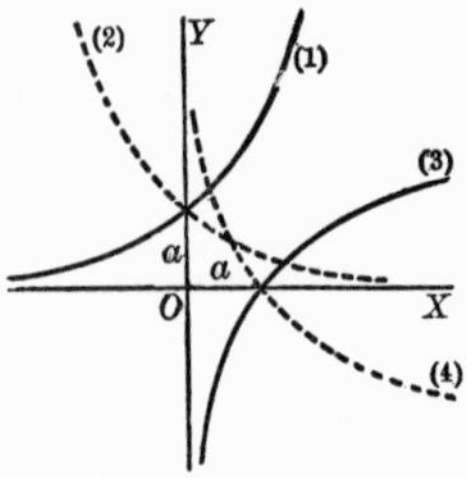

FIG. 81.

The equations $y = ae^{\pm nx}$ and $x = ae^{\pm ny}$ are special cases of above.

82 Oscillatory Wave of Decreasing Amplitude.

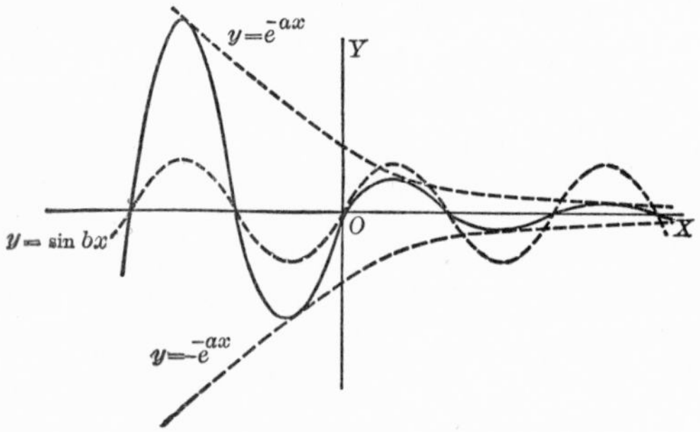

FIG. 82.

$$y = e^{-ax} \sin bx.$$

NOTE. The curve oscillates between $y = e^{-ax}$ and $y = -e^{-ax}$.

83 Catenary.

83 Catenary. Curve made by a chain or cord of uniform weight suspended freely between two points at the same level. [Fig. 83.] [For mensuration of catenary, see 49].

$$y = \frac{a}{2}\left(e^{\frac{x}{a}} + e^{-\frac{x}{a}}\right).$$

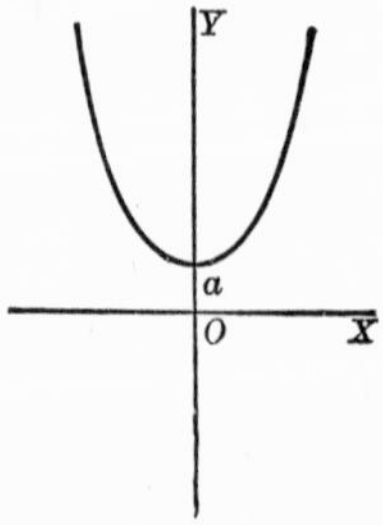

FIG. 83.

84 Cycloid. Curve described by a point on a circle which rolls along a fixed straight line. [Fig. 84]

$$\begin{cases} x = a(\phi - \sin\phi). \\ y = a(1 - \cos\phi). \end{cases}$$

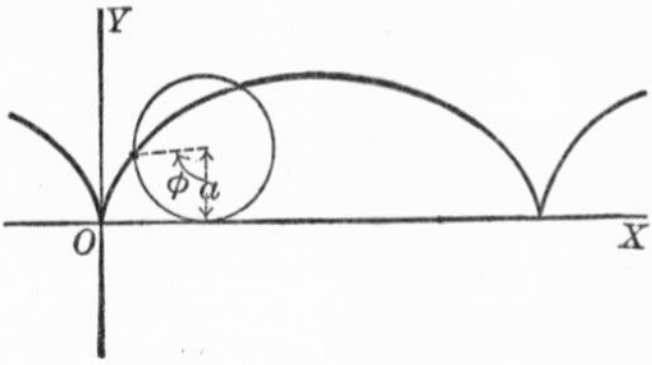

FIG. 84.

85 Epicycloid. Curve described by a point on a circle which rolls along the outside of a fixed circle. [Fig. 85]

$$\begin{cases} x = (a + b)\cos\phi - b\cos\left(\dfrac{a + b}{b}\phi\right). \\ y = (a + b)\sin\phi - b\sin\left(\dfrac{a + b}{b}\phi\right). \end{cases}$$

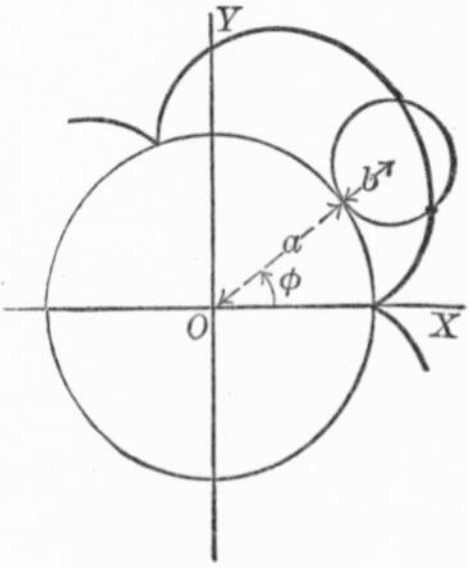

FIG. 85.

86 Cardioid. Epicycloid with radii of fixed and rolling circles equal.

$r = a(1 + \cos\theta)$. [Fig. 86]

$r = a(1 + \sin\theta)$. [Fig. 86 rotated through $+90°$]

$r = a(1 - \cos\theta)$. [Fig. 86 rotated through $+180°$]

$r = a(1 - \sin\theta)$. [Fig. 86 rotated through $-90°$]

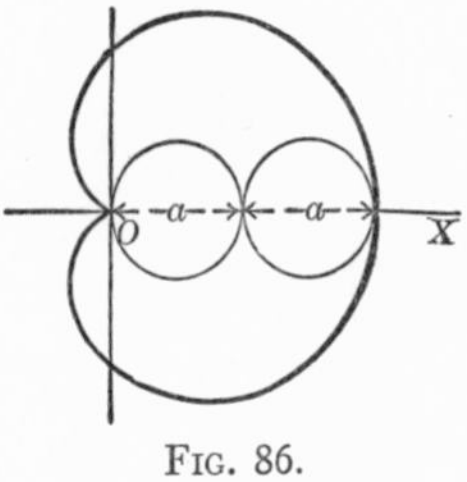

FIG. 86.

87 Hypocycloid. Curve described by a point on a circle which rolls along the inside of a fixed circle.

$$\begin{cases} x = (a - b)\cos\phi + b\cos\left(\dfrac{a - b}{b}\phi\right). \\ y = (a - b)\sin\phi - b\sin\left(\dfrac{a - b}{b}\phi\right). \end{cases}$$

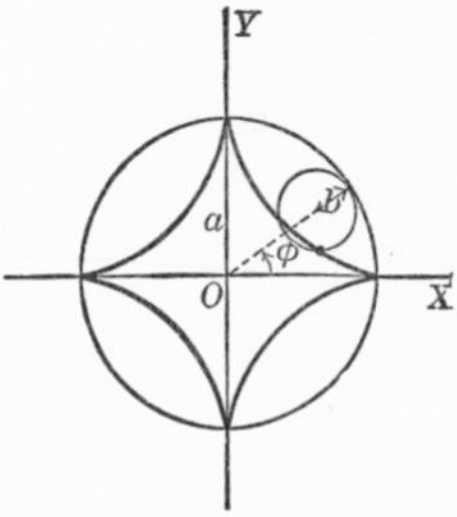

FIG. 88.

88 Hypocycloid of four cusps: radius of fixed circle equals four times the radius of the rolling circle. [Fig. 88]

$$x^{\frac{2}{3}} + y^{\frac{2}{3}} = a^{\frac{2}{3}}.$$

$$x = a\cos^3\phi, \quad y = a\sin^3\phi.$$

89 Involute of the Circle.

Curve described by the end of a string which is kept taut while being unwound from a circle. [Fig. 89]

$$\begin{cases} x = a\cos\phi + a\phi\sin\phi. \\ y = a\sin\phi - a\phi\cos\phi. \end{cases}$$

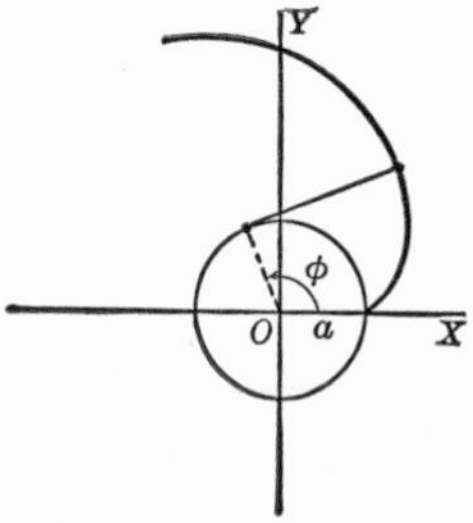

FIG. 89.

90 Lemniscate.

Locus of a point which moves so that the product of its distances from two fixed points (foci) is constant, or $PF' \times PF = a^2$.

$r^2 = 2a^2\cos 2\theta$. [Fig. 90]

$r^2 = 2a^2\sin 2\theta$. [Fig. 90 turned through 45°]

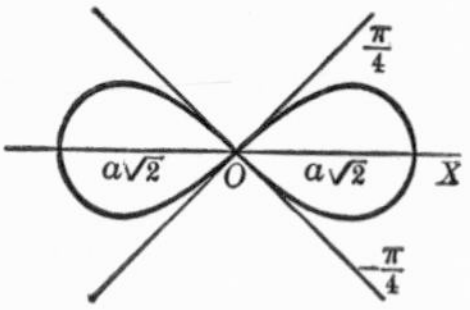

FIG. 90.

91 N-leaved Rose

(1) $r = a\sin n\theta$. [Fig. 91 (1)]

(2) $r = a\cos n\theta$. [Fig. 91 (2)]

There are **n** leaves if **n** is odd, **2 n** leaves if **n** is even.

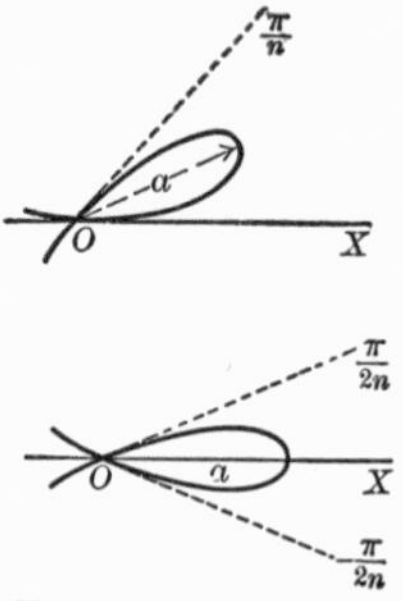

FIGS. 91 (1), 91 (2).

92 Spirals. [Fig. 92]

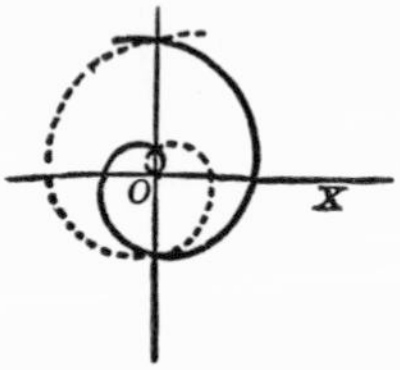

FIG. 92 (1).

(1) Archimedian.

$r = a\theta$.

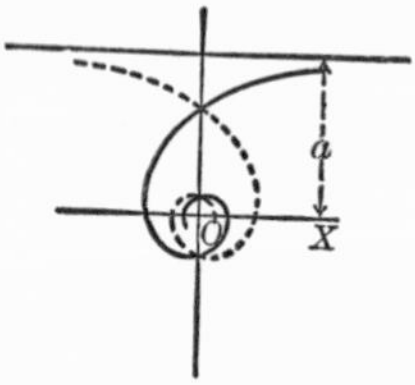

FIG. 92 (2).

(2) Hyperbolic.

$r = \frac{a}{\theta}$.

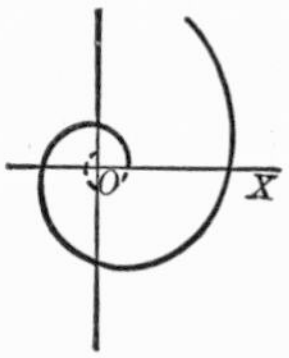

FIG. 92 (3).

(3) Logarithmic.

$r = e^{a\theta}$.

II. Solid

93 Coördinates

Let three mutually perpendicular planes, **XOY, YOZ, ZOX** (coördinate planes) meet in a point **O** (origin).

Rectangular system. The position of a point **P** (x, y, z) in space is fixed by its three distances **x, y,** and **z** from the three coördinate planes.

Cylindrical system. The position of any point **P** (r, θ, z) is fixed by **z**, its distance from the **XOY** plane, and by (r, θ), the polar coördinates of the projection of **P** in the **XOY** plane.

Relations connecting rectangular and cylindrical coördinates are the same as those given in 65.

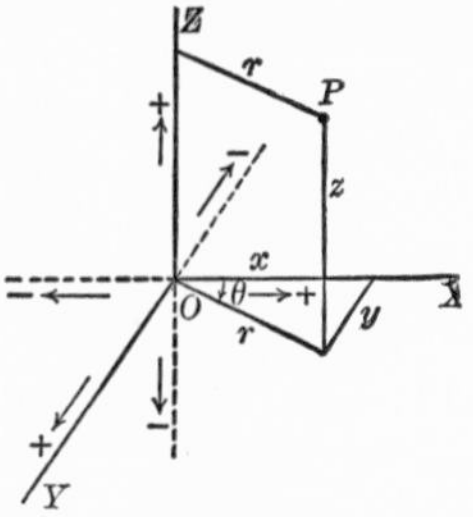

FIG. 93.

94 Points, Lines, and Planes

Distance (d) between two points $P_1(x_1, y_1, z_1)$ and $P_2(x_2, y_2, z_2)$,

$$d = \sqrt{(x_2 - x_1)^2 + (y_2 - y_1)^2 + (z_2 - z_1)^2}.$$

Direction cosines of a line (cosines of the angles α, β, γ which the line or any parallel line makes with the coördinate axes) are related by

$$\cos^2\alpha + \cos^2\beta + \cos^2\gamma = 1.$$

If $\cos\alpha : \cos\beta : \cos\gamma = a : b : c$,

then $\cos\alpha = \dfrac{a}{\sqrt{a^2 + b^2 + c^2}}, \cos\beta = \dfrac{b}{\sqrt{a^2 + b^2 + c^2}}, \cos\gamma = \dfrac{c}{\sqrt{a^2 + b^2 + c^2}}.$

Direction cosines of the line joining $P_1(x_1, y_1, z_1)$ and $P_2(x_2, y_2, z_2)$,

$$\cos\alpha : \cos\beta : \cos\gamma = x_2 - x_1 : y_2 - y_1 : z_2 - z_1.$$

Angle (θ) between two lines, whose direction angles are $\alpha_1, \beta_1, \gamma_1$ and $\alpha_2, \beta_2, \gamma_2$,

$$\cos\theta = \cos\alpha_1\cos\alpha_2 + \cos\beta_1\cos\beta_2 + \cos\gamma_1\cos\gamma_2.$$

Equation of a plane is of the first degree in x, y, and z,

$$Ax + By + Cz + D = 0,$$

where A, B, C are proportional to the direction cosines of a normal or perpendicular to the plane.

Angle between two planes is the angle between their normals.

Equations of a straight line are two equations of the first degree,

$$A_1x + B_1y + C_1z + D_1 = 0, \quad A_2x + B_2y + C_2z + D_2 = 0.$$

Equations of a straight line through the point $P_1(x_1, y_1, z_1)$ with direction cosines proportional to a, b, and c,

$$\frac{x - x_1}{a} = \frac{y - y_1}{b} = \frac{z - z_1}{c}.$$

95 Cylindrical Surfaces

The locus in space of an equation containing only two of the coördinates **x, y, z** is a cylindrical surface with its elements perpendicular to the plane of coördinates. Considered as a plane geometry equation, the equation represents the curve of intersection of the cylinder with the plane of the two coördinates.

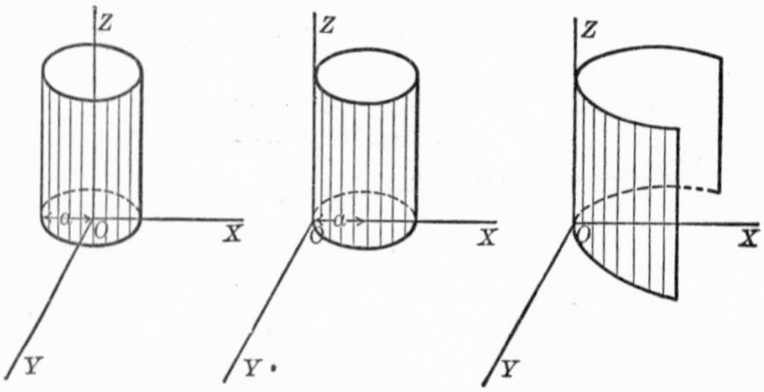

FIG. 95 (1). FIG. 95 (2). FIG. 95 (3).

Circular cylinders. [Fig. 95] [For mensuration see **53**]

(1) $\begin{cases} x^2 + y^2 = a^2. \\ r = a. \end{cases}$ (2) $\begin{cases} x^2 + y^2 = 2\,ax. \\ r = 2\,a \cos\theta. \end{cases}$

Parabolic cylinder (3) $y^2 = ax$.

96 Surfaces of Revolution

Equation of the surface of revolution obtained by revolving the plane curve $y = f(x)$ or $z = f(x)$ about **OX**,

$$y^2 + z^2 = [f(x)]^2.$$

Sphere (revolve circle $x^2 + y^2 = a^2$ about **OX**)

$$x^2 + y^2 + z^2 = a^2.$$ [For mensuration of sphere, see **57**]

Spheroid (revolve ellipse $\frac{x^2}{a^2} + \frac{y^2}{b^2} = 1$ about **OX**)

$$\frac{x^2}{a^2} + \frac{y^2 + z^2}{b^2} = 1$$ (prolate if $a > b$, oblate if $b > a$).

[For mensuration of ellipsoid, see **59**]

Cone (revolve line $y = mx$ about **OX**)

$$y^2 + z^2 = m^2x^2.$$ [For mensuration of cone, see **54**]

Paraboloid (revolve parabola $y^2 = ax$ about **OX**)

$$y^2 + z^2 = ax.$$ [For mensuration of paraboloid, see **60**]

97 Space Curves

A curve in space may be represented by two equations connecting the coördinates **x, y, z** of any point on the curve, or by three equations expressing the coördinates **x, y, z** in terms of a fourth variable or parameter.

Helix. Curve generated by a point moving on a cylinder so that the distance traversed parallel to the axis of the cylinder is proportional to the angle of rotation about the axis.

$$x = a \cos\theta,\ y = a \sin\theta,\ z = k\theta,$$ [Fig. 97]

where a = radius of cylinder, $2\,\pi k$ = pitch.

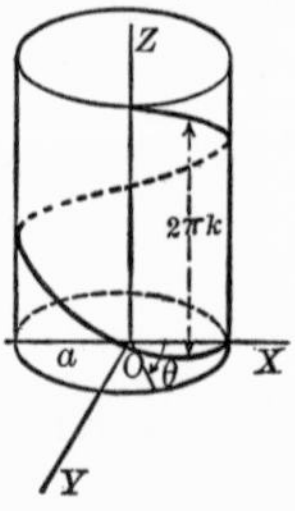

FIG. 97.

DIFFERENTIAL CALCULUS

98 Definition of Function. Notation

A variable y is said to be a function of another variable x, if, when x is given y is determined.

The symbols f(x), F(x), ϕ(x), etc., represent various functions of x.

The symbol f(a) represents the value of f(x) when x = a.

99 Definition of Derivative. Notation

Let y = f(x). If Δx is any increment (increase or decrease) given to x, and Δy is the corresponding increment in y, then the derivative of y with respect to x is the limit of the ratio of Δy to Δx as Δx approaches zero, that is

$$\frac{dy}{dx} = \lim_{\Delta x \doteq 0} \frac{\Delta y}{\Delta x} = \lim_{\Delta x \doteq 0} \frac{f(x + \Delta x) - f(x)}{\Delta x} = f'(x).$$

$$\frac{d^2y}{dx^2} = \frac{d}{dx}\left(\frac{dy}{dx}\right) = \frac{d}{dx} f'(x) = f''(x). \qquad \text{[2d derivative]}$$

$$\frac{d^3y}{dx^3} = \frac{d}{dx}\left(\frac{d^2y}{dx^2}\right) = \frac{d}{dx} f''(x) = f'''(x). \qquad \text{[3d derivative]}$$

$$\frac{d^ny}{dx^n} = \frac{d}{dx}\left(\frac{d^{n-1}y}{dx^{n-1}}\right) = \frac{d}{dx} f^{(n-1)}(x) = f^{(n)}(x). \qquad [n\text{th derivative}]$$

The symbols f′(a), f″(a), . . . , $f^{(n)}(a)$ represent the values of f′(x), f″(x), . . . , $f^{(n)}(x)$, respectively, when x = a.

100 Some Relations Among Derivatives

If x = f(y), then $\frac{dy}{dx} = 1 \div \frac{dx}{dy}$

If x = f(t), and y = F(t), then $\frac{dy}{dx} = \frac{dy}{dt} \div \frac{dx}{dt}$

If y = f(u), and u = F(x), then $\frac{dy}{dx} = \frac{dy}{du} \cdot \frac{du}{dx}$.

101 Table of Derivatives

Functions of x are represented by u and v, constants are represented by a, n, and e.

$$\frac{d}{dx}(x) = 1.$$

$$\frac{d}{dx}(a) = 0.$$

$$\frac{d}{dx}(u \pm v \pm \cdots) = \frac{du}{dx} \pm \frac{dv}{dx} \pm \cdots.$$

$$\frac{d}{dx}(au) = a\frac{du}{dx}.$$

$$\frac{d}{dx}(uv) = u\frac{dv}{dx} + v\frac{du}{dx}.$$

$$\frac{d}{dx}\left(\frac{u}{v}\right) = \frac{v\frac{du}{dx} - u\frac{dv}{dx}}{v^2}.$$

$$\frac{d}{dx}(u^n) = nu^{n-1}\frac{du}{dx}.$$

$$\frac{d}{dx}\log_a u = \frac{\log_a e}{u}\frac{du}{dx}.$$

$$\frac{d}{dx}\ln u = \frac{1}{u}\frac{du}{dx}.$$

$$\frac{d}{dx}a^u = a^u \ln a \frac{du}{dx}.$$

$$\frac{d}{dx}e^u = e^u\frac{du}{dx}.$$

$$\frac{d}{dx}u^v = vu^{v-1}\frac{du}{dx} + u^v \ln u \frac{dv}{dx}.$$

$$\frac{d}{dx}\sin u = \cos u \frac{du}{dx}.$$

$$\frac{d}{dx}\cos u = -\sin u \frac{du}{dx}.$$

$$\frac{d}{dx}\tan u = \sec^2 u \frac{du}{dx}.$$

$$\frac{d}{dx}\cot u = -\csc^2 u \frac{du}{dx}.$$

$$\frac{d}{dx}\sec u = \sec u \tan u \frac{du}{dx}.$$

$$\frac{d}{dx}\csc u = -\csc u \cot u \frac{du}{dx}.$$

$$\frac{d}{dx}\operatorname{vers} u = \sin u \frac{du}{dx}.$$

$$\frac{d}{dx}\sin^{-1} u = \frac{1}{\sqrt{1-u^2}}\frac{du}{dx} \quad \left(\text{where } \sin^{-1} u \text{ lies between } -\frac{\pi}{2} \text{ and } +\frac{\pi}{2}\right)$$

$$\frac{d}{dx}\cos^{-1} u = -\frac{1}{\sqrt{1-u^2}}\frac{du}{dx} \quad (\text{where } \cos^{-1} u \text{ lies between } 0 \text{ and } \pi).$$

$$\frac{d}{dx}\tan^{-1} u = \frac{1}{1+u^2}\frac{du}{dx}.$$

$$\frac{d}{dx}\cot^{-1} u = -\frac{1}{1+u^2}\frac{du}{dx}.$$

$$\frac{d}{dx}\sec^{-1} u = \frac{1}{u\sqrt{u^2-1}}\frac{du}{dx} \quad (\text{where } \sec^{-1} u \text{ lies between } 0 \text{ and } \pi).$$

$$\frac{d}{dx}\csc^{-1} u = -\frac{1}{u\sqrt{u^2-1}}\frac{du}{dx} \quad \left(\text{where } \csc^{-1} u \text{ lies between } -\frac{\pi}{2} \text{ and } +\frac{\pi}{2}\right)$$

$$\frac{d}{dx}\operatorname{vers}^{-1} u = \frac{1}{\sqrt{2u-u^2}}\frac{du}{dx} \quad (\text{where } \operatorname{vers}^{-1} u \text{ lies between } 0 \text{ and } \pi).$$

102 The nth Derivative of Certain Functions

$$\frac{d^n}{dx^n}e^{ax} = a^n e^{ax}.$$

$$\frac{d^n}{dx^n}a^x = (\ln a)^n a^x.$$

$$\frac{d^n}{dx^n}\ln x = \frac{(-1)^{n-1}\,\underline{|n-1}}{x^n}, \quad \underline{|n-1} = 1\cdot 2\cdot 3 \ldots (n-1).$$

$$\frac{d^n}{dx^n}\sin ax = a^n \sin\left(ax + \frac{n\pi}{2}\right).$$

$$\frac{d^n}{dx^n}\cos ax = a^n \cos\left(ax + \frac{n\pi}{2}\right).$$

103 Slope of a Curve. Tangent and Normal

The slope of the curve (slope of the tangent line to the curve) whose equation is $y = f(x)$ is

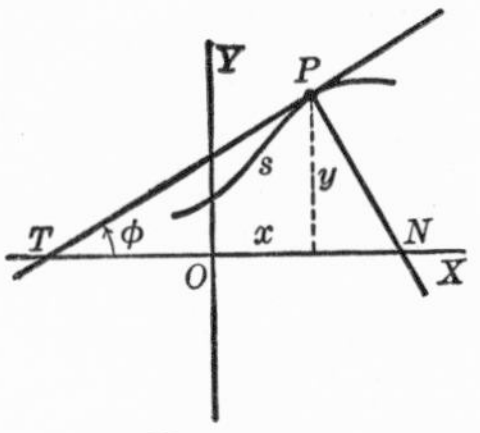

FIG. 103.

Slope $= m = \tan\phi = \dfrac{dy}{dx} = f'(x)$.

Slope at $x = x_1$ is $m_1 = f'(x_1)$.

Equation of tangent line at $P_1\,(x_1, y_1)$ is

$$y - y_1 = m_1\,(x - x_1).$$

Equation of normal at $P_1\,(x_1, y_1)$ is

$$y - y_1 = -\frac{1}{m_1}\,(x - x_1).$$

Angle (β) of intersection of two curves whose slopes at a common point are m_1 and m_2 is $\beta = \tan^{-1}\dfrac{m_2 - m_1}{1 + m_1 m_2}$.

104 Derivative of Length of Arc. Radius of Curvature

If s is the length of arc measured along the curve $y = f(x)$ from some fixed point to any point $P\,(x, y)$, and ϕ is the inclination of the tangent line at P to OX, then [Fig. 103]

$$\frac{dx}{ds} = \cos\phi = \frac{1}{\sqrt{1 + \left(\frac{dy}{dx}\right)^2}};\quad \frac{dy}{ds} = \sin\phi = \frac{1}{\sqrt{1 + \left(\frac{dx}{dy}\right)^2}};\quad \left(\frac{dx}{ds}\right)^2 + \left(\frac{dy}{ds}\right)^2 = 1.$$

Radius of curvature (ρ) at any point of the curve $y = f(x)$ or $r = f(\theta)$.

$$\rho = \frac{ds}{d\phi} = \frac{\left[1 + \left(\frac{dy}{dx}\right)^2\right]^{\frac{3}{2}}}{\left(\frac{d^2y}{dx^2}\right)} = \frac{\{1 + [f'(x)]^2\}^{\frac{3}{2}}}{f''(x)} = \frac{\left[r^2 + \left(\frac{dr}{d\theta}\right)^2\right]^{\frac{3}{2}}}{r^2 + 2\left(\frac{dr}{d\theta}\right)^2 - r\frac{d^2r}{d\theta^2}}.$$

ρ at $x = a$ is $\dfrac{\{1 + [f'(a)]^2\}^{\frac{3}{2}}}{f''(a)}$.

Curvature (k) at any point is $k = \dfrac{1}{\rho}$.

105 Maximum and Minimum Values of a Function

The $\begin{Bmatrix}\text{maximum}\\ \text{minimum}\end{Bmatrix}$ value of a function, $f(x)$, in an interval $x = a$ to $x = b$, is the value of the function which is $\begin{Bmatrix}\text{larger}\\ \text{smaller}\end{Bmatrix}$ than the values of the function in its immediate vicinity. Thus in [Fig. 105], the value of the function at M_1 and M_2 is a maximum, its value at m_1 and m_2 is a minimum.

Test for a maximum at $x = x_1$: $f'(x_1) = 0$ or ∞, and $f''(x_1) < 0$.

Test for a minimum at $x = x_1$: $f'(x_1) = 0$ or ∞, and $f''(x_1) > 0$.

If $f''(x_1) = 0$ or ∞, then for a maximum, $f'''(x_1) = 0$ or ∞, and $f^{IV}(x_1) < 0$,
for a minimum, $f'''(x_1) = 0$ or ∞, and $f^{IV}(x_1) > 0$,

and similarly if $f^{IV}(x_1) = 0$ or ∞, etc.

In a practical problem which suggests that the function, $f(x)$, has a maximum or has a minimum in an interval $x = a$ to $x = b$, merely equate $f'(x)$ to zero and solve for the required value of x. To find the largest or smallest values of a function, $f(x)$, in an interval $x = a$ to $x = b$, find also the values $f(a)$ and $f(b)$, for (see Fig. 105 at L and S) these may be the largest and smallest values although they are not maximum or minimum values.

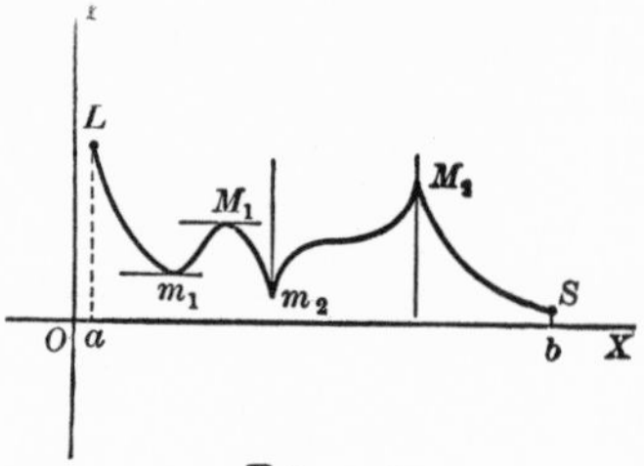

FIG. 105.

106 Points of Inflection of a Curve

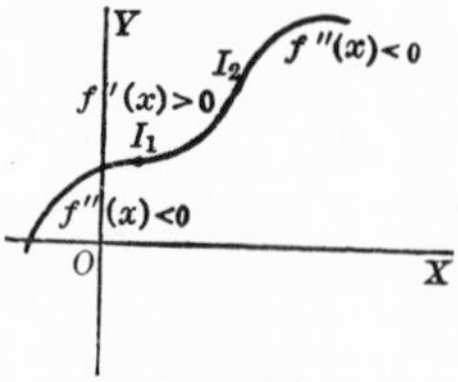

FIG. 106.

Wherever $f''(x) < 0$, the curve is concave down.
Wherever $f''(x) > 0$, the curve is concave up.

The curve is said to have a point of inflection at $x = x_1$ if $f''(x_1) = 0$ or ∞ and the curve is concave up on one side of $x = x_1$ and concave down on the other (see points I_1 and I_2 in Fig. 106).

107 Taylor's and Maclaurin's Theorems

Any $f(x)$ may, in general, be expanded into a **Taylor's Series.**

$$f(x) = f(a) + f'(a)\frac{x-a}{1} + f''(a)\frac{(x-a)^2}{2!} + f'''(a)\frac{(x-a)^3}{3!} + \cdots + f^{(n-1)}(a)\frac{(x-a)^{n-1}}{(n-1)!} + R_n,$$

where a is any quantity whatever so chosen that none of the expressions $f(a)$, $f'(a)$, $f''(a)$, . . . become infinite. If the series is to be used for the purpose of computing the approximate value of $f(x)$ for a given value of x, a should be chosen so that $(x - a)$ is numerically very small, and thus only a few terms of the series need be used. $R_n = f^{(n)}(x_1)\frac{(x-a)^n}{n!}$, where x_1 lies between a and x, is the remainder after n terms, and gives the limits between which the error lies in using n terms of the series for the value of the function. $n! = 1 \cdot 2 \cdot 3 \cdots n$.

If $a = 0$, the above series becomes **Maclaurin's Series.**

$$f(x) = f(0) + f'(0)\frac{x}{1} + f''(0)\frac{x^2}{2!} + f'''(0)\frac{x^3}{3!} + \ldots + f^{(n-1)}(0)\frac{x^{n-1}}{(n-1)!} + R_n.$$

This series may be used for purposes of computation when x is numerically very small.

Some Standard Series

The following series are obtained through expansions of the functions by Taylor's or Maclaurin's theorems. The expression in brackets following each series gives the region of convergence of the series, that is, the values of **x** for which the remainder, R_n, approaches zero as **n** increases, so that a number of terms of the series may be used for an approximation of the function. If the region of convergence is not indicated, it is to be understood that the series converges for all finite values of **x**. ($n! = 1 \cdot 2 \cdot 3 \cdots n$.)

108 Binomial Series

$$(a+x)^n = a^n + na^{n-1}x + \frac{n(n-1)}{2!}a^{n-2}x^2 + \frac{n(n-1)(n-2)}{3!}a^{n-3}x^3 + \cdots. \qquad [x^2 < a^2]$$

NOTE. The series consists of $(n+1)$ terms when **n** is a positive integer; the number of terms is infinite when **n** is a negative or fractional number.

$$(a-bx)^{-1} = \frac{1}{a}\left(1 + \frac{bx}{a} + \frac{b^2x^2}{a^2} + \frac{b^3x^3}{a^3} + \cdots\right). \qquad [b^2x^2 < a^2]$$

109 Exponential Series

$$a^x = 1 + x \ln a + \frac{(x \ln a)^2}{2!} + \frac{(x \ln a)^3}{3!} + \cdots.$$

$$e^x = 1 + x + \frac{x^2}{2!} + \frac{x^3}{3!} + \cdots.$$

$$\tfrac{1}{2}(e^x + e^{-x}) = 1 + \frac{x^2}{2!} + \frac{x^4}{4!} + \cdots.$$

$$\tfrac{1}{2}(e^x - e^{-x}) = x + \frac{x^3}{3!} + \frac{x^5}{5!} + \cdots.$$

$$e^{-x^2} = 1 - x^2 + \frac{x^4}{2!} - \frac{x^6}{3!} + \frac{x^8}{4!} - \cdots.$$

110 Logarithmic Series

$$\ln x = (x-1) - \tfrac{1}{2}(x-1)^2 + \tfrac{1}{3}(x-1)^3 - \cdots. \qquad [x \text{ between } 0 \text{ and } 2]$$

$$\ln x = \frac{x-1}{x} + \frac{1}{2}\left(\frac{x-1}{x}\right)^2 + \frac{1}{3}\left(\frac{x-1}{x}\right)^3 + \cdots. \qquad [x > \tfrac{1}{2}]$$

$$\ln x = 2\left[\frac{x-1}{x+1} + \frac{1}{3}\left(\frac{x-1}{x+1}\right)^3 + \frac{1}{5}\left(\frac{x-1}{x+1}\right)^5 + \cdots\right]. \qquad [x \text{ positive}]$$

$$\ln(1+x) = x - \frac{x^2}{2} + \frac{x^3}{3} - \frac{x^4}{4} + \cdots.$$

$$\ln(a+x) = \ln a + 2\left[\frac{x}{2a+x} + \frac{1}{3}\left(\frac{x}{2a+x}\right)^3 + \frac{1}{5}\left(\frac{x}{2a+x}\right)^5 + \cdots\right].$$

$$\left[\begin{array}{l} a \text{ positive} \\ x \text{ between } -a \text{ and } +\infty \end{array}\right]$$

$$\ln\left(\frac{1+x}{1-x}\right) = 2\left(x + \frac{x^3}{3} + \frac{x^5}{5} + \frac{x^7}{7} + \cdots\right). \quad [x^2 < 1]$$

$$\ln\left(\frac{x+1}{x-1}\right) = 2\left[\frac{1}{x} + \frac{1}{3}\left(\frac{1}{x^3}\right) + \frac{1}{5}\left(\frac{1}{x}\right)^5 + \frac{1}{7}\left(\frac{1}{x}\right)^7 + \cdots\right]. \quad [x^2 > 1]$$

$$\ln\left(\frac{x+1}{x}\right) = 2\left[\frac{1}{2x+1} + \frac{1}{3(2x+1)^3} + \frac{1}{5(2x+1)^5} + \cdots\right].$$

[x positive]

$$\ln(x + \sqrt{1+x^2}) = x - \frac{1}{2}\frac{x^3}{3} + \frac{1\cdot 3}{2\cdot 4}\frac{x^5}{5} - \frac{1\cdot 3\cdot 5}{2\cdot 4\cdot 6}\frac{x^7}{7} + \cdots. \quad [x^2 < 1]$$

111 Trigonometric Series

$$\sin x = x - \frac{x^3}{3!} + \frac{x^5}{5!} - \frac{x^7}{7!} + \cdots.$$

$$\cos x = 1 - \frac{x^2}{2!} + \frac{x^4}{4!} - \frac{x^6}{6!} + \cdots.$$

$$\tan x = x + \frac{x^3}{3} + \frac{2x^5}{15} + \frac{17x^7}{315} + \frac{62x^9}{2835} + \cdots. \quad \left[x^2 < \frac{\pi^2}{4}\right]$$

$$\sin^{-1} x = x + \frac{1}{2}\frac{x^3}{3} + \frac{1\cdot 3}{2\cdot 4}\frac{x^5}{5} + \frac{1\cdot 3\cdot 5}{2\cdot 4\cdot 6}\frac{x^7}{7} + \cdots. \quad [x^2 < 1]$$

$$\tan^{-1} x = x - \tfrac{1}{3}x^3 + \tfrac{1}{5}x^5 - \tfrac{1}{7}x^7 + \cdots. \quad [x^2 \leqq 1]$$

112 Logarithmic Trigonometric Series

$$\ln \sin x = \ln x - \frac{x^2}{6} - \frac{x^4}{180} - \frac{x^6}{2835} - \cdots. \quad [x^2 < \pi^2]$$

$$\ln \cos x = -\frac{x^2}{2} - \frac{x^4}{12} - \frac{x^6}{45} - \frac{17x^8}{2520} - \cdots. \quad \left[x^2 < \frac{\pi^2}{4}\right]$$

$$\ln \tan x = \ln x + \frac{x^2}{3} + \frac{7x^4}{90} + \frac{62x^6}{2835} + \cdots. \quad \left[x^2 < \frac{\pi^2}{4}\right]$$

113 Exponential Trigonometric Series

$$e^{\sin x} = 1 + x + \frac{x^2}{2!} - \frac{3x^4}{4!} - \frac{8x^5}{5!} + \frac{3x^6}{6!} + \cdots.$$

$$e^{\cos x} = e\left(1 - \frac{x^2}{2!} + \frac{4x^4}{4!} - \frac{31x^6}{6!} + \cdots\right).$$

$$e^{\tan x} = 1 + x + \frac{x^2}{2!} + \frac{3x^3}{3!} + \frac{9x^4}{4!} + \frac{37x^5}{5!} + \cdots. \quad \left[x^2 < \frac{\pi^2}{4}\right]$$

114 Approximations of Expressions Containing Small Terms

These may be derived from various infinite series given in 108–113. Some first approximations derived by neglecting all powers but the first of the small positive or negative quantity x = s are given below. The expression in brackets gives the next term beyond that which is used and by means of it the accuracy of the approximation may be estimated.

$\frac{1}{1+s} = 1 - s.$ $\quad [+s^2]$

$(1+s)^n = 1 + ns.$ $\quad \left[+\frac{n(n-1)}{2}s^2\right]$

$e^s = 1 + s.$ $\quad \left[+\frac{s^2}{2}\right]$

$\ln(1+s) = s.$ $\quad \left[-\frac{s^2}{2}\right]$

$\sin s = s.$ $\quad \left[-\frac{s^3}{6}\right]$

$\cos s = 1.$ $\quad \left[-\frac{s^2}{2}\right]$

$(1+s_1)(1+s_2) = (1+s_1+s_2)$ $\quad [+s_1s_2]$

The following expressions are some that may be approximated by $1 + s$, where s is a small positive or negative quantity and n is any number.

$\left(1+\frac{s}{n}\right)^n.$ $\quad e^s.$ $\quad 1 + \ln\sqrt{\frac{1+s}{1-s}}.$

$\sqrt[n]{1+ns}.$ $\quad 2 - e^{-s}.$ $\quad 1 + n\sin\frac{s}{n}.$

$\sqrt[n]{\frac{1+\frac{ns}{2}}{1-\frac{ns}{2}}}.$ $\quad 1 + n\ln\left(1+\frac{s}{n}\right).$ $\quad \cos\sqrt{-2s}.$

115 Evaluation of Indeterminate Forms [see Algebra, 2]

Let f(x) and F(x) be two functions of x, and let a be a value of x.

(1) If $\frac{f(a)}{F(a)} = \frac{0}{0}$ or $\frac{\infty}{\infty}$, use $\frac{f'(a)}{F'(a)}$ for the value of this fraction.

If $\frac{f'(a)}{F'(a)} = \frac{0}{0}$ or $\frac{\infty}{\infty}$, use $\frac{f''(a)}{F''(a)}$ for the value of this fraction, etc.

(2) If $f(a) \cdot F(a) = 0 \cdot \infty$ or if $f(a) - F(a) = \infty - \infty$, evaluate by changing the product or difference to the form $\frac{0}{0}$ or $\frac{\infty}{\infty}$ and use (1).

(3) If $f(a)^{F(a)} = 0^0$ or ∞^0 or 1^∞, then $f(a)^{F(a)} = e^{F(a)\cdot\ln f(a)}$, [Algebra, 8] and the exponent, being of the form $0 \cdot \infty$, may be evaluated by (2).

116 Differential of a Function

If $y = f(x)$ and Δx = increment in x, then the differential of x equals the increment of x, or $dx = \Delta x$; and the differential of y is the derivative of y multiplied by the differential of x, thus

$$dy = \frac{dy}{dx}\,dx = \frac{df(x)}{dx}\,dx = f'(x)\,dx,$$

and $\frac{dy}{dx} = dy \div dx$.

If $x = f_1(t)$ and $y = f_2(t)$, then $dx = f_1'(t)\,dt$, $dy = f_2'(t)\,dt$.

Every derivative formula has a corresponding differential formula; thus from the table 101, we have, for example,

$$d\,(uv) = u\,dv + v\,du;\quad d\,(\sin u) = \cos u\,du;\quad d\,(\tan^{-1} u) = \frac{du}{1+u^2},\ \text{etc.}$$

117 Functions of Several Variables. Partial Derivatives. Differentials

Let z be a function of two variables, $z = f(x, y)$, then its partial derivatives are

$$\frac{\partial z}{\partial x} = \frac{dz}{dx} \text{ when } y \text{ is kept constant.}$$

$$\frac{\partial z}{\partial y} = \frac{dz}{dy} \text{ when } x \text{ is kept constant.}$$

$$\frac{\partial^2 z}{\partial x^2} = \frac{\partial}{\partial x}\left(\frac{\partial z}{\partial x}\right);\quad \frac{\partial^2 z}{\partial y^2} = \frac{\partial}{\partial y}\left(\frac{\partial z}{\partial y}\right);\quad \frac{\partial^2 z}{\partial x\,\partial y} = \frac{\partial}{\partial x}\left(\frac{\partial z}{\partial y}\right) = \frac{\partial}{\partial y}\left(\frac{\partial z}{\partial x}\right) = \frac{\partial^2 z}{\partial y\,\partial x}.$$

Similarly, if $z = f(x, y, u, \cdots)$, then, for example,

$$\frac{\partial z}{\partial x} = \frac{dz}{dx} \text{ when } y, u, \ldots \text{ are kept constant.}$$

If $z = f(x, y, \cdots)$ and $x, y, \cdots$ are functions of a single variable, t,

$$\frac{dz}{dt} = \frac{\partial z}{\partial x}\frac{dx}{dt} + \frac{\partial z}{\partial y}\frac{dy}{dt} + \cdots.$$

If $z = f(x, y, \cdots)$, then $dz = \frac{\partial z}{\partial x}dx + \frac{\partial z}{\partial y}dy + \cdots$.

If $F(x, y, z, \cdots) = 0$, then $\frac{\partial F}{\partial x}dx + \frac{\partial F}{\partial y}dy + \frac{\partial F}{\partial z}dz + \cdots = 0$.

If $f(x, y) = 0$, then $\frac{dy}{dx} = -\frac{\partial f}{\partial x} \div \frac{\partial f}{\partial y}$.

118 Maxima and Minima of Functions of Two Variables

If $u = f(x, y)$, the values of x and y which make u a maximum or a minimum must satisfy the conditions

$$\frac{\partial u}{\partial x} = 0,\quad \frac{\partial u}{\partial y} = 0,\quad \left(\frac{\partial^2 u}{\partial x\,\partial y}\right)^2 < \left(\frac{\partial^2 u}{\partial x^2}\right)\left(\frac{\partial^2 u}{\partial y^2}\right).$$

A $\left\{\begin{matrix}\text{maximum}\\ \text{minimum}\end{matrix}\right\}$ also requires both $\frac{\partial^2 u}{\partial x^2}$ and $\frac{\partial^2 u}{\partial y^2}$ to be $\left\{\begin{matrix}\text{negative}\\ \text{positive}\end{matrix}\right\}$.

119 Space Curves. Surfaces (see Analytic Geometry, 95–97)

Let $x = f_1(t)$, $y = f_2(t)$, $z = f_3(t)$ be the equations of any space curve. The direction cosines of the tangent line to the curve at any point are proportional to dx, dy, and dz, or to $\frac{dx}{dt}$, $\frac{dy}{dt}$, and $\frac{dz}{dt}$.

Equations of tangent line at a point (x_1, y_1, z_1) are

$$\frac{x - x_1}{(dx)_1} = \frac{y - y_1}{(dy)_1} = \frac{z - z_1}{(dz)_1},$$ where $(dx)_1$ = value of dx at (x_1, y_1, z_1), etc.

Angle between two space curves is the angle between their tangent lines. (see Analytic Geometry, 94)

Let $F(x, y, z) = 0$ be the equation of a surface.

Direction cosines of the normal to the surface at any point are proportional to $\frac{\partial F}{\partial x}$, $\frac{\partial F}{\partial y}$, $\frac{\partial F}{\partial z}$.

Equations of the normal at any point (x_1, y_1, z_1) are

$$\frac{x - x_1}{\left(\frac{\partial F}{\partial x}\right)_1} = \frac{y - y_1}{\left(\frac{\partial F}{\partial y}\right)_1} = \frac{z - z_1}{\left(\frac{\partial F}{\partial z}\right)_1}.$$

Equation of the tangent plane at any point (x_1, y_1, z_1) is

$$(x - x_1)\left(\frac{\partial F}{\partial x}\right)_1 + (y - y_1)\left(\frac{\partial F}{\partial y}\right)_1 + (z - z_1)\left(\frac{\partial F}{\partial z}\right)_1 = 0,$$

where $\left(\frac{\partial F}{\partial x}\right)_1$ is the value of $\frac{\partial F}{\partial x}$ at the point (x_1, y_1, z_1), etc.

Angle between two surfaces is the angle between their normals.

INTEGRAL CALCULUS

120 Definition of Integral

F(x) is said to be the integral of f(x), if the derivative of F(x) is f(x), or the differential of F(x) is f(x) dx; in symbols:

$$F(x) = \int f(x)\,dx \text{ if } \frac{d\,F(x)}{dx} = f(x), \text{ or } d\,F(x) = f(x)\,dx.$$

In general: $\int f(x)\,dx = F(x) + C$, where C is an arbitrary constant.

121 Fundamental Theorems on Integrals

$$\int df(x) = f(x) + C.$$

$$d\int f(x)\,dx = f(x)\,dx.$$

$$\int [f_1(x) \pm f_2(x) \pm \cdots]\,dx = \int f_1(x)\,dx \pm \int f_2(x)\,dx \pm \cdots.$$

$\int a\,f(x)\,dx = a\int f(x)\,dx$, where a is any constant.

$\int u^n\,du = \frac{u^{n+1}}{n+1} + C \quad (n \neq -1)$; u is any function of x.

$$\int \frac{du}{u} = \ln u + C; \text{ u is any function of x.}$$

$$\int u\,dv = uv - \int v\,du; \text{ u and v are any functions of x.}$$

Table of Integrals

NOTE. In the following table, the constant of integration (C) is omitted but should be added to the result of every integration. The letter x represents any variable; the letter u represents any function of x; all other letters represent constants which may have any finite value unless otherwise indicated: $\ln = \log_e$; all angles are in radians.

Functions containing ax + b

122 $$\int (ax+b)^n\,dx = \frac{1}{a(n+1)}(ax+b)^{n+1}. \quad (n \neq -1)$$

123 $$\int \frac{dx}{ax+b} = \frac{1}{a}\ln(ax+b).$$

124 $$\int x(ax+b)^n\,dx = \frac{1}{a^2(n+2)}(ax+b)^{n+2} - \frac{b}{a^2(n+1)}(ax+b)^{n+1}.$$
$$(n \neq -1, -2)$$

125 $$\int \frac{x\,dx}{ax+b} = \frac{x}{a} - \frac{b}{a^2}\ln(ax+b).$$

126 $$\int \frac{x\,dx}{(ax+b)^2} = \frac{b}{a^2(ax+b)} + \frac{1}{a^2}\ln(ax+b).$$

127 $$\int x^2(ax+b)^n\,dx = \frac{1}{a^3}\left[\frac{(ax+b)^{n+3}}{n+3} - 2b\frac{(ax+b)^{n+2}}{n+2} + b^2\frac{(ax+b)^{n+1}}{n+1}\right]$$
$$(n \neq -1, -2, -3)$$

128 $$\int \frac{x^2\,dx}{ax+b} = \frac{1}{a^3}\left[\frac{1}{2}(ax+b)^2 - 2b(ax+b) + b^2\ln(ax+b)\right].$$

129 $$\int \frac{x^2\,dx}{(ax+b)^2} = \frac{1}{a^3}\left[(ax+b) - 2b\ln(ax+b) - \frac{b^2}{ax+b}\right].$$

130 $$\int \frac{x^2\,dx}{(ax+b)^3} = \frac{1}{a^3}\left[\ln(ax+b) + \frac{2b}{ax+b} - \frac{b^2}{2(ax+b)^2}\right].$$

131 $$\int x^m(ax+b)^n\,dx$$
$$= \frac{1}{a(m+n+1)}\left[x^m(ax+b)^{n+1} - mb\int x^{m-1}(ax+b)^n\,dx\right]$$
$$= \frac{1}{m+n+1}\left[x^{m+1}(ax+b)^n + nb\int x^m(ax+b)^{n-1}dx\right]. \quad \begin{pmatrix} m \text{ pos.} \\ m+n \\ +1 \neq 0 \end{pmatrix}$$

132 $$\int \frac{dx}{x(ax+b)} = \frac{1}{b}\ln\frac{x}{ax+b}.$$

133 $$\int \frac{dx}{x^2(ax+b)} = -\frac{1}{bx} + \frac{a}{b^2}\ln\frac{ax+b}{x}.$$

134 $$\int \frac{dx}{x(ax+b)^2} = \frac{1}{b(ax+b)} - \frac{1}{b^2}\ln\frac{ax+b}{x}.$$

135 $$\int \frac{dx}{x^2(ax+b)^2} = -\frac{b+2ax}{b^2x(ax+b)} + \frac{2a}{b^3}\ln\frac{ax+b}{x}.$$

136 $$\int \frac{dx}{x\sqrt{ax+b}} = \frac{1}{\sqrt{b}}\ln\frac{\sqrt{ax+b}-\sqrt{b}}{\sqrt{ax+b}+\sqrt{b}}. \quad \text{(b pos.)}$$

137 $$\int \frac{dx}{x\sqrt{ax+b}} = \frac{2}{\sqrt{-b}}\tan^{-1}\sqrt{\frac{ax+b}{-b}}. \quad \text{(b neg.)}$$

138 $$\int \frac{dx}{x(ax+b)^{\frac{n}{2}}} = \frac{2}{b(n-2)(ax+b)^{\frac{n}{2}-1}} + \frac{1}{b}\int\frac{dx}{x(ax+b)^{\frac{n}{2}-1}}. \quad \text{(n odd and pos.)}$$

139 $$\int \frac{\sqrt{ax+b}}{x}dx = 2\sqrt{ax+b} + \sqrt{b}\ln\frac{\sqrt{ax+b}-\sqrt{b}}{\sqrt{ax+b}+\sqrt{b}}. \quad \text{(b pos.)}$$

140 $$\int \frac{\sqrt{ax+b}}{x}dx = 2\sqrt{ax+b} - 2\sqrt{-b}\tan^{-1}\sqrt{\frac{ax+b}{-b}}. \quad \text{(b neg.)}$$

141 $$\int \frac{(ax+b)^{\frac{n}{2}}}{x}dx = \frac{2}{n}(ax+b)^{\frac{n}{2}} + b\int\frac{(ax+b)^{\frac{n}{2}-1}}{x}dx. \quad \text{(n odd and pos.)}$$

142 $$\int \frac{dx}{x^2\sqrt{ax+b}} = -\frac{\sqrt{ax+b}}{bx} - \frac{a}{2b\sqrt{b}}\ln\frac{\sqrt{ax+b}-\sqrt{b}}{\sqrt{ax+b}+\sqrt{b}}. \quad \text{(b pos.)}$$

143 $$\int \frac{dx}{x^2\sqrt{ax+b}} = -\frac{\sqrt{ax+b}}{bx} - \frac{a}{b\sqrt{-b}}\tan^{-1}\sqrt{\frac{ax+b}{-b}}. \quad \text{(b neg.)}$$

144 $$\int \frac{dx}{(ax+b)(px+q)} = \frac{1}{bp-aq}\ln\frac{px+q}{ax+b}. \quad (bp-aq\neq 0)$$

145 $$\int \frac{dx}{(ax+b)^2(px+q)} = \frac{1}{bp-aq}\left[\frac{1}{ax+b}+\frac{p}{bp-aq}\ln\frac{px+q}{ax+b}\right]. \quad (bp-aq\neq 0)$$

146 $$\int \frac{dx}{(ax+b)^n(px+q)^m} = \frac{1}{(m-1)(bp-aq)}\left[\frac{1}{(ax+b)^{n-1}(px+q)^{m-1}} - a(m+n-2)\int\frac{dx}{(ax+b)^n(px+q)^{m-1}}\right].$$
$$(m>1,\ n\ \text{pos.},\ bp-aq\neq 0)$$

147 $$\int \frac{x\,dx}{(ax+b)(px+q)} = \frac{1}{bp-aq}\left[\frac{b}{a}\ln(ax+b)-\frac{q}{p}\ln(px+q)\right]. \quad (bp-aq\neq 0)$$

148 $$\int \frac{x\,dx}{(ax+b)^2(px+q)} = \frac{1}{bp-aq}\left[-\frac{b}{a(ax+b)}-\frac{q}{bp-aq}\ln\frac{px+q}{ax+b}\right].$$
$$(bp-aq\neq 0)$$

149 $$\int \frac{px+q}{\sqrt{ax+b}}dx = \frac{2}{3a^2}(3aq-2bp+apx)\sqrt{ax+b}.$$

150 $$\int \frac{\sqrt{ax+b}}{px+q}dx = \frac{2\sqrt{ax+b}}{p} - \frac{2}{p}\sqrt{\frac{aq-bp}{p}}\tan^{-1}\sqrt{\frac{p(ax+b)}{aq-bp}}.$$
$$(p\ \text{pos.},\ aq>bp)$$

151 $$\int \frac{\sqrt{ax+b}}{px+q}dx = \frac{2\sqrt{ax+b}}{p} + \frac{1}{p}\sqrt{\frac{bp-aq}{p}} \ln \frac{\sqrt{p(ax+b)}-\sqrt{bp-aq}}{\sqrt{p(ax+b)}+\sqrt{bp-aq}}.$$
(p pos., $bp > aq$)

152 $$\int \frac{dx}{(px+q)\sqrt{ax+b}} = \frac{2}{\sqrt{p}\sqrt{aq-bp}} \tan^{-1}\sqrt{\frac{p(ax+b)}{aq-bp}}.$$ (p pos., $aq > bp$)

153 $$\int \frac{dx}{(px+q)\sqrt{ax+b}} = -\frac{1}{\sqrt{p}\sqrt{bp-aq}} \ln \frac{\sqrt{p(ax+b)}-\sqrt{bp-aq}}{\sqrt{p(ax+b)}+\sqrt{bp-aq}}.$$
(p pos., $bp > aq$)

154 $$\int \frac{\sqrt{px+q}}{\sqrt{ax+b}}dx = \frac{1}{a}\sqrt{ax+b}\sqrt{px+q} - \frac{bp-aq}{a\sqrt{ap}}$$
$$\ln\left(\sqrt{p(ax+b)} + \sqrt{a(px+q)}\right).$$ (a and p, same sign)
$$= \frac{1}{a}\sqrt{ax+b}\sqrt{px+q} - \frac{bp-aq}{a\sqrt{-ap}}\tan^{-1}\frac{\sqrt{-ap(ax+b)}}{a\sqrt{px+q}}$$
(a and p have opposite signs)
$$= \frac{1}{a}\sqrt{ax+b}\sqrt{px+q} + \frac{bp-aq}{2a\sqrt{-ap}}$$
$$\sin^{-1}\frac{2apx+aq+bp}{bp-aq}.$$ (a and p have opposite signs)

Functions containing $ax^2 + b$

155 $$\int \frac{dx}{ax^2+b} = \frac{1}{\sqrt{ab}}\tan^{-1}\left(x\sqrt{\frac{a}{b}}\right).$$ (a and b pos.)

156 $$\int \frac{dx}{ax^2+b} = \frac{1}{2\sqrt{-ab}} \ln \frac{x\sqrt{a}-\sqrt{-b}}{x\sqrt{a}+\sqrt{-b}}.$$ (a pos., b neg.)
$$= \frac{1}{2\sqrt{-ab}} \ln \frac{\sqrt{b}+x\sqrt{-a}}{\sqrt{b}-x\sqrt{-a}}.$$ (a neg., b pos.)

157 $$\int \frac{dx}{(ax^2+b)^n} = \frac{1}{2(n-1)b}\frac{x}{(ax^2+b)^{n-1}} + \frac{2n-3}{2(n-1)b}\int \frac{dx}{(ax^2+b)^{n-1}}$$ (n integ. > 1)

158 $$\int (ax^2+b)^n x\,dx = \frac{1}{2a}\frac{(ax^2+b)^{n+1}}{n+1}.$$ ($n \neq -1$)

159 $$\int \frac{x\,dx}{ax^2+b} = \frac{1}{2a}\ln(ax^2+b).$$

160 $$\int \frac{dx}{x(ax^2+b)} = \frac{1}{2b}\ln\frac{x^2}{ax^2+b}.$$

161 $$\int \frac{x^2\,dx}{ax^2+b} = \frac{x}{a} - \frac{b}{a}\int \frac{dx}{ax^2+b}.$$

162 $$\int \frac{x^2\,dx}{(ax^2+b)^n} = -\frac{1}{2(n-1)a}\frac{x}{(ax^2+b)^{n-1}} + \frac{1}{2(n-1)a}\int \frac{dx}{(ax^2+b)^{n-1}}.$$
(n integ. > 1)

163 $$\int \frac{dx}{x^2(ax^2+b)^n} = \frac{1}{b}\int \frac{dx}{x^2(ax^2+b)^{n-1}} - \frac{a}{b}\int \frac{dx}{(ax^2+b)^n}.$$ (n pos. integ.)

164 $\int \sqrt{ax^2+b}\,dx = \frac{x}{2}\sqrt{ax^2+b} + \frac{b}{2\sqrt{a}}\ln\left(x\sqrt{a}+\sqrt{ax^2+b}\right).$ (a pos.)

165 $\int \sqrt{ax^2+b}\,dx = \frac{x}{2}\sqrt{ax^2+b} + \frac{b}{2\sqrt{-a}}\sin^{-1}\left(x\sqrt{-\frac{a}{b}}\right).$ (a neg.)

166 $\int \frac{dx}{\sqrt{ax^2+b}} = \frac{1}{\sqrt{a}}\ln\left(x\sqrt{a}+\sqrt{ax^2+b}\right).$ (a pos.)

167 $\int \frac{dx}{\sqrt{ax^2+b}} = \frac{1}{\sqrt{-a}}\sin^{-1}\left(x\sqrt{-\frac{a}{b}}\right).$ (a neg.)

168 $\int \sqrt{ax^2+b}\,x\,dx = \frac{1}{3a}(ax^2+b)^{\frac{3}{2}}.$

169 $\int \frac{x\,dx}{\sqrt{ax^2+b}} = \frac{1}{a}\sqrt{ax^2+b}.$

170 $\int \frac{\sqrt{ax^2+b}}{x}\,dx = \sqrt{ax^2+b} + \sqrt{b}\ln\frac{\sqrt{ax^2+b}-\sqrt{b}}{x}.$ (b pos.)

171 $\int \frac{\sqrt{ax^2+b}}{x}\,dx = \sqrt{ax^2+b} - \sqrt{-b}\tan^{-1}\frac{\sqrt{ax^2+b}}{\sqrt{-b}}.$ (b neg.)

172 $\int \frac{dx}{x\sqrt{ax^2+b}} = \frac{1}{\sqrt{b}}\ln\frac{\sqrt{ax^2+b}-\sqrt{b}}{x}.$ (b pos.)

173 $\int \frac{dx}{x\sqrt{ax^2+b}} = \frac{1}{\sqrt{-b}}\sec^{-1}\left(x\sqrt{-\frac{a}{b}}\right).$ (b neg.)

174 $\int \sqrt{ax^2+b}\,x^2\,dx = \frac{x}{4a}(ax^2+b)^{\frac{3}{2}} - \frac{bx}{8a}\sqrt{ax^2+b}$

$- \frac{b^2}{8a\sqrt{a}}\ln\left(x\sqrt{a}+\sqrt{ax^2+b}\right).$ (a pos.)

175 $\int \sqrt{ax^2+b}\,x^2\,dx = \frac{x}{4a}(ax^2+b)^{\frac{3}{2}} - \frac{bx}{8a}\sqrt{ax^2+b}$

$- \frac{b^2}{8a\sqrt{-a}}\sin^{-1}\left(x\sqrt{\frac{-a}{b}}\right).$ (a neg.)

176 $\int \frac{x^2\,dx}{\sqrt{ax^2+b}} = \frac{x}{2a}\sqrt{ax^2+b} - \frac{b}{2a\sqrt{a}}\ln\left(x\sqrt{a}+\sqrt{ax^2+b}\right).$ (a pos.)

177 $\int \frac{x^2\,dx}{\sqrt{ax^2+b}} = \frac{x}{2a}\sqrt{ax^2+b} - \frac{b}{2a\sqrt{-a}}\sin^{-1}\left(x\sqrt{-\frac{a}{b}}\right).$ (a neg.)

178 $\int \frac{\sqrt{ax^2+b}}{x^2}\,dx = -\frac{\sqrt{ax^2+b}}{x} + \sqrt{a}\ln\left(x\sqrt{a}+\sqrt{ax^2+b}\right).$ (a pos.)

179 $\int \frac{\sqrt{ax^2+b}}{x^2}\,dx = -\frac{\sqrt{ax^2+b}}{x} - \sqrt{-a}\sin^{-1}\left(x\sqrt{-\frac{a}{b}}\right).$ (a neg.)

180 $\int \frac{dx}{x^2\sqrt{ax^2+b}} = -\frac{\sqrt{ax^2+b}}{bx}.$

181 $\int \frac{x^n\,dx}{\sqrt{ax^2+b}} = \frac{x^{n-1}\sqrt{ax^2+b}}{na} - \frac{(n-1)b}{na}\int \frac{x^{n-2}\,dx}{\sqrt{ax^2+b}}.$ (n pos.)

182 $\int x^n \sqrt{ax^2+b}\,dx = \frac{x^{n-1}(ax^2+b)^{\frac{3}{2}}}{(n+2)a} - \frac{(n-1)b}{(n+2)a}\int x^{n-2}\sqrt{ax^2+b}\,dx.$ (n pos.)

183 $\int \frac{\sqrt{ax^2+b}\,dx}{x^n} = -\frac{(ax^2+b)^{\frac{3}{2}}}{b(n-1)x^{n-1}} - \frac{(n-4)a}{(n-1)b}\int \frac{\sqrt{ax^2+b}}{x^{n-2}}dx.$ ($n > 1$)

184 $\int \frac{dx}{x^n\sqrt{ax^2+b}} = -\frac{\sqrt{ax^2+b}}{b(n-1)x^{n-1}} - \frac{(n-2)a}{(n-1)b}\int \frac{dx}{x^{n-2}\sqrt{ax^2+b}}.$ ($n > 1$)

185 $\int (ax^2+b)^{\frac{3}{2}}dx = \frac{x}{8}(2ax^2+5b)\sqrt{ax^2+b} + \frac{3b^2}{8\sqrt{a}}\ln\left(x\sqrt{a}+\sqrt{ax^2+b}\right).$ (a pos.)

186 $\int (ax^2+b)^{\frac{3}{2}}dx = \frac{x}{8}(2ax^2+5b)\sqrt{ax^2+b} + \frac{3b^2}{8\sqrt{-a}}\sin^{-1}\left(x\sqrt{-\frac{a}{b}}\right).$ (a neg.)

187 $\int \frac{dx}{(ax^2+b)^{\frac{3}{2}}} = \frac{x}{b\sqrt{ax^2+b}}.$

188 $\int (ax^2+b)^{\frac{3}{2}}x\,dx = \frac{1}{5a}(ax^2+b)^{\frac{5}{2}}.$

189 $\int \frac{x\,dx}{(ax^2+b)^{\frac{3}{2}}} = -\frac{1}{a\sqrt{ax^2+b}}.$

190 $\int \frac{x^2\,dx}{(ax^2+b)^{\frac{3}{2}}} = -\frac{x}{a\sqrt{ax^2+b}} + \frac{1}{a\sqrt{a}}\ln\left(x\sqrt{a}+\sqrt{ax^2+b}\right).$ (a pos.)

191 $\int \frac{x^2\,dx}{(ax^2+b)^{\frac{3}{2}}} = -\frac{x}{a\sqrt{ax^2+b}} + \frac{1}{a\sqrt{-a}}\sin^{-1}\left(x\sqrt{-\frac{a}{b}}\right).$ (a neg.)

192 $\int \frac{dx}{x(ax^n+b)} = \frac{1}{bn}\ln\frac{x^n}{ax^n+b}.$

193 $\int \frac{dx}{x\sqrt{ax^n+b}} = \frac{1}{n\sqrt{b}}\ln\frac{\sqrt{ax^n+b}-\sqrt{b}}{\sqrt{ax^n+b}+\sqrt{b}}.$ (b pos.)

194 $\int \frac{dx}{x\sqrt{ax^n+b}} = \frac{2}{n\sqrt{-b}}\sec^{-1}\sqrt{\frac{-ax^n}{b}}.$ (b neg.)

Functions containing $ax^2 + bx + c$

195 $\int \frac{dx}{ax^2+bx+c} = \frac{1}{\sqrt{b^2-4ac}}\ln\frac{2ax+b-\sqrt{b^2-4ac}}{2ax+b+\sqrt{b^2-4ac}}.$ ($b^2 > 4ac$)

196 $\int \frac{dx}{ax^2+bx+c} = \frac{2}{\sqrt{4ac-b^2}}\tan^{-1}\frac{2ax+b}{\sqrt{4ac-b^2}}.$ ($b^2 < 4ac$)

197 $\int \frac{dx}{ax^2+bx+c} = -\frac{2}{2ax+b}.$ ($b^2 = 4ac$)

198 $\int \frac{x\,dx}{ax^2+bx+c} = \frac{1}{2a}\ln(ax^2+bx+c) - \frac{b}{2a}\int\frac{dx}{ax^2+bx+c}.$

199 $\int \frac{x^2\,dx}{ax^2+bx+c} = \frac{x}{a} - \frac{b}{2a^2}\ln(ax^2+bx+c) + \frac{b^2-2ac}{2a^2}\int\frac{dx}{ax^2+bx+c}.$

200 $\int \frac{dx}{\sqrt{ax^2+bx+c}} = \frac{1}{\sqrt{a}}\ln\left(2ax+b+2\sqrt{a}\sqrt{ax^2+bx+c}\right).$ (a pos.)

201 $\int \frac{dx}{\sqrt{ax^2+bx+c}} = \frac{1}{\sqrt{-a}}\sin^{-1}\frac{-2ax-b}{\sqrt{b^2-4ac}}.$ (a neg.)

202 $\int \sqrt{ax^2+bx+c}\,dx = \frac{2ax+b}{4a}\sqrt{ax^2+bx+c} + \frac{4ac-b^2}{8a}\int\frac{dx}{\sqrt{ax^2+bx+c}}.$

203 $\int \frac{x\,dx}{\sqrt{ax^2+bx+c}} = \frac{\sqrt{ax^2+bx+c}}{a} - \frac{b}{2a}\int\frac{dx}{\sqrt{ax^2+bx+c}}.$

204 $\int \sqrt{ax^2+bx+c}\;x\,dx = \frac{(ax^2+bx+c)^{\frac{3}{2}}}{3a} - \frac{b}{2a}\int\sqrt{ax^2+bx+c}\,dx.$

205 $\int \frac{dx}{x\sqrt{ax^2+bx+c}} = -\frac{1}{\sqrt{c}}\ln\left(\frac{\sqrt{ax^2+bx+c}+\sqrt{c}}{x}+\frac{b}{2\sqrt{c}}\right).$ (c pos.)

206 $\int \frac{dx}{x\sqrt{ax^2+bx+c}} = \frac{1}{\sqrt{-c}}\sin^{-1}\frac{bx+2c}{x\sqrt{b^2-4ac}}.$ (c neg.)

207 $\int \frac{dx}{x\sqrt{ax^2+bx}} = -\frac{2}{bx}\sqrt{ax^2+bx}.$

208 $\int \frac{dx}{(ax^2+bx+c)^{\frac{3}{2}}} = -\frac{2(2ax+b)}{(b^2-4ac)\sqrt{ax^2+bx+c}}.$

Functions containing sin ax

209 $\int \sin u\,du = -\cos u.$ (u is any function of x)

210 $\int \sin ax\,dx = -\frac{1}{a}\cos ax.$

211 $\int \sin^2 ax\,dx = \frac{x}{2} - \frac{\sin 2ax}{4a}.$

212 $\int \sin^3 ax\,dx = -\frac{1}{a}\cos ax + \frac{1}{3a}\cos^3 ax.$

213 $\int \sin^4 ax\,dx = \frac{3}{8}x - \frac{1}{4a}\sin 2ax + \frac{1}{32a}\sin 4ax.$

214 $\int \sin^n ax\,dx = -\frac{\sin^{n-1}ax\cos ax}{na} + \frac{n-1}{n}\int\sin^{n-2}ax\,dx.$ (n pos. integ.)

215 $\int \frac{dx}{\sin ax} = \frac{1}{a}\ln\tan\frac{ax}{2} = \frac{1}{a}\ln(\csc ax - \cot ax).$

216 $\int \frac{dx}{\sin^2 ax} = -\frac{1}{a}\cot ax.$

217 $\int \frac{dx}{\sin^n ax} = -\frac{1}{a(n-1)}\frac{\cos ax}{\sin^{n-1}ax} + \frac{n-2}{n-1}\int\frac{dx}{\sin^{n-2}ax}.$ (n integ. > 1)

218 $\int \frac{dx}{1+\sin ax} = -\frac{1}{a}\tan\left(\frac{\pi}{4}-\frac{ax}{2}\right).$

219 $\int \frac{dx}{1 - \sin ax} = \frac{1}{a} \cot \left(\frac{\pi}{4} - \frac{ax}{2}\right).$

220 $\int \frac{dx}{b + c \sin ax} = \frac{-2}{a\sqrt{b^2 - c^2}} \tan^{-1}\left[\sqrt{\frac{b-c}{b+c}} \tan\left(\frac{\pi}{4} - \frac{ax}{2}\right)\right].$ $(b^2 > c^2)$

221 $\int \frac{dx}{b + c \sin ax} = \frac{-1}{a\sqrt{c^2 - b^2}} \ln \frac{c + b \sin ax + \sqrt{c^2 - b^2} \cos ax}{b + c \sin ax}.$ $(c^2 > b^2)$

222 $\int \sin ax \sin bx \, dx = \frac{\sin (a-b) x}{2(a-b)} - \frac{\sin (a+b) x}{2(a+b)}.$ $(a^2 \neq b^2)$

Functions containing cos ax

223 $\int \cos u \, du = \sin u.$ (u is any function of x)

224 $\int \cos ax \, dx = \frac{1}{a} \sin ax.$

$\int \sqrt{1 - \cos x}\, dx = \sqrt{2} \int \sin \frac{x}{2} dx.$

225 $\int \cos^2 ax \, dx = \frac{x}{2} + \frac{\sin 2ax}{4a}.$

$\int \sqrt{1 + \cos x}\, dx = \sqrt{2} \int \cos \frac{x}{2} dx.$

226 $\int \cos^3 ax \, dx = \frac{1}{a} \sin ax - \frac{1}{3a} \sin^3 ax.$

227 $\int \cos^4 ax \, dx = \frac{3}{8} x + \frac{1}{4a} \sin 2ax + \frac{1}{32a} \sin 4ax.$

228 $\int \cos^n ax \, dx = \frac{\cos^{n-1} ax \sin ax}{na} + \frac{n-1}{n} \int \cos^{n-2} ax \, dx.$ (n pos. integ.)

229 $\int \frac{dx}{\cos ax} = \frac{1}{a} \ln \tan\left(\frac{ax}{2} + \frac{\pi}{4}\right) = \frac{1}{a} \ln (\tan ax + \sec ax).$

230 $\int \frac{dx}{\cos^2 ax} = \frac{1}{a} \tan ax.$

231 $\int \frac{dx}{\cos^n ax} = \frac{1}{a(n-1)} \frac{\sin ax}{\cos^{n-1} ax} + \frac{n-2}{n-1} \int \frac{dx}{\cos^{n-2} ax}.$ (n integ. > 1)

232 $\int \frac{dx}{1 + \cos ax} = \frac{1}{a} \tan \frac{ax}{2}.$

233 $\int \frac{dx}{1 - \cos ax} = -\frac{1}{a} \cot \frac{ax}{2}.$

234 $\int \frac{dx}{b + c \cos ax} = \frac{2}{a\sqrt{b^2 - c^2}} \tan^{-1}\left(\sqrt{\frac{b-c}{b+c}} \tan \frac{ax}{2}\right).$ $(b^2 > c^2)$

235 $\int \frac{dx}{b + c \cos ax} = \frac{1}{a\sqrt{c^2 - b^2}} \ln \frac{c + b \cos ax + \sqrt{c^2 - b^2} \sin ax}{b + c \cos ax}.$ $(c^2 > b^2)$

236 $\int \cos ax \cos bx \, dx = \frac{\sin (a-b) x}{2(a-b)} + \frac{\sin (a+b) x}{2(a+b)}.$ $(a^2 \neq b^2)$

Functions containing sin ax and cos ax

237 $\int \sin ax \cos bx \, dx = -\frac{1}{2}\left[\frac{\cos (a-b) x}{a-b} + \frac{\cos (a+b) x}{a+b}\right].$ $(a^2 \neq b^2)$

238 $\int \sin^n ax \cos ax \, dx = \frac{1}{a(n+1)} \sin^{n+1} ax.$ $(n \neq -1).$

239 $\int \frac{\cos ax}{\sin ax} dx = \frac{1}{a} \ln \sin ax.$

240 $\int (b + c \sin ax)^n \cos ax\, dx = \frac{1}{ac(n+1)}(b + c \sin ax)^{n+1}. \quad (n \neq -1)$

241 $\int \frac{\cos ax\, dx}{b + c \sin ax} = \frac{1}{ac} \ln (b + c \sin ax).$

242 $\int \cos^n ax \sin ax\, dx = -\frac{1}{a(n+1)} \cos^{n+1} ax. \quad (n \neq -1).$

243 $\int \frac{\sin ax}{\cos ax} dx = -\frac{1}{a} \ln \cos ax.$

244 $\int (b + c \cos ax)^n \sin ax\, dx = -\frac{1}{ac(n+1)}(b + c \cos ax)^{n+1}. \quad (n \neq -1)$

245 $\int \frac{\sin ax}{b + c \cos ax} dx = -\frac{1}{ac} \ln (b + c \cos ax).$

246 $\int \frac{dx}{b \sin ax + c \cos ax} = \frac{1}{a\sqrt{b^2 + c^2}} \ln \left[\tan \frac{1}{2} \left(ax + \tan^{-1} \frac{c}{b} \right) \right].$

247 $\int \sin^2 ax \cos^2 ax\, dx = \frac{x}{8} - \frac{\sin 4ax}{32a}.$

248 $\int \frac{dx}{\sin ax \cos ax} = \frac{1}{a} \ln \tan ax.$

249 $\int \frac{dx}{\sin^2 ax \cos^2 ax} = \frac{1}{a}(\tan ax - \cot ax).$

250 $\int \frac{\sin^2 ax}{\cos ax} dx = \frac{1}{a}\left[-\sin ax + \ln \tan \left(\frac{ax}{2} + \frac{\pi}{4} \right) \right].$

251 $\int \frac{\cos^2 ax}{\sin ax} dx = \frac{1}{a}\left[\cos ax + \ln \tan \frac{ax}{2} \right].$

252 $\int \sin^m ax \cos^n ax\, dx = -\frac{\sin^{m-1} ax \cos^{n+1} ax}{a(m+n)}$

$+ \frac{m-1}{m+n} \int \sin^{m-2} ax \cos^n ax\, dx. \quad$ (m, n pos.)

253 $\int \sin^m ax \cos^n ax\, dx = \frac{\sin^{m+1} ax \cos^{n-1} ax}{a(m+n)}$

$+ \frac{n-1}{m+n} \int \sin^m ax \cos^{n-2} ax\, dx. \quad$ (m, n pos.)

254 $\int \frac{\cos^n ax}{\sin^m ax} dx = \frac{-\cos^{n+1} ax}{a(m-1)\sin^{m-1} ax}$

$+ \frac{m-n-2}{(m-1)} \int \frac{\cos^n ax}{\sin^{m-2} ax} dx. \quad (m, n \text{ pos.}, m \neq 1)$

255 $\int \frac{\sin^m ax}{\cos^n ax} dx = \frac{\sin^{m+1} ax}{a(n-1)\cos^{n-1} ax}$

$- \frac{m-n+2}{n-1} \int \frac{\sin^m ax}{\cos^{n-2} ax} dx. \quad (m, n \text{ pos.}, n \neq 1)$

256 $\int \frac{\sin^{2n} ax}{\cos ax} dx = \int \frac{(1 - \cos^2 ax)^n}{\cos ax} dx.$ (Expand, divide, and use 224–229)

257 $\int \frac{\cos^{2n} ax}{\sin ax} dx = \int \frac{(1 - \sin^2 ax)^n}{\sin ax} dx.$ (Expand, divide, and use 210–215)

258 $\int \frac{\sin^{2n+1} ax}{\cos ax} dx = \int \frac{(1 - \cos^2 ax)^n}{\cos ax} \sin ax\, dx.$

(Expand, divide, and use **242–243**)

259 $\int \frac{\cos^{2n+1} ax}{\sin ax} dx = \int \frac{(1 - \sin^2 ax)^n}{\sin ax} \cos ax\, dx.$

(Expand, divide, and use **238–239**)

Functions containing tan ax $\left(= \frac{1}{\cot ax}\right)$ or cot ax $\left(= \frac{1}{\tan ax}\right)$

260 $\int \tan u\, du = -\ln \cos u.$ (u is any function of **x**)

261 $\int \tan ax\, dx = -\frac{1}{a} \ln \cos ax.$

262 $\int \tan^2 ax\, dx = \frac{1}{a} \tan ax - x.$

263 $\int \tan^n ax\, dx = \frac{1}{a(n-1)} \tan^{n-1} ax - \int \tan^{n-2} ax\, dx.$ (n integ. > 1)

264 $\int \cot u\, du = \ln \sin u.$ (u is any function of **x**)

265 $\int \cot ax\, dx = \int \frac{dx}{\tan ax} = \frac{1}{a} \ln \sin ax.$

266 $\int \cot^2 ax\, dx = \int \frac{dx}{\tan^2 ax} = -\frac{1}{a} \cot ax - x.$

267 $\int \cot^n ax\, dx = \int \frac{dx}{\tan^n ax} = -\frac{1}{a(n-1)} \cot^{n-1} ax - \int \cot^{n-2} ax\, dx.$

(n integ. > 1)

268 $\int \frac{dx}{b + c \tan ax} = \int \frac{\cot ax\, dx}{b \cot ax + c} = \frac{1}{b^2 + c^2}\left[bx + \frac{c}{a} \ln (b \cos ax + c \sin ax)\right].$

269 $\int \frac{dx}{b + c \cot ax} = \int \frac{\tan ax\, dx}{b \tan ax + c} = \frac{1}{b^2 + c^2}\left[bx - \frac{c}{a} \ln (c \cos ax + b \sin ax)\right].$

270 $\int \frac{dx}{\sqrt{1 + \tan^2 ax}} = \frac{1}{a} \sin ax.$

271 $\int \frac{dx}{\sqrt{b + c \tan^2 ax}} = \frac{1}{a\sqrt{b - c}} \sin^{-1}\left(\sqrt{\frac{b - c}{b}} \sin ax\right).$ (b pos., $b^2 > c^2$)

Functions containing sec ax $\left(= \frac{1}{\cos ax}\right)$ or csc ax $\left(= \frac{1}{\sin ax}\right)$

272 $\int \sec u\, du = \ln (\sec u + \tan u) = \ln \tan \left(\frac{u}{2} + \frac{\pi}{4}\right).$ (u is any function of **x**)

273 $\int \sec ax\, dx = \frac{1}{a} \ln \tan \left(\frac{ax}{2} + \frac{\pi}{4}\right).$

274 $\int \sec^2 ax\, dx = \frac{1}{a} \tan ax.$

275 $\int \sec^n ax\, dx = \frac{1}{a(n-1)} \frac{\sin ax}{\cos^{n-1} ax} + \frac{n-2}{n-1} \int \sec^{n-2} ax\, dx.$ (n integ. > 1)

276 $\int \csc u\, du = \ln (\csc u - \cot u) = \ln \tan \frac{u}{2}.$ (u is any function of **x**)

277 $\int \csc ax\, dx = \frac{1}{a} \ln \tan \frac{ax}{2}.$

278 $\int \csc^2 ax\, dx = -\frac{1}{a} \cot ax.$

279 $\int \csc^n ax\, dx = -\frac{1}{a(n-1)} \frac{\cos ax}{\sin^{n-1} ax} + \frac{n-2}{n-1} \int \csc^{n-2} ax\, dx.$ (n integ. > 1)

Functions containing tan ax and sec ax or cot ax and csc ax

280 $\int \tan u \sec u\, du = \sec u.$ (u is any function of x)

281 $\int \tan ax \sec ax\, dx = \frac{1}{a} \sec ax.$

282 $\int \tan^n ax \sec^2 ax\, dx = \frac{1}{a(n+1)} \tan^{n+1} ax.$ $(n \neq -1)$

283 $\int \frac{\sec^2 ax\, dx}{\tan ax} = \frac{1}{a} \ln \tan ax.$

284 $\int \cot u \csc u\, du = -\csc u.$ (u is any function of x)

285 $\int \cot ax \csc ax\, dx = -\frac{1}{a} \csc ax.$

286 $\int \cot^n ax \csc^2 ax\, dx = -\frac{1}{a(n+1)} \cot^{n+1} ax.$ $(n \neq -1)$

287 $\int \frac{\csc^2 ax\, dx}{\cot ax} = -\frac{1}{a} \ln \cot ax.$

Inverse Trigonometric Functions

288 $\int \sin^{-1} ax\, dx = x \sin^{-1} ax + \frac{1}{a} \sqrt{1 - a^2x^2}.$

289 $\int \cos^{-1} ax\, dx = x \cos^{-1} ax - \frac{1}{a} \sqrt{1 - a^2x^2}.$

290 $\int \tan^{-1} ax\, dx = x \tan^{-1} ax - \frac{1}{2a} \ln (1 + a^2x^2).$

291 $\int \cot^{-1} ax\, dx = x \cot^{-1} ax + \frac{1}{2a} \ln (1 + a^2x^2).$

292 $\int \sec^{-1} ax\, dx = x \sec^{-1} ax - \frac{1}{a} \ln \left(ax + \sqrt{a^2x^2 - 1}\right).$

293 $\int \csc^{-1} ax\, dx = x \csc^{-1} ax + \frac{1}{a} \ln \left(ax + \sqrt{a^2x^2 - 1}\right).$

Algebraic and Trigonometric Functions

294 $\int x \sin ax\, dx = \frac{1}{a^2} \sin ax - \frac{1}{a} x \cos ax.$

295 $\int x^n \sin ax\, dx = -\frac{1}{a} x^n \cos ax + \frac{n}{a} \int x^{n-1} \cos ax\, dx.$ (n pos.)

296 $\int \frac{\sin ax\, dx}{x} = ax - \frac{(ax)^3}{3 \cdot 3!} + \frac{(ax)^5}{5 \cdot 5!} - \cdots.$

297 $\int x \cos ax\, dx = \frac{1}{a^2} \cos ax + \frac{1}{a} x \sin ax.$

298 $\int x^n \cos ax\, dx = \frac{1}{a} x^n \sin ax - \frac{n}{a} \int x^{n-1} \sin ax\, dx.$ (n pos.)

299 $\int \frac{\cos ax\, dx}{x} = \ln ax - \frac{(ax)^2}{2 \cdot 2!} + \frac{(ax)^4}{4 \cdot 4!} - \cdots.$

Exponential, Algebraic, Trigonometric, Logarithmic Functions

300 $\int b^u\, du = \frac{b^u}{\ln b}.$ (u is any function of x)

301 $\int e^u\, du = e^u.$ (u is any function of x)

302 $\int b^{ax}\, dx = \frac{b^{ax}}{a \ln b}.$

303 $\int e^{ax}\, dx = \frac{1}{a} e^{ax}.$

304 $\int \frac{dx}{b + ce^{ax}} = \frac{1}{ab} [ax - \ln (b + ce^{ax})].$

305 $\int \frac{e^{ax}\, dx}{b + ce^{ax}} = \frac{1}{ac} \ln (b + ce^{ax}).$

306 $\int \frac{dx}{be^{ax} + ce^{-ax}} = \frac{1}{a\sqrt{bc}} \tan^{-1}\left(e^{ax}\sqrt{\frac{b}{c}}\right).$ (b and c pos.)

307 $\int xb^{ax}\, dx = \frac{xb^{ax}}{a \ln b} - \frac{b^{ax}}{a^2 (\ln b)^2}.$

308 $\int xe^{ax}\, dx = \frac{e^{ax}}{a^2} (ax - 1).$

309 $\int x^n b^{ax}\, dx = \frac{x^n b^{ax}}{a \ln b} - \frac{n}{a \ln b} \int x^{n-1} b^{ax}\, dx.$ (n pos.)

310 $\int x^n e^{ax}\, dx = \frac{1}{a} x^n e^{ax} - \frac{n}{a} \int x^{n-1} e^{ax}\, dx.$ (n pos.)

311 $\int \frac{e^{ax}}{x}\, dx = \ln x + ax + \frac{(ax)^2}{2 \cdot 2!} + \frac{(ax)^3}{3 \cdot 3!} + \cdots.$

312 $\int \frac{e^{ax}}{x^n}\, dx = \frac{1}{n-1} \left[-\frac{e^{ax}}{x^{n-1}} + a \int \frac{e^{ax}}{x^{n-1}}\, dx \right].$ (n integ. > 1)

313 $\int e^{ax} \ln x\, dx = \frac{1}{a} e^{ax} \ln x - \frac{1}{a} \int \frac{e^{ax}}{x}\, dx.$

314 $\int e^{ax} \sin bx\, dx = \frac{e^{ax}}{a^2 + b^2} (a \sin bx - b \cos bx).$

315 $\int e^{ax} \cos bx\, dx = \frac{e^{ax}}{a^2 + b^2} (a \cos bx + b \sin bx).$

316 $$\int xe^{ax} \sin bx\, dx = \frac{xe^{ax}}{a^2 + b^2} (a \sin bx - b \cos bx) - \frac{e^{ax}}{(a^2 + b^2)^2} [(a^2 - b^2) \sin bx - 2ab \cos bx].$$

317 $$\int xe^{ax}\cos bx\,dx = \frac{xe^{ax}}{a^2+b^2}(a\cos bx + b\sin bx) - \frac{e^{ax}}{(a^2+b^2)^2}[(a^2-b^2)\cos bx + 2ab\sin bx].$$

318 $$\int \ln ax\,dx = x\ln ax - x.$$

319 $$\int (\ln ax)^n\,dx = x(\ln ax)^n - n\int(\ln ax)^{n-1}\,dx. \quad (n \text{ pos.})$$

320 $$\int x^n \ln ax\,dx = x^{n+1}\left[\frac{\ln ax}{n+1} - \frac{1}{(n+1)^2}\right]. \quad (n \neq -1)$$

321 $$\int \frac{(\ln ax)^n}{x}\,dx = \frac{(\ln ax)^{n+1}}{n+1}. \quad (n \neq -1)$$

322 $$\int \frac{dx}{x\ln ax} = \ln(\ln ax).$$

323 $$\int \frac{dx}{\ln ax} = \frac{1}{a}\left[\ln(\ln ax) + \ln ax + \frac{(\ln ax)^2}{2\cdot 2!} + \frac{(\ln ax)^3}{3\cdot 3!} + \cdots\right].$$

324 $$\int \sin(\ln ax)\,dx = \frac{x}{2}[\sin(\ln ax) - \cos(\ln ax)].$$

325 $$\int \cos(\ln ax)\,dx = \frac{x}{2}[\sin(\ln ax) + \cos(\ln ax)].$$

Some Definite Integrals

326 $$\int_0^a \sqrt{a^2-x^2}\,dx = \frac{\pi a^2}{4}.$$

327 $$\int_0^a \sqrt{2ax-x^2}\,dx = \frac{\pi a^2}{4}.$$

328 $$\int_0^\infty \frac{dx}{ax^2+b} = \frac{\pi}{2\sqrt{ab}}. \quad (a \text{ and } b \text{ pos.})$$

329 $$\int_0^{\sqrt{\frac{b}{a}}} \frac{dx}{ax^2+b} = \int_{\sqrt{\frac{b}{a}}}^\infty \frac{dx}{ax^2+b} = \frac{\pi}{4\sqrt{ab}}. \quad (a \text{ and } b \text{ pos.})$$

330 $$\int_0^{\frac{\pi}{2}} \sin^n ax\,dx = \int_0^{\frac{\pi}{2}} \cos^n ax\,dx = \frac{1\cdot 3\cdot 5\cdot \ldots (n-1)}{2\cdot 4\cdot 6\cdot \ldots n}\,\frac{\pi}{2a}. \quad (n, \text{ pos. even integ.})$$

331 $$\int_0^{\frac{\pi}{2}} \sin^n ax\,dx = \int_0^{\frac{\pi}{2}} \cos^n ax\,dx = \frac{2\cdot 4\cdot 6\cdot \ldots (n-1)}{1\cdot 3\cdot 5\cdot \ldots n}\,\frac{1}{a}. \quad (n, \text{ pos. odd integ.})$$

332 $$\int_0^\pi \sin ax \sin bx\,dx = \int_0^\pi \cos ax\cos bx\,dx = 0. \quad (a \neq b)$$

333 $$\int_0^\pi \sin^2 ax\,dx = \int_0^\pi \cos^2 ax\,dx = \frac{\pi}{2}.$$

334 $$\int_0^\infty e^{-ax^2}\,dx = \frac{1}{2}\sqrt{\frac{\pi}{a}}.$$

335 $$\int_0^\infty x^n e^{-ax}\,dx = \frac{n!}{a^{n+1}}. \quad (n \text{ pos. integ.})$$

336 Definition and Approximate Value of the Definite Integral

If $f(x)$ is continuous from $x = a$ to $x = b$ inclusive, and this interval is divided into n equal parts by the points $a, x_1, x_2, \ldots x_{n-1}, b$ such that $\Delta x = (b - a) \div n$, then the definite integral of $f(x)\,dx$ between the limits $x = a$ to $x = b$ is

$$\int_a^b f(x)\,dx = \lim_{n \doteq \infty} [f(a)\,\Delta x + f(x_1)\,\Delta x + f(x_2)\,\Delta x + \cdots + f(x_{n-1})\,\Delta x].$$

$$= \left[\int f(x)\,dx\right]_a^b = \left[F(x)\right]_a^b = F(b) - F(a).$$

If $y_0, y_1, y_2, \ldots, y_{n-1}, y_n$ are the values of $f(x)$ when $x = a, x_1, x_2, \ldots, x_{n-1}, b$ respectively, and if $h = (b - a) \div n$, then approximate values of this definite integral are given by the Trapezoidal, Durand's, and Simpson's Rules on page 18.

337 Some Fundamental Theorems on Definite Integrals

$$\int_a^b [f_1(x) + f_2(x) + \cdots]\,dx = \int_a^b f_1(x)\,dx + \int_a^b f_2(x)\,dx + \cdots.$$

$$\int_a^b k\,f(x)\,dx = k\int_a^b f(x)\,dx. \quad (k \text{ is any constant})$$

$$\int_a^b f(x)\,dx = -\int_b^a f(x)\,dx.$$

$$\int_a^b f(x)\,dx = \int_a^c f(x)\,dx + \int_c^b f(x)\,dx.$$

$$\int_a^b f(x)\,dx = (b - a)\,f(x_1), \text{ where } x_1 \text{ lies between } a \text{ and } b.$$

$$\int_a^\infty f(x)\,dx = \lim_{b \doteq \infty}\int_a^b f(x)\,dx.$$

Some Applications of the Definite Integral

338 Plane Area

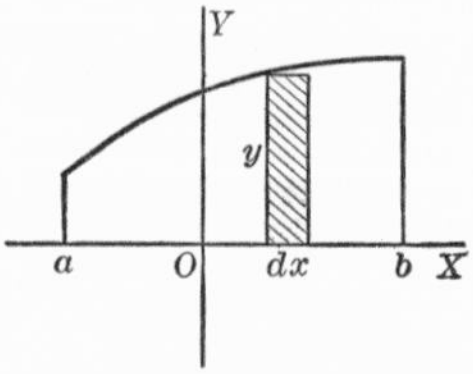

FIG. 338 (a).

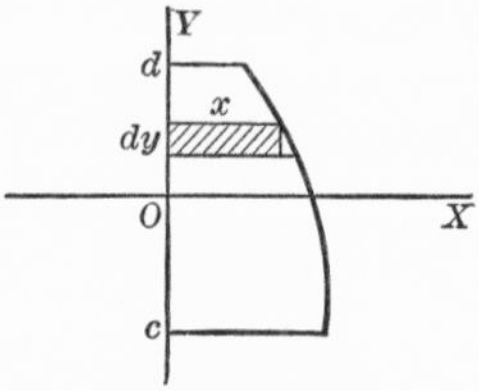

FIG. 338 (b).

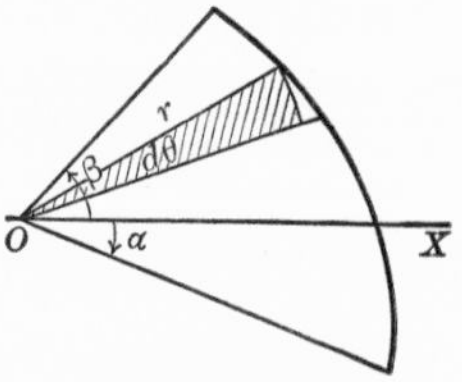

FIG. 338 (d).

(a) Area (A) bounded by the curve $y = f(x)$, the axis OX, and the ordinates $x = a$, $x = b$.

$$dA = y\,dx, \qquad A = \int_a^b f(x)\,dx.$$

(b) Area (A) bounded by the curve $x = f(y)$, the axis **OY**, and the abscissas $y = c$, $y = d$.

$$dA = x\,dy, \qquad A = \int_c^d f(y)\,dy.$$

(c) Area (A) bounded by the curve $x = f_1(t)$, $y = f_2(t)$, the axis **OX**, and $t = a$, $t = b$.

$$dA = y\,dx, \qquad A = \int_a^b f_2(t)\,f_1'(t)\,dt.$$

(d) Area (A) bounded by the curve $r = f(\theta)$ and two radii $\theta = \alpha$, $\theta = \beta$.

$$dA = \tfrac{1}{2}\,r^2 d\theta, \qquad A = \tfrac{1}{2}\int_\alpha^\beta [f(\theta)]^2\,d\theta.$$

339 Length of Arc

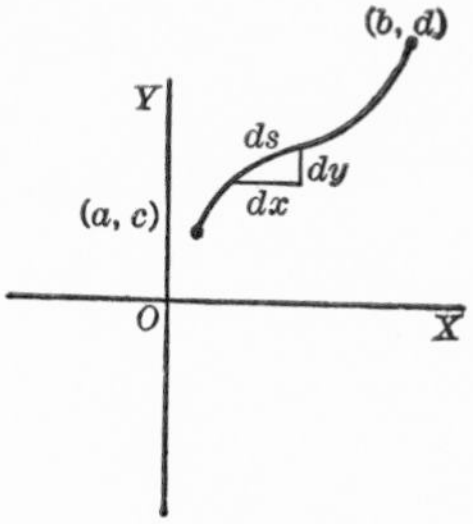

FIG. 339 (a).

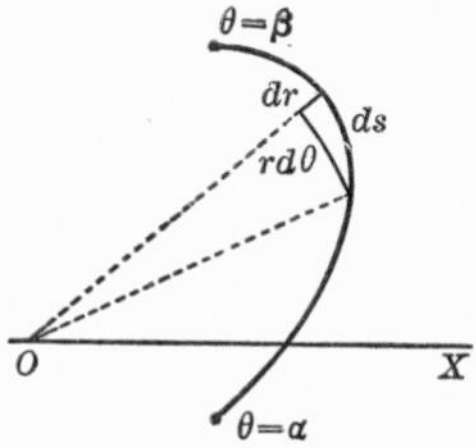

FIG. 339 (c).

(a) Length (s) of arc of curve $f(x, y) = 0$ from the point (a, c) to the point (b, d).

$$ds = \sqrt{(dx)^2 + (dy)^2}, \qquad s = \int_a^b \sqrt{1 + \left(\frac{dy}{dx}\right)^2}\,dx = \int_c^d \sqrt{1 + \left(\frac{dx}{dy}\right)^2}\,dy.$$

(b) Length (s) of arc of curve $x = f_1(t)$, $y = f_2(t)$ from $t = a$ to $t = b$.

$$ds = \sqrt{(dx)^2 + (dy)^2}, \qquad s = \int_a^b \sqrt{\left(\frac{dx}{dt}\right)^2 + \left(\frac{dy}{dt}\right)^2}\,dt.$$

(c) Length (s) of arc of curve $r = f(\theta)$ from $\theta = \alpha$ to $\theta = \beta$

$$ds = \sqrt{(dr)^2 + (r\,d\theta)^2}, \qquad s = \int_\alpha^\beta \sqrt{r^2 + \left(\frac{dr}{d\theta}\right)^2}\,d\theta.$$

(d) Length (s) of arc of space curve $x = f_1(t)$, $y = f_2(t)$, $z = f_3(t)$ from $t = a$ to $t = b$.

$$ds = \sqrt{(dx)^2 + (dy)^2 + (dz)^2}, \qquad s = \int_a^b \sqrt{\left(\frac{dx}{dt}\right)^2 + \left(\frac{dy}{dt}\right)^2 + \left(\frac{dz}{dt}\right)^2}\,dt.$$

340 Volume of Revolution

(a) Volume (V) of revolution generated by revolving about the line $y = k$ the area enclosed by the curve $y = f(x)$, the ordinates $x = a$, $x = b$, and the line $y = k$.

$$dV = \pi R^2\,dx = \pi (y - k)^2\,dx,$$

$$V = \pi \int_a^b [f(x) - k]^2\,dx.$$

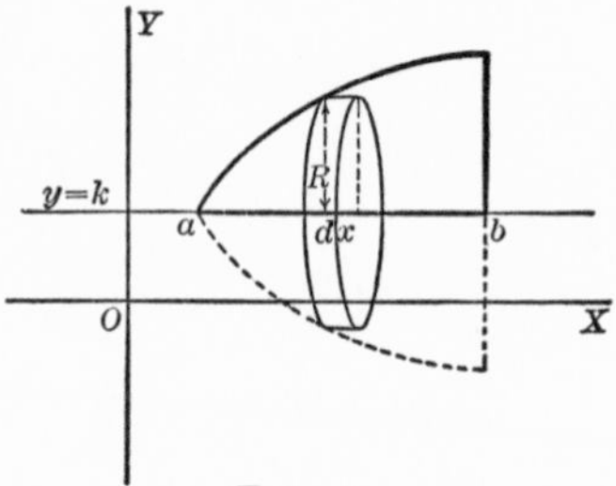

FIG. 340.

(b) Volume (V) of revolution generated by revolving about the line $x = k$ the area enclosed by the curve $x = f(y)$, the abscissas $y = c$, $y = d$, and the line $x = k$.

$$dV = \pi R^2 dy = \pi (x - k)^2\,dy, \qquad V = \pi \int_c^d [f(y) - k]^2\,dy.$$

341 Area of Surface of Revolution

(a) Area (S) of surface of revolution generated by revolving the arc of the curve $f(x, y) = 0$ from the point (a, c) to the point (b, d).

About $y = k$: $dS = 2\pi R\,ds$,

$$S = 2\pi \int_a^b (y - k) \sqrt{1 + \left(\frac{dy}{dx}\right)^2}\,dx.$$

About $x = k$: $dS = 2\pi R\,ds$,

$$S = 2\pi \int_c^d (x - k) \sqrt{1 + \left(\frac{dx}{dy}\right)^2}\,dy.$$

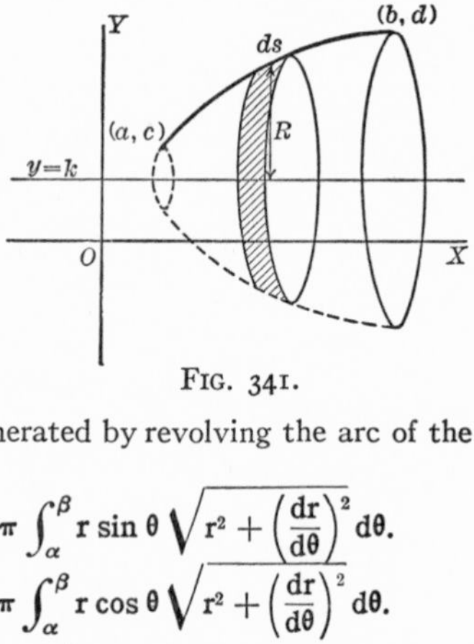

FIG. 341.

(b) Area (S) of surface of revolution generated by revolving the arc of the curve $r = f(\theta)$ from $\theta = \alpha$ to $\theta = \beta$.

About OX: $dS = 2\pi R\,ds$, $\qquad S = 2\pi \int_\alpha^\beta r \sin\theta \sqrt{r^2 + \left(\frac{dr}{d\theta}\right)^2}\,d\theta.$

About OY: $dS = 2\pi R\,ds$, $\qquad S = 2\pi \int_\alpha^\beta r \cos\theta \sqrt{r^2 + \left(\frac{dr}{d\theta}\right)^2}\,d\theta.$

342 Volume by Parallel Sections

Volume (V) of a solid generated by moving a plane section of area A_x perpendicular to OX from $x = a$ to $x = b$.

$$dV = A_x\,dx, \qquad V = \int_a^b A_x\,dx,$$

where A_x must be expressed as a function of x.

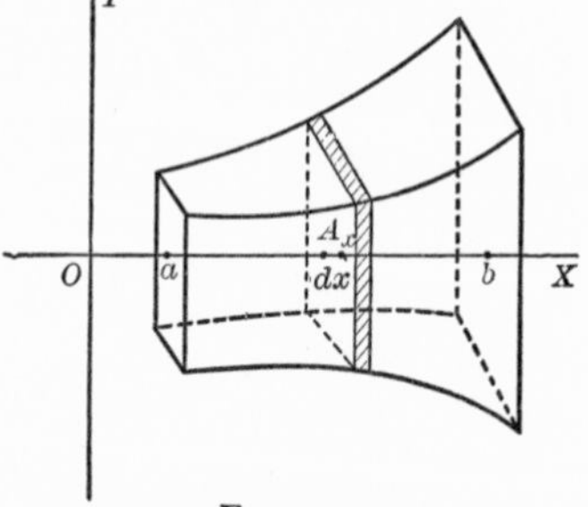

FIG. 342.

343 Mass

Mass (m) of constant or variable density (δ).

$$dm = \delta\,dA \quad \text{or} \quad \delta\,ds \quad \text{or} \quad \delta\,dV \quad \text{or} \quad \delta\,dS, \qquad m = \int dm,$$

where dA, ds, dV, dS are the elements of area, length, volume, surface in 338-342, and δ = mass per unit element.

344 Moment

Moment (M) of a mass (m).

$$\text{About OX:}\quad M_x = \int y\,dm = \int r \sin\theta\,dm.$$

$$\text{About OY:}\quad M_y = \int x\,dm = \int r \cos\theta\,dm.$$

$$\text{About O:}\quad M_0 = \int \sqrt{x^2 + y^2}\,dm = \int r\,dm.$$

345 Moment of Inertia

Moment of inertia (J) of a mass (m).

$$\text{About OX:}\quad J_x = \int y^2\,dm = \int r^2 \sin^2\theta\,dm.$$

$$\text{About OY:}\quad J_y = \int x^2\,dm = \int r^2 \cos^2\theta\,dm.$$

$$\text{About O:}\quad J_0 = \int (x^2 + y^2)\,dm = \int r^2\,dm.$$

346 Center of Gravity

Coördinates ($\bar{x}$, $\bar{y}$) of the center of gravity of a mass (m).

$$\bar{x} = \frac{\int x\,dm}{\int dm}, \qquad \bar{y} = \frac{\int y\,dm}{\int dm}.$$

Note. The center of gravity of the element of area may be taken at its mid-point. In the above equations **x** and **y** are the coördinates of the center of gravity of the element.

347 Work

Work (W) done in moving a particle from s = a to s = b against a force whose component in the direction of motion is F_s.

$$dW = F_s\,ds, \qquad W = \int_a^b F_s\,ds,$$

where F_s must be expressed as a function of s.

348 Pressure

Pressure (p) against an area vertical to the surface of the liquid and between depths **a** and **b.**

$$dp = wyx\,dy, \qquad p = \int_a^b wyx\,dy,$$

where **w** = weight of liquid per unit volume, **y** = depth beneath surface of liquid of a horizontal element of area, and **x** = length of horizontal element of area; **x** must be expressed in terms of **y.**

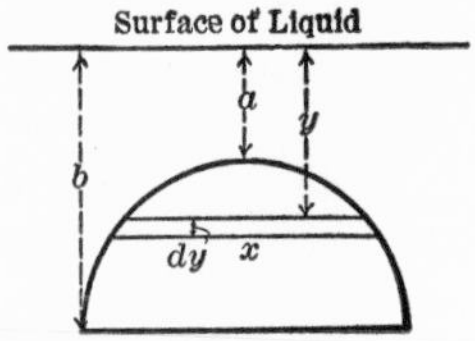

FIG. 348.

349 Center of Pressure

The depth ($\bar{y}$) of the center of pressure against an area **vertical to the** surface of the liquid and between depths **a** and **b.**

$$\bar{y} = \frac{\int_a^b y\,dp}{\int_a^b dp}. \qquad \text{(for dp see 348)}$$

DIFFERENTIAL EQUATIONS

350 Definitions and Notation

A **differential equation** is an equation involving differentials or **derivatives.**

The **order** of a differential equation is the same as that of the derivative of highest order which it contains.

The **degree** of a differential equation is the same as the power to which the derivative of highest order in the equation is raised, that derivative entering the equation free from radicals.

The **solution** of a differential equation is the relation involving only the variables (but not their derivatives) and arbitrary constants, consistent with the given differential equation.

The most **general solution** of a differential equation of the **n**th order contains **n** arbitrary constants. If particular values are assigned to these arbitrary constants, the solution is called a particular solution.

Notation. **M** and **N** denote functions of **x** and **y**; **X** denotes a function of **x** alone or a constant, **Y** denotes a function of **y** alone or a constant; $C, C_1, C_2, \ldots, C_n$ denote arbitrary constants of integration, **a, b, k, l, m, n,** . . . denote given constants.

Equations of First Order and First Degree. $M\,dx + N\,dy = 0$

351 Variables Separable: $X_1Y_1\,dx + X_2Y_2\,dy = 0.$

Solution:

$$\int \frac{X_1}{X_2}\,dx + \int \frac{Y_2}{Y_1}\,dy = C.$$

352 Homogeneous Equation: $dy - f\left(\frac{y}{x}\right)dx = 0.$

Solution: $$x = Ce^{\int \frac{dv}{f(v)-v}} \quad \text{and} \quad v = \frac{y}{x}.$$

NOTE. Here, **M** ÷ **N** can be written in a form such that **x** and **y** occur only in the combination **y** ÷ **x**; this can always be done if every term in **M** and **N** is of the same degree in **x** and **y**.

353 Linear Equation: $dy + (X_1y - X_2)\,dx = 0.$

Solution: $$y = e^{-\int X_1\,dx}\left(\int X_2 e^{\int X_1\,dx}dx + C\right).$$

NOTE. A similar solution exists for $dx + (Y_1 x - Y_2)\,dy = 0.$

354 Exact Equation: $M\,dx + N\,dy = 0$, where $\frac{\partial M}{\partial y} = \frac{\partial N}{\partial x}.$

Solution: $$\int M\,dx + \int\left[N - \frac{\partial}{\partial y}\int M\,dx\right]dy = C,$$

where **y** is constant when integrating with respect to **x**.

355 Non-exact Equation: $M\,dx + N\,dy = 0$, where $\frac{\partial M}{\partial y} \neq \frac{\partial N}{\partial x}.$

Solution: The equation may be made exact by multiplying by an integrating factor $\mu(x, y)$. The form of this factor is readily recognized in a large number of cases. Then solve by 354.

Certain Special Equations of the Second Order. $\frac{d^2y}{dx^2} = f\left(x, y, \frac{dy}{dx}\right)$

356 Equation: $$\frac{d^2y}{dx^2} = X.$$

Solution: $$y = x\int X\,dx - \int xX\,dx + C_1x + C_2.$$

357 Equation: $$\frac{d^2y}{dx^2} = Y.$$

Solution: $$x = \int \frac{dy}{\sqrt{2\int Y\,dy + C_1}} + C_2.$$

358 Equation: $$\frac{d^2y}{dx^2} = f\left(\frac{dy}{dx}\right).$$

Solution: $$x = \int \frac{dp}{f(p)} + C_1 \quad \text{and} \quad y = \int \frac{p\,dp}{f(p)} + C_2.$$

From these two equations eliminate $p = \frac{dy}{dx}$ if necessary.

359 Equation: $$\frac{d^2y}{dx^2} = f\left(x, \frac{dy}{dx}\right).$$

Solution: Place $\frac{dy}{dx} = p$ and $\frac{d^2y}{dx^2} = \frac{dp}{dx}$, thus bringing the equation into the form $\frac{dp}{dx} = f(x, p)$. This is of the first order and may be solved for **p** by **351–355**. Then replace **p** by $\frac{dy}{dx}$ and integrate for **y**.

360 Equation: $$\frac{d^2y}{dx^2} = f\left(y, \frac{dy}{dx}\right).$$

Solution: Place $\frac{dy}{dx} = p$ and $\frac{d^2y}{dx^2} = p\frac{dp}{dy}$, thus bringing the equation into the form $p\frac{dp}{dy} = f(y, p)$. This is of the first order and may be solved for **p** by **351–355**. Then replace **p** by $\frac{dy}{dx}$ and integrate for **y**.

Linear Equations of Physics. Second Order with Constant Coefficients. $\frac{d^2x}{dt^2} + 2l\frac{dx}{dt} \pm k^2x = f(t)$

361 Equation: $$\frac{d^2x}{dt^2} - k^2x = 0.$$

Solution: $$x = C_1e^{kt} + C_2e^{-kt}.$$

362 Equation of Simple Harmonic Motion: $\frac{d^2x}{dt^2} + k^2x = 0.$

Solution: This may be written in the following forms:

(a) $x = C_1e^{kt\sqrt{-1}} + C_2e^{-kt\sqrt{-1}}.$

(b) $x = C_1 \cos kt + C_2 \sin kt.$

(c) $x = C_1 \sin (kt + C_2).$

(d) $x = C_1 \cos (kt + C_2).$

363 Equation of Harmonic Motion with Constant Disturbing Force: $$\frac{d^2x}{dt^2} + k^2x = a.$$

Solution: $$x = C_1 \cos kt + C_2 \sin kt + \frac{a}{k^2},$$

or $$x = C_1 \sin (kt + C_2) + \frac{a}{k^2}.$$

364 Equation of Forced Vibration

(a) $\frac{d^2x}{dt^2} + k^2x = a \cos nt + b \sin nt$, where $n \neq k$.

Solution: $x = C_1 \cos kt + C_2 \sin kt + \frac{1}{k^2 - n^2}(a \cos nt + b \sin nt)$.

(b) $\frac{d^2x}{dt^2} + k^2x = a \cos kt + b \sin kt.$

Solution: $x = C_1 \cos kt + C_2 \sin kt + \frac{t}{2k}(a \sin kt - b \cos kt)$.

365 Equation of Damped Vibration: $\frac{d^2x}{dt^2} + 2l\frac{dx}{dt} + k^2x = 0.$

Solution: If $l^2 = k^2$, $x = e^{-lt}(C_1 + C_2t)$.

If $l^2 > k^2$, $x = e^{-lt}\left(C_1e^{\sqrt{l^2-k^2}\,t} + C_2e^{-\sqrt{l^2-k^2}\,t}\right)$.

If $l^2 < k^2$, $x = e^{-lt}\left(C_1 \cos \sqrt{k^2 - l^2}\,t + C_2 \sin \sqrt{k^2 - l^2}\,t\right)$

or $x = C_1e^{-lt} \sin \left(\sqrt{k^2 - l^2}\,t + C_2\right)$.

366 Equation of Damped Vibration with Constant Disturbing Force:

$$\frac{d^2x}{dt^2} + 2l\frac{dx}{dt} + k^2x = a.$$

Solution:

$$x = x_1 + \frac{a}{k^2},$$

where x_1 is the solution of equation 365.

367 General Equation: $\frac{d^2x}{dt^2} + 2l\frac{dx}{dt} + k^2x = f(t) = T.$

Solution:

$$x = x_1 + I,$$

where x_1 is the solution of equation 365, and I is given by

(a) $l^2 = k^2$, $I = e^{-lt}\left[t\int e^{lt}\,T\,dt - \int e^{lt}\,T\,t\,dt\right]$.

(b) $l^2 > k^2$, $I = \frac{1}{\alpha - \beta}\left[e^{\alpha t}\int e^{-\alpha t}\,T\,dt - e^{\beta t}\int e^{-\beta t}\,T\,dt\right]$,

where $\alpha = -1 + \sqrt{l^2 - k^2}$, $\beta = -1 - \sqrt{l^2 - k^2}$.

(c) $l^2 < k^2$, $I = \frac{e^{\alpha t}}{\beta}\left[\sin \beta t\int e^{-\alpha t} \cos \beta t\,T\,dt - \cos \beta t\int e^{-\alpha t} \sin \beta t\,T\,dt\right]$,

where $\alpha = -1$, $\beta = \sqrt{k^2 - l^2}$.

NOTE. I may also be found by the method indicated in 369.

Linear Equations with Constant Coefficients: nth Order

368 Equation

$$a_n\frac{d^nx}{dt^n} + a_{n-1}\frac{d^{n-1}x}{dt^{n-1}} + a_{n-2}\frac{d^{n-2}x}{dt^{n-2}} + \cdots + a_1\frac{dx}{dt} + a_0x = 0.$$

Solution: Let $D = \alpha_1, \alpha_2, \alpha_3, \ldots, \alpha_n$ be the n roots of the auxiliary algebraic equation $a_nD^n + a_{n-1}D^{n-1} + a_{n-2}D^{n-2} + \cdots + a_1D + a_0 = 0$.

(a) If all roots are real and distinct,

$$x = C_1 e^{\alpha_1 t} + C_2 e^{\alpha_2 t} + \cdots + C_n e^{\alpha_n t}.$$

(b) If 2 roots are equal: $\alpha_1 = \alpha_2$, the rest real and distinct,

$$x = e^{\alpha_1 t}(C_1 + C_2 t) + C_3 e^{\alpha_3 t} + \cdots + C_n e^{\alpha_n t}.$$

(c) If p roots are equal: $\alpha_1 = \alpha_2 = \cdots = \alpha_p$, the rest real and distinct,

$$x = e^{\alpha_1 t}(C_1 + C_2 t + C_3 t^2 + \cdots + C_p t^{p-1}) + \cdots + C_n e^{\alpha_n t}.$$

(d) If 2 roots are conjugate imaginary: $\alpha_1 = \beta + \gamma\sqrt{-1}$, $\alpha_2 = \beta - \gamma\sqrt{-1}$

$$x = e^{\beta t}(C_1 \cos \gamma t + C_2 \sin \gamma t) + C_3 e^{\alpha_3 t} + \cdots + C_n e^{\alpha_n t}.$$

(e) If there is a pair of conjugate imaginary double roots:

$$\alpha_1 = \beta + \gamma\sqrt{-1} = \alpha_2, \qquad \alpha_3 = \beta - \gamma\sqrt{-1} = \alpha_4,$$

$$x = e^{\beta t}[(C_1 + C_2 t)\cos \gamma t + (C_3 + C_4 t)\sin \gamma t] + \cdots + C_n e^{\alpha_n t}.$$

369 Equation

$$a_n \frac{d^n x}{dt^n} + a_{n-1}\frac{d^{n-1}x}{dt^{n-1}} + \cdots + a_1 \frac{dx}{dt} + a_0 x = f(t).$$

Solution: $x = x_1 + I,$

where x_1 is the solution of equation 368, and where I may be found by the following method.

Let $f(t) = T_1 + T_2 + T_3 + \cdots$. Find the 1st, 2d, 3d, ... derivatives of these terms. If $\tau_1, \tau_2, \tau_3, \ldots \tau_n$ are the resulting expressions which have different functional form (disregarding constant coefficients), assume

$$I = A\tau_1 + B\tau_2 + C\tau_3 + \cdots + K\tau_k + \cdots + N\tau_n.$$

NOTE. Thus, if $T = a \sin nt + bt^2 e^{kt}$, all possible successive derivatives of $\sin nt$ and $t^2 e^{kt}$ give terms of the form: $\sin nt$, $\cos nt$, e^{kt}, te^{kt}, $t^2 e^{kt}$, hence assume $I = A \sin nt + B \cos nt + Ce^{kt} + Dte^{kt} + Et^2 e^{kt}$.

Substitute this value of I for x in the given equation, expand, equate coefficients of like terms in the left and right members of the equation, and solve for A, B, C, ... N.

NOTE. If a root, α_k, occurring m times, of the algebraic equation in D (see 368) gives rise to a term of the form τ_k in x_1, then the corresponding term in the assumed value of I is $Kt^m\tau_k$.

370 Simultaneous Equations

$$\begin{cases} a_n \dfrac{d^n x}{dt^n} + b_m \dfrac{d^m y}{dt^m} + \cdots + a_1 \dfrac{dx}{dt} + b_1 \dfrac{dy}{dt} + a_0 x + b_0 y = f_1(t). \\ c_k \dfrac{d^k x}{dt^k} + g_l \dfrac{d^l y}{dt^l} + \cdots + c_1 \dfrac{dx}{dt} + g_1 \dfrac{dy}{dt} + c_0 x + g_0 y = f_2(t). \end{cases}$$

Solution: Write the equations in the form:

$$\begin{cases} (a_n D^n + \cdots + a_1 D + a_0)\,x + (b_m D^m + \cdots + b_1 D + b_0)\,y = f_1(t), \\ (c_k D^k + \cdots + c_1 D + c_0)\,x + (g_l D^l + \cdots + g_1 D + g_0)\,y = f_2(t), \end{cases}$$

where $$D = \frac{d}{dt}, \ldots, D^i = \frac{d^i}{dt^i}, \ldots.$$

Regarding this set of equations as a pair of simultaneous algebraic equations in **x** and **y**, eliminate **y** and **x** in turn, getting two linear differential equations of the form 369 whose solutions are

$$x = x_1 + I_1, \qquad y = y_1 + I_2.$$

Substitute these values of **x** and **y** in the original equations, equate coefficients of like terms, and thus express the arbitrary constants in y_1, say, in terms of those in x_1.

Partial Differential Equations

371 Equation of Oscillation: $\dfrac{\partial^2 y}{\partial t^2} = a^2 \dfrac{\partial^2 y}{\partial x^2}.$

Solution:
$$y = \sum_{i=1}^{i=\infty} C_i e^{(x+at)\alpha_i} + \sum_{i=1}^{i=\infty} C_i' e^{(x-at)\alpha_i},$$

where C_i, C_i', α_i are arbitrary constants.

372 Equation of Thermodynamics: $\dfrac{\partial u}{\partial t} = a^2 \dfrac{\partial^2 u}{\partial x^2}.$

Solution:
$$u = \sum_{i=1}^{i=\infty} C_i e^{\alpha_i x} e^{a^2 \alpha_i^2 t},$$

where C_i and α_i are arbitrary constants.

373 Equation of Laplace or Condition of Continuity of Incompressible Liquids: $\dfrac{\partial^2 u}{\partial x^2} + \dfrac{\partial^2 u}{\partial y^2} = 0.$

Solution:
$$u = \sum_{i=1}^{i=\infty} C_i e^{(x+y\sqrt{-1})\alpha_i} + \sum_{i=1}^{i=\infty} C_i' e^{(x-y\sqrt{-1})\alpha_i},$$

where C_i, C_i', α_i are arbitrary constants.

COMPLEX QUANTITIES

374 Definition and Representation of a Complex Quantity

If $z = x + jy$, where $j = \sqrt{-1}$ and **x** and **y** are real, **z** is called a complex quantity. **z** is completely determined by **x** and **y**.

If **P(x, y)** is a point in the plane (Fig. 374) then the segment **OP** in magnitude and direction is said to represent the complex quantity $z = x + jy$.

If θ is the angle from **OX** to **OP** and **r** is the length of **OP**, then

$$z = x + jy = r(\cos\theta + j\sin\theta) = re^{j\theta},$$

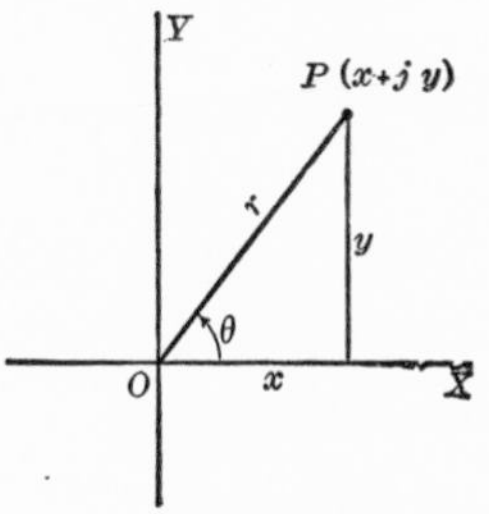

FIG. 374.

where $\theta = \tan^{-1}\dfrac{y}{x}$, $r = +\sqrt{x^2 + y^2}$, and **e** is the base of natural logarithms. $x + jy$ and $x - jy$ are called conjugate complex quantities.

375 Properties of Complex Quantities

Let z, z_1, z_2 represent complex quantities, then:

Sum or Difference: $z_1 \pm z_2 = (x_1 \pm x_2) + j(y_1 \pm y_2)$.

Product: $z_1 \cdot z_2 = r_1 r_2 [\cos(\theta_1 + \theta_2) + j \sin(\theta_1 + \theta_2)]$

$$= r_1 r_2 e^{j(\theta_1+\theta_2)} = (x_1 x_2 - y_1 y_2) + j(x_1 y_2 + x_2 y_1).$$

Quotient: $\frac{z_1}{z_2} = \frac{r_1}{r_2}[\cos(\theta_1 - \theta_2) + j \sin(\theta_1 - \theta_2)]$

$$= \frac{r_1}{r_2} e^{j(\theta_1-\theta_2)} = \frac{x_1 x_2 + y_1 y_2}{x_2^2 + y_2^2} + j \frac{x_2 y_1 - x_1 y_2}{x_2^2 + y_2^2}.$$

Power: $z^n = r^n [\cos n\theta + j \sin n\theta] = r^n e^{jn\theta}$.

Root: $\sqrt[n]{z} = \sqrt[n]{r}\left[\cos \frac{\theta + 2k\pi}{n} + j \sin \frac{\theta + 2k\pi}{n}\right] = \sqrt[n]{r}\, e^{j\frac{\theta+2k\pi}{n}}$,

where **k** takes in succession the values 0, 1, 2, 3, . . . , $n - 1$.

Equation: If $z_1 = z_2$, then $x_1 = x_2$ and $y_1 = y_2$.

Periodicity: $z = r(\cos\theta + j \sin\theta) = r[\cos(\theta + 2k\pi) + j \sin(\theta + 2k\pi)]$,

or $z = re^{j\theta} = re^{j(\theta+2k\pi)}$ and $e^{j2k\pi} = 1$, where **k** is any integer

Exponential-Trigonometric Relations:

$$e^{jz} = \cos z + j \sin z, \quad e^{-jz} = \cos z - j \sin z,$$

$$\cos z = \frac{1}{2}(e^{jz} + e^{-jz}), \quad \sin z = \frac{1}{2j}(e^{jz} - e^{-jz}).$$

VECTORS

376 Definition and Graphical Representation of a Vector

A vector (**V**) is a quantity which is completely specified by a magnitude and a direction. A scalar (s) is a quantity which is completely specified by a magnitude.

The vector (**V**) may be represented geometrically by the segment $\overrightarrow{OA}$, the length of OA signifying the magnitude of **V** and the arrow carried by OA signifying the direction of **V**.

The segment $\overrightarrow{AO}$ represents the vector $-$**V**.

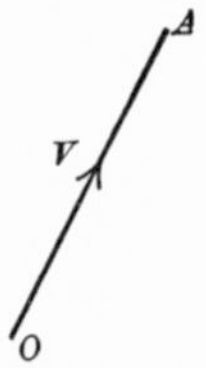

FIG. 376.

377 Graphical Summation of Vectors

If V_1, V_2 are two vectors, their graphical sum, $V = V_1 + V_2$, is formed by drawing the vector $V_1 = \overrightarrow{OA}$ from any point **O**, and the vector $V_2 = \overrightarrow{AB}$ from the end of V_1, and joining O and B; then $V = \overrightarrow{OB}$. Also $V_1 + V_2 = V_2 + V_1$ and $V_1 + V_2 - V = 0$ (Fig. 377a).

Similarly, if $V_1, V_2, V_3, \ldots V_n$ are any number of vectors drawn so that the initial point of one is the end point of the preceding one, then their graphical sum, $\mathbf{V} = \mathbf{V}_1 + \mathbf{V}_2 + \ldots + \mathbf{V}_n$, is the vector joining the initial point of $\mathbf{V}_1$ with the end point of $\mathbf{V}_n$ (Fig. 377b).

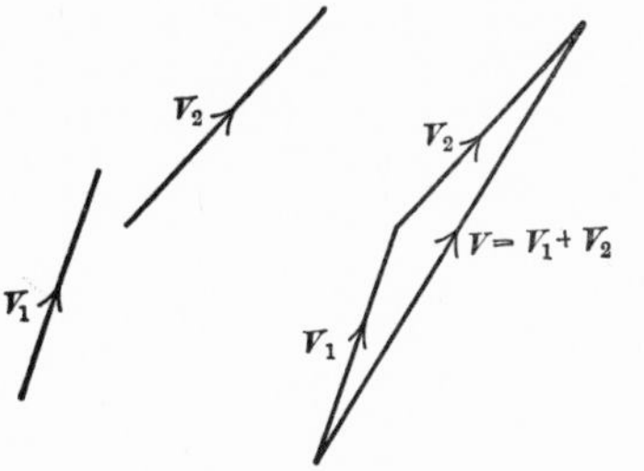

FIG. 377 (a).

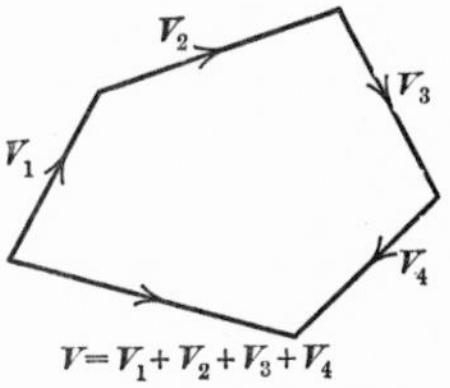

FIG. 377 (b).

378 Components of a Vector. Analytic Representation

A vector (**V**) considered as lying in the **xy** coördinate plane is completely determined by its horizontal and vertical components **x** and **y**. If **i** and **j** represent vectors of unit magnitude along **OX** and **OY** respectively, and a and b are the magnitudes of the components **x** and **y**, then **V** may be represented by $\mathbf{V} = a\mathbf{i} + b\mathbf{j}$, its magnitude by $|\mathbf{V}| = +\sqrt{a^2 + b^2}$, and its direction by $\alpha = \tan^{-1}\frac{b}{a}$.

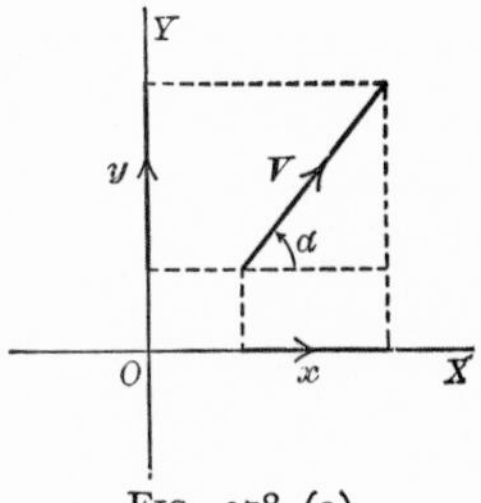

FIG. 378 (a).

A vector (**V**) considered as lying in space is completely determined by its components **x**, **y**, and **z** along three mutually perpendicular lines **OX**, **OY**, and **OZ**, directed as in Fig. 378. If **i**, **j**, **k** represent vectors of unit magnitude along **OX**, **OY**, **OZ** respectively, and a, b, c are the magnitudes of the components **x**, **y**, **z** respectively, then **V** may be represented by $\mathbf{V} = a\mathbf{i} + b\mathbf{j} + c\mathbf{k}$, its magnitude by $|\mathbf{V}| = +\sqrt{a^2 + b^2 + c^2}$, and its direction by $\cos\alpha : \cos\beta : \cos\gamma = a : b : c$.

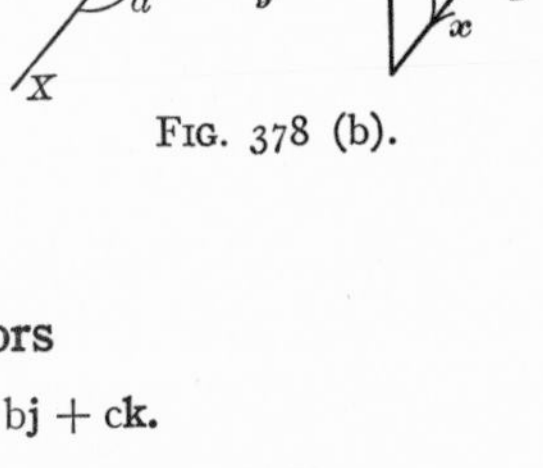

FIG. 378 (b).

Properties of Vectors

$$\mathbf{V} = a\mathbf{i} + b\mathbf{j} \quad \text{or} \quad \mathbf{V} = a\mathbf{i} + b\mathbf{j} + c\mathbf{k}.$$

379 Vector sum (V) of any number of vectors, $\mathbf{V}_1, \mathbf{V}_2, \mathbf{V}_3, \ldots$.

$$\mathbf{V} = \mathbf{V}_1 + \mathbf{V}_2 + \mathbf{V}_3 + \cdots = (a_1 + a_2 + a_3 + \cdots)\mathbf{i} + (b_1 + b_2 + \cdots)\mathbf{j} + (c_1 + c_2 + c_3 + \cdots)\mathbf{k}.$$

380 Product of a vector (V) by a scalar (s)

$s\mathbf{V} = (sa)\,\mathbf{i} + (sb)\,\mathbf{j} + (sc)\,\mathbf{k}$.

$(s_1 + s_2)\,\mathbf{V} = s_1\mathbf{V} + s_2\mathbf{V}$; $(\mathbf{V}_1 + \mathbf{V}_2)\,s = \mathbf{V}_1 s + \mathbf{V}_2 s$.

NOTE. sV has the same direction as V and its magnitude is s times the magnitude of V.

381 Scalar product of 2 vectors: $\mathbf{V}_1 \cdot \mathbf{V}_2$.

$\mathbf{V}_1 \cdot \mathbf{V}_2 = |\mathbf{V}_1|\,|\mathbf{V}_2| \cos\phi$, where ϕ is the angle between $\mathbf{V}_1$ and $\mathbf{V}_2$.

$\mathbf{V}_1 \cdot \mathbf{V}_2 = \mathbf{V}_2 \cdot \mathbf{V}_1$; $\mathbf{V}_1 \cdot \mathbf{V}_1 = |\mathbf{V}_1|^2$.

$(\mathbf{V}_1 + \mathbf{V}_2) \cdot \mathbf{V}_3 = \mathbf{V}_1 \cdot \mathbf{V}_3 + \mathbf{V}_2 \cdot \mathbf{V}_3$;

$(\mathbf{V}_1 + \mathbf{V}_2) \cdot (\mathbf{V}_3 + \mathbf{V}_4) = \mathbf{V}_1 \cdot \mathbf{V}_3 + \mathbf{V}_1 \cdot \mathbf{V}_4 + \mathbf{V}_2 \cdot \mathbf{V}_3 + \mathbf{V}_2 \cdot \mathbf{V}_4$.

$\mathbf{i}\cdot\mathbf{i} = \mathbf{j}\cdot\mathbf{j} = \mathbf{k}\cdot\mathbf{k} = 1$; $\mathbf{i}\cdot\mathbf{j} = \mathbf{j}\cdot\mathbf{k} = \mathbf{k}\cdot\mathbf{i} = 0$.

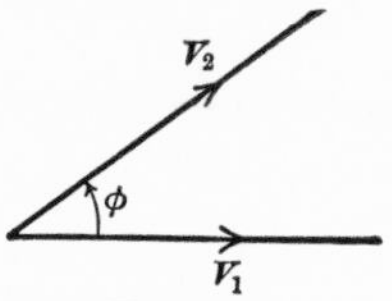

FIG. 381.

In plane: $\mathbf{V}_1 \cdot \mathbf{V}_2 = a_1a_2 + b_1b_2$; **in space:** $\mathbf{V}_1 \cdot \mathbf{V}_2 = a_1a_2 + b_1b_2 + c_1c_2$.

NOTE. The scalar product of two vectors $\mathbf{V}_1 \cdot \mathbf{V}_2$ is a scalar quantity and may physically be represented by the work done by a constant force of magnitude $|\mathbf{V}_1|$ on a unit particle moving through a distance $|\mathbf{V}_2|$, where ϕ is the angle between the line of force and the direction of motion.

382 Vector product of 2 vectors: $\mathbf{V}_1 \times \mathbf{V}_2$.

$\mathbf{V}_1 \times \mathbf{V}_2 = \mathbf{l}\,|\mathbf{V}_1|\,|\mathbf{V}_2| \sin\phi$, where ϕ is the angle from $\mathbf{V}_1$ to $\mathbf{V}_2$ and $\mathbf{l}$ is a unit vector perpendicular to the plane of the vectors $\mathbf{V}_1$ and $\mathbf{V}_2$ and so directed that a right-handed screw driven in the direction of $\mathbf{l}$ would carry $\mathbf{V}_1$ into $\mathbf{V}_2$.

$\mathbf{V}_1 \times \mathbf{V}_2 = -\mathbf{V}_2 \times \mathbf{V}_1$; $\mathbf{V}_1 \times \mathbf{V}_1 = 0$.

$(\mathbf{V}_1 + \mathbf{V}_2) \times \mathbf{V}_3 = \mathbf{V}_1 \times \mathbf{V}_3 + \mathbf{V}_2 \times \mathbf{V}_3$;

$\mathbf{V}_1 \times (\mathbf{V}_2 \times \mathbf{V}_3) = \mathbf{V}_2\,(\mathbf{V}_1 \cdot \mathbf{V}_3) - \mathbf{V}_3\,(\mathbf{V}_1 \cdot \mathbf{V}_2)$.

$\mathbf{V}_1 \cdot (\mathbf{V}_2 \times \mathbf{V}_3) = \mathbf{V}_2 \cdot (\mathbf{V}_3 \times \mathbf{V}_1) = \mathbf{V}_3 \cdot (\mathbf{V}_1 \times \mathbf{V}_2)$;

$(\mathbf{V}_1 + \mathbf{V}_2) \times (\mathbf{V}_3 + \mathbf{V}_4) = \mathbf{V}_1 \times \mathbf{V}_3 + \mathbf{V}_1 \times \mathbf{V}_4 + \mathbf{V}_2 \times \mathbf{V}_3 + \mathbf{V}_2 \times \mathbf{V}_4$.

$\mathbf{i} \times \mathbf{i} = \mathbf{j} \times \mathbf{j} = \mathbf{k} \times \mathbf{k} = 0$; $\mathbf{i} \times \mathbf{j} = \mathbf{k}$; $\mathbf{j} \times \mathbf{k} = \mathbf{i}$; $\mathbf{k} \times \mathbf{i} = \mathbf{j}$.

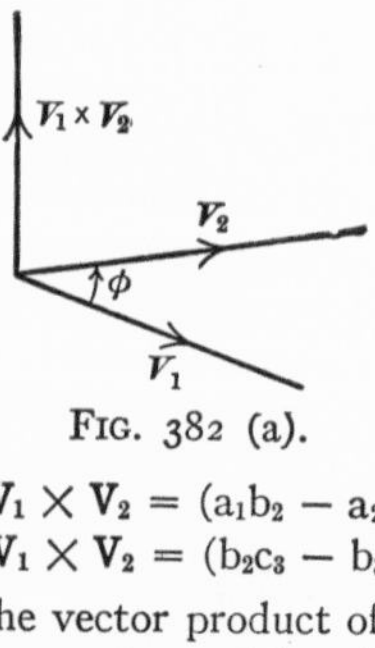

FIG. 382 (a).

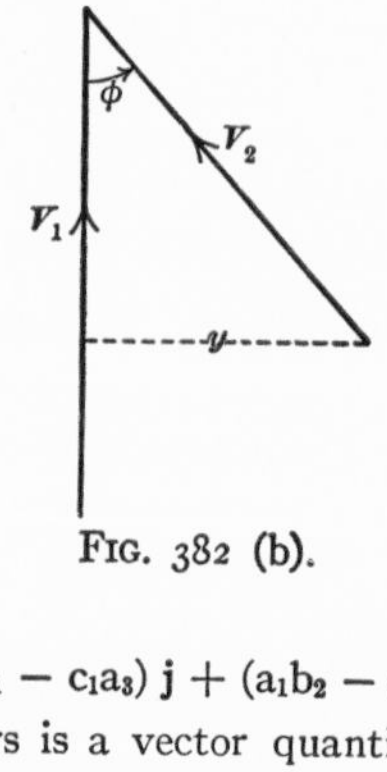

FIG. 382 (b).

In plane: $\mathbf{V}_1 \times \mathbf{V}_2 = (a_1b_2 - a_2b_1)\,\mathbf{k}$.

In space: $\mathbf{V}_1 \times \mathbf{V}_2 = (b_2c_3 - b_3c_2)\,\mathbf{i} + (c_3a_1 - c_1a_3)\,\mathbf{j} + (a_1b_2 - a_2b_1)\,\mathbf{k}$.

NOTE. The vector product of two vectors is a vector quantity and may physically be represented by the moment of a force $\mathbf{V}_1$ about a point **O** placed so that the moment arm is $\mathbf{y} = |\mathbf{V}_2| \sin\phi$ (see Fig. 382 b).

HYPERBOLIC FUNCTIONS

383 Definitions of Hyperbolic Functions. (See Table, p. 290.)

Hyperbolic sine (sinh) $x = \frac{1}{2}(e^x - e^{-x})$; $\quad \operatorname{csch} x = \frac{1}{\sinh x}$

Hyperbolic cosine (cosh) $x = \frac{1}{2}(e^x + e^{-x})$; $\quad \operatorname{sech} x = \frac{1}{\cosh x}$

Hyperbolic tangent (tanh) $x = \frac{e^x - e^{-x}}{e^x + e^{-x}}$; $\quad \coth x = \frac{1}{\tanh x}$

where e = base of natural logarithms.

NOTE. The circular or ordinary trigonometric functions were defined with reference to a circle; in a similar manner, the hyperbolic functions may be defined with reference to a hyperbola. In the above definitions the hyperbolic functions are abbreviations for certain exponential functions.

384 Graphs of Hyperbolic Functions (a) $y = \sinh x$; (b) $y = \cosh x$; (c) $y = \tanh x$.

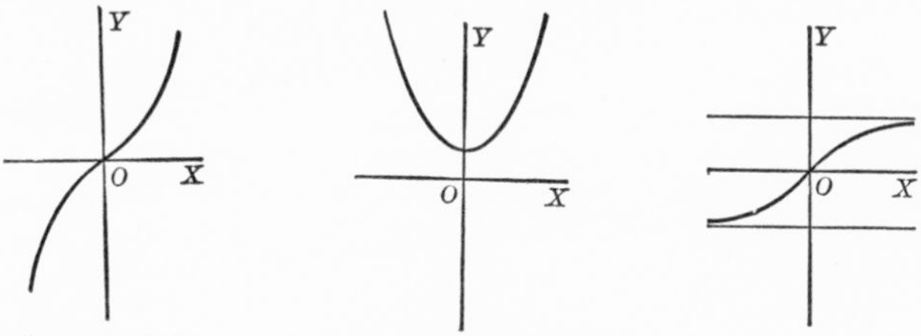

FIG. 384 (a). FIG. 384 (b). FIG. 384 (c).

385 Some Relations among Hyperbolic Functions

$\sinh 0 = 0$, $\quad \cosh 0 = 1$, $\quad \tanh 0 = 0$.

$\sinh \infty = \infty$, $\quad \cosh \infty = \infty$, $\quad \tanh \infty = 1$.

$\sinh(-x) = -\sinh x$, $\quad \cosh(-x) = \cosh x$, $\quad \tanh(-x) = -\tanh x$.

$\cosh^2 x - \sinh^2 x = 1$, $\quad \operatorname{sech}^2 x + \tanh^2 x = 1$, $\quad \operatorname{csch}^2 x - \coth^2 x = -1$.

$\sinh 2x = 2 \sinh x \cosh x$, $\quad \cosh 2x = \cosh^2 x + \sinh^2 x$.

$2\sinh^2 \frac{x}{2} = \cosh x - 1$, $\quad 2\cosh^2 \frac{x}{2} = \cosh x + 1$.

$\sinh(x \pm y) = \sinh x \cosh y \pm \cosh x \sinh y$.

$\cosh(x \pm y) = \cosh x \cosh y \pm \sinh x \sinh y$.

$\tanh(x \pm y) = \frac{\tanh x \pm \tanh y}{1 \pm \tanh x \tanh y}$.

386 Hyperbolic Functions of Pure Imaginary and Complex Quantities

$\sinh jy = j \sin y$; $\quad \cosh jy = \cos y$; $\quad \tanh jy = j \tan y$.

$\sinh(x + jy) = \sinh x \cos y + j \cosh x \sin y$.

$\cosh(x + jy) = \cosh x \cos y + j \sinh x \sin y$.

$\sinh(x + 2j\pi) = \sinh x; \quad \cosh(x + 2j\pi) = \cosh x.$
$\sinh(x + j\pi) = -\sinh x; \quad \cosh(x + j\pi) = -\cosh x.$
$\sinh(x + \frac{1}{2}j\pi) = j\cosh x; \quad \cosh(x + \frac{1}{2}j\pi) = j\sinh x.$

387 Inverse or Anti-Hyperbolic Functions

If $x = \sinh y$, then y is the anti-hyperbolic sine of x or $y = \sinh^{-1} x$.

$$\sinh^{-1} x = \ln\left(x + \sqrt{x^2 + 1}\right); \qquad \operatorname{csch}^{-1} x = \sinh^{-1}\frac{1}{x}.$$

$$\cosh^{-1} x = \ln\left(x + \sqrt{x^2 - 1}\right); \qquad \operatorname{sech}^{-1} x = \cosh^{-1}\frac{1}{x}.$$

$$\tanh^{-1} x = \frac{1}{2}\ln\frac{1 + x}{1 - x}; \qquad \coth^{-1} x = \tanh^{-1}\frac{1}{x}.$$

388 Derivatives of Hyperbolic Functions

$$\frac{d}{dx}\sinh x = \cosh x; \quad \frac{d}{dx}\cosh x = \sinh x; \quad \frac{d}{dx}\tanh x = \operatorname{sech}^2 x.$$

$$\frac{d}{dx}\coth x = -\operatorname{csch}^2 x; \quad \frac{d}{dx}\operatorname{sech} x = -\operatorname{sech} x \tanh x; \quad \frac{d}{dx}\operatorname{csch} x = -\operatorname{csch} x \coth x.$$

$$\frac{d}{dx}\sinh^{-1} x = \frac{1}{\sqrt{x^2+1}}; \quad \frac{d}{dx}\cosh^{-1} x = \frac{1}{\sqrt{x^2 - 1}}; \quad \frac{d}{dx}\tanh^{-1} x = \frac{1}{1 - x^2}.$$

$$\frac{d}{dx}\coth^{-1} x = -\frac{1}{x^2-1}; \quad \frac{d}{dx}\operatorname{sech}^{-1} x = -\frac{1}{x\sqrt{1 - x^2}}; \quad \frac{d}{dx}\operatorname{csch}^{-1} x = -\frac{1}{x\sqrt{x^2+1}}.$$

389 Some Integrals Leading to Hyperbolic Functions

$$\int \sinh x\,dx = \cosh x; \quad \int \cosh x\,dx = \sinh x; \quad \int \tanh x\,dx = \ln\cosh x.$$

$$\int \coth x\,dx = \ln\sinh x; \quad \int \operatorname{sech} x\,dx = \sin^{-1}(\tanh x); \quad \int \operatorname{csch} x\,dx = \ln\tanh\frac{x}{2}.$$

$$\int \frac{dx}{\sqrt{x^2+a^2}} = \sinh^{-1}\frac{x}{a}; \quad \int \frac{dx}{\sqrt{x^2-a^2}} = \cosh^{-1}\frac{x}{a}; \quad \int \frac{dx}{a^2-x^2} = \frac{1}{a}\tanh^{-1}\frac{x}{a}. \quad (x < a)$$

$$\int \frac{dx}{x\sqrt{a^2 + x^2}} = -\frac{1}{a}\sinh^{-1}\frac{a}{x}; \qquad \int \frac{dx}{x\sqrt{a^2 - x^2}} = -\frac{1}{a}\cosh^{-1}\frac{a}{x};$$

$$\int \frac{dx}{x^2 - a^2} = -\frac{1}{a}\tanh^{-1}\frac{a}{x}. \quad (x > a)$$

$$\int \sqrt{x^2 - a^2}\,dx = \frac{x}{2}\sqrt{x^2 - a^2} - \frac{a^2}{2}\cosh^{-1}\frac{x}{a}.$$

$$\int \sqrt{x^2 + a^2}\,dx = \frac{x}{2}\sqrt{x^2 + a^2} + \frac{a^2}{2}\sinh^{-1}\frac{x}{a}.$$

390 Expansions of Hyperbolic Functions into Series

$$\sinh x = x + \frac{x^3}{3!} + \frac{x^5}{5!} + \cdots.$$

$$\cosh x = 1 + \frac{x^2}{2!} + \frac{x^4}{4!} + \cdots.$$

$$\tanh x = x - \frac{x^3}{3} + \frac{2x^5}{15} - \frac{17x^7}{315} + \cdots.$$

$$\sinh^{-1} x = x - \frac{1}{2}\frac{x^3}{3} + \frac{1\cdot 3}{2\cdot 4}\frac{x^5}{5} - \frac{1\cdot 3\cdot 5}{2\cdot 4\cdot 6}\frac{x^7}{7} + \cdots \quad (x < 1)$$

$$\sinh^{-1} x = \ln 2x + \frac{1}{2}\frac{1}{2x^2} - \frac{1\cdot 3}{2\cdot 4}\frac{1}{4x^4} + \frac{1\cdot 3\cdot 5}{2\cdot 4\cdot 6}\frac{1}{6x^6} - \cdots \quad (x > 1)$$

$$\cosh^{-1} x = \ln 2x - \frac{1}{2}\frac{1}{2x^2} - \frac{1\cdot 3}{2\cdot 4}\frac{1}{4x^4} - \frac{1\cdot 3\cdot 5}{2\cdot 4\cdot 6}\frac{1}{6x^6} - \cdots.$$

$$\tanh^{-1} x = x + \frac{x^3}{3} + \frac{x^5}{5} + \frac{x^7}{7} + \cdots.$$

391 The Catenary. (For definition, see 83)

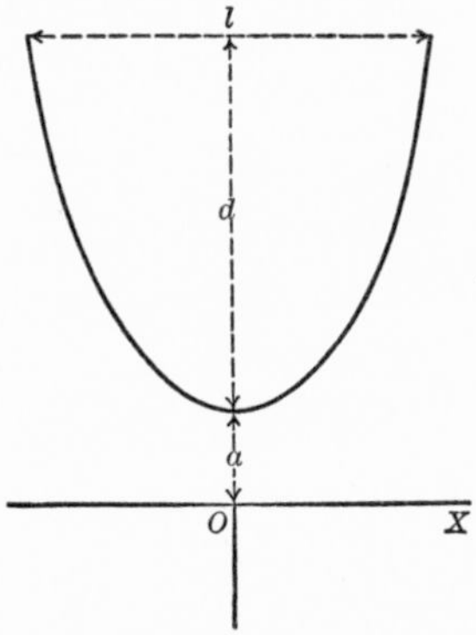

FIG. 391.

Equation: $$y = \frac{a}{2}\left(e^{\frac{x}{a}} + e^{-\frac{x}{a}}\right) = a \cosh \frac{x}{a}.$$

If the width of the span is l and the sag is d, then the length of the arc (s) is found by means of the equations:

$$\cosh z = \frac{2d}{l} z + 1, \qquad s = \frac{l}{z}\sinh z,$$

where z is to be found approximately by means of the table, p. 290, from the first of these equations and this value substituted in the second.

If s and l are known, d may be found similarly by means of

$$\sinh z = \frac{s}{l} z, \qquad d = \frac{l}{2z}(\cosh z - 1).$$

MECHANICS

KINEMATICS

Rectilinear Motion

Velocity (**v**) of a particle which moves uniformly **s** feet in **t** seconds.

392 $$v = \frac{s}{t} \text{ feet per second.}$$

NOTE. The velocity (**v**) of a moving particle at any instant equals $\frac{ds}{dt}$. The speed of a moving particle equals the magnitude of its velocity but has no direction.

Acceleration (**a**) of a particle whose velocity increases uniformly **v** feet per second in **t** seconds.

393 $$a = \frac{v}{t} \text{ feet per second per second.}$$

NOTE. The acceleration (**a**) of a moving particle at any instant equals $\frac{dv}{dt}$ or $\frac{d^2s}{dt^2}$. The acceleration (**g**) of a falling body in vacuo at sea level and latitude 45 degrees equals 32.17 feet per second per second.

Velocity (v_t) at the end of **t** seconds acquired by a particle having an initial velocity of v_0 feet per second and a uniform acceleration of **a** feet per second per second.

394 $$v_t = v_0 + at \text{ feet per second.}$$

NOTE. **a** is negative if the initial velocity and the acceleration act in opposite directions.

Space (**s**) traversed in **t** seconds by a particle having an initial velocity of v_0 feet per second and a uniform acceleration of **a** feet per second per second.

395 $$s = v_0t + \tfrac{1}{2}at^2 \text{ feet.}$$

Space (s) required for a particle with an initial velocity of v_0 feet per second and a uniform acceleration of a feet per second per second to reach a velocity of v_t feet per second.

396 $$s = \frac{v_t^2 - v_0^2}{2a} \text{ feet.}$$

Velocity (v_t) acquired, in travelling s feet, by a particle having an initial velocity of v_0 feet per second and a uniform acceleration of a feet per second per second.

397 $$v_t = \sqrt{v_0^2 + 2as} \text{ feet per second.}$$

Time (t) required for a particle having an initial velocity of v_0 feet per second and a uniform acceleration of a feet per second per second to travel s feet.

398 $$t = \frac{-v_0 + \sqrt{v_0^2 + 2as}}{a} \text{ seconds.}$$

Uniform acceleration (a) required to move a particle, with an initial velocity of v_0 feet per second, s feet in t seconds.

399 $$a = \frac{2(s - v_0 t)}{t^2} \text{ feet per second per second.}$$

Circular Motion

Angular velocity (ω) of a particle moving uniformly through θ radians in t seconds.

400 $$\omega = \frac{\theta}{t} \text{ radians per second.}$$

NOTE. The angular velocity (ω) of a moving particle at any instant equals $\frac{d\theta}{dt}$.

Normal acceleration (a) toward the center of its path of a particle moving uniformly with v feet per second tangential velocity and r feet radius of curvature of path.

401 $$a = \frac{v^2}{r} \text{ feet per second per second.}$$

NOTE. The tangential acceleration of a particle moving with constant speed in a circular path is zero.

Angular acceleration (α) of a particle whose angular velocity increases uniformly ω radians per second in t seconds.

402 $$\alpha = \frac{\omega}{t} \text{ radians per second per second.}$$

NOTE. The angular acceleration (α) of a moving particle at any instant equals $\frac{d\omega}{dt}$ or $\frac{d^2\theta}{dt^2}$.

Angular velocity (ω_t) at the end of t seconds acquired by a particle having an initial angular velocity of ω_0 radians per second and a uniform angular acceleration of α radians per second per second.

403 $$\omega_t = \omega_0 + \alpha t \text{ radians per second.}$$

Angle (θ) subtended in t seconds by a particle having an initial angular velocity of ω_0 radians per second and a uniform angular acceleration of α radians per second per second.

404 $$\theta = \omega_0 t + \tfrac{1}{2} \alpha t^2 \text{ radians.}$$

Angle (θ) subtended by a particle with an initial angular velocity of ω_0 radians per second and a uniform angular acceleration of α radians per second per second in acquiring an angular velocity of ω_t radians per second.

405 $$\theta = \frac{\omega_t^2 - \omega_0^2}{2\,\alpha} \text{ radians.}$$

Angular velocity (ω_t) acquired in subtending θ radians by a particle having an initial angular velocity of ω_0 radians per second and a uniform angular acceleration of α radians per second per second.

406 $$\omega_t = \sqrt{\omega_0^2 + 2\,\alpha\theta} \text{ radians per second.}$$

Time (t) required for a particle having an initial angular velocity of ω_0 radians per second and a uniform angular acceleration of α radians per second per second to subtend θ radians.

407 $$t = \frac{-\omega_0 + \sqrt{\omega_0^2 + 2\,\alpha\theta}}{\alpha} \text{ seconds.}$$

Uniform angular acceleration (α) required for a particle with an initial angular velocity of ω_0 radians per second to subtend θ radians in t seconds.

408 $$\alpha = \frac{2\,(\theta - \omega_0 t)}{t^2} \text{ radians per second per second.}$$

Velocity (v) of a particle **r** feet from the axis of rotation in a body making **n** revolutions per second.

409 $$v = 2\pi rn \text{ feet per second.}$$

Velocity (v) of a particle **r** feet from the axis of rotation in a body rotating with an angular velocity of **ω** radians per second.

410 $$v = \omega r \text{ feet per second.}$$

Angular velocity (ω) of a body making **n** revolutions per second.

411 $$\omega = 2\pi n \text{ radians per second.}$$

Path of a Projectile*

Horizontal component of velocity (v_x) of a particle having an initial velocity of v_0 feet per second in a direction making an angle of **β** degrees with the horizontal.

412 $$v_x = v_0 \cos \beta \text{ feet per second.}$$

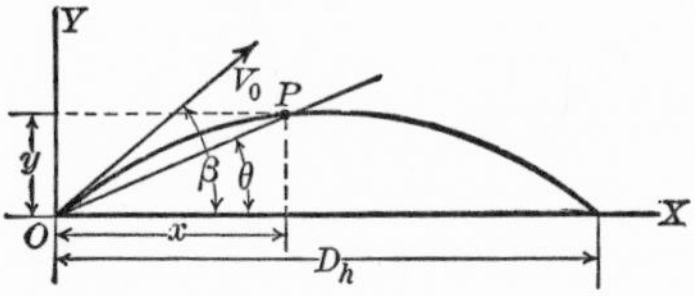

FIG. 412.

Horizontal distance (x) travelled in **t** seconds by a particle having an initial velocity of v_0 feet per second at **β** degrees with the horizontal and a uniform downward acceleration of **a** feet per second per second.

413 $$x = v_0 t \cos \beta \text{ feet.}$$

Vertical component of velocity (v_y) at the end of **t** seconds of a particle having an initial velocity of v_0 feet per second at **β** degrees with the horizontal and a uniform downward acceleration of **a** feet per second per second.

414 $$v_y = v_0 \sin \beta - at \text{ feet per second.}$$

Vertical distance (y) travelled in **t** seconds by a particle having an initial velocity of v_0 feet per second at **β** degrees with the

* Friction of the air is neglected throughout.

horizontal and a uniform downward acceleration of **a** feet per second per second.

415 $$y = v_0 t \sin \beta - \tfrac{1}{2} a t^2 \text{ feet.}$$

Time (t_v) to reach the highest point of the path of a particle having an initial velocity of v_0 feet per second at β degrees with the horizontal and a uniform downward acceleration of **a** feet per second per second.

416 $$t_v = \frac{v_0 \sin \beta}{a} \text{ seconds.}$$

Vertical distance (d_v) from the horizontal to the highest point of the path of a particle having an initial velocity of v_0 feet per second at β degrees with the horizontal and a uniform downward acceleration of **a** feet per second per second.

417 $$d_v = \frac{v_0^2 \sin^2 \beta}{2\,a} \text{ feet.}$$

Velocity (v) at the end of **t** seconds of a particle having an initial velocity of v_0 feet per second at β degrees with the horizontal and a uniform downward acceleration of **a** feet per second per second.

418 $$v = \sqrt{v_x^2 + v_y^2} = \sqrt{v_0^2 - 2\,v_0 a t \sin \beta + a^2 t^2} \text{ feet per second.}$$

Time (t_h) to reach the same horizontal as at start for a particle having an initial velocity of v_0 feet per second at β degrees with the horizontal and a uniform downward acceleration of **a** feet per second per second.

419 $$t_h = \frac{2\,v_0 \sin \beta}{a} \text{ seconds.}$$

Horizontal distance (d_h) travelled by a particle having an initial velocity of v_0 feet per second at β degrees with the horizontal and a uniform downward acceleration of **a** feet per second per second in returning to the same horizontal as at start.

420 $$d_h = \frac{v_0^2 \sin 2\,\beta}{a} \text{ feet.}$$

Time (t) to reach any point **P** for a particle having an initial velocity of v_0 feet per second at β degrees with the horizontal and a uniform downward acceleration of **a** feet per second per second, if a line through **P** and the point of starting makes **θ** degrees with the horizontal.

421 $$t = \frac{2\,v_0 \sin(\beta - \theta)}{a \cos\theta} \text{ seconds.}$$

Harmonic Motion

Simple harmonic motion is the motion of the projection, on the diameter of a circle, of a particle moving with constant speed around the circumference of the circle. **Amplitude** is one-half the projection of the path of the particle or equal to the radius of the circle. **Frequency** is the number of complete oscillations per unit time.

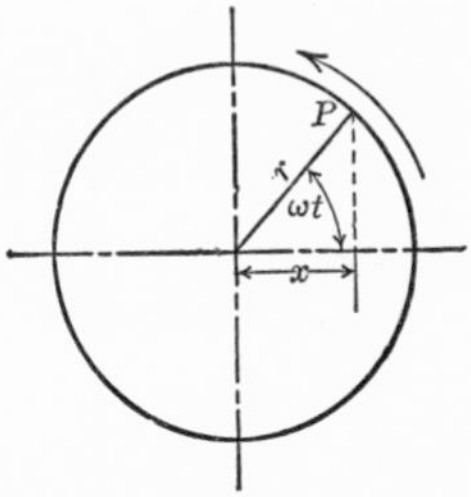

FIG. 422.

Displacement (x) from the center **t** seconds after starting, of the projection on the diameter, of a particle moving with a uniform angular velocity of **ω** radians per second about a circle **r** feet in radius.

422 $$x = r \cos \omega t \text{ feet.}$$

Velocity (v) **t** seconds after starting, of the projection on the diameter, of a particle moving with a uniform angular velocity of **ω** radians per second about a circle **r** feet in radius.

423 $$v = -\omega r \sin \omega t \text{ feet per second.}$$

Acceleration (a) **t** seconds after starting, of the projection on the diameter, of a particle moving with a uniform angular velocity of **ω** radians per second about a circle **r** feet in radius.

424 $$a = -\omega^2 r \cos \omega t = -\omega^2 x \text{ feet per second per second.}$$

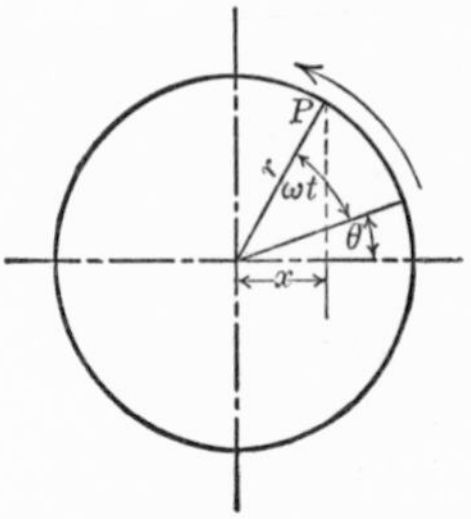

FIG. 424.

NOTE. If the time (t) is reckoned from a position displaced by **θ** radians from the horizontal (called lead if positive and lag if negative) the formulas become: $x = r\cos(\omega t + \theta)$ feet, $v = -\omega r \sin(\omega t + \theta)$ feet per second and $a = -\omega^2 r \cos(\omega t + \theta)$ feet per second per second.

RELATIONS OF MASS AND SPACE

Mass

Mass (**m**) of a body weighing **w** pounds.

425 $$m = \frac{w}{g} \text{ pounds (grav.).}$$

NOTE. The mass (**m**) of a body may be measured by its weight (**w**), designated " pounds (abs.)" etc., or by its weight (**w**) divided by the acceleration due to gravity (**g**), designated " pounds (grav.)" etc. In this text the latter unit is used throughout.

Center of Gravity

Center of gravity of a body or system of bodies is that point through which the resultant of the weights of the component particles passes, whatever position be given the body or system.

NOTE. The center of mass of a body is the same as the center of gravity. The center of gravity of a line, surface or volume is obtained by considering it to be the center of gravity of a slender rod, thin plate or homogeneous body and is often called the centroid.

Moment (**M**) of a body of weight (**w**), or of mass (**m**), about a plane if **x** is the perpendicular distance from the center of gravity of the body to the plane.

426 $$M = wx \quad \text{or} \quad M = mx.$$

Statical moment (S) of an area (A), about an axis X if **x** is the perpendicular distance from the center of gravity of the area to the axis.

427 $$S = Ax.$$

NOTE. The statical moment of an area about an axis through its center of gravity is zero.

Distances (x_0, y_0, z_0) from each of three coördinate planes (X, Y, Z) to the center of gravity or mass of a system of bodies, if Σw is the sum of their weights or Σm is the sum of their masses and Σwx, Σwy, Σwz or Σmx, Σmy, Σmz are the algebraic sums of moments of the separate bodies about the X, Y and Z planes.

428 $$x_0 = \frac{\Sigma wx}{\Sigma w} = \frac{\Sigma mx}{\Sigma m}.$$
$$y_0 = \frac{\Sigma wy}{\Sigma w} = \frac{\Sigma my}{\Sigma m}.$$
$$z_0 = \frac{\Sigma wz}{\Sigma w} = \frac{\Sigma mz}{\Sigma m}.$$

Distances (x_0, y_0, z_0) from each of three coördinate planes to the center of gravity of a volume, if Σv is the sum of the component volumes and Σvx, Σvy and Σvz are the algebraic sums of the moments of these component volumes about the X, Y and Z planes.

429 $$x_0 = \frac{\Sigma vx}{\Sigma v}. \quad y_0 = \frac{\Sigma vy}{\Sigma v}. \quad z_0 = \frac{\Sigma vz}{\Sigma v}.$$

Distances (x_0, y_0) from each of two coördinate axes to the center of gravity of an area, if ΣA is the sum of the component areas and ΣAx and ΣAy are the algebraic sums of the moments of these component areas about the X and Y axes.

430 $$x_0 = \frac{\Sigma Ax}{\Sigma A}. \quad y_0 = \frac{\Sigma Ay}{\Sigma A}.$$

NOTE. The general method of finding the center of gravity of an irregular area is to divide it into component areas, the centers of gravity of which may be calculated or determined from the table on page 78; then find the sum of statical moments of the component areas about some convenient axis and divide by the total area to obtain the distance from that axis to the center of gravity of the whole area. In numerical problems it is often convenient to take the axis of reference through the center of gravity of one of the component areas thereby eliminating the moment of that area and simplifying the numerical work.

Moment of Inertia of Plane Areas

Moment of inertia (J) of an area about an axis is the sum of the products of the component areas into the square of their distances from the axis (ΣAx^2).

431 $$J = \Sigma Ax^2.$$

NOTE. In general an expression for moment of inertia involves the use of the calculus, the area being considered as divided into differential areas **dA.** $J_x = \int y^2\,dA$ and $J_y = \int x^2\,dA$. The unit of moment of inertia of an area is inches, feet, etc., to the fourth power.

Moment of inertia (J_x) of an area **A** about any axis in terms of the moment of inertia J_0 about a parallel axis through the center of gravity of the area, if x_0 is the distance between the two axes.

432 $$J_x = J_0 + Ax_0^2.$$

Radius of gyration (K) of an area **A** from an axis about which the moment of inertia is **J**.

433 $$K = \sqrt{\frac{J}{A}}.$$

Radius of gyration (K_x) of an area **A** about any axis in terms of the radius of gyration K_0 about a parallel axis through the center of gravity of the area, if x_0 is the distance between the two axes.

434 $$K_x^2 = K_0^2 + x_0^2.$$

Product of inertia (U) of an area with respect to two rectangular coördinate axes is the sum of the products of the component areas into the product of their distances from the two axes (ΣAxy).

435 $$U = \Sigma Axy.$$

NOTE. Product of inertia, like moment of inertia, is generally expressed by use of the calculus:

$$U = \int xy\, dA.$$

In case one of the axes is an axis of symmetry the product of inertia is zero.

Product of inertia (U_{xy}) of an area **A** about any two rectangular axes in terms of the product of inertia U_0 about two parallel rectangular axes through the center of gravity of the area, if x_0 and y_0 are the distances between these two sets of axes.

436 $$U_{xy} = U_0 + Ax_0y_0.$$

Moment of inertia ($J_{x'}$ and $J_{y'}$) and product of inertia ($U_{x'y'}$) of an area **A** about each of two rectangular coördinate axes (X′ and Y′) in terms of the moments and product of inertia (J_x, J_y, U_{xy}) about two other rectangular coördinate axes making an angle α with X′ and Y′.

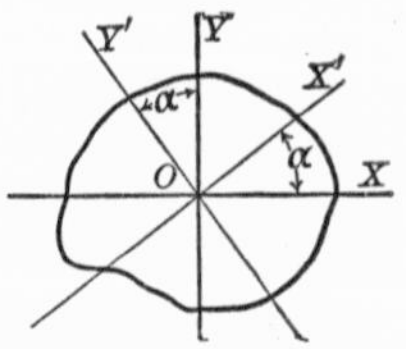

FIG. 437.

437 $$J_{x'} = J_y \sin^2 \alpha + J_x \cos^2 \alpha - 2\,U_{xy} \cos \alpha \sin \alpha.$$
$$J_{y'} = J_y \cos^2 \alpha + J_x \sin^2 \alpha + 2\,U_{xy} \cos \alpha \sin \alpha.$$
$$U_{x'y'} = (J_x - J_y) \cos \alpha \sin \alpha + U_{xy} (\cos^2 \alpha - \sin^2 \alpha).$$

Principal axes of an area are those axes, through any point, about one of which the moment of inertia is a maximum, the moment of inertia about the other being a minimum. The axes are at right angles to each other.

Angle (α) between the rectangular coördinate axes X and Y, about which the moments and products of inertia are J_x, J_y and U_{xy}, and the principal axes through the point of intersection of X and Y.

438 $$\tan 2\alpha = \frac{2\,U_{xy}}{J_y - J_x}.$$

NOTE. An axis of symmetry is a principal axis. The product of inertia about principal axes is zero. If J_y and J_x are moments of inertia about principal axes the equations for the moments of inertia about rectangular axes making an angle α with these principal axes are: $J_{x'} = J_y \sin^2 \alpha + J_x \cos^2 \alpha$ and $J_{y'} = J_y \cos^2 \alpha + J_x \sin^2 \alpha$. The sum of the moments of inertia about rectangular coördinate axes is a constant for all pairs of axes intersecting at the same point, i.e., $J_x + J_y = J_{x'} + J_{y'}$.

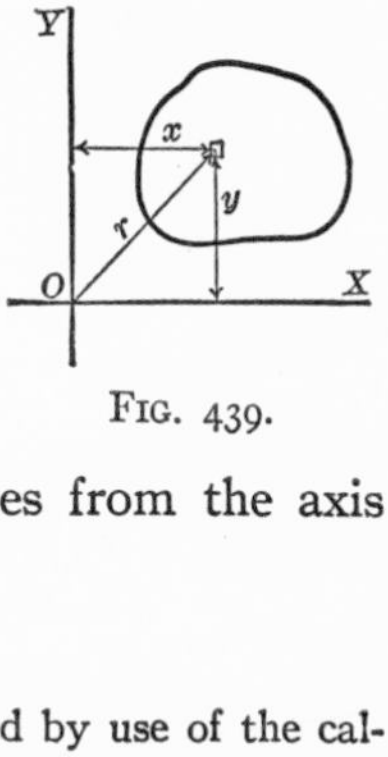

FIG. 439.

Polar moment of inertia (J_p) of an area is the moment of inertia about an axis perpendicular to the plane of the area and is equal to the sum of the products of the component areas into the squares of their distances from the axis (ΣAr^2).

439 $$J_p = \Sigma Ar^2.$$

NOTE. Polar moment of inertia is generally expressed by use of the calculus: $J_p = \int r^2\, dA$.

Polar moment of inertia (J_p) of an area A in terms of the moments of inertia J_x and J_y about two rectangular coördinate axes intersecting on the polar axis.

440 $$J_p = J_x + J_y.$$

Properties of Various Plane Sections

Section	Distance to center of gravity, x	Moment of inertia, J*	Radius of gyration, K
Square 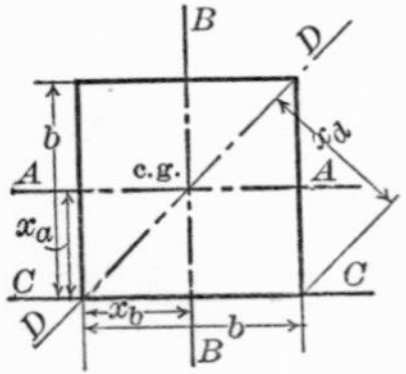	$x_a = x_b = \frac{b}{2}$. $x_d = \frac{b}{\sqrt{2}}$.	$J_{AA} = J_{BB} = J_{DD} = \frac{b^4}{12}$. $J_{CC} = \frac{b^4}{3}$. $J_p = \frac{b^4}{6}$.	$K_{AA} = K_{BB} = K_{DD} = \frac{b}{\sqrt{12}}$ $= 0.289\, b$. $K_{CC} = \frac{b}{\sqrt{3}} = 0.577\, b$.
Hollow Square 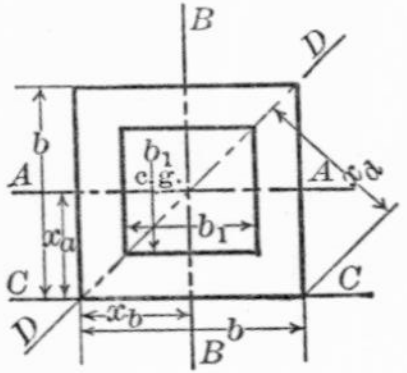	$x_a = x_b = \frac{b}{2}$. $x_d = \frac{b}{\sqrt{2}}$.	$J_{AA} = J_{BB} = J_{DD} = \frac{b^4 - b_1^4}{12}$. $J_{CC} = \frac{b^4}{3} - \frac{b_1^2\,(3\,b^2 + b_1^2)}{12}$ $J_p = \frac{b^4 - b_1^4}{6}$.	$K_{AA} = K_{BB} = K_{DD} = \sqrt{\frac{b^2 + b_1^2}{12}}$ $= 0.289\,\sqrt{b^2 + b_1^2}$.

* J_p, polar moment of inertia, refers to an axis through the center of gravity.

Properties of Various Plane Sections (*Continued*)

Section	Distance to center of gravity, x	Moment of inertia, J *	Radius of gyration, K
Rectangle 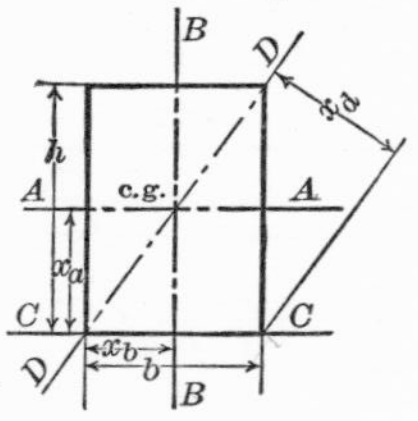	$x_a = \frac{h}{2}$. $x_b = \frac{b}{2}$. $x_d = \frac{bh}{\sqrt{b^2+h^2}}$.	$J_{AA} = \frac{bh^3}{12}$. $J_{BB} = \frac{hb^3}{12}$. $J_{CC} = \frac{bh^3}{3}$. $J_{DD} = \frac{b^3h^3}{6(b^2+h^2)}$. $J_p = \frac{bh^3+hb^3}{12}$.	$K_{AA} = \frac{h}{\sqrt{12}} = 0.289\,h$. $K_{BB} = \frac{b}{\sqrt{12}} = 0.289\,b$. $K_{CC} = \frac{h}{\sqrt{3}} = 0.577\,h$. $K_{DD} = \frac{bh}{\sqrt{6(b^2+h^2)}}$.
Rectangle 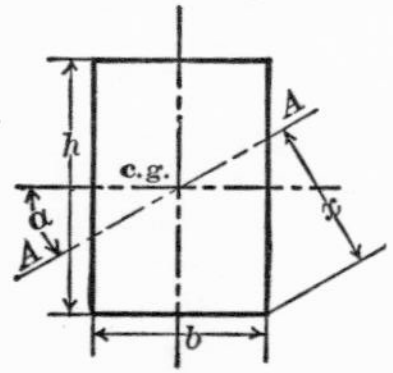	$x = \frac{b \sin\alpha + h\cos\alpha}{2}$.	$J_{AA} = \frac{bh(b^2\sin^2\alpha + h^2\cos^2\alpha)}{12}$.	$K_{AA} = \sqrt{\frac{b^2\sin^2\alpha + h^2\cos^2\alpha}{12}}$.

* J_p, polar moment of inertia, refers to an axis through the cente· of gravity.

Properties of Various Plane Sections (*Continued*)

Section	Distance to center of gravity, x	Moment of inertia, J*	Radius of gyration, K
Hollow Rectangle 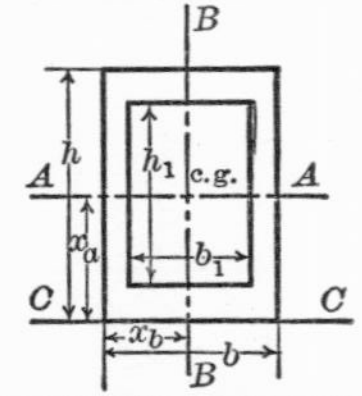	$x_a = \frac{h}{2}$. $x_b = \frac{b}{2}$.	$J_{AA} = \frac{bh^3 - b_1h_1^3}{12}$. $J_{BB} = \frac{hb^3 - h_1b_1^3}{12}$. $J_{CC} = \frac{bh^3}{3} - \frac{b_1h_1}{12}(3h^2 + h_1^2)$.	$K_{AA} = \sqrt{\frac{bh^3 - b_1h_1^3}{12(bh - b_1h_1)}}$. $K_{BB} = \sqrt{\frac{hb^3 - h_1b_1^3}{12(hb - h_1b_1)}}$.
Triangle 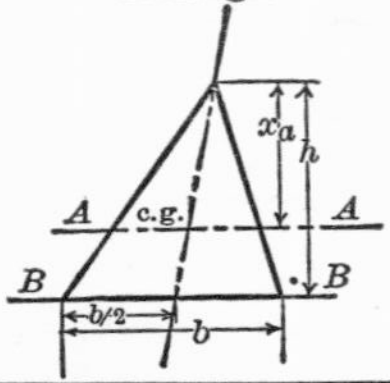	$x_a = \frac{2}{3}h$.	$J_{AA} = \frac{bh^3}{36}$. $J_{BB} = \frac{bh^3}{12}$.	$K_{AA} = \frac{h}{\sqrt{18}} = 0.236\,h$. $K_{BB} = \frac{h}{\sqrt{6}} = 0.408\,h$.
Trapezoid 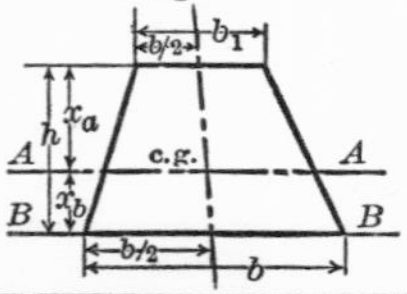	$x_a = \frac{h(b_1 + 2b)}{3(b + b_1)}$. $x_b = \frac{h(b + 2b_1)}{3(b + b_1)}$.	$J_{AA} = \frac{h^3(b^2 + 4bb_1 + b_1^2)}{36(b + b_1)}$. $J_{BB} = \frac{h^3(b + 3b_1)}{12}$.	$K_{AA} = \frac{h\sqrt{2(b^2 + 4bb_1 + b_1^2)}}{6(b + b_1)}$. $K_{BB} = \frac{h\sqrt{b + 3b_1}}{\sqrt{6}(b + b_1)}$.

* J_p, polar moment of inertia, refers to an axis through the center of gravity.

Properties of Various Plane Sections (*Continued*)

Section	Distance to center of gravity, x	Moment of inertia, J*	Radius of gyration, K
Circle 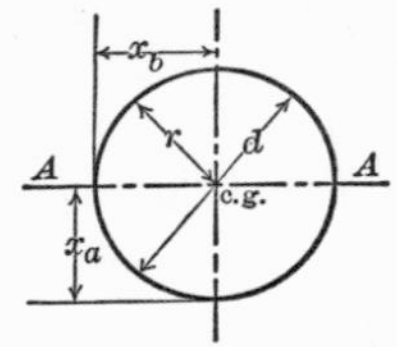	$x_a = x_b = \frac{d}{2} = r.$	$J_{AA} = \frac{\pi d^4}{64} = 0.0491\, d^4 = \frac{\pi r^4}{4} = 0.7854\, r^4.$ $J_p = \frac{\pi r^4}{2}.$	$K_{AA} = \frac{d}{4} = \frac{r}{2}.$
Hollow circle 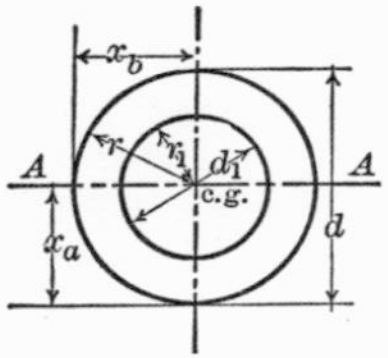	$x_a = x_b = \frac{d}{2} = r.$	$J_{AA} = \frac{\pi (d^4 - d_1^4)}{64} = 0.0491\,(d^4 - d_1^4) = \frac{\pi (r^4 - r_1^4)}{4} = 0.7854\,(r^4 - r_1^4).$ $J_p = \frac{\pi (r^4 - r_1^4)}{2}.$	$K_{AA} = \frac{\sqrt{d^2 + d_1^2}}{4} = \frac{\sqrt{r^2 + r_1^2}}{2}.$
Semi-circle 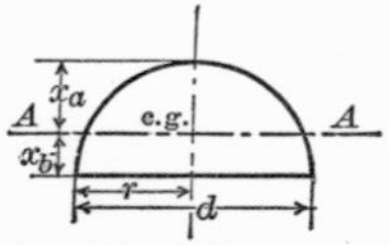	$x_a = \frac{d\,(3\pi - 4)}{6\pi} = 0.288\, d = 0.576\, r.$ $x_b = \frac{2d}{3\pi} = 0.212\, d = \frac{4r}{3\pi} = 0.424\, r.$	$J_{AA} = \frac{d^4 (9\pi^2 - 64)}{1152\,\pi} = 0.00686\, d^4 = 0.1098\, r^4.$	$K_{AA} = \frac{d}{12\,\pi} \sqrt{(9\pi^2 - 64)} = 0.132\, d.$

* J_p, polar moment of inertia, refers to an axis through the center of gravity.

Properties of Various Plane Sections (*Continued*)

Section	Distance to center of gravity, x	Moment of inertia, J*	Radius of gyration, K
Hollow Half Circle 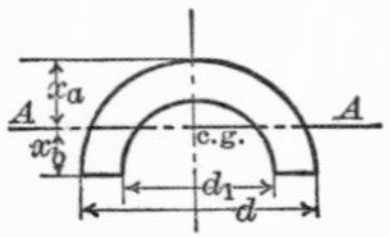	$x_b = \frac{2(d^3 - d_1^3)}{3\pi(d^2 - d_1^2)}$.	$J_{AA} = \frac{\pi(d^4 - d_1^4)}{128} - \frac{4(d^3 - d_1^3)^2}{72\pi(d^2 - d_1^2)}$.	$K_{AA} = \sqrt{\frac{(d^4 - d_1^4)}{16(d^2 - d_1^2)} - \frac{4(d^3 - d_1^3)^2}{9\pi^2(d^2 - d_1^2)^2}}$.
Circular Segment 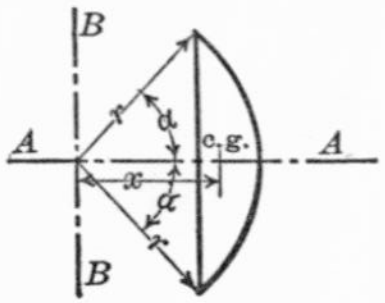	$x = \frac{2}{3}\frac{r^3 \sin^3 \alpha}{A}$. [$A = \frac{1}{2} r^2 (2\alpha - \sin 2\alpha)$ where first α is in radians]	$J_{AA} = \frac{1}{4}Ar^2\left[1 - \frac{2}{3}\frac{\sin^3 \alpha \cos \alpha}{\alpha - \sin \alpha \cos \alpha}\right]$. $J_{BB} = \frac{1}{4}Ar^2\left[1 + \frac{2 \sin^3 \alpha \cos \alpha}{\alpha - \sin \alpha \cos \alpha}\right]$.	$K = \sqrt{\frac{J}{A}}$.
Circular Sector 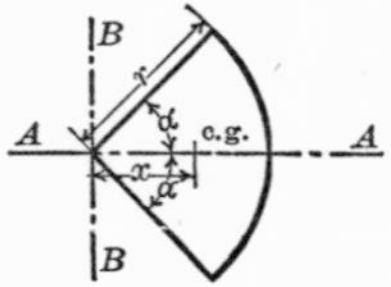	$x = \frac{2}{3}\frac{r \sin \alpha}{\alpha}$.	$J_{AA} = \frac{1}{4}Ar^2\left(1 - \frac{\sin \alpha \cos \alpha}{\alpha}\right)$. $J_{BB} = \frac{1}{4}Ar^2\left(1 + \frac{\sin \alpha \cos \alpha}{\alpha}\right)$.	$K = \sqrt{\frac{J}{A}}$.

* J_p, polar moment of inertia, refers to an axis through the center of gravity.

Properties of Various Plane Sections (*Continued*)

Section	Distance to center of gravity, x	Moment of inertia, J*	Radius of gyration, K
Parabolic Segment 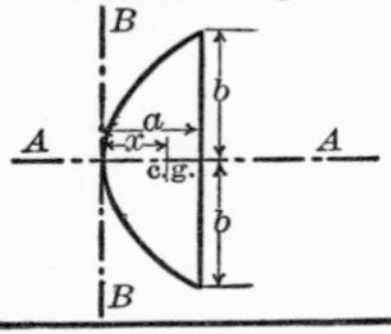	$x = \frac{3}{5}a$. (for half segment, $y = \frac{3}{8}b$)	$J_{AA} = \frac{4}{15}ab^3$. $J_{BB} = \frac{4}{7}ba^3$.	$K_{AA} = \frac{b}{\sqrt{5}} = 0.447\,b$ $K_{BB} = a\sqrt{\frac{3}{7}} = 0.654\,a$.
Ellipse 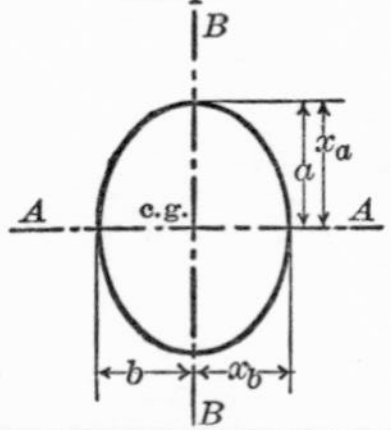	$x_a = a$. $x_b = b$.	$J_{AA} = \frac{\pi a^3 b}{4} = 0.7854\,a^3b$. $J_{BB} = \frac{\pi b^3 a}{4} = 0.7854\,ab^3$. $J_p = \frac{\pi ab\,(a^2 + b^2)}{4}$.	$K_{AA} = \frac{a}{2}$. $K_{BB} = \frac{b}{2}$.
Elliptical Ring 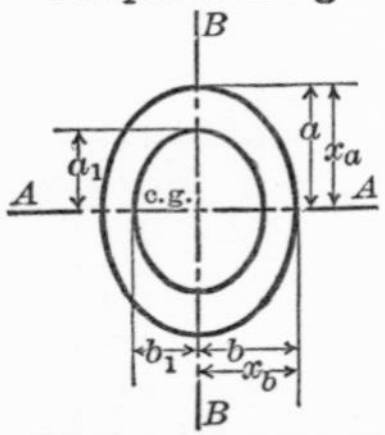	$x_a = a$. $x_b = b$.	$J_{AA} = \frac{\pi}{4}(a^3b - a_1^3b_1)$ $= 0.7854\,(a^3b - a_1^3b_1)$. $J_{BB} = \frac{\pi}{4}(b^3a - b_1^3a_1)$ $= 0.7854\,(b^3a - b_1^3a_1)$.	$K_{AA} = \frac{1}{2}\sqrt{\frac{a^3b - a_1^3b_1}{ab - a_1b_1}}$. $K_{BB} = \frac{1}{2}\sqrt{\frac{b^3a - b_1^3a_1}{ba - b_1a_1}}$.

* J_p, polar moment of inertia, refers to an axis through the center of gravity.

Properties of Various Plane Sections (*Continued*)

Section	Distance to center of gravity, x	Moment of inertia, J*	Radius of gyration, K
Equal Angle 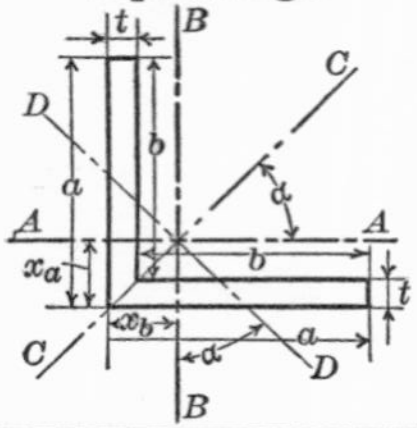	$x_a = x_b = \frac{a^2 + (a - t)\,t}{2\,(2\,a - t)}$. $[\alpha = 45°]$	$J_{AA} = \frac{t\,(a - x)^3 + ax^3 - a\,(x - t)^3}{3}$. $J_{BB} = J_{AA}$. $J_{CC} = \frac{bt^3 + b^3t + 3\,a^2bt + t^4}{12}$. $J_{DD} = \frac{bt^3 + b^3t + 3\,bt\,(a - 4x + 2t)^2 + t^4 + 6t^2\,(2x - t)^2}{12}$.	$K = \sqrt{\frac{J}{A}}$.
Unequal Angle 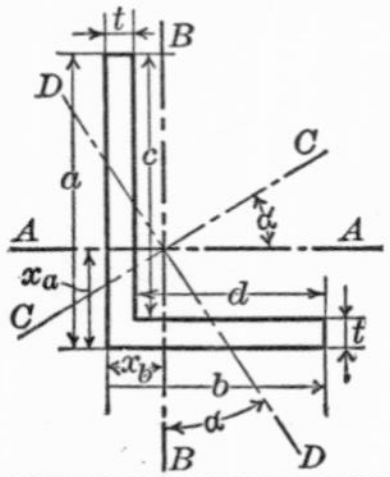	$x_a = \frac{t\,(b + 2\,c) + c^2}{2\,(b + c)}$. $x_b = \frac{t\,(2\,d + a) + d^2}{2\,(a + d)}$. $\tan 2\,d = \frac{t(2x_b - t)a(a - 2x_a) + d(2x_a - t)(b + t - 2x_b)}{2\,(J_{AA} - J_{BB})}$	$J_{AA} = \frac{t\,(a - x_a)^3 + bx_a^3 - d\,(x_a - t)^3}{3}$. $J_{BB} = \frac{t\,(b - x_b)^3 + ax_b^3 - c\,(x_b - t)^3}{3}$. $J_{CC} = \frac{J_{AA}\cos^2\alpha - J_{BB}\sin^2\alpha}{\cos 2\,\alpha}$. $J_{DD} = \frac{J_{BB}\cos^2\alpha - J_{AA}\sin^2\alpha}{\cos 2\,\alpha}$.	$K = \sqrt{\frac{J}{A}}$.
I-Beam 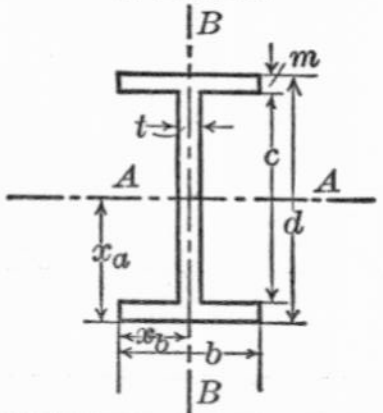	$x_a = \frac{d}{2}$. $x_b = \frac{b}{2}$.	$J_{AA} = \frac{bd^3 - c^3\,(b - t)}{12}$. $J_{BB} = \frac{2\,mb^3 + ct^3}{12}$.	$K_{AA} = \sqrt{\frac{bd^3 - c^3\,(b - t)}{12\,[bd - c\,(b - t)]}}$. $K_{BB} = \sqrt{\frac{2\,mb^3 + ct^3}{12\,[bd - c\,(b - t)]}}$.

Properties of Various Plane Sections (*Continued*)

Section	Distance to center of gravity, x	Moment of inertia, J*	Radius of gyration, K
Standard I-Beam 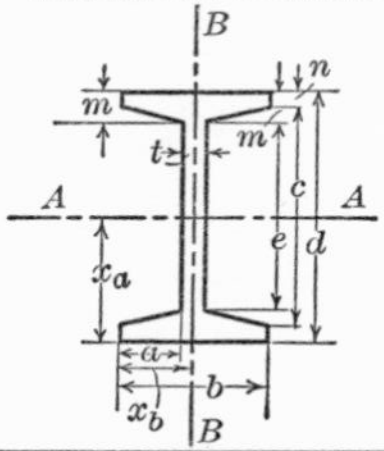	$x_a = \frac{d}{2}$. $x_b = \frac{b}{2}$.	$J_{AA} = \frac{bd^3 - \frac{a}{4(m-n)}(c^4 - e^4)}{12}$. $J_{BB} = \frac{2nb^3 + et^3 + \frac{m-n}{4a}(b^4 - t^4)}{12}$.	$K = \sqrt{\frac{J}{A}}$.
Channel 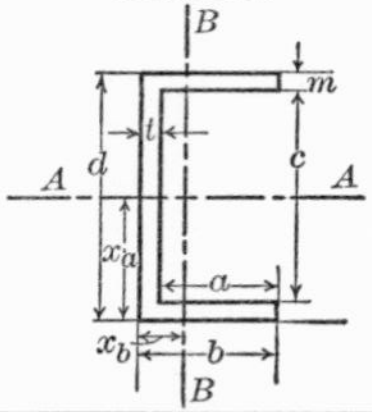	$x_a = \frac{d}{2}$. $x_b = \frac{\frac{dt^2}{2} + 2am\left(t + \frac{a}{2}\right)}{dt + 2am}$.	$J_{AA} = \frac{bd^3 - ac^3}{12}$. $J_{BB} = \frac{dx_b^3 - d(x_b - t)^3 + 2m(b - x_b)^3}{3}$.	$K_{AA} = \sqrt{\frac{bd^3 - ac^3}{12(bd - ac)}}$. $K_{BB} = \sqrt{\frac{dx_b^3 - d(x_b - t)^3 + 2m(b - x_b)^3}{3(bd - ac)}}$.
Standard Channel 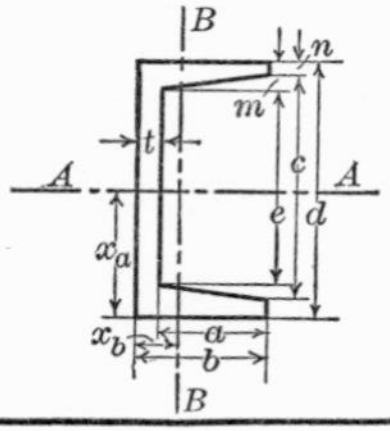	$x_a = \frac{d}{2}$. $x_b = \frac{\frac{dt^2}{2} + 2am\left(t + \frac{a}{2}\right) + a(m-n)\left(t + \frac{a}{3}\right)}{dt + a(m+n)}$.	$J_{AA} = \frac{bd^3 - \frac{a}{8(m-n)}(c^4 - e^4)}{12}$. $J_{BB} = \frac{2nb^3 + et^3 + \frac{m-n}{2a}(b^4 - t^4)}{3} - [dt + a(m+n)]\,x_b^2$.	$K = \sqrt{\frac{J}{A}}$.

* J_n, polar moment of inertia, refers to an axis through the center of gravity

Properties of Various Plane Sections (*Concluded*)

Section	Distance to center of gravity, x	Moment of inertia, J *	Radius of gyration, K
Zee 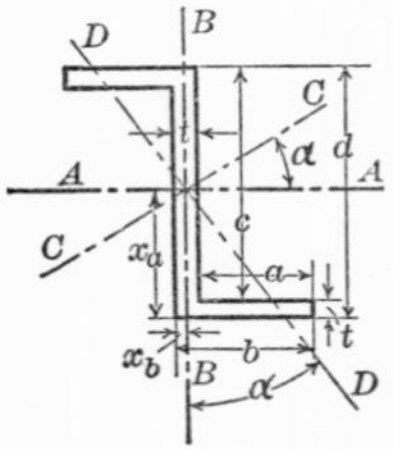	$x_a = \frac{d}{2}$. $x_b = \frac{t}{2}$. $\tan 2\alpha = \frac{(dt - t^2)(b^2 - bt)}{J_{AA} - J_{BB}}$.	$J_{AA} = \frac{bd^3 - a(d - 2t)^3}{12}$. $J_{BB} = \frac{d(b+a)^3 - 2a^3c - 6b^2ac}{12}$. $J_{CC} = \frac{J_{AA}\cos^2\alpha - J_{BB}\sin^2\alpha}{\cos 2\alpha}$. $J_{DD} = \frac{J_{BB}\cos^2\alpha - J_{AA}\sin^2\alpha}{\cos 2\alpha}$.	$K = \sqrt{\frac{J}{A}}$.
Tee 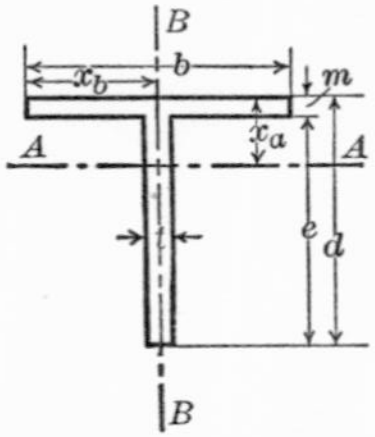	$x_a = \frac{\frac{bm^2}{2} + et\left(\frac{e}{2} + m\right)}{bm + et}$. $x_b = \frac{b}{2}$.	$J_{AA} = \frac{bx_a^3 + t(d - x_a)^3 - (b - t)(x_a - m)^3}{3}$. $J_{BB} = \frac{mb^3 + et^3}{12}$.	$K_{AA} = \sqrt{\frac{bx_a^3 + t(d - x_a)^3 - (b - t)(x_a - m)^3}{3(bm + et)}}$. $K_{BB} = \sqrt{\frac{mb^3 + et^3}{12(bm + et)}}$.

* J_p, polar moment of inertia, refers to an axis through the center of gravity.

Moment of Inertia of Bodies

Moment of inertia (J_m) of a body about an axis, in terms of the mass, is the sum of the products of the component masses into the squares of their distances from the axis (Σmr^2).

441 $$J_m = \Sigma mr^2.$$

Moment of inertia (J) of a body about an axis, in terms of the weight, is the sum of the products of the component weights into the squares of their distances from the axis (Σwr^2).

442 $$J = \Sigma wr^2.$$

Moment of inertia (J_m) in terms of the mass for a case where the moment of inertia in terms of the weight is **J**.

443 $$J_m = \frac{J}{g}.$$

NOTE. The moment of inertia of a body is generally expressed by the calculus. $J_m = \int r^2\,dm$. $J = \int r^2\,dw$. The unit of moment of inertia of solid is pound-feet², etc.

Moment of inertia (J_x) of a body of weight **W** about any axis in terms of the moment of inertia (J_0) about a parallel axis through the center of gravity of the body, if x_0 is the distance between the axes.

444 $$J_x = J_0 + Wx_0^2.$$

Radius of gyration (K) of a body of weight **W** from an axis about which the moment of inertia is **J**.

445 $$K = \sqrt{\frac{J}{W}}.$$

Moment of inertia (J_m), in terms of the mass, of a body of weight **W** about an axis for which the radius of gyration is **K**.

446 $$J_m = \frac{W}{g}K^2.$$

Product of inertia (U or U_m) of a body with respect to two coördinate planes is the sum of the products of the component weights (or masses) into the products of their distances from these planes (Σwxy or Σmxy).

447 $$U = \Sigma wxy \qquad U_m = \Sigma mxy.$$

NOTE. The product of inertia of a body is generally expressed by the calculus. $U = \int xy\,dw$. $U_m = \int xy\,dm$.

Moment of inertia (J) with respect to the axis V′V in terms of the moments of inertia J_x, J_y and J_z with respect to the axes X′X, Y′Y and Z′Z and the products of inertia U_{xy}, U_{xz} and U_{yz} with respect to the planes Y_{oy} and X_{ox}, the planes Y_{oz} and X_{ox} and the planes X_{oz} and X_{oy} respectively, where V′V passes through the origin of these three axes and makes the angles α, β and γ with the axes X′X, Y′Y and Z′Z respectively.

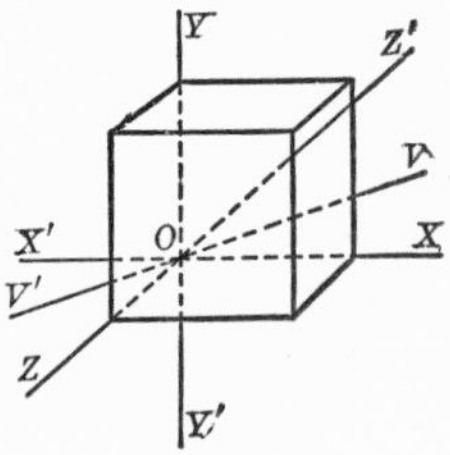

FIG. 448.

448 $$J = J_x \cos^2\alpha + J_y \cos^2\beta + J_z \cos^2\gamma - 2\,U_{xy}\cos\alpha\cos\beta - 2\,U_{xz}\cos\alpha\cos\gamma - 2\,U_{yz}\cos\beta\cos\gamma.$$

Principal axes of a body are those three rectangular axes through any point, about one of which the moment of inertia is a maximum and about another a minimum, the moment of inertia about the third axis being intermediate in value. **Principal planes** are the planes perpendicular to the principal axes. The products of inertia with respect to the principal planes are zero.

Properties of Various Solids *

Solids	Moment of inertia, J	Radius of gyration, K
Straight Rod 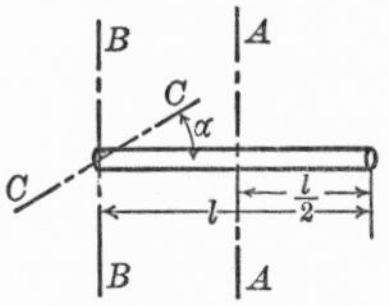	$J_{AA} = \frac{1}{12} Wl^2.$ $J_{BB} = \frac{1}{3} Wl^2.$ $J_{CC} = \frac{1}{3} Wl^2 \sin^2 \alpha.$	$K_{AA} = \frac{l}{\sqrt{12}}.$ $K_{BB} = \frac{l}{\sqrt{3}}.$ $K_{CC} = l\sqrt{\frac{\sin \alpha}{3}}.$
Rod bent into a Circular Arc 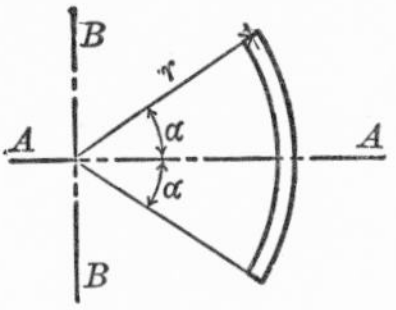	$J_{AA} =$ $\frac{1}{2} Wr^2 \left[1 - \frac{\sin \alpha \cos \alpha}{\alpha}\right].$ $J_{BB} =$ $\frac{1}{2} Wr^2 \left[1 + \frac{\sin \alpha \cos \alpha}{\alpha}\right].$	$K_{AA} =$ $r\sqrt{\frac{1}{2}\left(1 - \frac{\sin \alpha \cos \alpha}{\alpha}\right)}.$ $K_{BB} =$ $r\sqrt{\frac{1}{2}\left(1 + \frac{\sin \alpha \cos \alpha}{\alpha}\right)}.$
Cube 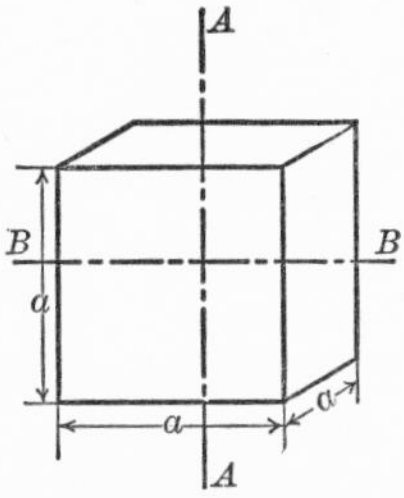	$J_{AA} = J_{BB} = \frac{1}{6} Wa^2.$	$K_{AA} = K_{BB} = \frac{a}{\sqrt{6}}.$
Rectangular Prism 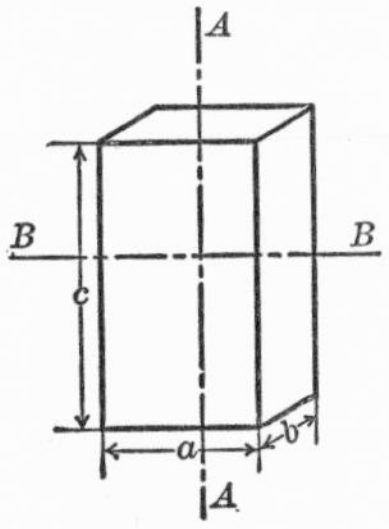	$J_{AA} = \frac{1}{12} W (a^2 + b^2).$ $J_{BB} = \frac{1}{12} W (b^2 + c^2).$	$K_{AA} = \sqrt{\frac{a^2 + b^2}{12}}.$ $K_{BB} = \sqrt{\frac{b^2 + c^2}{12}}.$

* All axes pass through the center of gravity unless otherwise noted. $J_m = \frac{J}{g}$. W = total weight of the body.

Properties of Various Solids* (*Continued*)

Solids	Moments of inertia J	Radius of gyration, K
Right Circular Cylinder 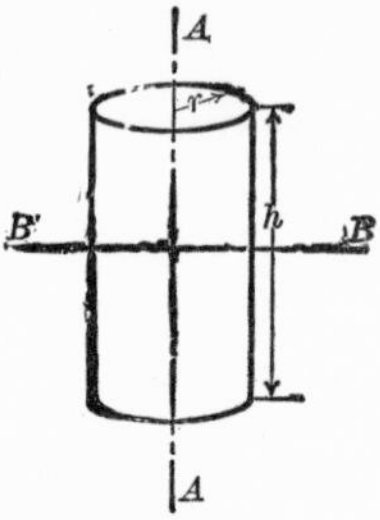	$J_{AA} = \frac{1}{2} Wr^2.$ $J_{BB} = \frac{1}{12} W(3r^2 + h^2).$	$K_{AA} = \frac{r}{\sqrt{2}}.$ $K_{BB} = \sqrt{\frac{3r^2 + h^2}{12}}.$
Hollow Right Circular Cylinder 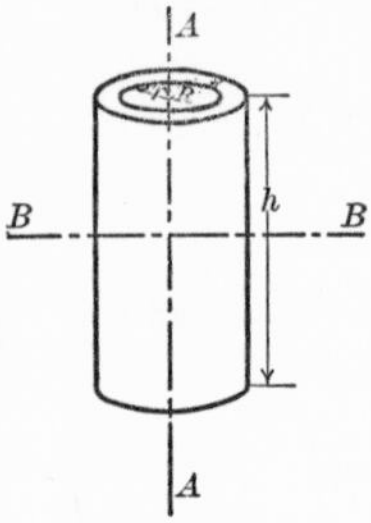	$J_{AA} = \frac{1}{2} W(R^2 + r^2).$ $J_{BB} = \frac{1}{4} W\left(R^2 + r^2 + \frac{h^2}{3}\right).$	$K_{AA} = \sqrt{\frac{R^2 + r^2}{2}}.$ $K_{BB} = \sqrt{\frac{3R^2 + 3r^2 + h^2}{12}}.$
Thin Hollow Cylinder 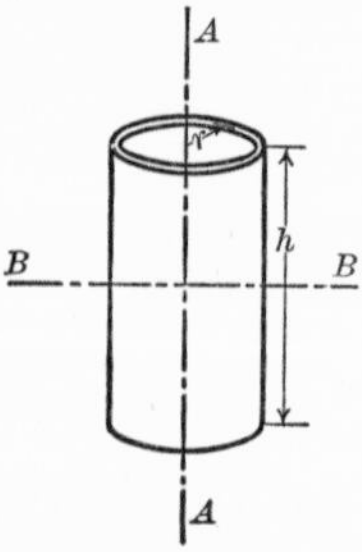	$J_{AA} = Wr^2.$ $J_{BB} = \frac{W}{2}\left(r^2 + \frac{h^2}{6}\right).$	$K_{AA} = r.$ $K_{BB} = \sqrt{\frac{6r^2 + h^2}{12}}.$

* All axes pass through the center of gravity unless otherwise noted. $J_m = \frac{J}{g}$. W = total weight of the body.

Properties of Various Solids* (*Continued*)

Solids	Moments of inertia, J	Radius of gyration, K
Elliptical Cylinder 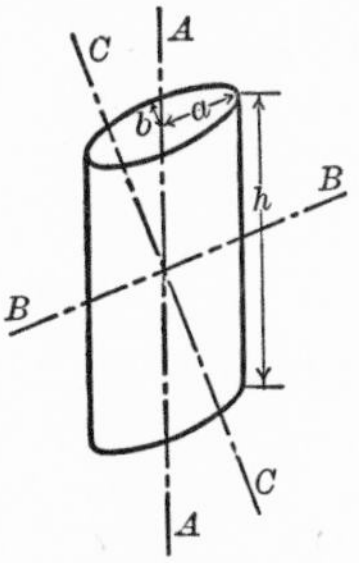	$J_{AA} = \frac{1}{4} W (a^2 + b^2)$. $J_{BB} = \frac{1}{12} W (3\,b^2 + h^2)$. $J_{CC} = \frac{1}{12} W (3\,a^2 + h^2)$.	$K_{AA} = \sqrt{\frac{a^2 + b^2}{2}}$. $K_{BB} = \sqrt{\frac{3\,b^2 + h^2}{12}}$. $K_{CC} = \sqrt{\frac{3\,a^2 + h^2}{12}}$.
Sphere 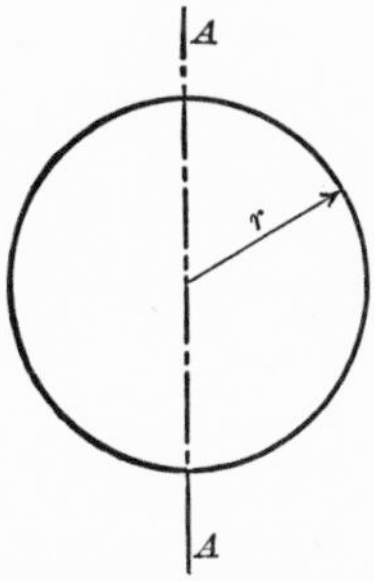	$J_{AA} = \frac{2}{5} Wr^2$.	$K_{AA} = \frac{2\,r}{\sqrt{10}}$.
Hollow Sphere 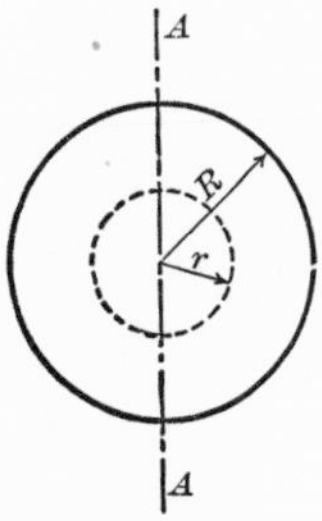	$J_{AA} = \frac{2}{5} W \frac{R^5 - r^5}{R^3 - r^3}$.	$K_{AA} = \sqrt{\frac{2}{5}\left(\frac{R^5 - r^5}{R^3 - r^3}\right)}$.

* All axes pass through the center of gravity unless otherwise noted. $J_m = \frac{J}{g}$. W = total weight of the body.

Properties of Various Solids* (*Continued*)

Solids	Moment of inertia, J	Radius of gyration, K
Thin Hollow Sphere 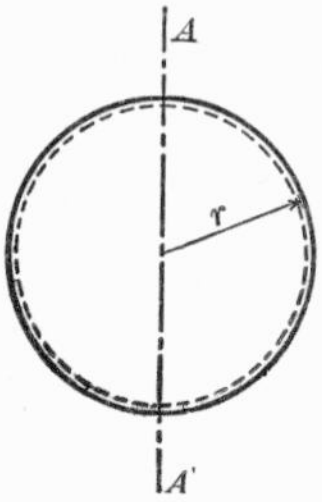	$J_{AA} = \frac{2}{3} Wr^2.$	$K_{AA} = \frac{2r}{\sqrt{6}}.$
Ellipsoid 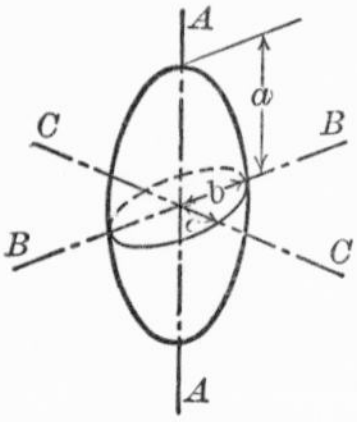	$J_{AA} = \frac{1}{5} W (b^2 + c^2).$ $J_{BB} = \frac{1}{5} W (a^2 + c^2).$ $J_{CC} = \frac{1}{5} W (a^2 + b^2).$	$K_{AA} = \sqrt{\frac{b^2 + c^2}{5}}.$ $K_{BB} = \sqrt{\frac{a^2 + c^2}{5}}.$ $K_{CC} = \sqrt{\frac{a^2 + b^2}{5}}.$
Torus 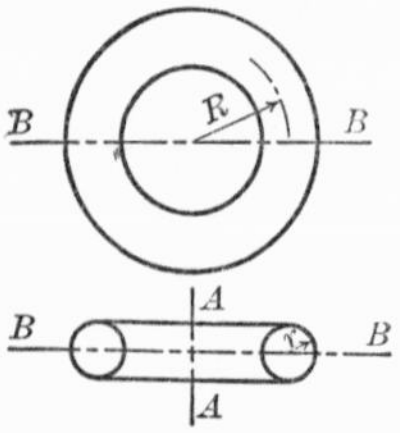	$J_{AA} = W (R^2 + \frac{3}{4} r^2).$ $J_{BB} = W \left(\frac{R^2}{2} + \frac{5}{8} r^2\right).$	$K_{AA} = \frac{1}{2} \sqrt{4 R^2 + 3 r^2}.$ $K_{BB} = \sqrt{\frac{4 R^2 + 5 r^2}{8}}.$

* All axes pass through the center of gravity unless otherwise noted. $J_m = \frac{J}{g}$. W = total weight of the body.

Properties of Various Solids * (*Continued*)

Solids	Distance to center of gravity, x	Moment of inertia, J	Radius of gyration, K
Right Rectangular Pyramid 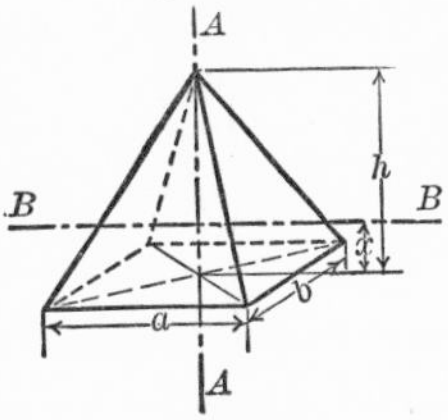	$x = \frac{h}{4}$.	$J_{AA} = \frac{1}{20} W (a^2 + b^2)$. $J_{BB} = \frac{1}{20} W \left(b^2 + \frac{3h^2}{4}\right)$.	$K_{AA} = \sqrt{\frac{a^2 + b^2}{20}}$. $K_{BB} = \sqrt{\frac{1}{80}(4b^2 + 3h^2)}$.
Right Circular Cone 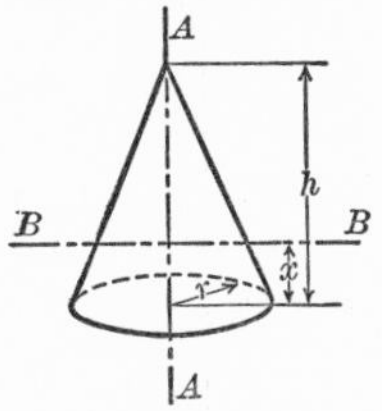	$\bar{x} = \frac{h}{4}$.	$J_{AA} = \frac{3}{10} W r^2$. $J_{BB} = \frac{3}{20} W \left(r^2 + \frac{h^2}{4}\right)$.	$K_{AA} = \frac{3r}{\sqrt{30}}$. $K_{BB} = \sqrt{\frac{3}{80}(4r^2 + h^2)}$.
Frustum of a Cone 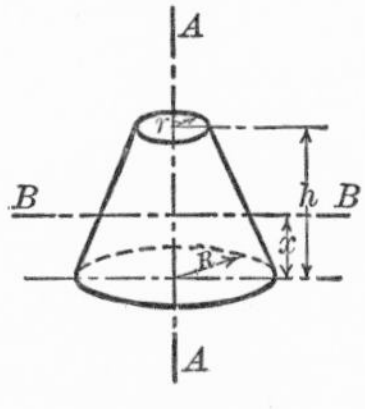	$x = \frac{h(R^2 + 2Rr + 3r^2)}{4(R^2 + Rr + r^2)}$	$J_{AA} = \frac{3}{10} W \frac{(R^5 - r^5)}{(R^3 - r^3)}$.	$K_{AA} = \sqrt{\frac{3}{10} \frac{(R^5 - r^5)}{(R^3 - r^3)}}$.
Paraboloid 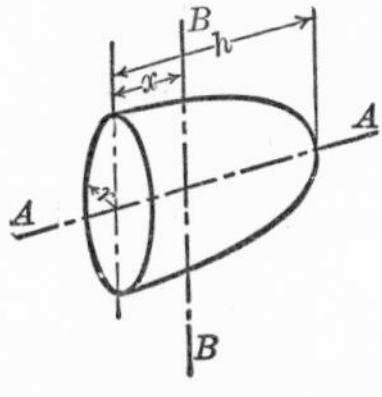	$x = \frac{1}{3} h$.	$J_{AA} = \frac{1}{3} W r^2$. $J_{BB} = \frac{1}{18} W (3r^2 + h^2)$.	$K_{AA} = \frac{r}{\sqrt{3}}$. $K_{BB} = \sqrt{\frac{1}{18}(3r^2 + h^2)}$.

* All axes pass through the center of gravity unless otherwise noted. $J_m = \frac{J}{g}$. W = total weight of the body.

Properties of Various Solids * (*Concluded*)

Solids	Distance to center of gravity, x	Moment of inertia, J	Radius of gyration, K
Spherical Sector 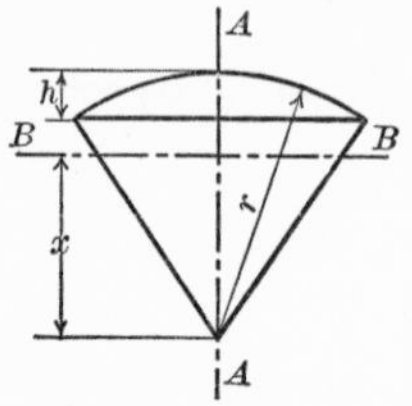	$x = \frac{3}{8}(2r - h).$	$J_{AA} = \frac{1}{5}W(3rh - h^2).$	$K_{AA} = \sqrt{\frac{3rh - h^2}{5}}.$
Spherical Segment 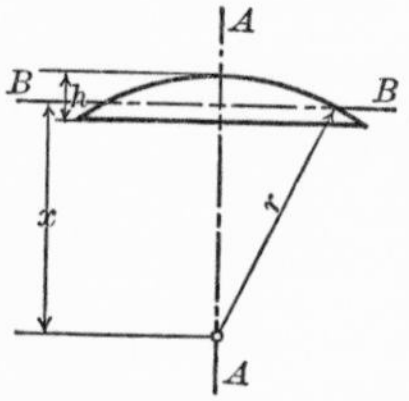	$x = \frac{3}{4}\frac{(2r - h)^2}{(3r - h)}.$ For half sphere $x = \frac{3}{8}r.$	$J_{AA} = W\left(r^2 - \frac{3rh}{4} + \frac{3h^2}{20}\right)\frac{2h}{3r - h}.$	$K_{AA} = \sqrt{\frac{J}{W}}.$

* All axes pass through the center of gravity unless otherwise noted. $J_m = \frac{J}{g}$. W = total weight of the body.

KINETICS

Translation

Three laws of motion. (1) A body remains in a state of rest or of uniform motion except under the action of some unbalanced force. (2) A single force acting on a body causes it to move with accelerated motion in the direction of the force. The acceleration is directly proportional to the force and inversely proportional to the mass of the body. (3) To every action there is an equal and opposite reaction.

Force (F) imparting an acceleration of **a** feet per second per second to a mass of **m** pounds (grav.).

449 $$F = ma \text{ pounds.}$$

NOTE. In terms of the weight w, $F = \frac{w}{g}a.$

Impulse (I) of a force of **F** pounds acting for **t** seconds.

450 $$I = Ft \text{ pound-seconds.}$$

Momentum ($\mathfrak{M}$) of a body of **m** pounds (grav.) mass moving with a velocity of **v** feet per second.

451 $$\mathfrak{M} = mv \text{ pound(grav.)-feet per second.}$$

Force (F) required to change the velocity of a mass of **m** pounds (grav.) from v_1 feet per second to v_2 feet per second in **t** seconds.

452 $$F = \frac{m(v_1 - v_2)}{t} \text{ pounds.}$$

NOTE. The change in momentum of a body during any time interval equals the impulse of the force acting on the body for that time.

Work (W) done by a force of **F** pounds acting through a distance of **s** feet.

453 $$W = Fs \text{ foot-pounds.}$$

NOTE. If the force is variable, $W = \int_0^s F\,ds$.

Power (P) required to do W foot-pounds of work at a constant rate in **t** seconds.

454 $$P = \frac{W}{t} \text{ foot-pounds per second.}$$

Potential energy (W), referred to a certain datum, of a body of **w** pounds weight and at an elevation of **h** feet above the datum.

455 $$W = wh \text{ foot-pounds.}$$

Kinetic energy (W) of a body of **m** pounds (grav.) mass having a velocity of translation of **v** feet per second.

456 $$W = \frac{mv^2}{2} \text{ foot-pounds.}$$

Force (F) required to change the velocity of a mass of **m** pounds (grav.) from v_1 feet per second to v_2 feet per second in **s** feet.

457 $$F = \frac{m(v_1^2 - v_2^2)}{2s} \text{ pounds.}$$

NOTE. The change in kinetic energy of the body equals the work done on the body.

Force (F) required to move a mass of **m** pounds (grav.) in a circular path of **r** feet radius with a constant speed of **v** feet per second.

458 $$F = \frac{mv^2}{r} \text{ pounds.}$$

NOTE. The above force acts along the normal to the path of the body toward the center of curvature, and is called the centripetal or deviating force. The reaction to this force along the normal to the path of the body away from the center of curvature is called the centrifugal force.

Rotation

Torque or moment (T) about the axis of rotation imparting an angular acceleration of α radians per second per second to a body with a mass moment of inertia of J_m pound(grav.)-feet squared about the axis of rotation.

459 $$T = J_m\alpha \text{ pound-feet.}$$

NOTE. In terms of the weight, **w** pounds, of the body and its radius of gyration, **K** feet, about the axis of rotation, $T = \frac{w}{g}K^2\alpha$ pound-feet.

Angular impulse (I_a) of a torque of **T** pound-feet acting for **t** seconds.

460 $$I_a = Tt \text{ pound-feet-seconds.}$$

Angular momentum ($\mathcal{M}_a$) of a body with a mass moment of inertia of J_m pound(grav.)-feet squared about the axis of rotation and an angular velocity of ω radians per second.

461 $$\mathcal{M}_a = J_m\omega \text{ pound(grav.)-feet squared per second.}$$

NOTE. The angular momentum of a body is sometimes called its moment of momentum. The angular momentum of a body moving in a plane perpendicular to the axis of rotation is given by $\mathcal{M}_a = \mathcal{M}r$ pound(grav.)-feet squared per second where $\mathcal{M}$ equals the momentum of the body in pound (grav.)-feet per second, and **r** equals the perpendicular distance in feet from the line of direction of the momentum to the axis of rotation.

Torque (T) required to change the angular velocity of a body of mass moment of inertia of J_m pound(grav.)-feet squared about the axis of rotation from ω_1 radians per second to ω_2 radians per second in **t** seconds.

462 $$T = \frac{J_m(\omega_1 - \omega_2)}{t} \text{ pound-feet.}$$

NOTE. The change in angular momentum of a body is equal to the angular impulse.

Work (W) done by a torque of **T** pound-feet acting through an angle of **θ** radians.

463 $$W = T\theta \text{ foot-pounds.}$$

NOTE. If the torque is variable, $W = \int_0^\theta T\,d\theta$. The work done by a torque of T pound-feet in N revolutions is given by $W = T\,2\pi N$ foot-pounds.

Kinetic energy (W) of a body which has an angular velocity of **ω** radians per second and a mass moment of inertia of J_m pound(grav.)-feet squared about the axis of rotation.

464 $$W = \frac{J_m\omega^2}{2} \text{ foot-pounds.}$$

NOTE. In terms of the weight, w pounds, of the body and its radius of gyration, K feet, about the axis of rotation, $W = \frac{wK^2\omega^2}{2g}$ foot-pounds.

Torque (T) required to change the angular velocity of a body of mass moment of inertia of J_m pound(grav.)-feet squared about the axis of rotation from ω_1 radians per second to ω_2 radians per second, the torque acting through an angle of **θ** radians.

465 $$T = \frac{J_m(\omega_1^2 - \omega_2^2)}{2\theta} \text{ pound-feet.}$$

NOTE. The change in kinetic energy of a body equals the work done on the body.

Center of percussion with respect to the axis of rotation is the point through which the line of action of the resultant of all the external forces acting on the rotating body passes.

Distance (l) from the axis of rotation to the center of percussion of a body with a mass moment of inertia of J_m pound (grav.)-feet squared about the axis of rotation, **m** pounds (grav.) mass and x_0 feet between the axis and the center of gravity.

466 $$l = \frac{J_m}{x_0 m} \text{ feet.}$$

NOTE. In terms of the radius of gyration K, $l = \frac{K^2}{x_0}$.

General Formulas for Rotation about a Fixed Axis

Assume a body **AB** rotating about the axis $Z'Z$. Let $\mathbf{m}$ = mass of the body; α = angular acceleration at any instant; ω = angular velocity at any instant and $\mathbf{x_0}, \mathbf{y_0}, \mathbf{z_0}$ = the coördinates of the center of gravity of the body.

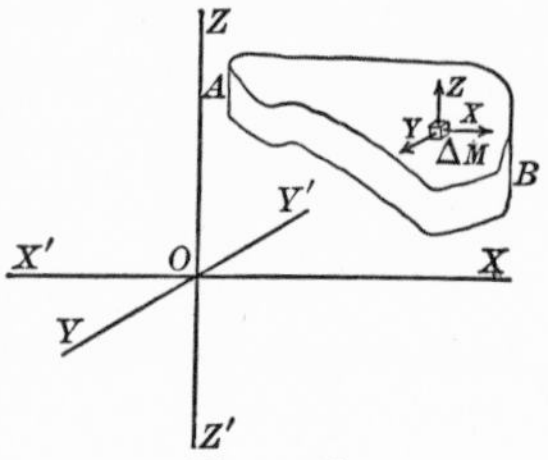

FIG. 466.

Considering the forces and motions of the small particles (as Δm) of which it may be composed, if ΣX, ΣY, ΣZ = the sums of the components of the forces parallel to the axes $X'X$, $Y'Y$, $Z'Z$ respectively; ΣT_x, ΣT_y, ΣT_z = the sums of the torques about the axes $X'X$, $Y'Y$, $Z'Z$ respectively; ΣJ_m = the moment of inertia of the mass about the axis $Z'Z$; ΣU_{xz_m}, ΣU_{yz_m} = the products of inertia of mass with respect to the planes **YOZ** and **XOY** and the planes **XOZ** and **XOY** respectively.

$$\left.\begin{aligned} \Sigma X &= -\alpha y_0 m - \omega^2 x_0 m \\ \Sigma Y &= +\alpha x_0 m - \omega^2 y_0 m \\ \Sigma Z &= 0 \end{aligned}\right\} \qquad \left.\begin{aligned} \Sigma T_x &= -\alpha U_{xz_m} + \omega^2 U_{yz_m} \\ \Sigma T_y &= -\alpha U_{yz_m} - \omega^2 U_{xz_m} \\ \Sigma T_z &= \alpha J_m. \end{aligned}\right\}$$

Analogy of Formulas for Translation and Rotation

Translation		Rotation	
Force.	$F = ma$	Torque.	$T = J_m\alpha$
Impulse.	$I = Ft$	Angular impulse. .	$I_a = Tt$
Momentum. . . .	$\mathfrak{M} = mv$	Angular momentum	$\mathfrak{M}_a = J_m\omega$
Change of momentum.	$m(v_1 - v_0) = Ft$	Change of angular momentum.	$J_m(\omega_1 - \omega_0) = Tt$
Work.	$W = Fs$	Work.	$W = T\theta$
Kinetic energy.	$W = \frac{1}{2} mv^2$	Kinetic energy. . .	$W = \frac{1}{2} J_m\omega^2$
Change of kinetic energy.	$\frac{1}{2} m(v_1^2 - v_0^2) = F(s_1 - s_0)$	Change of kinetic energy.	$\frac{1}{2} J_m(\omega_1^2 - \omega_0^2) = T(\theta_1 - \theta_0)$

Translation and Rotation

Work (W) done on a body by a force of **F** pounds having a torque of **T** pound-feet about the center of gravity of the body in moving the body **s** feet and causing it to rotate through an angle of θ radians.

467 $$W = Fs + T\theta.$$

Kinetic energy (W) of a body of **m** pounds (grav.) mass, with a mass moment of inertia of J_m pound(grav.)-feet squared about its center of gravity and having a velocity of translation of v feet per second and an angular velocity of ω radians per second.

468 $$W = \frac{1}{2}mv^2 + \frac{1}{2}\omega^2 J_m \text{ foot-pounds.}$$

NOTE. If the body weighs w pounds and has K feet radius of gyration about the center of gravity, $W = \frac{1}{2}\frac{w}{g}v^2 + \frac{1}{2}\frac{w}{g}K^2\omega^2$.

Kinetic energy developed in a body during any displacement is equal to the external work done upon it.

469 $$Fs + T\theta = \frac{1}{2}mv^2 + \frac{1}{2}\omega^2 J_m \text{ foot-pounds.}$$

Instantaneous axis. Any plane motion may be considered as a rotation about an axis which may be constantly changing to successive parallel positions. This axis at any instant is called the instantaneous axis.

NOTE. If the velocities, at any instant, of two points in a body are known the instantaneous axis passes through the intersection of the perpendiculars to the lines of motion of these two points.

Distance (l) from the instantaneous axis to the center of percussion of a body of m pounds (grav.) mass and mass moment of inertia of J_m pound(grav.)-feet squared about its center of gravity, for a position of the instantaneous axis of x_0 feet distance from the center of gravity of the body.

470 $$l = \frac{J_m}{x_0 m} + x_0.$$

Velocity of translation (v_c) of the center of gravity of a body having an angular velocity of ω radians per second about the instantaneous axis which is x_0 feet from the center of gravity.

471 $$v_c = \omega x_0 \text{ feet per second.}$$

Kinetic energy (W) of a body with a mass moment of inertia of J'_m pound(grav.)-feet squared about the instantaneous axis and an angular velocity of ω radians per second about the instantaneous axis.

472 $$W = \frac{1}{2}\omega^2 J'_m \text{ foot-pounds.}$$

Pendulum

The imaginary pendulum conceived as a material point suspended by a weightless cord is called a simple pendulum. A real pendulum is called a compound pendulum.

Time (t) of oscillation (from a maximum deflection to the right to a maximum deflection to the left) of a simple pendulum l feet in length.

473 $$t = \pi\sqrt{\frac{l}{g}} \text{ seconds (for small vibrations).}$$

NOTE. An approximate expression for all arcs is $t = \pi\sqrt{\frac{l}{g}}\left(1 + \frac{h}{8l}\right)$, where h is the vertical distance between the highest and lowest points of the path.

Length (l) of a simple seconds pendulum (one whose time of oscillation is one second).

474 $$l = \frac{g}{\pi^2} \text{ feet.}$$

Time (t) of oscillation of a compound pendulum of **K** feet radius of gyration with respect to the axis of suspension and **l** feet length from the axis of suspension to the center of gravity of the pendulum.

475 $$t = \pi\sqrt{\frac{K^2}{lg}} \text{ (for small vibrations).}$$

Distance (d) from the center of suspension to the center of oscillation, of a compound pendulum, of **K** feet radius of gyration about the center of suspension, the distance from the center of suspension to the center of gravity being **l** feet.

476 $$d = \frac{K^2}{l} \text{ feet.}$$

NOTE. The time of oscillation, for a small vibration, about an axis through the center of suspension is the same as that of a small vibration about a parallel axis through the center of oscillation.

Tension (T) in the cord of a conical pendulum with a weight of **W** pounds and **l** feet length of cord, rotating with **n** revolutions per second.

477 $$T = \frac{Wl4\pi^2n^2}{g} \text{ pounds.}$$

NOTE. In terms of the angular velocity ω radians per second; $T = \frac{Wl\omega^2}{g}$ pounds.

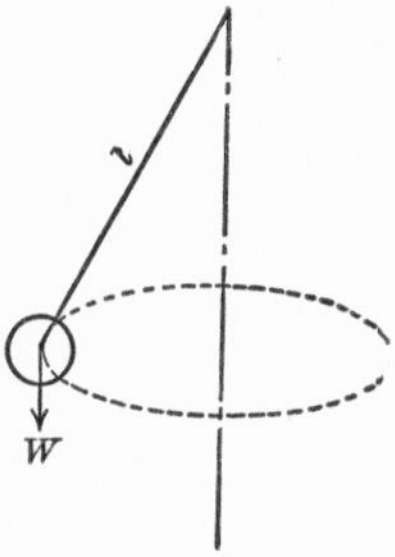

FIG. 477.

Time (t) of oscillation of a simple cycloidal pendulum swinging on the arc of a cycloid described by a circle of **r** feet radius.

478 $$t = 2\pi\sqrt{\frac{r}{g}} \text{ seconds.}$$

Prony Brake

Power (P) indicated by a Prony brake when the perpendicular distance from the center of the pulley to the direction of a force of **F** pounds applied at the end of the brake arm is **l** feet and the pulley revolves at a speed of **S** revolutions per minute.

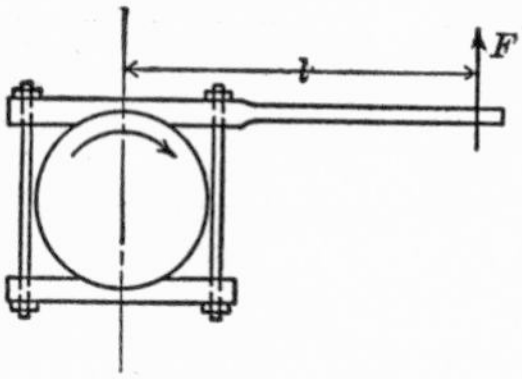

FIG. 479.

479 $$P = 1.903\, lSF \times 10^{-4} \text{ horse-power.}$$

NOTE. The torque of the pulley equals **lF** pound-feet. If **l** is made 5 feet 3 inches, $P = \frac{SF}{1000}$ horse-power.

Friction

Static friction is the force, in addition to that overcoming inertia, required to set in motion one body in contact with another.

Coefficient of static friction (f) between two bodies, when **N** is the normal pressure between them and **F** is the corresponding static friction. [**N** and **F** in the same units]

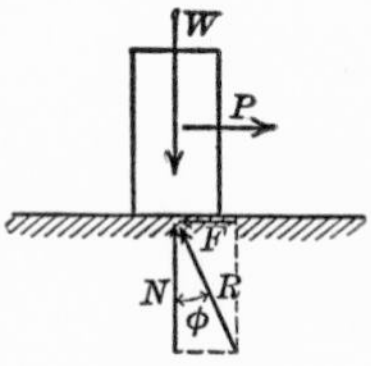

FIG. 480.

480 $$f = \frac{F}{N}.$$

Resultant force (R) between two bodies starting from relative rest with a normal pressure of **N** pounds and a static friction of **F** pounds between them.

481 $$R = \sqrt{F^2 + N^2} \text{ pounds.}$$

Angle of static friction (ϕ) for two surfaces with a normal pressure **N** and a static friction **F** between them. [**N** and **F** in the same units]

482 $$\tan \phi = \frac{F}{N} = f.$$

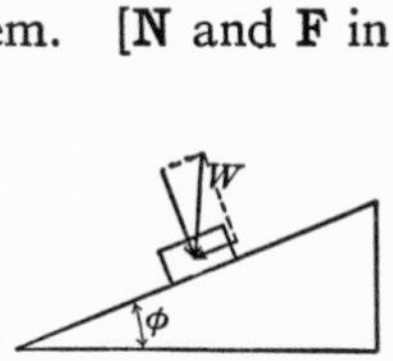

FIG. 482.

NOTE. The angle of repose is the angle of inclination of the surface of one body at which the other body will begin to slide along it, under the action of its own weight. The angle of repose (ϕ) is equal to the angle of static friction.

Coefficients of Friction

Materials	Condition	Sliding friction		Static friction	
		ϕ	f	ϕ	f
Cast-iron on cast-iron or bronze	wet	$17\frac{1}{4}^\circ$	0.31	....	....
Cast-iron on cast-iron or bronze	greased	$4\frac{1}{2}^\circ$–$5\frac{3}{4}^\circ$	0.08–0.10	....	0.16
Cast-iron on oak (fibers parallel)	dry	$16\frac{3}{4}^\circ$–$26\frac{1}{2}^\circ$	0.30–0.50	....	....
Cast-iron on oak (fibers parallel)	wet	$12\frac{1}{2}^\circ$	0.22	33°	0.65
Cast-iron on oak (fibers parallel	greased	$10\frac{3}{4}^\circ$	0.19	....	....
Earth on earth	......	....	....	14°–45°	0.25–1.0
Earth on earth (clay)	damp	....	....	45°	1.0
Earth on earth (clay)	wet	....	....	$17\frac{1}{4}^\circ$	0.31
Hemp-rope on rough wood	dry	$26\frac{1}{2}^\circ$	0.50	$26\frac{1}{2}^\circ$–$38\frac{3}{4}^\circ$	0.50–0.80
Hemp-rope on polished wood	dry	....	....	$18\frac{1}{4}^\circ$	0.33
Leather on oak	dry	$16\frac{3}{4}^\circ$–$26\frac{1}{2}^\circ$	0.30–0.50	$26\frac{1}{2}^\circ$–31°	0.50–0.60
Leather on cast-iron	dry	$29\frac{1}{4}^\circ$	0.56	$16\frac{3}{4}^\circ$–$26\frac{1}{2}^\circ$	0.30–0.50
Oak on oak (fibers parallel)	dry	$25\frac{3}{4}^\circ$	0.48	$31\frac{3}{4}^\circ$	0.62
Oak on oak (fibers crossed)	dry	$18\frac{3}{4}^\circ$	0.34	$22\frac{1}{4}^\circ$	0.54
Oak on oak (fibers crossed)	wet	14°	0.25	$35\frac{1}{4}^\circ$	0.71
Oak on oak (fibers perpendicular)	dry	$10\frac{3}{4}^\circ$	0.19	$23\frac{1}{4}^\circ$	0.43
Steel on ice	dry	....	0.014	....	0.027
Steel on steel	dry	vel. 10 ft. per sec.	0.09	$8\frac{1}{2}^\circ$	0.15
		vel. 100 ft. per sec.	0.03		
Stone masonry on concrete	dry	....	....	$37\frac{1}{4}^\circ$	0.76
Stone masonry on undisturbed ground	dry	....	....	33°	0.65
Stone masonry on undisturbed ground	wet	....	....	$16\frac{3}{4}^\circ$	0.30
Wrought-iron on wrought-iron	dry	$23\frac{3}{4}^\circ$	0.44	....	....
Wrought-iron on wrought-iron	greased	$4\frac{1}{2}^\circ$–$5\frac{3}{4}^\circ$	0.08–0.10	....	0.11
Wrought-iron on cast-iron or bronze	dry	$10\frac{1}{4}^\circ$	0.18	$10\frac{3}{4}^\circ$	0.19
Wrought-iron on cast-iron or bronze	greased	......	....	....	0.07–0.08

Sliding friction is the force, in addition to that overcoming inertia, required to maintain relative motion between two bodies.

NOTE. Laws of sliding friction. (1) For moderate pressures the friction is proportional to the normal pressure between the surfaces. (2) For moderate pressures the friction is independent of the extent of the surface in contact. (3) At low velocities the friction is independent of the velocity of rubbing. The friction decreases as the velocity increases. (4) Sliding friction is usually less than static friction.

Coefficient of sliding friction (f) between two bodies when **N** is the normal pressure between them and **F** is the corresponding sliding friction. [**N** and **F** in the same units]

483
$$f = \frac{F}{N}.$$

Angle of sliding friction (ϕ) for two surfaces with a normal pressure **N** and a sliding friction **F** between them. [**N** and **F** in the same units]

NOTE. See formula 482. The angle of sliding friction is the angle of inclination of the surface of one body, at which the motion of another body sliding upon it will be maintained. The angle of sliding friction is in general less than the angle of static friction.

Applications of Principles of Friction

Inclined plane. Let **W** = weight in pounds of a body sliding on the plane, α = angle of inclination of plane, β = angle between force **F** and plane, ϕ = angle of repose, f = coefficient of friction ($\tan \phi = f$), and **F** = force applied to the body along the line of action indicated.

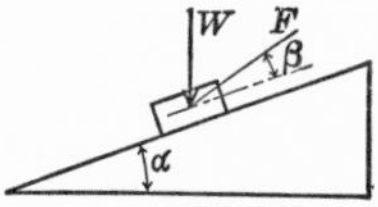

FIG. 484.

484 (a) Force (F) to prevent slipping. ($\alpha > \phi$)

$$F = W \frac{\sin(\alpha - \phi)}{\cos(\beta + \phi)} \text{ pounds.}$$

(b) Force (F) to start the body up the plane. ($\alpha > \phi$)

$$F = W \frac{\sin(\alpha + \phi)}{\cos(\beta - \phi)} \text{ pounds.}$$

(c) Force (F) to start the body down the plane. ($\alpha < \phi$)

$$F = W \frac{\sin(\phi - \alpha)}{\cos(\beta + \phi)} \text{ pounds.}$$

Wedge. Let **W** = force in pounds opposing motion, α = angle of inclination of sides of wedge, ϕ = angle of friction, and **F** = force applied to wedge.

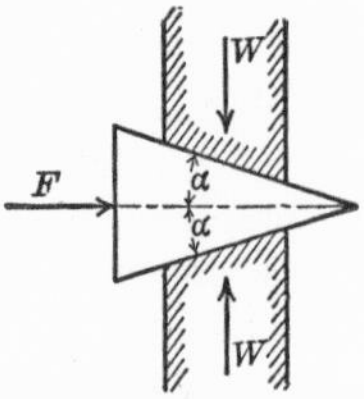

FIG. 485.

485 (a) Force (F) to push wedge.

$$F = 2\,W \tan(\alpha + \phi) \text{ pounds.}$$

(b) Force (F) to draw wedge ($\alpha > \phi$).

$$F = 2\,W \tan(\phi - \alpha) \text{ pounds.}$$

Square threaded screw. Let **r** = mean radius of screw, **p** = pitch of screw, α = angle of pitch $\left(\tan\alpha = \frac{p}{2\pi r}\right)$, **F** = force applied to screw at end of arm **a**, **W** = total weight in pounds to be moved and ϕ = angle of friction. [**r** and **a** in same units]

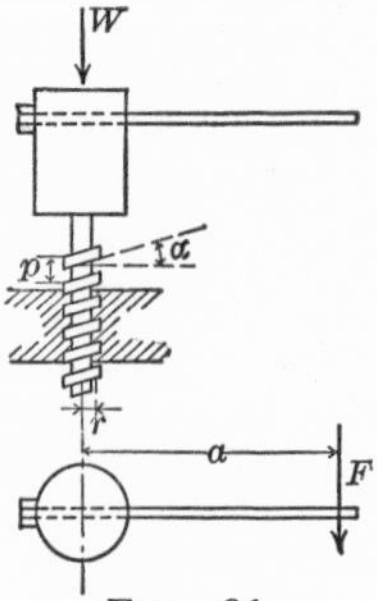

FIG. 486.

486 (a) Force (F) to lower screw.

$$F = \frac{Wr(\tan\phi - \tan\alpha)}{a} \text{ pounds (approx.).}$$

(b) Force **F** to raise screw.

$$F = \frac{Wr(\tan\phi + \tan\alpha)}{a} \text{ pounds (approx.).}$$

Sharp threaded screw. Let **r** = mean radius of screw, α = angle of pitch, β = angle between faces of the screw, **F** = force in pounds applied to screw at end of arm **a**, **W** = total weight in pounds to move, and ϕ = angle of friction. [**r** and **a** in same units]

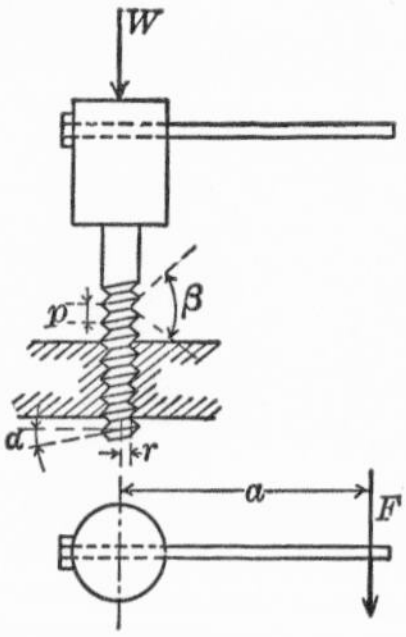

FIG. 487.

487 (a) Force (F) to lower screw.

$$F = \frac{Wr}{a}\left(\frac{\tan\phi\cos\alpha}{\cos\frac{\beta}{2}} - \tan\alpha\right) \text{ pounds (approx.).}$$

(b) Force (F) to raise screw.

$$F = \frac{Wr}{a}\left(\frac{\tan\phi\cos\alpha}{\cos\frac{\beta}{2}} + \tan\alpha\right) \text{ pounds (approx.).}$$

Pivot Friction

f = coefficient of friction. W = load in pounds.
T = torque of friction about the axis of the shaft.
r = radius in inches. n = revolutions per second.

Type of Pivot	Torque T in pound-inches	Power P lost by friction in ft.-lbs. per second
Shafts and Journals 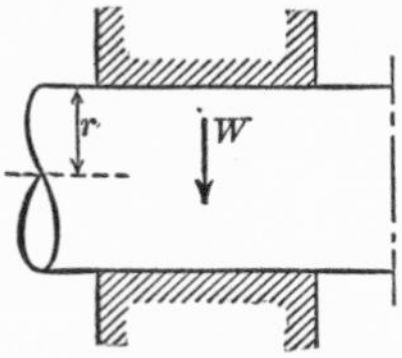 (180° bearing)	$T = fWr.$	$P = \frac{2\pi n}{12} fWr.$
Flat Pivot 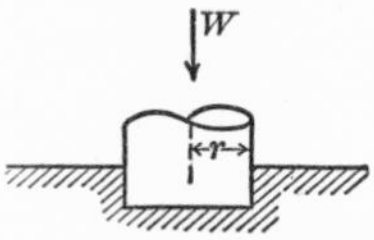	$T = \frac{2}{3} fWr.$	$P = \frac{4\pi n}{3 \times 12} fWr.$
Collar-bearing 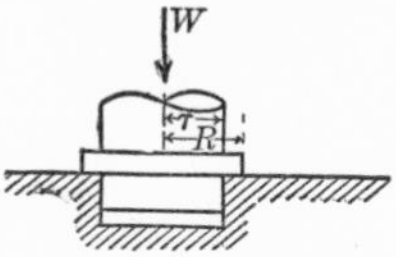	$T = \frac{2}{3} fW \frac{R^3 - r^3}{R^2 - r^2}.$	$P = \frac{4\pi n}{3 \times 12} fw \frac{R^3 - r^3}{R^2 - r^2}.$
Conical Pivot 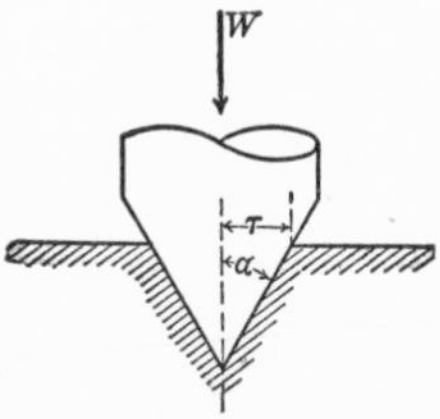	$T = \frac{2}{3} fW \frac{r}{\sin \alpha}.$	$P = \frac{4\pi nfWr}{3 \times 12 \sin \alpha}.$
Truncated-cone Pivot 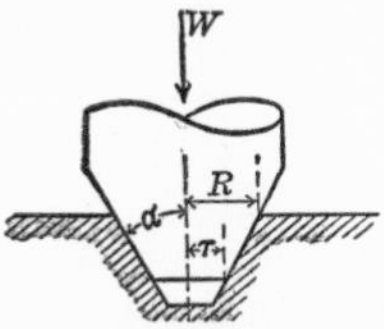	$T = \frac{2}{3} fW \frac{(R^3 - r^3)}{(R^2 - r^2) \sin \alpha}.$	$P = \frac{4\pi nfW (R^3 - r^3)}{3 \times 12 (R^2 - r^2) \sin \alpha}.$

Rolling Friction

Coefficient of rolling friction (**c**) of a wheel with a load of **W** pounds and with **r** inches radius, moved at a uniform speed by a force of **F** pounds applied at its center.

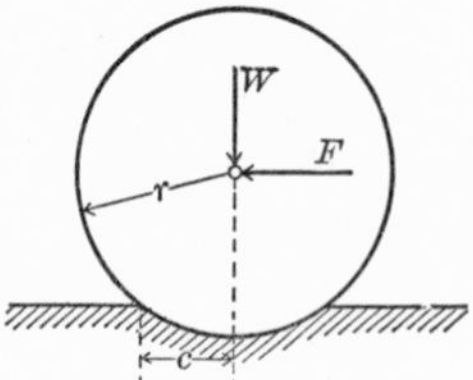

FIG. 488.

488 $$c = \frac{Fr}{W} \text{ inches.}$$

NOTE. Coefficients of rolling friction.
Lignum vitæ roller on oak track.................c = 0.019 inches.
Elm roller on oak track.........................c = 0.032 inches.
Iron on iron (and steel on steel)...............c = 0.020 inches.

Belt Friction

Ratio $\left(\frac{F_1}{F_2}\right)$ of the pull F_1 on the driving side of a belt to the pull F_2 on the driven side of the belt, when slipping is impending, in terms of the coefficient of friction **f** and the angle of contact α, in radians. [$\epsilon = 2.718$]

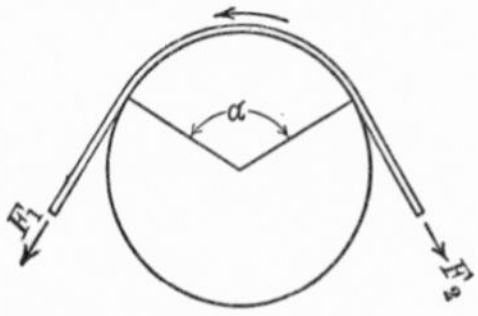

FIG. 489.

489 $$\frac{F_1}{F_2} = \epsilon^{f\alpha}.$$

NOTE. Mean values of f are as follows:
Leather on wood (somewhat oily)........................ 0.47
Leather on cast iron (somewhat oily)................... 0.28
Leather on cast iron (moist)........................... 0.38
Hemp-rope on iron drum................................. 0.25
Hemp-rope on wooden drum............................... 0.40
Hemp-rope on polished wood............................. 0.33
Hemp-rope on rough wood................................ 0.50

Values of $\frac{F_1}{F_2}$ (Slipping impending)

$\frac{\alpha}{2\pi}$	f = 0.25	f = 0.33	f = 0.40	f = 0.50	$\frac{\alpha}{2\pi}$	f = 0.25	f = 0.33	f = 0.40	f = 0.50
0.1	1.17	1.23	1.29	1.37	0.6	2.57	3.47	4.52	6.59
0.2	1.37	1.51	1.65	1.87	0.7	3.00	4.27	5.81	9.00
0.3	1.60	1.86	2.13	2.57	0.8	3.51	5.25	7.47	12.34
0.4	1.87	2.29	2.73	3.51	0.9	4.11	6.46	9.60	16.90
0.425	1.95	2.41	2.91	3.80	1.0	4.81	7.95	12.35	23.14
0.45	2.03	2.54	3.10	4.11	1.5	10.55	22.42	43.38	111.2
0.475	2.11	2.68	3.30	4.45	2.0	23.14	63.23	152.4	535.5
0.5	2.19	2.82	3.51	4.81	2.5	50.75	178.5	535.5	2,576
0.525	2.28	2.97	3.74	5.20	3.0	111.3	502.9	1881	12,392
0.55	2.37	3.31	3.98	5.63	3.5	244.2	1418	6611	59,610

Impact *

Common velocity (v'), after direct central impact, of two inelastic bodies of mass m_1 and m_2 and initial velocities v_1 and v_2 respectively.

490 $$v' = \frac{m_1 v_1 + m_2 v_2}{m_1 + m_2}.$$

Final velocities (v_1' **and** v_2'), after direct central impact, of two perfectly elastic bodies of mass m_1 and m_2 and initial velocities v_1 and v_2 respectively.

491 $$\begin{cases} v_1' = \dfrac{m_1 v_1 - m_2 v_1 + 2\, m_2 v_2}{m_1 + m_2} \\ v_2' = \dfrac{m_2 v_2 - m_1 v_2 + 2\, m_1 v_1}{m_1 + m_2}. \end{cases}$$

Final velocities (v_1' **and** v_2'), after direct central impact, of two partially but equally inelastic bodies of mass m_1 and m_2 and initial velocities v_1 and v_2 respectively and constant **e** depending on the elasticity of bodies.

492 $$\begin{cases} v_1' = \dfrac{m_1 v_1 + m_2 v_2 - e m_2 (v_1 - v_2)}{m_1 + m_2} \\ v_2' = \dfrac{m_1 v_1 + m_2 v_2 - e m_1 (v_2 - v_1)}{m_1 + m_2}. \end{cases}$$

NOTE. $e = \sqrt{\frac{H}{h}}$ where **H** is the height of rebound of a sphere dropped from a height **h** on to a horizontal surface of a rigid mass. If the bodies are inelastic $e = 0$, and if bodies are perfectly elastic $e = 1$.

STATICS

Components of a force F (F_x **and** F_y) parallel to two rectangular axes **X′X** and **Y′Y**, the axis **X′X** making an angle α with the force **F**.

493 $F_x = F \cos \alpha, \qquad F_y = F \sin \alpha.$

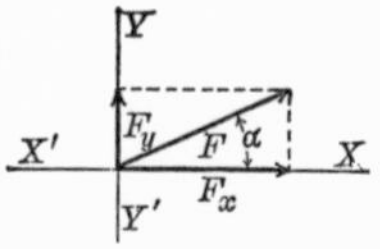

FIG. 493.

Moment or torque (**M**) of a force of **F** pounds about a given point, the perpendicular distance from the point to the direction of the force being **d** feet.

494 $$M = Fd \text{ pound-feet.}$$

* m_1 and m_2, v_1 and v_2 in the same units.

NOTE. A couple is formed by two equal, opposite, parallel forces acting in the same plane but not in the same straight line. The moment (M) of a couple of two forces, each of F pounds, with a perpendicular distance of d feet between them is Fd pound-feet. The moment, about any point, of the resultant of several forces, lying in the same plane, is the algebraic sum of the moments of the separate forces about that point.

Resultant force (R) of two forces, F_1 and F_2, which make an angle α with each other, the angle between the resultant force R and the force F_1 being θ.

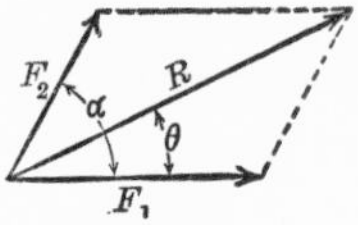

FIG. 495.

495 $$R = \sqrt{F_1^2 + F_2^2 + 2\,F_1F_2\cos\alpha}.$$

496 $$\tan\theta = \frac{F_2\sin\alpha}{F_1 + F_2\cos\alpha}, \quad \text{or,} \quad \sin\theta = \frac{F_2\sin\alpha}{R}.$$

Parallelogram of forces. The resultant force (R) of two forces F_1 and F_2 is represented in magnitude and direction by the diagonal lying between those two sides of a parallelogram which represent F_1 and F_2 in magnitude and direction.

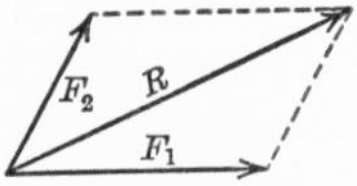

FIG. 496.

Triangle of forces. The resultant force (R) of two forces F_1 and F_2 is represented in magnitude and direction by the third side of a triangle in which the other two sides represent F_1 and F_2 in magnitude and direction.

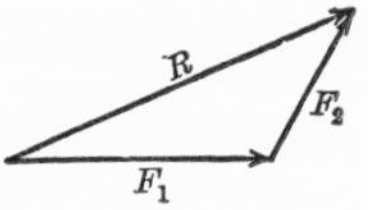

FIG. 496a.

Resultant force (R) of three forces F_1, F_2 and F_3 mutually at right angles to each other and not lying in the same plane, the angles between the resultant force R and the forces F_1, F_2 and F_3 being α, β and γ respectively.

497 $$R = \sqrt{F_1^2 + F_2^2 + F_3^2}.$$

498 $$\cos\alpha = \frac{F_1}{R}, \quad \cos\beta = \frac{F_2}{R}, \quad \cos\gamma = \frac{F_3}{R}.$$

NOTE. If three forces not in the same plane are not mutually at right angles to each other, the resultant force may be found by formula 504.

Parallelopiped of forces. The resultant force (**R**) of three forces $\mathbf{F_1}$, $\mathbf{F_2}$ and $\mathbf{F_3}$, not lying in the same plane, is represented in magnitude and direction by the diagonal lying between those three sides of a parallelopiped which represent $\mathbf{F_1}$, $\mathbf{F_2}$ and $\mathbf{F_3}$ in magnitude and direction.

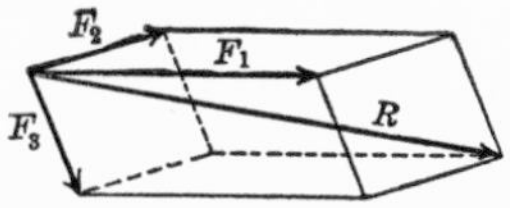

FIG. 498.

Resultant force (**R**) of several forces lying in the same plane, if $\Sigma\mathbf{F_x}$ and $\Sigma\mathbf{F}$ are the algebraic sums of the components of the forces parallel to two rectangular axes **X′X** and **Y′Y**, the angle between the resultant force and the axis **X′X** being $\boldsymbol{\alpha}$.

499 $$R = \sqrt{(\Sigma F_x)^2 + (\Sigma F_y)^2}.$$

500 $$\tan\alpha = \frac{\Sigma F_y}{\Sigma F_x}, \quad \sin\alpha = \frac{\Sigma F_y}{R}, \quad \cos\alpha = \frac{\Sigma F_x}{R}.$$

Perpendicular distance (**d**) from a given point to the resultant force (**R**) of several forces lying in the same plane, if $\Sigma\mathbf{M}$ is the algebraic sum of the moments, about that point, of the separate forces.

501 $$d = \frac{\Sigma M}{R}.$$

NOTE. The resultant of several parallel forces is the algebraic sum of the forces (ΣF). If $\Sigma F = 0$ the resultant is a couple whose moment is $\Sigma\mathbf{M}$.

Force Polygon. The resultant force (**R**) of several forces $\mathbf{F_1}$, $\mathbf{F_2}$ . . . $\mathbf{F_n}$, lying in the same plane, is represented in magnitude and direction by the closing side of a polygon in which the remaining sides represent the forces $\mathbf{F_1}$, $\mathbf{F_2}$ . . . $\mathbf{F_n}$ in magnitude and direction.

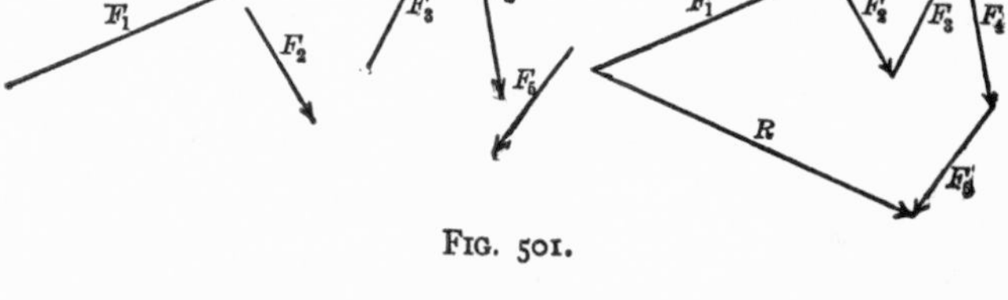

FIG. 501.

NOTE. The arrows indicate the directions of the forces and for the given forces they must point in the same way around the polygon, but for the resultant force in the opposite direction or leading from the starting point of the first force to the end point of the last force.

Moment (**M**) of a force **F**, about a line, is the product of the rectangular component of the force perpendicular to the line (the other component being parallel to the line) into the perpendicular distance between the line and this rectangular component, or the force **F** may be resolved into three rectangular components, one parallel and the other two perpendicular to the line, as in Fig. 502. The moment of the force about each axis is then obtained as follows:

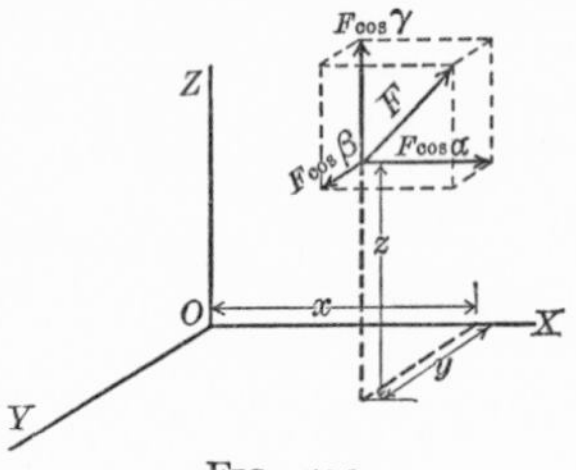

FIG. 502.

502
$$M_x = yF\cos\gamma - zF\cos\beta.$$
$$M_y = zF\cos\alpha - xF\cos\gamma.$$
$$M_z = xF\cos\beta - yF\cos\alpha.$$

Resultant force (**R**) of several parallel forces, not lying in the same plane, is the algebraic sum (**ΣF**) of the forces.

NOTE. If $\Sigma F = 0$, the resultant is a couple whose moments are ΣM_x, ΣM_y, etc.

Perpendicular distances (d_x) **and** (d_y) from each of two axes **X′X** and **Y′Y** to the resultant force (**R**) of several parallel forces, not lying in the same plane, if ΣM_x and ΣM_y are the algebraic sums of the moments of the separate forces about the axes **X′X** and **Y′Y** respectively.

503
$$d_x = \frac{\Sigma M_x}{R}, \qquad d_y = \frac{\Sigma M_y}{R}.$$

Resultant force (**R**) and direction (**α, β, γ**) of the resultant force of several forces, not lying in the same plane, if ΣF_x, ΣF_y and ΣF_z are the algebraic sums of the components parallel to three rectangular axes **X′X, Y′Y** and **Z′Z**, and **α**, **β** and **γ** are the angles which the resultant force makes with the axes **X′X, Y′Y** and **Z′Z** respectively.

504 $$R = \sqrt{(\Sigma F_x)^2 + (\Sigma F_y)^2 + (\Sigma F_z)^2}.$$

505 $$\cos \alpha = \frac{\Sigma F_x}{R}, \quad \cos \beta = \frac{\Sigma F_y}{R}, \quad \cos \gamma = \frac{\Sigma F_z}{R}.$$

Resultant couple (M) and direction ($\alpha_m, \beta_m, \gamma_m$) of the axis of the resultant couple of several forces, not acting in the same plane, if ΣM_x, ΣM_y and ΣM_z are the algebraic sums of the moments about three rectangular axes **X′X, Y′Y** and **Z′Z** and α_m, β_m and γ_m are the angles which the moment axis of the resultant couple makes with the axes **X′X, Y′Y** and **Z′Z** respectively.

506 $$M = \sqrt{(\Sigma M_x)^2 + (\Sigma M_y)^2 + (\Sigma M_z)^2}.$$

507 $$\cos \alpha_m = \frac{\Sigma M_x}{\Sigma M}, \quad \cos \beta_m = \frac{\Sigma M_y}{\Sigma M}, \quad \cos \gamma_m = \frac{\Sigma M_z}{\Sigma M}.$$

NOTE. In general the resultant of several non-parallel forces, not in the same plane, is not a single force, but by the use of the above principles the system may be reduced to a single force and a couple.

Conditions of equilibrium of several forces, lying in the same plane, if ΣF_x and ΣF_y are the algebraic sums of the components parallel to two axes **X′X** and **Y′Y** and ΣM is the algebraic sum of the moments of the forces about any point.

508 $$\Sigma F_x = 0, \quad \Sigma F_y = 0, \quad \Sigma M = 0.$$

Conditions of equilibrium of several forces, not lying in the same plane, if ΣF_x, ΣF_y and ΣF_z are the algebraic sums of the components parallel to three axes **X′X, Y′Y** and **Z′Z** which intersect at a common point but do not lie in the same plane, and ΣM_x, ΣM_y and ΣM_z are the algebraic sums of the moments of the forces about these three axes.

509 $$\Sigma F_x = 0, \quad \Sigma F_y = 0, \quad \Sigma F_z = 0.$$

510 $$\Sigma M_x = 0, \quad \Sigma M_y = 0, \quad \Sigma M_z = 0.$$

Stresses in Framed Structures *

Pratt Truss. Two live loads of 10 tons each as shown in Fig. 508a.

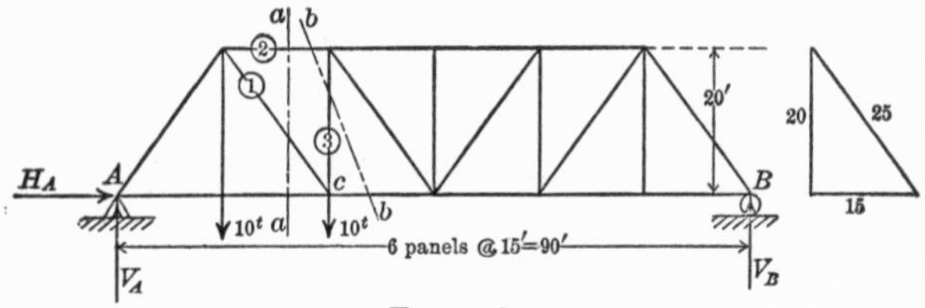

FIG. 508a.

(a) Reactions (use conditions of equilibrium, formula 508).

By $\Sigma M = 0$, $\Sigma M_A = 0 = 10 \times 15 + 10 \times 30 - V_B \times 90$, $V_B = 5$ tons.

By $\Sigma F_y = 0$, $20 - V_B = V_A$, $V_A = 15$ tons.

By $\Sigma F_x = 0$, $H_A = 0$. (note that a roller is used at **B**, fixing the reaction there in a vertical direction)

(b) Stresses in bars.

To find the stress in a bar consider a plane (cutting the bar in question) to divide the truss into two parts; remove one part and replace the portion of the bars which are removed by their stresses which may now be treated as outer forces. These stresses are found by applying the equations of equilibrium. It is essential that only three of the bars which are cut shall have unknown stresses.

NOTE. If tension is called positive and all unknown stresses are assumed to be tension stresses, a positive sign for the result indicates tension and a negative sign compression.

Bar ①. Truss cut by plane **aa**. Consider left portion.

Let $V_{①}$ = the vertical component of $S_{①}$, the stress in bar ①.

By $\Sigma F_y = 0$, $-V_A + 10 + V_{①} = 0$, $V_{①} = 5$, $S_{①} = \frac{25}{20} \times 5 =$ 6.25 tons tension.

Bar ②. Truss cut by plane **aa**. Take moments about joint **c**.

By $\Sigma M = 0$, $\Sigma M_c = 0 = V_A \times 30 - 10 \times 15 + S_{②} \times 20$.

$$S_{②} = \frac{-450 + 150}{20} = -15 = 15 \text{ tons compression.}$$

Bar ③. Truss cut by plane **bb**.

By $\Sigma F_y = 0$, $-V_A + 20 + S_{③} = 0$, $S_{③} = -5 = 5$ tons compression below cut or 5 tons tension above cut.

* Due to live loads only. Weight of structure is neglected.

Roof Truss. Two live loads of 3 tons each as shown in Fig. 508b.

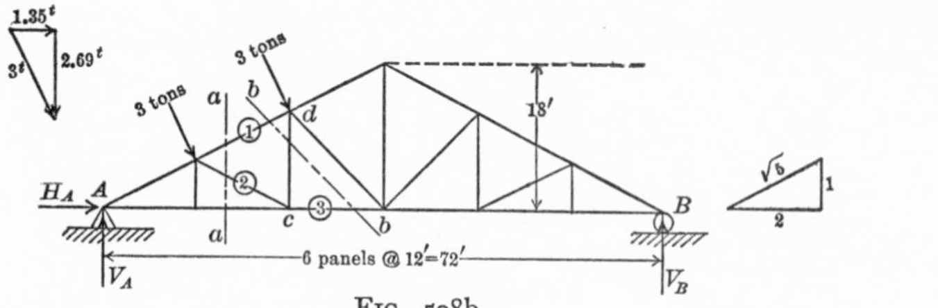

FIG. 508b.

(**a**) Reactions (use conditions of equilibrium formula 508).

By $\Sigma M=0$, $\Sigma M_A=0=3\times13.4+3\times26.8-V_B\times72$,

$V_B=1.67$ tons.

By $\Sigma F_y=0$, $2\times2.69-V_B-V_A=0$, $V_A=3.71$ tons.

By $\Sigma F_x=0$, $2.70+H_A=0$, $H_A=-2.70$ tons, *i.e.*, acting to the left. (note that a roller is used at **B**, fixing the reaction in a vertical direction)

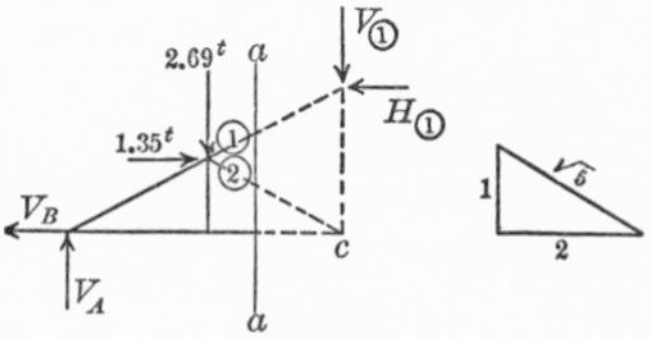

FIG. 508c.

(**b**) Stresses in bars. (See **b** under Pratt Truss.)

Bar ①. Truss cut by plane **aa.** Consider left portion. Take moments about joint **c.** Let $H_{①}$ = horizontal component of $S_{①}$.

By $\Sigma M=0$, $\Sigma M_c=0=V_A\times24-2.69\times12+1.35\times6+H_{①}\times12$.

$$H_{①}=-5.34,\quad S_{①}=\frac{\sqrt{5}}{2}\times5.34=5.96 \text{ tons compression.}$$

Bar ②. Truss cut by plane **aa.** Take moments about **A.**

Let $V_{②}$ = vertical component of $S_{②}$.

By $\Sigma M=0$, $\Sigma M_A=0=3\times13.4+V_{②}\times24$, $V_{②}=-1.67$ tons.

$$S_{②}=\sqrt{5}\times1.67=3.73 \text{ tons compression.}$$

Bar ③. Truss cut by plane **bb.** Consider right portion, as fewer loads lie to the right of cutting plane.

Take moments about joint **d.**

By $\Sigma M=0$, $\Sigma M_d=0=-V_B\times48+S_{③}\times12$, $S_{③}=6.68$ tons tension.

PROPERTIES OF MATERIALS

Intensity of stress is the stress per unit area, usually expressed in pounds per square inch. The simple term, Stress, is often used to indicate intensity of stress.

Ultimate stress is the greatest stress which can be produced in a body before rupture occurs.

Allowable stress or **working stress** is the intensity of stress which the material of a structure or a machine is designed to resist.

Factor of safety is a factor by which the ultimate stress is divided to obtain the allowable stress.

Elastic limit is the maximum intensity of stress to which a material may be subjected and return to its original shape upon the removal of the stress.

NOTE. For stresses below the elastic limit the deformations are directly proportional to the stresses producing them: that is, Hooke's Law holds for stresses below the elastic limit.

Yield point is the intensity of stress beyond which the change in length increases rapidly with little if any increase in stress.

Modulus of elasticity is the ratio of stress to the strain, for stresses below the elastic limit.

NOTE. Modulus of elasticity may also be defined as the stress which would produce a change of length of a bar equal to the original length of the bar, assuming the material to retain its elastic properties up to that point.

Properties of Common Materials

Material	Wt., lbs. per cu. ft.	Ultimate strength in pounds per sq. in.				Elastic limit in lbs. per sq. in.		Modulus of elasticity in pounds per sq. in.	
		Tension	Bending	Compression	Shear	Tension	Compression	Tension and Compression	Shear
Cast iron....	450	20000	35000	90000	18000	6000	20000	15000000	6000000
Wrought iron	480	50000	60000	50000	40000	25000	25000	28000000	12000000
Struct. steel.	490	60000	80000	60000	50000	36000	36000	30000000	13000000
Concrete....	150	300	...	2500	1000	...	1000	2000000	......
Yellow pine..	40	9000	8000	*7000 / 700	*400 / 1500	3000	....	1500000	4000000
White oak...	48	10000	8000	*6000 / 2000	*600 / 4000	...	...	1500000	4000000

* Parallel to the grain and across the grain respectively.

Poisson's ratio is the ratio of the relative change of diameter of a bar to its unit change of length under an axial load which does not stress it beyond the elastic limit.

NOTE. Poisson's ratio is usually denoted by $\frac{1}{m}$. It varies for different materials but is usually about $\frac{1}{4}$.

Intensity of stress (**f**) due to a force of **P** pounds producing tension, compression or shear on an area of **A** square inches, over which it is uniformly distributed.

511 $$f = \frac{P}{A} \text{ pounds per sq. in.}$$

Modulus of elasticity (E) of a bar of **A** square inches cross-sectional area and **l** inches length, which undergoes a change of length of **d** inches under an axial load of **P** pounds.

512 $$E = \frac{Pl}{Ad} \text{ pounds per sq. in.}$$

NOTE. The load must be such as to produce an intensity of stress below the elastic limit. If **f** is the intensity of stress produced and **e** the ratio of change of length to total length, $E = \frac{f}{e}$ and $e = \frac{f}{E}$.

Change of length (d) of a bar of **A** square inches cross-sectional area, **l** inches length, and **E** pounds per square inch modulus of elasticity of material, due to an axial load of **P** pounds.

513 $$d = \frac{Pl}{AE} \text{ inches.}$$

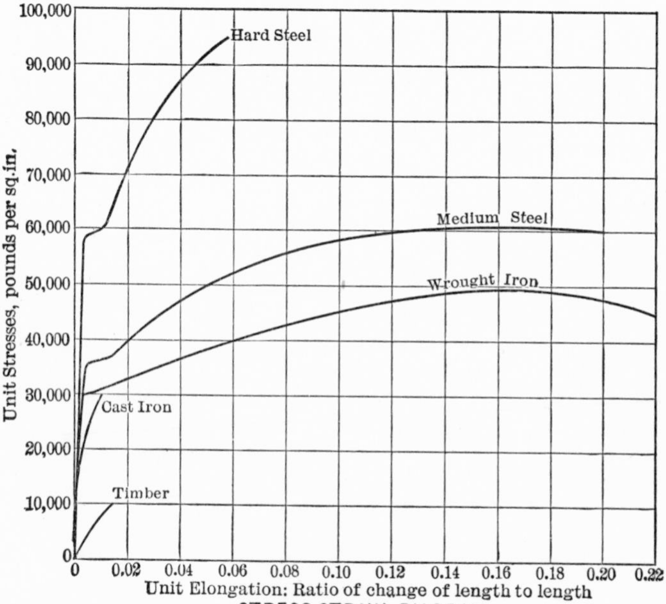

STRESS-STRAIN DIAGRAM

FIG. 513.

NOTE. Stress-strain diagrams show the relation of the intensities of stress of a material to the corresponding strains or deformations.

RIVETED JOINTS

Shearing strength (r_s) of a rivet **d** inches in diameter, with an allowable stress in shear of f_s pounds per square inch.

514 $$r_s = \frac{\pi d^2}{4} f_s \text{ pounds.}$$

Bearing strength (r_b of a rivet **d** inches in diameter, with an allowable stress in bearing of f_b pounds per square inch, against a plate **t** inches in thickness.

515 $$r_b = dtf_b \text{ pounds.}$$

Total stress (**r**) on each of **n** rivets resisting a pull or thrust of **P** pounds.

516 $$r = \frac{P}{n} \text{ pounds.}$$

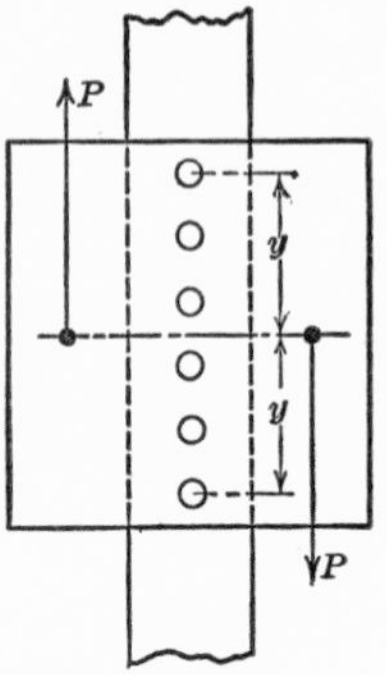

FIG. 517.

Total stress (r_m) on the most stressed rivet of a group of rivets resisting the action of a couple of **M** inch-pounds, if **y** is the distance in inches from the center of gravity of the group of rivets to the outermost rivet and Σy^2 is the sum of the squares of the distances from the center of gravity of the group to each of the rivets.

517 $$r_m = \frac{My}{\Sigma y^2} \text{ pounds.}$$

Resistance to moment (**M**) of a group of rivets, if the distance of the outermost rivet from the center of gravity of the group is **y** inches and the sum of the squares of the distances from the center of gravity of the group to each of the rivets is Σy^2 and **r** is the total allowable stress on a rivet.

518 $$M = \frac{r\Sigma y^2}{y} \text{ inch-pounds.}$$

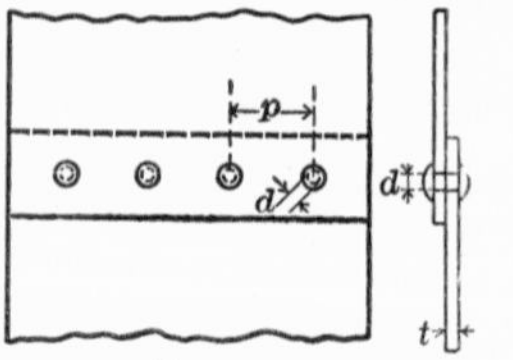

FIG. 519.

Resistance to tearing (**T**) between rivets, of a plate **t** inches in thickness in which rivets of **d** inches diameter are placed with **p** inches pitch, if the allowable intensity of stress of the plate in tension is f_t pounds per square inch.

519 $$T = t(p - d) f_t \text{ pounds.}$$

Efficiency of a riveted joint is the ratio of the least strength of the joint to the tensile strength of the solid plate.

Strength of Various Types of Riveted Joints

f_s = allowable shearing stress in pounds per square inch.
f_b = allowable bearing stress in pounds per square inch.
f_t = allowable tension stress in pounds per square inch.
d = diameter of rivet in inches.
t = thickness of plate in inches.
p = pitch of inner row of rivets in inches.
P = pitch of outer row of rivets in inches.
t_c = thickness of cover plates in inches.

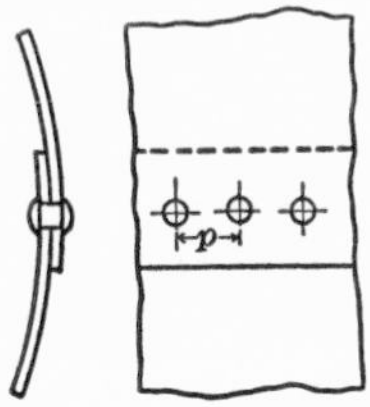

Single-riveted Lap Joint

(1) Shearing one rivet $= \frac{\pi d^2}{4} f_s$.
(2) Tearing plate between rivets $= (p - d)\, t f_t$.
(3) Crushing of rivet or plate $= d t f_b$.

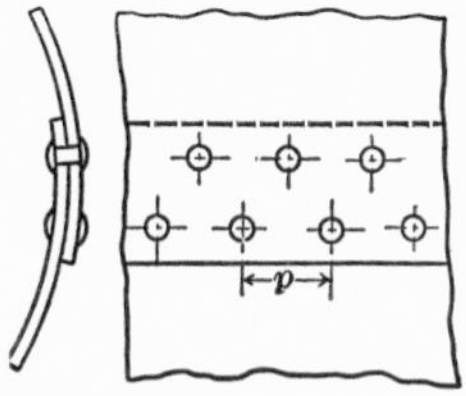

Double-riveted Lap Joint

(1) Shearing two rivets $= \frac{2\, \pi d^2}{4} f_s$.
(2) Tearing between two rivets $= (p - d)\, t f_t$.
(3) Crushing in front of rivets $= 2\, d t f_b$.

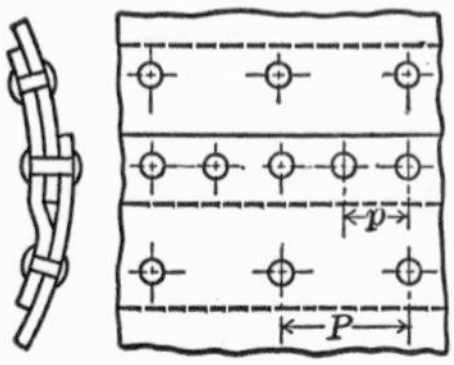

Single-riveted Lap Joint with inside Cover-Plate

(1) Tearing between outer row of rivets $= (P - d)\, t f_t$.
(2) Tearing between inner row of rivets and shearing outer row of rivets $= (P - 2\, d)\, t f_t + \frac{\pi d^2}{4} f_s$.
(3) Shearing three rivets $= \frac{3\, \pi d^2}{4} f_s$.
(4) Crushing in front of three rivets $= 3\, t d f_b$.
(5) Tearing at inner row of rivets and crushing in front of one rivet in outer row $= (P - 2\, d)\, t f_t + t d f_b$.

Strength of Various Types of Riveted Joints (*Continued*)

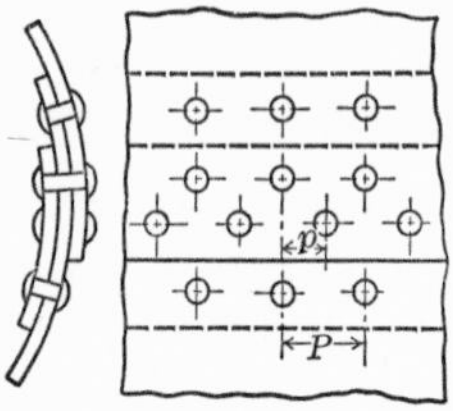

Double-riveted Lap Joint with inside Cover-Plate

(1) Tearing at outer row of rivets $= (P - d)\,tf_t$.

(2) Shearing four rivets $= \frac{4\,\pi d^2}{4} f_s$.

(3) Tearing at inner row and shearing outer row of rivets $= (P - 1\frac{1}{2}\,d)\,tf_t + \frac{\pi d^2}{4} f_s$.

(4) Crushing in front of four rivets $= 4\,tdf_b$.

(5) Tearing at inner row of rivets and crushing in front of one rivet $= (P - 1\frac{1}{2}\,d)\,tf_t + tdf_b$.

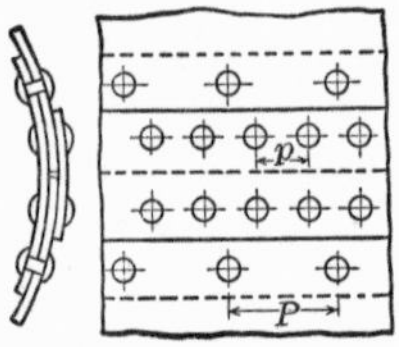

Double-riveted Butt-Joint

(1) Tearing at outer row of rivets $= (P - d)\,tf_t$.

(2) Shearing two rivets in double shear and one in single shear $= \frac{5\,\pi d^2}{4} f_s$.

(3) Tearing at inner row of rivets and shearing one of the outer row of rivets $= (P - 2\,d)\,tf_t + \frac{\pi d^2}{4} f_s$.

(4) Crushing in front of three rivets $= 3\,tdf_b$.

(5) Crushing in front of two rivets and shearing one rivet $= 2\,tdf_b + \frac{\pi d^2}{4} f_s$.

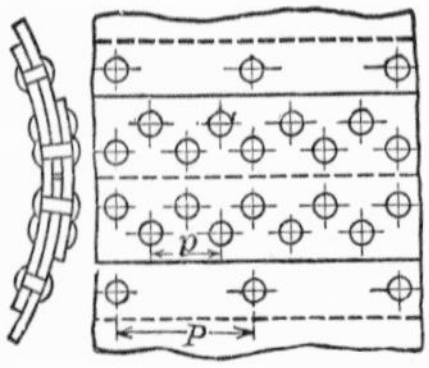

Triple-riveted Butt-Joint

(1) Tearing at outer row of rivets $= (P - d)\,tf_t$.

(2) Shearing four rivets in double shear and one in single shear $= \frac{9\,\pi d^2}{4} f_s$.

(3) Tearing at middle row of rivets and shearing one rivet $= (P - 2\,d)\,tf_t + \frac{\pi d^2}{4} f_s$.

(4) Crushing in front of four rivets and shearing one rivet $= 4\,dtf_b + \frac{\pi d^2}{4} f_s$.

(5) Crushing in front of five rivets $= 4\,dtf_b + dt_bf_b$.

BEAMS

Vertical shear at any section of a beam is equal to the algebraic sum of all the vertical forces on one side of the section. The shear is positive when the part of the beam to the left of the section tends to move upward under the action of the resultant of the vertical forces.

NOTE. In the study of beams, the reactions must be treated as applied loads and included in shear and moment. A section is always taken as cut by a plane normal to the axis of the beam. In all cases vertical means normal to the axis.

Bending moment at any section of a beam is equal to the algebraic sum of the moments, about the center of gravity of the section, of all the forces on one side of the section. Moment which causes compression in the upper fibers of a beam is positive.

NOTE. The maximum moment occurs at a section where the shear is zero. A curve of shears or of moments is a curve the ordinate to which at any section shows the value of the shear or moment at that section.

Moment and Shear Curves for a Simple Beam with a Uniformly Distributed Load.

Moment and Shear Curves for a Simple Beam with Concentrated Loads.

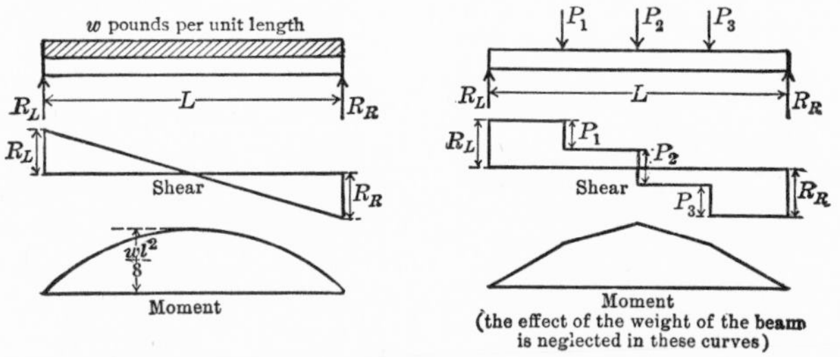

FIG. 520.

Neutral plane of a beam is the plane which undergoes no change in length due to the bending and along which the direct stress is zero. The fibers on one side of the neutral plane are stressed in tension and on the other side in compression and the intensities of these stresses in homogeneous beams are directly proportional to the distances of the fibers from the neutral plane.

NOTE. The neutral axis at any section in a beam subject to bending only passes through the center of gravity of that section.

Neutral axis at any section of a beam is the line formed by the intersection of the neutral plane and the section.

Elastic curve of a beam is the curve formed by the neutral plane when the beam deflects due to bending.

Equation of the elastic curve of a beam of **J** inches^4 moment of inertia and a modulus of elasticity of the material of **E** pounds per square inch, if **x** and **y** in inches are the abscissa and ordinate respectively of a point on the neutral axis referred to rectangular coördinates through the points of support and **M** is the moment in inch pounds at that point

520 $$M = EJ\frac{d^2y}{dx^2}$$

NOTE. The equation of the elastic curve is used to find the slope and deflection of a beam under loading. A single integration gives the slope, integrating twice gives the deflection; in each case, however, the proper value of the constant of integration must be determined.

BEAMS UNDER VARIOUS LOADINGS

Beam, loading and moment curve	Reactions	Bending moment	Deflection
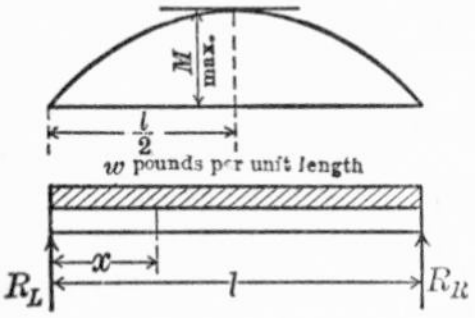	$R_L = R_R = \frac{wl}{2}$.	$M_x = \frac{wlx}{2} - \frac{wx^2}{2}$. $M_{max} = \frac{wl^2}{8}$.	$d_{max} = \frac{5\ wl^4}{384\ EJ}$.
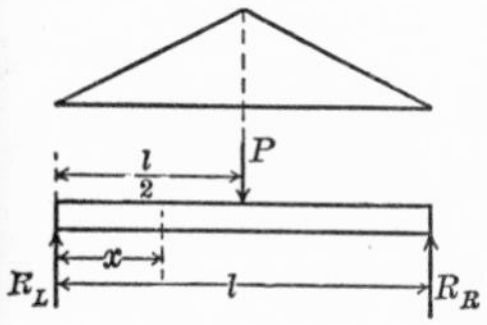	$R_L = R_R = \frac{P}{2}$.	$M_x = \frac{Px}{2}$. $M_{max} = \frac{Pl}{4}$.	$d_{max} = \frac{Pl^3}{48\ EJ}$.
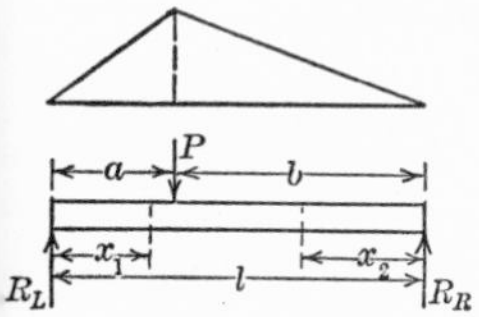	$R_L = \frac{Pb}{l}$. $R_R = \frac{Pa}{l}$.	$M_{x_1} = \frac{Pbx_1}{l}$. $M_{x_2} = \frac{Pax_2}{l}$. $M_{max} = \frac{Pab}{l}$.	$d_{max} = \frac{Pab(2a+b)\sqrt{3b(2a+b)}}{27\ EJl}$.
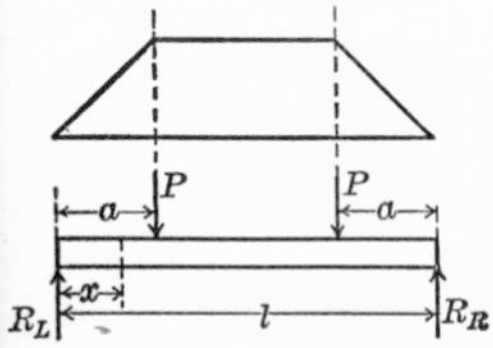	$R_L = R_R = P$.	$M_x = Px$. $M_{max} = Pa$.	$d_{max} = \frac{Pa}{6\ EJ}\left(\frac{3}{4}\ l^2 - a^2\right)$.
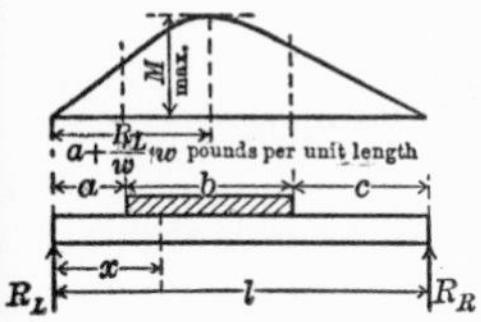	$R_L = \frac{wb\ (2\ c + b)}{2\ l}$. $R_R = \frac{wb\ (2\ a + b)}{2\ l}$.	$M_x = R_Lx - \frac{w\ (x-a)^2}{2}$. $M_{max} = R_L\left[a + \frac{R_L}{2\ w}\right]$.	

Beams Under Various Loadings (*Continued*)

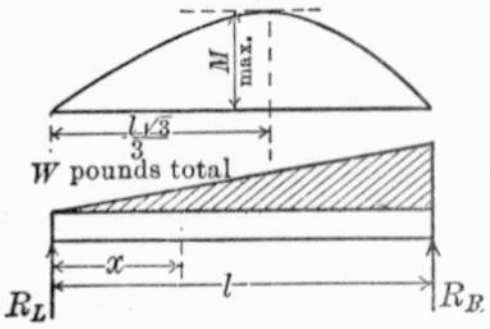	$R_L = \frac{1}{3} W.$ $R_R = \frac{2}{3} W.$	$M_x = \frac{Wx}{3}\left(1 - \frac{x^2}{l^2}\right).$ $M_{max} = \frac{2\,Wl}{9\sqrt{3}}.$	$d_{max} = \frac{0.013044\,Wl^3}{EJ}.$
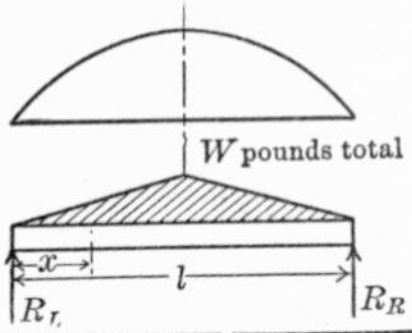	$R_L = R_R = \frac{W}{2}.$	$M_x = W_x\left(\frac{1}{2} - \frac{2\,x^2}{3\,l^2}\right).$ $M_{max} = \frac{Wl}{6}.$ (at center)	$d_{max} = \frac{Wl^3}{60\,EJ}.$
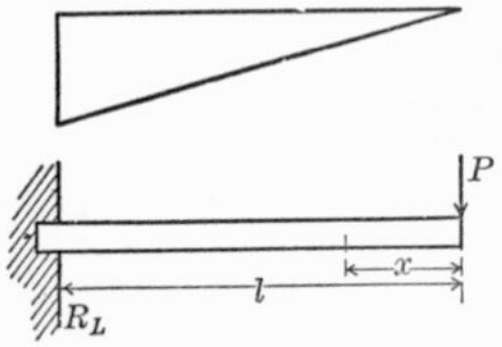	$R_L = P.$	$M_x = Px.$ $M_{max} = Pl.$	$d_{max} = \frac{Pl^3}{3\,EJ}.$
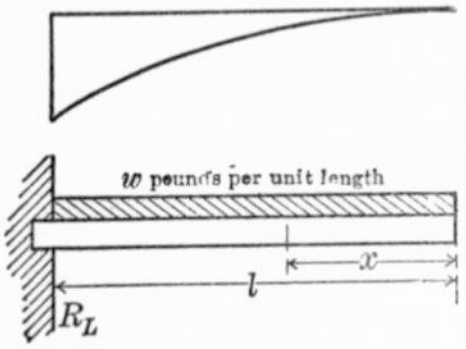	$R_L = wl.$	$M_x = \frac{wx^2}{2}.$ $M_{max} = \frac{wl^2}{2}.$	$d_{max} = \frac{wl^4}{8\,EJ}.$
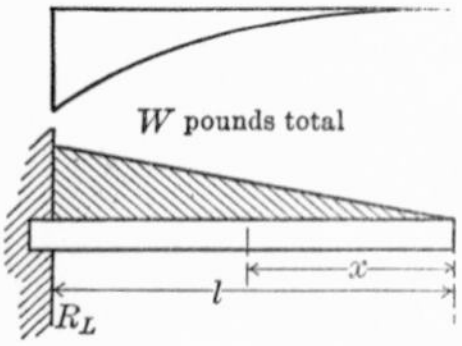	$R_L = W.$	$M_x = \frac{Wx^3}{3\,l^2}.$ $M_{max} = \frac{Wl}{3}.$	$d_{max} = \frac{Wl^3}{15\,EJ}.$
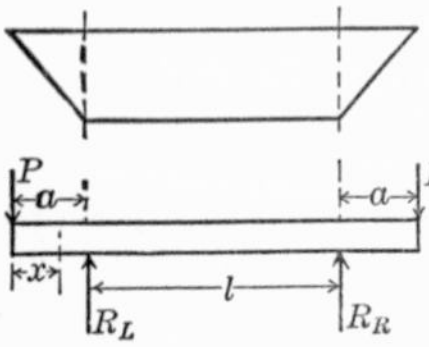	$R_L = R_R = P.$	$M_x = Px.$ $M_{max} = Pa.$	$d_{end} = \frac{Pa^2(2\,a + 3\,l)}{6\,EJ}.$ $d_{center} = -\frac{Pal^2}{8\,EJ}.$

Beams Under Various Loadings (*Concluded*)

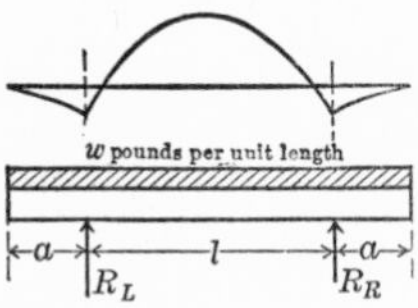	$R_L = R_R = \frac{w(l + 2a)}{2}$.	M at R_L and $R_R = \frac{wa^2}{2}$. $M_{center} = \frac{w(l^2 - 4a^2)}{8}$.	
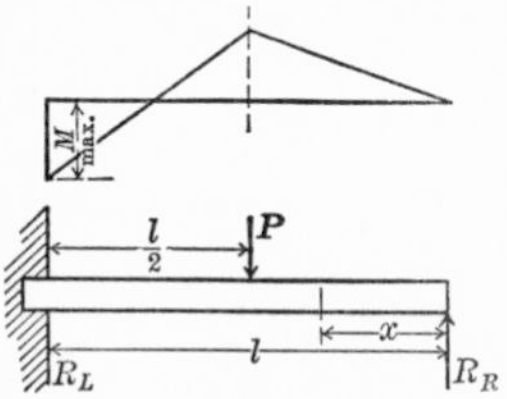	$R_L = \frac{11}{16} P$. $R_R = \frac{5}{16} P$.	$M_x = \frac{5}{16} Px$. $M_{max} = \frac{3}{16} Pl$.	$d_{max} = \sqrt{\frac{1}{5}} \frac{Pl^3}{48 EJ}$. at $x = l\sqrt{\frac{1}{5}}$.
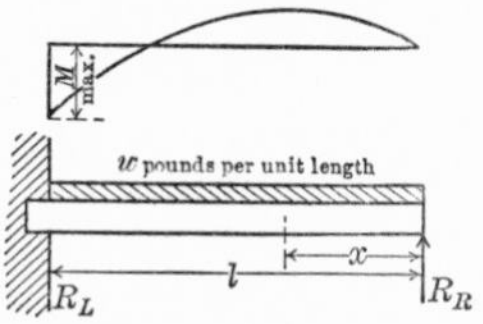	$R_L = \frac{5}{8} wl$. $R_R = \frac{3}{8} wl$.	$M_x = \frac{wlx}{2}\left(\frac{3}{4} - \frac{x}{l}\right)$. $M_{max} = \frac{wl^2}{8}$. (at R_L)	$d_{center} = \frac{wl^4}{192 EJ}$. $d_{max} = \frac{wl^4}{185 EJ}$. at $x = 0.4215 l$
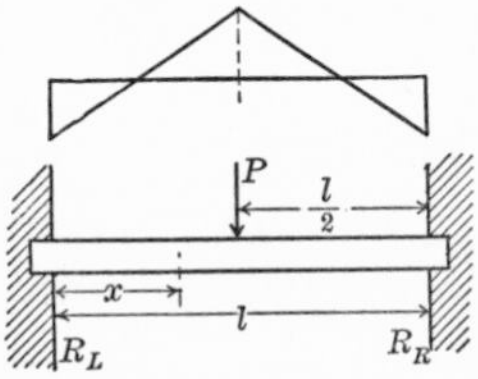	$R_L = R_R = \frac{P}{2}$.	$M_x = \frac{Pl}{2}\left(\frac{x}{l} - \frac{1}{4}\right)$. $M_{max} = \frac{Pl}{8}$. (at supports) $M_{max} = \frac{Pl}{8}$. (at center)	$d_{max} = \frac{Pl^3}{192 EJ}$.
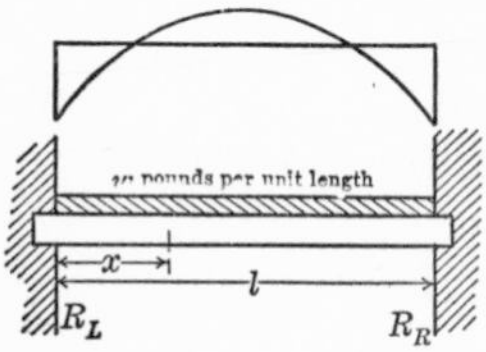	$R_L = R_R = \frac{wl}{2}$.	$M_x = \frac{wl^2}{2}\left(\frac{1}{6} - \frac{x}{l} + \frac{x^2}{l^2}\right)$. $M_{max} = \frac{wl^2}{12}$. (at supports)	$d_{max} = \frac{wl^4}{384 EJ}$.

Three moment equation gives the ratio between the moments M_a, M_b and M_c at three consecutive points of support (**a, b** and **c**) on a beam continuous over three or more supports.

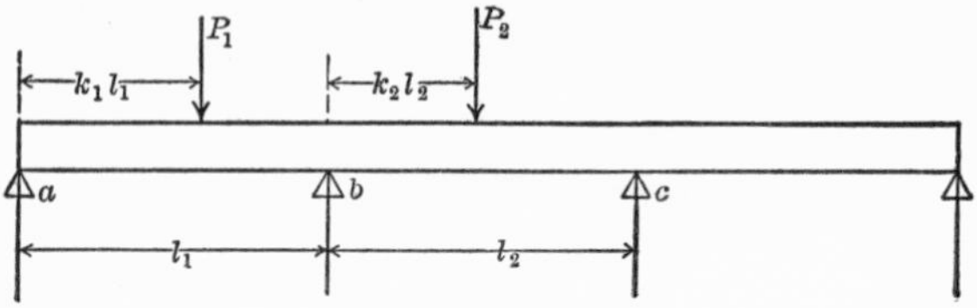

FIG. 521.

Case I. Concentrated loads. (See Fig. 521.)

521 $$M_a l_1 + 2\,M_b\,(l_1 + l_2) + M_c l_2 = P_1 l_1^2\,(k_1^3 - k_1) + P_2 l_2^2\,(3\,k_2^2 - k_2^3 - 2\,k_2).$$

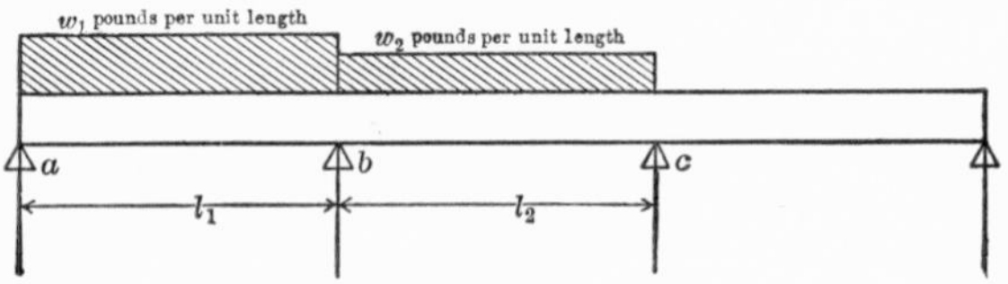

FIG. 522.

Case II. Uniformly distributed load. (See Fig. 522.)

522 $$M_a l_1 + 2\,M_b\,(l_1 + l_2) + M_c l_2 = -\tfrac{1}{4}\,w_1 l_1^3 - \tfrac{1}{4}\,w_2 l_2^3.$$

Intensity of stress (**f**) in tension or compression on a fiber **y** inches distant from the center of gravity of a section of a beam with **J** inches⁴ moment of inertia, due to a bending moment of **M** pound-inches.

523 $$f = \frac{My}{J} \text{ pounds per sq. in.}$$

Intensity of stress (**f**) on the outer fiber of a rectangular beam **h** inches in depth and **b** inches in breadth, due to a bending moment of **M** pound-inches.

524 $$f = \frac{6\,M}{bh^2} \text{ pounds per sq. in.}$$

Intensity of stress (**f**) in a fiber **y** inches distant from the center of gravity of a section of a beam of **A** square inches area and **J** inches⁴ moment of inertia, due to a direct load (parallel to axis of beam) of **P** pounds and a bending moment of **M** pound-inches.

525 $$f = \frac{P}{A} \pm \frac{My}{J} \text{ pounds per sq. in.}$$

Graphical representation of stress distribution in a beam.

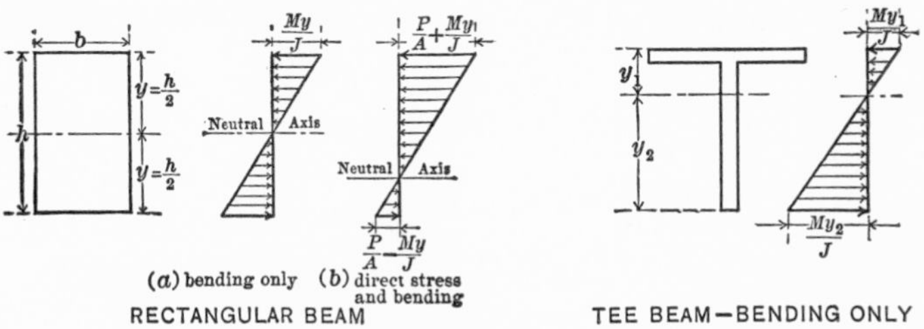

FIG. 525.

Maximum moment (M) which can be carried by a beam with **J** inches^4 moment of inertia and **y** inches greatest distance from center of gravity to outer fiber, without exceeding an intensity of stress of **f** pounds per square inch in the outer fiber.

526 $$M = \frac{fJ}{y} \text{ pound-inches.}$$

Section modulus (S) of a section of a beam with **J** inches^4 moment of inertia and **y** inches distance from center of gravity to outer fiber.

527 $$S = \frac{J}{y} \text{ inches}^3.$$

Intensity of stress (f) on the outer fiber of a beam of section modulus of S inches^3, due to a bending moment of **M** pound-inches.

528 $$f = \frac{M}{S} \text{ pounds per sq. in.}$$

Intensity of longitudinal shear (s) along a plane XX at the section of a beam where the total vertical shear is S pounds, if **J** inches^4 is the moment of inertia of the total section about its center of gravity axis, **b** the width of the beam at plane XX and **Q** inches^3 the statical moment, taken

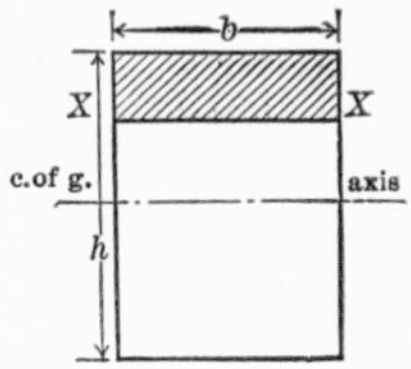

FIG. 529.

about the center of gravity axis, of that portion of the section which lies outside of the axis XX.

529 $$s = \frac{SQ}{bJ} \text{ pounds per sq. in.}$$

NOTE. The maximum intensity of shear always occurs at the center of gravity of the section of a beam.

Maximum intensity of shear (s) in a rectangular beam A square inches in area at a section where the total vertical shear is S pounds.

530 $$s = \frac{3}{2}\frac{S}{A} \text{ pounds per sq. in.}$$

NOTE. The intensity of vertical shear is equal to that of the longitudinal shear acting at right angles to it. The intensity of vertical shear is obtained by the formula $s = \frac{SQ}{bJ}$.

Properties of Standard I Beams*

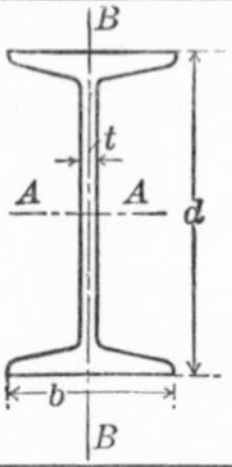

Depth of beam, d inches	Weight per foot, w pounds	Area of section, A inches²	Width of flange, b inches	Thickness of web, t inches	Axis A–A			Axis B–B		
					Moment of inertia, J inches⁴	Section modulus, S inches³	Radius of gyration, r inches	Moment of inertia, J inches⁴	Section modulus, S inches³	Radius of gyration, r inches
24	120.0	35.13	8.048	.798	3010.8	250.9	9.26	84.9	21.1	1.56
	115.0	33.67	7.987	.737	2940.5	245.0	9.35	82.8	20.7	1.57
	110.0	32.18	7.925	.675	2869.1	239.1	9.44	80.6	20.3	1.58
	105.9	30.98	7.875	.625	2811.5	234.3	9.53	78.9	20.0	1.60
24	100.0	29.25	7.247	.747	2371.8	197.6	9.05	48.4	13.4	1.29
	95.0	27.79	7.186	.686	2301.5	191.8	9.08	47.0	13.0	1.30
	90.0	26.30	7.124	.624	2230.1	185.8	9.21	45.5	12.8	1.32
	85.0	24.84	7.063	.563	2159.8	180.0	9.33	44.2	12.5	1.33
	79.9	23.33	7.000	.500	2087.2	173.9	9.46	42.9	12.2	1.36
20	100.0	29.20	7.273	.873	1648.3	164.8	7.51	52.4	14.4	1.34
	95.0	27.74	7.200	.800	1599.7	160.0	7.59	50.5	14.0	1.35
	90.0	26.26	7.126	.726	1550.3	155.0	7.68	48.7	13.7	1.36
	85.0	24.80	7.053	.653	1501.7	150.2	7.78	47.0	13.3	1.38
	81.4	23.74	7.000	.600	1466.3	146.6	7.86	45.8	13.1	1.39
20	75.0	21.90	6.391	.641	1263.5	126.3	7.60	30.1	9.4	1.17
	70.0	20.42	6.317	.567	1214.2	121.4	7.71	28.9	9.2	1.19
	65.4	19.08	6.250	.500	1169.5	116.9	7.83	27.9	8.9	1.21

* Manufactured by the Carnegie-Illinois Steel Company, Pittsburgh, Pa.

Properties of Standard I Beams* (*Continued*)

Depth of beam, d inches	Weight per foot, w pounds	Area of section, A inches²	Width of flange, b inches	Thickness of web, t inches	Axis A–A			Axis B–B		
					Moment of inertia, J inches⁴	Section modulus, S inches³	Radius of gyration, r inches	Moment of inertia, J inches⁴	Section modulus, S inches³	Radius of gyration, r inches
18	70.0	20.46	6.251	.711	917.5	101.9	6.70	24.5	7.8	1.09
	65.0	18.98	6.169	.629	877.7	97.5	6.80	23.4	7.6	1.11
	60.0	17.50	6.087	.547	837.8	93.1	6.92	22.3	7.3	1.13
	54.7	15.94	6.000	.460	795.5	88.4	7.07	21.2	7.1	1.15
15	75.0	21.85	6.278	.868	687.2	91.6	5.61	30.6	9.8	1.18
	70.0	20.38	6.180	.770	659.6	87.9	5.69	28.8	9.3	1.19
	65.0	18.91	6.082	.672	632.1	84.3	5.78	27.2	8.9	1.20
	60.8	17.68	6.000	.590	609.0	81.2	5.87	26.0	8.7	1.21
15	55.0	16.06	5.738	.648	508.7	67.8	5.63	17.0	5.9	1.03
	50.0	14.59	5.640	.550	481.1	64.2	5.74	16.0	5.7	1.05
	45.0	13.12	5.542	.452	453.6	60.5	5.88	15.0	5.4	1.07
	42.9	12.49	5.500	.410	441.8	58.9	5.95	14.6	5.3	1.08
12	55.0	16.04	5.600	.810	319.3	53.2	4.46	17.3	6.2	1.04
	50.0	14.57	5.477	.687	301.6	50.3	4.55	16.0	5.8	1.05
	45.0	13.10	5.355	.565	284.1	47.3	4.66	14.8	5.5	1.06
	40.8	11.84	5.250	.460	268.9	44.8	4.77	13.8	5.3	1.08
12	35.0	10.20	5.078	.428	227.0	37.8	4.72	10.0	3.9	0.99
	31.8	9.26	5.000	.350	215.8	36.0	4.83	9.5	3.8	1.01
10	40.0	11.69	5.091	.741	158.0	31.6	3.68	9.4	3.7	0.90
	35.0	10.22	4.944	.594	145.8	29.2	3.78	8.5	3.4	0.91
	30.0	8.75	4.797	.447	133.5	26.7	3.91	7.6	3.2	0.93
	25.4	7.38	4.660	.310	122.1	24.4	4.07	6.9	3.0	0.97
9	35.0	10.22	4.764	.724	111.3	24.7	3.30	7.3	3.0	0.84
	30.0	8.76	4.601	.561	101.4	22.5	3.40	6.4	2.8	0.85
	25.0	7.28	4.437	.397	91.4	20.3	3.54	5.6	2.5	0.88
	21.8	6.32	4.330	.290	84.9	18.9	3.67	5.2	2.4	0.90
8	25.5	7.43	4.262	.532	68.1	17.0	3.03	4.7	2.2	0.80
	23.0	6.71	4.171	.441	64.2	16.0	3.09	4.4	2.1	0.81
	20.5	5.97	4.079	.349	60.2	15.1	3.18	4.0	2.0	0.82
	18.4	5.34	4.000	.270	56.9	14.2	3.26	3.8	1.9	0.84
7	20.0	5.83	3.860	.450	41.9	12.0	2.68	3.1	1.6	0.74
	17.5	5.09	3.755	.345	38.9	11.1	2.77	2.9	1.6	0.76
	15.3	4.43	3.660	.250	36.2	10.4	2.86	2.7	1.5	0.78
6	17.25	5.02	3.565	.465	26.0	8.7	2.28	2.3	1.3	0.68
	14.75	4.29	3.443	.343	23.8	7.9	2.36	2.1	1.2	0.69
	12.5	3.61	3.330	.230	21.8	7.3	2.46	1.8	1.1	0.72
5	14.75	4.29	3.284	.494	15.0	6.0	1.87	1.7	1.0	0.63
	12.25	3.56	3.137	.347	13.5	5.4	1.95	1.4	0.91	0.63
	10.0	2.87	3.000	.210	12.1	4.8	2.05	1.2	0.82	0.65
4	10.5	3.05	2.870	.400	7.1	3.5	1.52	1.00	0.70	0.57
	9.5	2.76	2.796	.326	6.7	3.3	1.56	0.91	0.65	0.58
	8.5	2.46	2.723	.253	6.3	3.2	1.60	0.83	0.61	0.58
	7.7	2.21	2.660	.190	6.0	3.0	1.64	0.77	0.58	0.59
3	7.5	2.17	2.509	.349	2.9	1.9	1.15	0.59	0.47	0.52
	6.5	1.88	2.411	.251	2.7	1.8	1.19	0.51	0.43	0.52
	5.7	1.64	2.330	.170	2.5	1.7	1.23	0.46	0.40	0.53

* Manufactured by the Carnegie-Illinois Steel Company, Pittsburgh, Pa.

Properties of Standard Angles with Equal Legs*

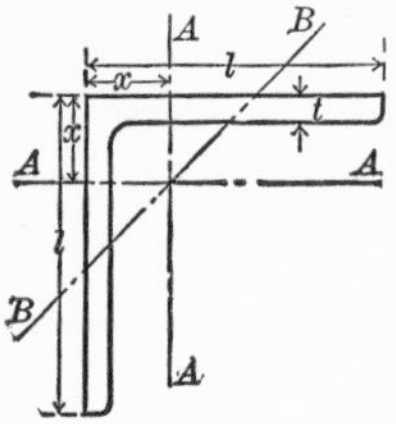

Size, l inches	Thickness, t inches	Weight per foot, w pounds	Area of section, A inches²	Axis A–A: Moment of inertia, J inches⁴	Axis A–A: Section modulus, S inches³	Axis A–A: Radius of gyration, r inches	Axis A–A: Distance from back of angle to center of gravity, x inches	Axis B–B: Minimum radius of gyration, r inches
8×8	$1\frac{1}{8}$	56.9	16.73	98.0	17.5	2.42	2.41	1.55
	$1\frac{1}{16}$	54.0	15.87	93.5	16.7	2.43	2.39	1.56
	1	51.0	15.00	89.0	15.8	2.44	2.37	1.56
	$\frac{15}{16}$	48.1	14.12	84.3	14.9	2.44	2.34	1.56
	$\frac{7}{8}$	45.0	13.23	79.6	14.0	2.45	2.32	1.56
	$\frac{13}{16}$	42.0	12.34	74.7	13.1	2.46	2.30	1.57
	$\frac{3}{4}$	38.9	11.44	69.7	12.2	2.47	2.28	1.57
	$\frac{11}{16}$	35.8	10.53	64.6	11.2	2.48	2.25	1.58
	$\frac{5}{8}$	32.7	9.61	59.4	10.3	2.49	2.23	1.58
	$\frac{9}{16}$	29.6	8.68	54.1	9.3	2.50	2.21	1.58
	$\frac{1}{2}$	26.4	7.75	48.6	8.4	2.51	2.19	1.58
6×6	$1\frac{1}{16}$	39.6	11.62	37.2	9.0	1.79	1.89	1.16
	1	37.4	11.00	35.5	8.6	1.80	1.86	1.16
	$\frac{15}{16}$	35.3	10.37	33.7	8.1	1.80	1.84	1.16
	$\frac{7}{8}$	33.1	9.73	31.9	7.6	1.81	1.82	1.17
	$\frac{13}{16}$	31.0	9.09	30.1	7.2	1.82	1.80	1.17
	$\frac{3}{4}$	28.7	8.44	28.2	6.7	1.83	1.78	1.17
	$\frac{11}{16}$	26.5	7.78	26.2	6.2	1.83	1.75	1.17
	$\frac{5}{8}$	24.2	7.11	24.2	5.7	1.84	1.73	1.17
	$\frac{9}{16}$	21.9	6.43	22.1	5.1	1.85	1.71	1.18
	$\frac{1}{2}$	19.6	5.75	19.9	4.6	1.86	1.68	1.18
	$\frac{7}{16}$	17.2	5.06	17.7	4.1	1.87	1.66	1.19
	$\frac{3}{8}$	14.9	4.36	15.4	3.5	1.88	1.64	1.19
	*$\frac{5}{16}$	12.6	3.66	13.0	3.0	1.89	1.61	1.19
5×5	1	30.6	9.00	19.6	5.8	1.48	1.61	0.96
	$\frac{15}{16}$	28.9	8.50	18.7	5.5	1.48	1.59	0.96
	$\frac{7}{8}$	27.2	7.98	17.8	5.2	1.49	1.57	0.96
	$\frac{13}{16}$	25.4	7.47	16.8	4.9	1.50	1.55	0.97
	$\frac{3}{4}$	23.6	6.94	15.7	4.5	1.50	1.52	0.97
	$\frac{11}{16}$	21.8	6.40	14.7	4.2	1.51	1.50	0.97
	$\frac{5}{8}$	20.0	5.86	13.6	3.9	1.52	1.48	0.97
	$\frac{9}{16}$	18.1	5.31	12.4	3.5	1.53	1.46	0.98
	$\frac{1}{2}$	16.2	4.75	11.3	3.2	1.54	1.43	0.98
	$\frac{7}{16}$	14.3	4.18	10.0	2.8	1.55	1.41	0.98
	$\frac{3}{8}$	12.3	3.61	8.7	2.4	1.56	1.39	0.99
	*$\frac{5}{16}$	10.3	3.03	7.4	2.0	1.56	1.36	0.99
	*$\frac{1}{4}$	8.3	2.44	6.0	1.6	1.57	1.34	0.99
4×4	$\frac{13}{16}$	19.9	5.84	8.1	3.0	1.18	1.29	0.77
	$\frac{3}{4}$	18.5	5.44	7.7	2.8	1.19	1.27	0.77
	$\frac{11}{16}$	17.1	5.03	7.2	2.6	1.19	1.25	0.77
	$\frac{5}{8}$	15.7	4.61	6.7	2.4	1.20	1.23	0.77
	$\frac{9}{16}$	14.3	4.18	6.1	2.2	1.21	1.21	0.78
	$\frac{1}{2}$	12.8	3.75	5.6	2.0	1.22	1.18	0.78
	$\frac{7}{16}$	11.3	3.31	5.0	1.8	1.23	1.16	0.78
	$\frac{3}{8}$	9.8	2.86	4.4	1.5	1.23	1.14	0.79
	$\frac{5}{16}$	8.2	2.40	3.7	1.3	1.24	1.12	0.79
	*$\frac{1}{4}$	6.6	1.94	3.0	1.0	1.25	1.09	0.79
	*$\frac{3}{16}$	5.0	1.46	2.3	0.80	1.26	1.07	0.79

* Manufactured by the Carnegie-Illinois Steel Company, Pittsburgh, Pa.

Properties of Standard Angles with Equal Legs* *(Continued)*

Size, l inches	Thickness, t inches	Weight per foot, w pounds	Area of section, A inches²	Axis A–A: Moment of inertia, J inches⁴	Axis A–A: Section modulus, S inches³	Axis A–A: Radius of gyration, r inches	Axis A–A: Distance from back of angle to center of gravity, x inches	Axis B–B: Minimum radius of gyration, r inches
3½×3½	13/16	17.1	5.03	5.3	2.3	1.02	1.17	0.67
	3/4	16.0	4.69	5.0	2.1	1.03	1.15	0.67
	11/16	14.8	4.34	4.7	2.0	1.04	1.12	0.67
	5/8	13.6	3.98	4.3	1.8	1.04	1.10	0.68
	9/16	12.4	3.62	4.0	1.6	1.05	1.08	0.68
	1/2	11.1	3.25	3.6	1.5	1.06	1.06	0.68
	7/16	9.8	2.87	3.3	1.3	1.07	1.04	0.68
	3/8	8.5	2.48	2.9	1.2	1.07	1.01	0.69
	5/16	7.2	2.09	2.5	0.98	1.08	0.99	0.69
	1/4	5.8	1.69	2.0	0.79	1.09	0.97	0.69
	*3/16	4.4	1.28	1.6	0.61	1.10	0.95	0.69
3×3	5/8	11.5	3.36	2.6	1.3	0.88	0.98	0.57
	9/16	10.4	3.06	2.4	1.2	0.89	0.95	0.58
	1/2	9.4	2.75	2.2	1.1	0.90	0.93	0.58
	7/16	8.3	2.43	2.0	0.95	0.91	0.91	0.58
	3/8	7.2	2.11	1.8	0.83	0.91	0.89	0.58
	5/16	6.1	1.78	1.5	0.71	0.92	0.87	0.59
	1/4	4.9	1.44	1.2	0.58	0.93	0.84	0.59
	*3/16	3.71	1.09	0.96	0.44	0.94	0.82	0.59
	*1/8	2.48	0.73	0.66	0.30	0.96	0.80	0.59
2½×2½	1/2	7.7	2.25	1.2	0.73	0.74	0.81	0.47
	7/16	6.8	2.00	1.1	0.65	0.75	0.78	0.48
	3/8	5.9	1.73	0.98	0.57	0.75	0.76	0.48
	5/16	5.0	1.47	0.85	0.48	0.76	0.74	0.49
	1/4	4.1	1.19	0.70	0.39	0.77	0.72	0.49
	3/16	3.07	0.90	0.55	0.30	0.78	0.69	0.49
	1/8	2.08	0.61	0.38	0.20	0.79	0.67	0.50
2×2	7/16	5.3	1.56	0.54	0.40	0.59	0.66	0.39
	3/8	4.7	1.36	0.48	0.35	0.59	0.64	0.39
	5/16	3.92	1.15	0.42	0.30	0.60	0.61	0.39
	1/4	3.19	0.94	0.35	0.25	0.61	0.59	0.39
	3/16	2.44	0.71	0.28	0.19	0.62	0.57	0.40
	1/8	1.65	0.48	0.19	0.13	0.63	0.55	0.40
1¾×1¾	3/8	3.99	1.17	0.31	0.26	0.51	0.57	0.34
	5/16	3.39	1.00	0.27	0.23	0.52	0.55	0.34
	1/4	2.77	0.81	0.23	0.19	0.53	0.53	0.34
	3/16	2.12	0.62	0.18	0.14	0.54	0.51	0.35
	1/8	1.44	0.42	0.13	0.10	0.55	0.48	0.35
1½×1½	3/8	3.35	0.98	0.19	0.19	0.44	0.51	0.29
	5/16	2.86	0.84	0.16	0.16	0.44	0.49	0.29
	1/4	2.34	0.69	0.14	0.13	0.45	0.47	0.29
	3/16	1.80	0.53	0.11	0.10	0.46	0.44	0.29
	1/8	1.23	0.36	0.08	0.07	0.46	0.42	0.30
1¼×1¼	5/16	2.33	0.68	0.09	0.11	0.36	0.42	0.24
	1/4	1.92	0.56	0.08	0.09	0.37	0.40	0.24
	3/16	1.48	0.43	0.06	0.07	0.38	0.38	0.24
	1/8	1.01	0.40	0.04	0.05	0.38	0.35	0.25
1×1	1/4	1.49	0.44	0.04	0.06	0.29	0.34	0.19
	5/16	1.16	0.34	0.03	0.04	0.30	0.32	0.19
	1/8	0.80	0.23	0.02	0.03	0.31	0.30	0.19

* Manufactured by the Carnegie-Illinois Steel Company, Pittsburgh, Pa.

Properties of Standard Angles with Unequal Legs*

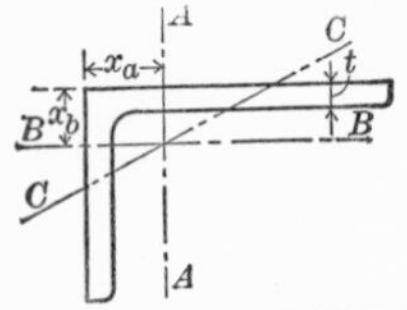

Size, inches	Thickness, t inches	Weight per foot, w pounds	Area of section, A inches²	Axis A–A: Moment of inertia, J inches⁴	Axis A–A: Section modulus, S inches³	Axis A–A: Radius of gyration, r inches	Axis A–A: Distance from back of angle to center of gravity, x_a ins.	Axis B–B: Moment of inertia, J inches⁴	Axis B–B: Section modulus, S inches³	Axis B–B: Radius of gyration, r inches	Axis B–B: Distance from back of angle to center of gravity, x_b ins.	Axis C–C: Minimum radius of gyration, r inches
8×6	1 1/8	49.3	14.48	88.9	16.8	2.48	2.70	42.5	9.9	1.71	1.70	1.28
	1 1/16	46.8	13.75	84.9	15.9	2.48	2.68	40.7	9.4	1.72	1.68	1.28
	1	44.2	13.00	80.8	15.1	2.49	2.65	38.8	8.9	1.73	1.65	1.28
	15/16	41.7	12.25	76.6	14.3	2.50	2.63	36.8	8.4	1.73	1.63	1.28
	7/8	39.1	11.48	72.3	13.4	2.51	2.61	34.9	7.9	1.74	1.61	1.28
	13/16	36.5	10.72	67.9	12.5	2.52	2.59	32.8	7.4	1.75	1.59	1.29
	3/4	33.8	9.94	63.4	11.7	2.53	2.56	30.7	6.9	1.76	1.56	1.29
	11/16	31.2	9.15	58.8	10.8	2.54	2.54	28.6	6.4	1.77	1.54	1.29
	5/8	28.5	8.36	54.1	9.9	2.54	2.52	26.3	5.9	1.77	1.52	1.30
	9/16	25.7	7.56	49.3	8.9	2.55	2.50	24.0	5.3	1.78	1.50	1.30
	1/2	23.0	6.75	44.3	8.0	2.56	2.47	21.7	4.8	1.79	1.47	1.30
	7/16	20.2	5.93	39.2	7.1	2.57	2.45	19.3	4.2	1.80	1.45	1.30
8×4	1	37.4	11.00	69.6	14.1	2.52	3.05	11.6	3.9	1.03	1.05	0.85
	15/16	35.3	10.37	66.1	13.3	2.52	3.02	11.1	3.7	1.03	1.02	0.85
	7/8	33.1	9.73	62.4	12.5	2.53	3.00	10.5	3.5	1.04	1.00	0.85
	13/16	31.0	9.09	58.7	11.7	2.54	2.98	10.0	3.3	1.05	0.98	0.85
	3/4	28.7	8.44	54.9	10.9	2.55	2.95	9.4	3.1	1.05	0.95	0.85
	11/16	26.5	7.78	51.0	10.0	2.56	2.93	8.7	2.8	1.06	0.93	0.85
	5/8	24.2	7.11	46.9	9.2	2.56	2.91	8.1	2.6	1.07	0.91	0.86
	9/16	21.9	6.43	42.8	8.4	2.58	2.88	7.4	2.4	1.07	0.88	0.86
	1/2	19.6	5.75	38.5	7.5	2.59	2.86	6.7	2.2	1.08	0.86	0.86
	7/16	17.2	5.06	34.1	6.6	2.60	2.83	6.0	1.9	1.09	0.83	0.87
7×4	1	34.0	10.00	47.7	10.8	2.18	2.60	11.2	3.9	1.06	1.10	0.85
	15/16	32.1	9.44	45.4	10.3	2.19	2.58	10.7	3.7	1.07	1.08	0.86
	7/8	30.2	8.86	42.9	9.7	2.20	2.55	10.2	3.5	1.07	1.05	0.86
	13/16	28.2	8.28	40.4	9.0	2.21	2.53	9.6	3.2	1.08	1.03	0.86
	3/4	26.2	7.69	37.8	8.4	2.22	2.51	9.1	3.0	1.09	1.01	0.86
	11/16	24.2	7.09	35.1	7.8	2.23	2.49	8.5	2.8	1.09	0.99	0.86
	5/8	22.1	6.49	32.4	7.1	2.24	2.46	7.8	2.6	1.10	0.96	0.86
	9/16	20.0	5.88	29.6	6.5	2.24	2.44	7.2	2.4	1.11	0.94	0.87
	1/2	17.9	5.25	26.7	5.8	2.25	2.42	6.5	2.1	1.11	0.92	0.87
	7/16	15.8	4.63	23.7	5.1	2.26	2.39	5.8	1.9	1.12	0.89	0.88
	3/8	13.6	3.99	20.6	4.4	2.27	2.37	5.1	1.6	1.13	0.87	0.88
6×4	1	30.6	9.00	30.8	8.0	1.85	2.17	10.8	3.8	1.09	1.17	0.85
	15/16	28.9	8.50	29.3	7.6	1.86	2.14	10.3	3.6	1.10	1.14	0.85
	7/8	27.2	7.98	27.7	7.2	1.86	2.12	9.8	3.4	1.11	1.12	0.86
	13/16	25.4	7.47	26.1	6.7	1.87	2.10	9.2	3.2	1.11	1.10	0.86
	3/4	23.6	6.94	24.5	6.2	1.88	2.08	8.7	3.0	1.12	1.08	0.86
	11/16	21.8	6.40	22.8	5.8	1.89	2.06	8.1	2.8	1.13	1.06	0.86
	5/8	20.0	5.86	21.1	5.3	1.90	2.03	7.5	2.5	1.13	1.03	0.86
	9/16	18.1	5.31	19.3	4.8	1.90	2.01	6.9	2.3	1.14	1.01	0.87
	1/2	16.2	4.75	17.4	4.3	1.91	1.99	6.3	2.1	1.15	0.99	0.87
	7/16	14.3	4.18	15.5	3.8	1.92	1.96	5.6	1.8	1.16	0.96	0.87
	3/8	12.3	3.61	13.5	3.3	1.93	1.94	4.9	1.6	1.17	0.94	0.88
	*5/16	10.3	3.03	11.4	2.8	1.94	1.92	4.2	1.4	1.17	0.92	0.88
6×3½	1	28.9	8.50	29.2	7.8	1.85	2.26	7.2	2.9	0.92	1.01	0.74
	15/16	27.3	8.03	27.8	7.4	1.86	2.24	6.9	2.7	0.93	0.99	0.74
	7/8	25.7	7.55	26.4	7.0	1.87	2.22	6.6	2.6	0.93	0.97	0.75
	13/16	24.0	7.06	24.9	6.6	1.88	2.20	6.2	2.4	0.94	0.95	0.75
	3/4	22.4	6.56	23.3	6.1	1.89	2.18	5.8	2.3	0.94	0.93	0.75
	11/16	20.6	6.06	21.7	5.6	1.89	2.15	5.5	2.1	0.95	0.90	0.75
	5/8	18.9	5.55	20.1	5.2	1.90	2.13	5.1	1.9	0.96	0.88	0.75

* Manufactured by the Carnegie-Illinois Steel Company, Pittsburgh, Pa.

Properties of Standard Angles with Unequal Legs* (*Continued*)

Size, inches	Thickness, t inches	Weight per foot, w pounds	Area of section, A inches²	Axis A–A: Moment of inertia, J inches⁴	Axis A–A: Section modulus, S inches³	Axis A–A: Radius of gyration, r inches	Axis A–A: Distance from back of angle to center of gravity, x_a ins.	Axis B–B: Moment of inertia, J inches⁴	Axis B–B: Section modulus, S inches³	Axis B–B: Radius of gyration, r inches	Axis B–B: Distance from back of angle to center of gravity, x_b ins.	Axis C–C: Minimum radius of gyration, r inches
6×3½	9/16	17.1	5.03	18.4	4.7	1.91	2.11	4.7	1.8	0.96	0.86	0.75
	1/2	15.3	4.50	16.6	4.2	1.92	2.08	4.3	1.6	0.97	0.83	0.76
	7/16	13.5	3.97	14.8	3.7	1.93	2.06	3.8	1.4	0.98	0.81	0.76
	3/8	11.7	3.42	12.9	3.3	1.94	2.04	3.3	1.2	0.99	0.79	0.77
	*5/16	9.8	2.87	10.9	2.7	1.95	2.01	2.9	1.0	1.00	0.76	0.77
	*1/4	7.9	2.31	8.9	2.2	1.96	1.99	2.5	0.91	1.01	0.74	0.77
5×4	3/4	21.1	6.19	14.6	4.4	1.54	1.66	8.5	2.9	1.14	1.16	0.78
	11/16	19.5	5.72	13.6	4.1	1.54	1.64	7.9	2.7	1.15	1.14	0.78
	5/8	17.8	5.23	12.4	3.7	1.55	1.62	7.3	2.5	1.16	1.12	0.78
	9/16	16.2	4.75	11.6	3.4	1.56	1.60	6.6	2.2	1.17	1.10	0.79
	1/2	14.5	4.25	10.5	3.1	1.57	1.57	6.0	2.0	1.18	1.07	0.79
	7/16	12.8	3.75	9.3	2.7	1.58	1.55	5.3	1.8	1.19	1.05	0.79
	3/8	11.0	3.23	8.1	2.3	1.59	1.53	4.7	1.6	1.20	1.03	0.80
5×3½	7/8	22.7	6.67	15.7	4.9	1.53	1.79	6.2	2.5	0.96	1.04	0.75
	13/16	21.3	6.25	14.8	4.6	1.54	1.77	5.9	2.4	0.97	1.02	0.75
	3/4	19.8	5.81	13.9	4.3	1.55	1.75	5.6	2.2	0.98	1.00	0.75
	11/16	18.3	5.37	13.0	4.0	1.56	1.72	5.2	2.1	0.98	0.97	0.75
	5/8	16.8	4.92	12.0	3.7	1.56	1.70	4.8	1.9	0.99	0.95	0.75
	9/16	15.2	4.47	11.0	3.3	1.57	1.68	4.4	1.7	1.00	0.93	0.75
	1/2	13.6	4.00	10.0	3.0	1.58	1.66	4.0	1.6	1.01	0.91	0.75
	7/16	12.0	3.53	8.9	2.6	1.59	1.63	3.6	1.4	1.01	0.88	0.76
	3/8	10.4	3.05	7.8	2.3	1.60	1.61	3.2	1.2	1.02	0.86	0.76
	5/16	8.7	2.56	6.6	1.9	1.61	1.59	2.7	1.0	1.03	0.84	0.76
	*1/4	7.0	2.06	5.4	1.6	1.61	1.56	2.2	0.83	1.04	0.81	0.76
5×3	13/16	19.9	5.84	14.0	4.5	1.55	1.86	3.7	1.7	0.80	0.86	0.64
	3/4	18.5	5.44	13.2	4.2	1.55	1.84	3.5	1.6	0.80	0.84	0.64
	11/16	17.1	5.03	12.3	3.9	1.56	1.82	3.3	1.5	0.81	0.82	0.64
	5/8	15.7	4.61	11.4	3.5	1.57	1.80	3.1	1.4	0.81	0.80	0.64
	9/16	14.3	4.18	10.4	3.2	1.58	1.77	2.8	1.3	0.82	0.77	0.65
	1/2	12.8	3.75	9.5	2.9	1.59	1.75	2.6	1.1	0.83	0.75	0.65
	7/16	11.3	3.31	8.4	2.6	1.60	1.73	2.3	1.0	0.84	0.73	0.65
	3/8	9.8	2.86	7.4	2.2	1.61	1.70	2.0	0.89	0.84	0.70	0.65
	5/16	8.2	2.40	6.3	1.9	1.61	1.68	1.8	0.75	0.85	0.68	0.66
	*1/4	6.6	1.94	5.1	1.5	1.62	1.66	1.4	0.61	0.86	0.66	0.66
4×3½	13/16	18.5	5.43	7.8	2.9	1.19	1.36	5.5	2.3	1.01	1.11	0.72
	3/4	17.3	5.06	7.3	2.8	1.20	1.34	5.2	2.1	1.01	1.09	0.72
	11/16	16.0	4.68	6.9	2.6	1.21	1.32	4.9	2.0	1.02	1.07	0.72
	5/8	14.7	4.30	6.4	2.4	1.22	1.29	4.5	1.8	1.03	1.04	0.72
	9/16	13.3	3.90	5.9	2.1	1.23	1.27	4.2	1.7	1.03	1.02	0.72
	1/2	11.9	3.50	5.3	1.9	1.23	1.25	3.8	1.5	1.04	1.00	0.72
	7/16	10.6	3.09	4.8	1 7	1.24	1.23	3.4	1.3	1.05	0.98	0.72
	3/8	9.1	2.67	4.2	1.5	1.25	1.21	3.0	1.2	1.06	0.96	0.73
	5/16	7.7	2.25	3.6	1.3	1.26	1.18	2.6	1.0	1.07	0.93	0.73
	*1/4	6.1	1.81	2.9	1.0	1.27	1.16	2.1	0.81	1.07	0.91	0.73
	*3/16	4.6	1.36	2.2	0.78	1.29	1.13	1 6	0.62	1.09	0.89	0.73
4×3	13/16	17.1	5.03	7.3	2.9	1.21	1.44	3.5	1.7	0.83	0.94	0.64
	3/4	16.0	4.69	6.9	2.7	1.22	1.42	3.3	1.6	0.84	0.92	0.64
	11/16	14.8	4.34	6.5	2.5	1.22	1.39	3.1	1.5	0.84	0.89	0.64
	5/8	13.6	3.98	6.0	2.3	1.23	1.37	2.9	1.4	0.85	0.87	0.64
	9/16	12.4	3.62	5.6	2.1	1.24	1.35	2.7	1.2	0.86	0.85	0.64
	1/2	11.1	3.25	5.0	1.9	1.25	1.33	2.4	1.1	0.86	0.83	0.64
	7/16	9.8	2.87	4.5	1.7	1.25	1.30	2.2	1.0	0.87	0.80	0.64
	3/8	8.5	2.48	4.0	1.5	1.26	1.38	1.9	0.87	0.88	0.78	0.64
	5/16	7.2	2.09	3.4	1.2	1.27	1.26	1.7	0.74	0.89	0.76	0.65
	1/4	5.8	1.69	2.8	1.0	1.28	1.24	1.4	0.60	0.89	0.74	0.65
	*3/16	4.4	1.28	2.1	0.77	1.29	1.21	1.1	0.46	0.91	0.71	0.65
3½×3	13/16	15.8	4.62	5.0	2.2	1.04	1.23	3.3	1.7	0.85	0.98	0.62
	3/4	14.7	4.31	4.7	2.1	1.04	1.21	3.1	1.5	0.85	0.96	0.62
	11/16	13.6	4.00	4.4	1.9	1.05	1.19	3.0	1.4	0.86	0.94	0.62
	5/8	12.5	3.67	4.1	1.8	1.06	1.17	2.8	1.3	0.87	0.92	0.62
	9/16	11.4	3.34	3.8	1.6	1.07	1.15	2.5	1.2	0.87	0.90	0.62
	1/2	10.2	3.00	3.5	1.5	1.07	1.13	2.3	1.1	0.88	0.88	0.62

* Manufactured by the Carnegie-Illinois Steel Company, Pittsburgh, Pa.

Properties of Standard Angles with Unequal Legs* (*Continued*)

Size, inches	Thickness, t inches	Weight per foot, w pounds	Area of section, A inches²	Axis A–A: Moment of inertia, J inches⁴	Axis A–A: Section modulus, S inches³	Axis A–A: Radius of gyration, r inches	Axis A–A: Distance from back of angle to center of gravity, x_a ins.	Axis B–B: Moment of inertia, J inches⁴	Axis B–B: Section modulus, S inches³	Axis B–B: Radius of gyration, r inches	Axis B–B: Distance from back of angle to center of gravity, x_b ins.	Axis C–C: Minimum radius of gyration, r inches
3½×3	7/16	9.1	2.65	3.1	1.3	1.08	1.10	2.1	0.98	0.89	0.85	0.62
	3/8	7.9	2.30	2.7	1.1	1.09	1.08	1.8	0.85	0.90	0.83	0.62
	3/16	6.6	1.93	2.3	0.96	1.10	1.06	1.6	0.72	0.90	0.81	0.63
	1/4	5.4	1.56	1.9	0.78	1.11	1.04	1.3	0.58	0.91	0.79	0.63
	*3/16	4.0	1.18	1.5	0.59	1.12	1.01	1.0	0.45	0.93	0.76	0.63
3½×2½	11/16	12.5	3.65	4.1	1.9	1.06	1.27	1.7	0.99	0.69	0.77	0.53
	5/8	11.5	3.36	3.8	1.7	1.07	1.25	1.6	0.92	0.69	0.75	0.53
	9/16	10.4	3.06	3.6	1.6	1.08	1.23	1.5	0.84	0.70	0.73	0.53
	1/2	9.4	2.75	3.2	1.4	1.09	1.20	1.4	0.76	0.70	0.70	0.53
	7/16	8.3	2.43	2.9	1.3	1.09	1.18	1.2	0.68	0.71	0.68	0.54
	3/8	7.2	2.11	2.6	1.1	1.10	1.16	1.1	0.59	0.72	0.66	0.54
	5/16	6.1	1.78	2.2	0.93	1.11	1.14	0.94	0.50	0.73	0.64	0.54
	1/4	4.9	1.44	1.8	0.75	1.12	1.11	0.78	0.41	0.74	0.61	0.54
	*3/16	3.7	1.09	1.4	0.58	1.13	1.09	0.60	0.32	0.74	0.59	0.55
3×2½	9/16	9.5	2.78	2.3	1.2	0.91	1.02	1.4	0.82	0.72	0.77	0.52
	1/2	8.5	2.50	2.1	1.0	0.91	1.00	1.3	0.74	0.72	0.75	0.52
	7/16	7.6	2.21	1.9	0.93	0.92	0.98	1.2	0.66	0.73	0.73	0.52
	3/8	6.6	1.92	1.7	0.81	0.93	0.96	1.0	0.58	0.74	0.71	0.52
	5/16	5.6	1.62	1.4	0.69	0.94	0.93	0.90	0.49	0.74	0.68	0.53
	1/4	4.5	1.31	1.2	0.56	0.95	0.91	0.74	0.40	0.75	0.66	0.53
	*3/16	3.39	1.00	0.91	0.43	0.95	0.89	0.58	0.31	0.76	0.64	0.53
3×2	1/2	7.7	2.25	1.9	1.0	0.92	1.08	0.67	0.47	0.55	0.58	0.43
	7/16	6.8	2.00	1.7	0.89	0.93	1.06	0.61	0.42	0.55	0.56	0.43
	3/8	5.9	1.73	1.5	0.78	0.94	1.04	0.54	0.37	0.56	0.54	0.43
	5/16	5.0	1.47	1.3	0.66	0.95	1.02	0.47	0.32	0.57	0.52	0.43
	1/4	4.1	1.19	1.1	0.54	0.95	0.99	0.39	0.26	0.57	0.49	0.43
	*3/16	3.07	0.90	0.84	0.41	0.97	0.97	0.31	0.20	0.58	0.47	0.44
2½×2	1/2	6.8	2.00	1.1	0.70	0.75	0.88	0.64	0.46	0.56	0.63	0.42
	7/16	6.1	1.78	1.0	0.62	0.76	0.85	0.58	0.41	0.57	0.60	0.42
	3/8	5.3	1.55	0.91	0.55	0.77	0.83	0.51	0.36	0.58	0.58	0.42
	5/16	4.5	1.31	0.79	0.47	0.78	0.81	0.45	0.31	0.58	0.56	0.42
	1/4	3.62	1.06	0.65	0.38	0.78	0.79	0.37	0.25	0.59	0.54	0.42
	3/16	2.75	0.81	0.51	0.29	0.79	0.76	0.29	0.20	0.60	0.51	0.43
	1/8	1.86	0.55	0.35	0.20	0.80	0.74	0.20	0.13	0.61	0.49	0.43
	3/8	4.7	1.36	0.82	0.52	0.78	0.92	0.22	0.20	0.40	0.42	0.32
2½×1½	5/16	3.92	1.15	0.71	0.44	0.79	0.90	0.19	0.17	0.41	0.40	0.32
	1/4	3.19	0.94	0.59	0.36	0.79	0.88	0.16	0.14	0.41	0.38	0.32
	3/16	2.44	0.72	0.46	0.28	0.80	0.85	0.13	0.11	0.42	0.35	0.33
2×1½	3/8	3.99	1.17	0.43	0.34	0.61	0.71	0.21	0.20	0.42	0.46	0.32
	5/16	3.39	1.00	0.38	0.29	0.62	0.69	0.18	0.17	0.42	0.44	0.32
	1/4	2.77	0.81	0.32	0.24	0.62	0.66	0.15	0.14	0.43	0.41	0.32
	3/16	2.12	0.62	0.25	0.18	0.63	0.64	0.12	0.11	0.44	0.39	0.32
	1/8	1.44	0.42	0.17	0.13	0.64	0.62	0.09	0.08	0.45	0.37	0.33
1¾×1¼	1/4	2.34	0.69	0.20	0.18	0.54	0.60	0.09	0.10	0.35	0.35	0.27
	3/16	1.80	0.53	0.16	0.14	0.55	0.58	0.07	0.08	0.36	0.33	0.27
	1/8	1.23	0.36	0.11	0.09	0.56	0.56	0.05	0.05	0.37	0.31	0.27

* Manufactured by the Carnegie-Illinois Steel Company, Pittsburgh, Pa.

Properties of Standard Channels*

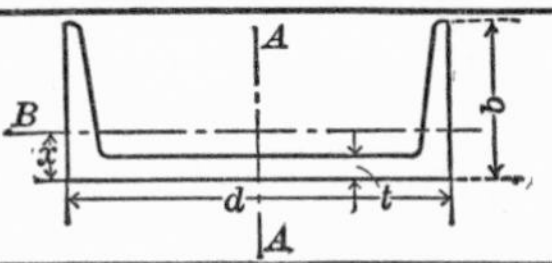

Depth of channel, d inches	Weight per foot, w pounds	Area of section, A inches²	Width of flange, b inches	Thickness of web, t inches	Axis A–A: Moment of inertia, J inches⁴	Axis A–A: Section modulus, S inches³	Axis A–A: Radius of gyration, r inches	Axis B–B: Moment of inertia, J inches⁴	Axis B–B: Section modulus, S inches³	Axis B–B: Radius of gyration, r inches	Distance from back of web to center of gravity, x inches
18	58.0	16.98	4.200	.700	670.7	74.5	6.29	18.5	5.6	1.04	0.88
	51.9	15.18	4.100	.600	622.1	69.1	6.40	17.1	5.3	1.06	0.87
	45.8	13.38	4.000	.500	573.5	63.7	6.55	15.8	5.1	1.09	0.89
	42.7	12.48	3.950	.450	549.2	61.0	6.64	15.0	4.9	1.10	0.90
15	55.0	16.11	3.814	.814	429.0	57.2	5.16	12.1	4.1	0.87	0.82
	50.0	14.64	3.716	.716	401.4	53.6	5.24	11.2	3.8	0.87	0.80
	45.0	13.17	3.618	.618	373.9	49.8	5.33	10.3	3.6	0.88	0.79
	40.0	11.70	3.520	.520	346.3	46.2	5.44	9.3	3.4	0.89	0.78
	35.0	10.23	3.422	.422	318.7	42.5	5.58	8.4	3.2	0.91	0.79
	33.9	9.90	3.400	.400	312.6	41.7	5.62	8.2	3.2	0.91	0.79
13	50.0	14.66	4.412	.787	312.9	48.1	4.62	16.7	4.9	1.07	0.98
	45.0	13.18	4.298	.673	292.0	44.9	4.71	15.3	4.6	1.08	0.97
	40.0	11.71	4.185	.560	271.4	41.7	4.82	13.9	4.3	1.09	0.97
	37.0	10.82	4.117	.492	258.9	39.8	4.89	13.0	4.2	1.10	0.98
	35.0	10.24	4.072	.447	250.7	38.6	4.95	12.5	4.0	1.10	0.99
	31.8	9.30	4.000	.375	237.5	36.5	5.05	11.6	3.9	1.11	1.01
12	40.0	11.73	3.415	.755	196.5	32.8	4.09	6.6	2.5	0.75	0.72
	35.0	10.26	3.292	.632	178.8	29.8	4.18	5.9	2.3	0.76	0.69
	30.0	8.79	3.170	.510	161.2	26.9	4.28	5.2	2.1	0.77	0.68
	25.0	7.32	3.047	.387	143.5	23.9	4.43	4.5	1.9	0.79	0.68
	20.7	6.03	2.940	.280	128.1	21.4	4.61	3.9	1.7	0.81	0.70
10	35.0	10.27	3.180	.820	115.2	23.0	3.34	4.6	1.9	0.67	0.69
	30.0	8.80	3.033	.673	103.0	20.6	3.42	4.0	1.7	0.67	0.65
	25.0	7.33	2.886	.526	90.7	18.1	3.52	3.4	1.5	0.68	0.62
	20.0	5.86	2.739	.379	78.5	15.7	3.66	2.8	1.3	0.70	0.61
	15.3	4.47	2.600	.240	66.9	13.4	3.87	2.3	1.2	0.72	0.64
9	25.0	7.33	2.812	.612	70.5	15.7	3.10	3.0	1.4	0.64	0.61
	20.0	5.86	2.648	.448	60.6	13.5	3.22	2.4	1.2	0.65	0.59
	15.0	4.39	2.485	.285	50.7	11.3	3.40	1.9	1.0	0.67	0.59
	13.4	3.89	2.430	.230	47.3	10.5	3.49	1.8	0.97	0.67	0.61
8	21.25	6.23	2.619	.579	47.6	11.9	2.77	2.20	1.10	0.60	0.59
	18.75	5.49	2.527	.487	43.7	10.9	2.82	2.00	1.00	0.60	0.57
	16.25	4.76	2.435	.395	39.8	9.9	2.89	1.80	0.94	0.61	0.56
	13.75	4.02	2.343	.303	35.8	9.0	2.99	1.50	0.86	0.62	0.56
	11.50	3.36	2.260	.220	32.3	8.1	3.10	1.30	0.79	0.63	0.58
7	19.75	5.79	2.509	.629	33.1	9.4	2.39	1.80	0.96	0.56	0.58
	17.25	5.05	2.404	.524	30.1	8.6	2.44	1.60	0.86	0.56	0.55
	14.75	4.32	2.299	.419	27.1	7.7	2.51	1.40	0.79	0.57	0.53
	12.25	3.58	2.194	.314	24.1	6.9	2.59	1.20	0.71	0.58	0.53
	9.80	2.85	2.090	.210	21.1	6.0	2.72	0.98	0.63	0.59	0.55
6	15.50	4.54	2.279	.559	19.5	6.5	2.07	1.30	0.73	0.53	0.55
	13.00	3.81	2.157	.437	17.3	5.8	2.13	1.10	0.65	0.53	0.52
	10.50	3.07	2.034	.314	15.1	5.0	2.22	0.87	0.57	0.53	0.50
	8.20	2.39	1.920	.200	13.0	4.3	2.34	0.70	0.50	0.54	0.52
5	11.50	3.36	2.032	.472	10.4	4.1	1.76	0.82	0.54	0.49	0.51
	9.00	2.63	1.885	.325	8.8	3.5	1.83	0.64	0.45	0.49	0.48
	6.70	1.95	1.750	.190	7.4	3.0	1.95	0.48	0.38	0.50	0.49
4	7.25	2.12	1.720	.320	4.5	2.3	1.47	0.44	0.35	0.46	0.46
	6.25	1.82	1.647	.247	4.1	2.1	1.50	0.38	0.32	0.45	0.46
	5.40	1.56	1.580	.180	3.8	1.9	1.56	0.32	0.29	0.45	0.46
3	6.00	1.75	1.596	.356	2.1	1.4	1.08	0.31	0.27	0.42	0.46
	5.00	1.46	1.498	.258	1.8	1.2	1.12	0.25	0.24	0.41	0.44
	4.10	1.19	1.410	.170	1.6	1.1	1.17	0.20	0.21	0.41	0.44

* Manufactured by the Carnegie-Illinois Steel Company, Pittsburgh, Pa.

Square and Round Steel Bars
Weights*, Areas and Circumferences
1/16 TO 15/16

Thickness or Diameter in Inches	Weight in Pounds				Area in Sq. Inches		Circumference
	Square		Round				
	One Inch Long	One Foot Long	One Inch Long	One Foot Long	Square	Round	Circumference
1/16	.001	.013	.001	.010	.0039	.0031	.1964
5/64	.002	.021	.001	.016	.0061	.0048	.2454
3/32	.002	.030	.002	.023	.0088	.0069	.2945
7/64	.003	.041	.003	.032	.0120	.0094	.3436
1/8	.004	.053	.004	.042	.0156	.0123	.3927
9/64	.006	.067	.004	.053	.0198	.0155	.4418
5/32	.007	.083	.005	.065	.0244	.0192	.4909
11/64	.008	.100	.007	.079	.0295	.0232	.5400
3/16	.010	.120	.008	.094	.0352	.0276	.5891
13/64	.012	.140	.009	.110	.0413	.0324	.6381
7/32	.014	.163	.011	.128	.0479	.0376	.6872
15/64	.016	.187	.012	.147	.0549	.0431	.7363
1/4	.018	.212	.014	.167	.0625	.0491	.7854
17/64	.020	.240	.016	.188	.0706	.0554	.8345
9/32	.022	.269	.018	.211	.0791	.0621	.8836
19/64	.025	.300	.020	.235	.0881	.0692	.9327
5/16	.028	.332	.022	.261	.0977	.0767	.9818
21/64	.031	.366	.024	.288	.1077	.0846	1.0308
11/32	.033	.402	.026	.316	.1182	.0928	1.0799
23/64	.037	.439	.029	.345	.1292	.1014	1.1290
3/8	.040	.478	.031	.376	.1406	.1104	1.1781
25/64	.043	.519	.034	.407	.1526	.1198	1.2272
13/32	.047	.561	.037	.441	.1650	.1296	1.2763
27/64	.050	.605	.040	.475	.1780	.1398	1.3254
7/16	.054	.651	.043	.511	.1914	.1503	1.3745
29/64	.058	.698	.046	.548	.2053	.1613	1.4235
15/32	.062	.747	.049	.587	.2197	.1726	1.4726
31/64	.066	.798	.052	.627	.2346	.1843	1.5217
1/2	.071	.850	.056	.668	.2500	.1963	1.5708
33/64	.075	.904	.060	.710	.2659	.2088	1.6199
17/32	.080	.960	.063	.754	.2822	.2217	1.6690
35/64	.085	1.017	.067	.799	.2991	.2349	1.7181
9/16	.090	1.076	.070	.845	.3164	.2485	1.7672
37/64	.095	1.136	.074	.893	.3342	.2625	1.8162
19/32	.100	1.199	.078	.941	.3525	.2769	1.8653
39/64	.105	1.263	.083	.992	.3713	.2916	1.9144
5/8	.111	1.328	.087	1.043	.3906	.3068	1.9635
41/64	.116	1.395	.091	1.096	.4104	.3223	2.0126
21/32	.122	1.464	.096	1.150	.4307	.3382	2.0617
43/64	.128	1.535	.100	1.205	.4514	.3545	2.1108
11/16	.134	1.607	.105	1.262	.4727	.3712	2.1599
45/64	.140	1.681	.110	1.320	.4944	.3883	2.2089
23/32	.146	1.756	.115	1.380	.5166	.4057	2.2580
47/64	.153	1.834	.120	1.440	.5393	.4236	2.3071
3/4	.159	1.913	.125	1.502	.5625	.4418	2.3562
13/16	.187	2.245	.147	1.763	.6602	.5185	2.5526
7/8	.217	2.603	.170	2.044	.7656	.6013	2.7489
15/16	.249	2.988	.196	2.347	.8789	.6903	2.9453

* One cubic foot of steel weighs 490 lbs.

1″ TO 3 15/16

Square and Round Steel Bars
Weights*, Areas and Circumferences

Thickness or Diameter in Inches	Weight in Pounds				Area in Sq. Inches		Circumference
	Square		Round		Square	Round	
	One Inch Long	One Foot Long	One Inch Long	One Foot Long			
1″	.28	3.400	.22	2.670	1.0000	.7854	3.1416
1/16	.32	3.838	.25	3.015	1.1289	.8866	3.3380
1/8	.36	4.303	.28	3.380	1.2656	.9940	3.5343
3/16	.40	4.795	.31	3.766	1.4102	1.1075	3.7306
1/4	.44	5.313	.35	4.172	1.5625	1.2272	3.9270
5/16	.49	5.857	.38	4.600	1.7227	1.3530	4.1234
3/8	.54	6.428	.42	5.049	1.8906	1.4849	4.3197
7/16	.58	7.026	.46	5.518	2.0664	1.6230	4.5161
1/2	.64	7.650	.50	6.008	2.2500	1.7671	4.7124
9/16	.69	8.301	.54	6.519	2.4414	1.9175	4.9088
5/8	.75	8.978	.59	7.051	2.6406	2.0739	5.1051
11/16	.81	9.682	.63	7.604	2.8477	2.2365	5.3015
3/4	.87	10.41	.68	8.178	3.0625	2.4053	5.4978
13/16	.94	11.17	.73	8.773	3.2852	2.5802	5.6942
7/8	1.00	11.95	.78	9.388	3.5156	2.7612	5.8905
15/16	1.06	12.76	.84	10.02	3.7539	2.9483	6.0869
2″	1.13	13.60	.89	10.68	4.0000	3.1416	6.2832
1/16	1.21	14.46	.95	11.36	4.2539	3.3410	6.4796
1/8	1.28	15.35	1.01	12.06	4.5156	3.5466	6.6759
3/16	1.36	16.27	1.07	12.78	4.7852	3.7583	6.8723
1/4	1.43	17.21	1.13	13.52	5.0625	3.9761	7.0686
5/16	1.52	18.18	1.19	14.28	5.3477	4.2000	7.2650
3/8	1.60	19.18	1.26	15.06	5.6406	4.4301	7.4613
7/16	1.68	20.20	1.32	15.87	5.9414	4.6664	7.6577
1/2	1.77	21.25	1.39	16.69	6.2500	4.9087	7.8540
9/16	1.86	22.33	1.46	17.53	6.5664	5.1573	8.0504
5/8	1.95	23.43	1.54	18.40	6.8906	5.4119	8.2467
11/16	2.05	24.56	1.61	19.29	7.2227	5.6727	8.4431
3/4	2.14	25.71	1.69	20.19	7.5625	5.9396	8.6394
13/16	2.24	26.90	1.76	21.12	7.9102	6.2126	8.8358
7/8	2.34	28.10	1.84	22.07	8.2656	6.4918	9.0321
15/16	2.44	29.34	1.92	23.04	8.6289	6.7771	9.2285
3″	2.55	30.60	2.01	24.03	9.0000	7.0686	9.4248
1/16	2.66	31.89	2.09	25.05	9.3789	7.3662	9.6212
1/8	2.77	33.20	2.18	26.08	9.7656	7.6699	9.8175
3/16	2.88	34.55	2.26	27.13	10.160	7.9798	10.014
1/4	2.99	35.92	2.35	28.21	10.563	8.2958	10.210
5/16	3.11	37.31	2.44	29.30	10.973	8.6179	10.407
3/8	3.23	38.73	2.53	30.42	11.391	8.9462	10.603
7/16	3.35	40.18	2.63	31.55	11.816	9.2806	10.799
1/2	3.47	41.65	2.73	32.71	12.250	9.6211	10.996
9/16	3.60	43.15	2.82	33.89	12.691	9.9678	11.192
5/8	3.72	44.68	2.92	35.09	13.141	10.321	11.388
11/16	3.85	46.23	3.03	36.31	13.598	10.680	11.585
3/4	3.98	47.82	3.13	37.55	14.063	11.045	11.781
13/16	4.12	49.42	3.23	38.81	14.535	11.416	11.977
7/8	4.25	51.05	3.34	40.10	15.016	11.793	12.174
15/16	4.39	52.71	3.45	41.40	15.504	12.177	12.370

* One cubic foot of steel weighs 490 lbs.

Properties of American Standard Yard Lumber and Timber Sizes

Nominal Size	American Standard Dressed Size	Area of Section $A = bd$	Wt. per Lineal foot	Moment of Inertia $I = \frac{bd^3}{12}$	Section Modulus $S = \frac{bd^2}{6}$
In.	In.	Sq. In.	Lbs.		
2×4	1⅝×3⅝	5.89	1.6	6.45	3.56
2×6	1⅝×5⅝	9.14	2.5	24.10	8.57
2×8	1⅝×7½	12.19	3.4	57.13	15.32
2×10	1⅝×9½	15.44	4.3	116.09	24.44
2×12	1⅝×11½	18.69	5.2	205.94	35.82
2×14	1⅝×13½	23.62	6.5	333.15	49.36
2×16	1⅝×15½	25.18	7.0	504.24	65.07
2×18	1⅝×17½	28.43	7.9	725.71	82.94
2×20	1⅝×19½	31.69	8.8	1004.05	102.98
3×4	2⅝×3⅝	9.51	2.6	10.42	5.75
3×6	2⅝×5⅝	14.76	4.2	38.93	13.84
3×8	2⅝×7½	19.68	5.7	92.28	24.60
3×10	2⅝×9½	24.93	7.2	187.55	39.48
3×12	2⅝×11½	30.18	8.8	332.69	57.86
3×14	2⅝×13½	35.43	10.3	538.21	79.73
3×16	2⅝×15½	40.68	11.3	814.60	105.11
3×18	2⅝×17½	45.94	12.8	1172.36	133.98
3×20	2⅝×19½	51.19	14.2	1622.00	166.36
4×4	3⅝×3⅝	13.14	3.6	14.38	7.94
4×6	3⅝×5⅝	20.39	5.7	53.76	19.11
4×8	3⅝×7½	27.18	7.5	127.44	33.98
4×10	3⅝×9½	34.43	9.6	258.99	54.52
4×12	3⅝×11½	41.68	11.6	459.42	79.90
4×14	3⅝×13½	48.93	13.6	743.23	110.11
4×16	3⅝×15½	56.18	15.6	1124.90	145.15
4×18	3⅝×17½	63.43	17.6	1618.96	185.02
4×20	3⅝×19½	70.69	19.6	2239.88	229.73
6×6	5½×5½	30.25	8.4	76.25	27.73
6×8	5½×7½	41.25	11.4	193.35	51.56
6×10	5½×9½	52.25	14.5	392.96	82.73
6×12	5½×11½	63.25	17.5	697.06	121.23
6×14	5½×13½	74.25	20.6	1127.66	167.06
6×16	5½×15½	85.25	23.6	1706.76	220.22
6×18	5½×17½	96.25	26.7	2456.36	280.73
6×20	5½×19½	107.25	29.8	3398.46	348.56
6×22	5½×21½	118.25	32.8	4555.05	423.73
8×8	7½×7½	56.25	15.6	263.67	70.31
8×10	7½×9½	71.25	19.8	535.85	112.81
8×12	7½×11½	86.25	23.9	950.55	165.31
8×14	7½×13½	101.25	28.0	1537.73	227.81
8×16	7½×15½	116.25	32.0	2327.42	300.31
8×18	7½×17½	131.25	36.4	2249.60	382.81
8×20	7½×19½	146.25	40.6	4634.30	475.31
8×22	7½×21½	161.25	44.8	6211.48	577.81
8×24	7½×23½	176.25	48.9	8111.17	690.31
10×10	9½×9½	90.25	25.0	678.75	142.89
10×12	9½×11½	109.25	30.3	1204.01	209.39
10×14	9½×13½	128.25	35.6	1947.78	288.56
10×16	9½×15½	147.25	40.9	2948.04	380.39
10×18	9½×17½	166.25	46.1	4242.80	484.89

Nominal Size	American Standard Dressed Size	Area of Section $A = bd$	Wt. per Lineal foot	Moment of Inertia $I = \frac{bd^3}{12}$	Section Modulus $S = \frac{bd^2}{6}$
In.	In.	Sq. In.	Lbs.		
10×20	9½×19½	185.25	51.4	5870.05	602.06
10×22	9½×21½	204.25	56.7	7867.81	731.89
10×24	9½×23½	223.25	62.0	10274.06	874.39
10×26	9½×25½	242.25	67.3	13126.81	1029.56
10×28	9½×27½	261.25	72.5	16465.24	1197.39
10×30	9½×29½	280.25	77.8	20323.79	1377.89
12×12	11½×11½	132.25	36.7	1457.50	253.47
12×14	11½×13½	155.25	43.1	2357.85	349.31
12×16	11½×15½	178.25	49.5	3568.70	460.48
12×18	11½×17½	201.25	55.9	5136.49	586.98
12×20	11½×19½	224.25	62.3	7105.90	728.81
12×22	11½×21½	247.25	68.7	9524.24	885.98
12×24	11½×23½	270.25	75.0	12437.08	1058.47
12×26	11½×25½	293.25	81.4	15890.42	1246.31
12×28	11½×27½	316.25	87.8	19932.58	1449.47
12×30	11½×29½	339.25	94.2	24602.61	1667.97
14×14	13½×13½	182.25	50.6	2767.92	410.06
14×16	13½×15½	209.25	58.1	4189.36	540.56
14×18	13½×17½	236.25	65.6	6029.29	689.06
14×20	13½×19½	263.25	73.1	8341.73	855.56
14×22	13½×21½	290.25	80.6	11180.67	1040.06
14×24	13½×23½	317.25	88.1	14600.10	1242.56
14×26	13½×25½	344.25	95.6	18654.04	1463.06
14×28	13½×27½	371.25	103.1	23398.73	1701.56
14×30	13½×29½	398.25	110.6	28881.42	1958.06
16×16	15½×15½	240.25	66.7	4809.98	620.64
16×18	15½×17½	271.25	75.3	6922.49	791.14
16×20	15½×19½	302.25	83.9	9577.50	982.31
16×22	15½×21½	333.25	92.5	12837.00	1194.14
16×24	15½×23½	364.25	101.2	16763.00	1426.64
16×26	15½×25½	395.25	109.8	21417.50	1679.81
16×28	15½×27½	426.25	118.4	26863.78	1953.64
16×30	15½×29½	457.25	127.0	33159.98	2248.14
18×18	17½×17½	306.25	85.0	7815.73	893.23
18×20	17½×19½	341.25	94.8	10813.33	1109.06
18×22	17½×21½	376.25	104.5	14493.43	1348.23
18×24	17½×23½	411.25	114.2	18926.02	1610.72
18×26	17½×25½	446.25	123.9	24181.11	1896.56
18×28	17½×27½	481.25	133.7	30331.62	2205.72
18×30	17½×29½	516.25	143.4	37438.79	2538.22
20×20	19½×19½	380.25	105.6	12049.49	1235.81
20×22	19½×21½	419.25	116.4	16149.86	1502.31
20×24	19½×23½	458.25	127.3	21089.04	1794.81
20×26	19½×25½	497.25	138.1	26944.73	2113.31
20×28	19½×27½	536.25	148.9	33798.17	2457.81
20×30	19½×29½	575.25	159.8	41717.61	2828.31
24×24	23½×23½	522.25	153.4	25414.96	2162.97
24×26	23½×25½	599.25	166.4	32471.80	2546.81
24×28	23½×27½	646.25	179.5	40731.06	2916.97
24×30	23½×29½	693.25	192.5	50274.98	3408.47

The weights given above are based on assumed average weight of 40 lbs. per cubic foot

Safe Load in Pounds per Square Inch of Cross-Sectional Area
Square and Rectangular Timber Columns
Dry Locations

Species of Lumber	American Standard Grade	*Ratio of Length to Least Dimension (l/d)										
		10& less	l/d 12	l/d 14	l/d 16	l/d 18	l/d 20	l/d 25	l/d 30	l/d 35	l/d 40	l/d 50
		lbs.	lbs.	lbs.	lbs.	lbs.	lbs.	lbs.	lbs.	lbs.	lbs.	lbs.
Ash, Commercial White	Select	1100	1076	1055	1023	978	913	658	457	336	257	164
	Common	880	868	857	840	818	784	647				
Cedar, Western Red; Fir, Balsam	Select	700	686	674	656	629	592	438	304	224	171	110
	Common	560	553	547	538	524	505	425				
Cedar, Northern and Southern White	Select	550	540	530	516	496	468	351	244	179	137	88
	Common	440	435	430	423	412	398	338				
Chestnut; Pine, Northern White, Idaho White, Sugar, Calif. White, and Pondosa	Select	750	733	718	695	663	617	438	304	224	171	110
	Common	600	591	583	572	556	532	434				
Cypress, Southern; Larch, Western	Select	1100	1063	1030	981	909	810	526	365	268	206	132
	Common	880	861	843	818	781	729					
Douglas Fir (Coast Region); Pine, Southern Yellow; Beech; Birch, Yellow and Sweet; Maple, Sugar	**Dense Select	1285	1251	1222	1176	1112	1022	702	487	358	274	175
	Select	1175	1149	1127	1093	1045	975	702				
	Common	880	870	861	847	826	796	675				
Douglas Fir (Rky. Mtn. Region); Spruce, Red, White, Sitka; Norway Pine; Alaska Cedar; Elm, Slippery and White; Sycamore; Gum, Red and Black; Tupelo	Select	800	786	774	753	726	688	526	365	268	206	132
	Common	640	632	627	617	602	582	500				
Hemlock, West Coast	Select	900	885	872	852	823	783	614	426	313	240	153
	Common	720	712	706	696	680	660	573				
Hemlock, Eastern; Fir, Commercial White	Select	700	689	678	664	641	611	482	335	246	188	121
	Common	560	554	549	542	530	515	449				
Oak, White and Red	Select	1000	982	967	943	908	860	658	457	336	257	164
	Common	800	790	783	771	753	728	625				
Redwood	Select	1000	972	947	910	856	781	526	268	365	206	132
	Common	800	786	773	754	726	688					
Spruce, Engelmann	Select	600	586	574	556	530	494	351	244	179	137	88
	Common	480	473	466	457	444	426	347				
Tamarack	Select	1000	976	955	923	877	817	570	396	291	223	142
	Common	800	788	777	761	737	706	566				

SAFE LOADS in compression parallel to grain for timber columns shall not exceed in pounds per square inch the values given in the above table for the respective species, grade, and ratio of unsupported length to least dimension, (*l*/*d*).

No column shall be used in which the unsupported length is more than 50 times the least diameter. *l* and *d* must be figured in the same unit of measurement.

RESULTANT OF SHEARING AND DIRECT STRESSES

Resultant intensity (p′) of normal stress and (s′) of shearing stress on a plane inclined $\alpha°$ to the horizontal at a point in the beam where the intensity of the horizontal and vertical shearing stresses is **s** pounds per square inch, the intensity of the stress normal to the vertical plane is p_x pounds per square inch, and that normal to the horizontal plane is p_y pounds per square inch.

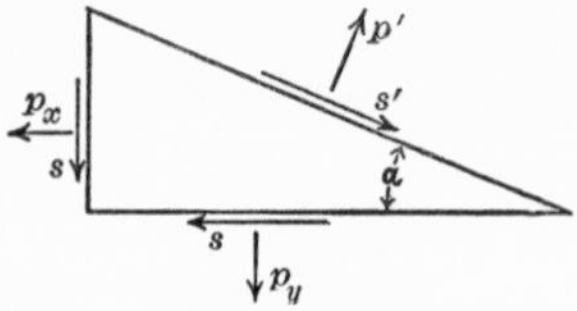

FIG. 531.

531 $$p' = \frac{p_x + p_y}{2} + \frac{p_y - p_x}{2}\cos 2\alpha + s \sin 2\alpha \text{ pounds per sq. in.}$$

532 $$s' = \frac{p_x - p_y}{2}\sin 2\alpha + s \cos 2\alpha \text{ pounds per sq. in.}$$

Angle (α) made with the horizontal by the plane on which the maximum intensity of normal stress occurs.

533 $$\tan 2\alpha = \frac{2s}{p_y - p_x}.$$

Maximum and minimum intensities of normal stress.

534 $$p'_{\substack{max\\min}} = \frac{p_x + p_y}{2} \pm \frac{1}{2}\sqrt{4s^2 + (p_x - p_y)^2} \text{ pounds per sq. in.}$$

NOTE. The maximum and minimum normal stresses are called principal stresses and occur on planes which are at right angles to each other and on each of which the shearing stress is zero.

Angle (α) made with the horizontal by the plane on which the maximum intensity of shear occurs.

535 $$\tan 2\alpha = \frac{p_x - p_y}{2s}.$$

NOTE. The planes of the maximum and minimum shearing stresses are inclined at 45° to the planes of maximum and minimum normal stresses.

Maximum and minimum intensities of shearing stress.

536 $$s'_{\substack{max\\min}} = \pm\frac{1}{2}\sqrt{4s^2 + (p_x - p_y)^2} \text{ pounds per sq. in.}$$

COLUMNS

Euler's formula for the ultimate average intensity of stress (f) on a column l inches in length, with a least radius of gyration of r inches and of material of E pounds per square inch modulus of elasticity. f should not exceed the elastic limit.

537 Column with end rounded $f = \pi^2 E \left(\frac{r}{l}\right)^2$ pounds per sq. in.

538 Column with ends fixed $f = 4\pi^2 E \left(\frac{r}{l}\right)^2$ pounds per sq. in.

539 Column with one end fixed and one end rounded $f = \frac{9}{4}\pi^2 E \left(\frac{r}{l}\right)^2$ pounds per sq. in.

Gordon Formula for allowable average intensity of stress (f) on a column l inches in length, with a least radius of gyration of r inches and a maximum allowable compression stress of f_c pounds per square inch on the material.

540 $$f = \frac{f_c}{1 + \frac{1}{c}\left(\frac{l}{r}\right)^2} \text{ pounds per sq. in.}$$

NOTE. The following values of **c** are commonly used for steel columns.

Column with ends rounded	9,000
Column with ends fixed	20,000
Column with one end fixed and one end rounded	36,000

Pin-ended columns are generally considered to have ends rounded.

Straight-line formula for the allowable average intensity of stress (f) in a column l inches in length, with a least radius of gyration of r inches and a maximum allowable compression stress of f_c pounds per square inch on the material.

541 $$f = f_c - c\left(\frac{l}{r}\right) \text{ pounds per sq. in.}$$

NOTE. The American Railway Engineering and Maintenance of Way Association gives the following formula in its specifications. $f = 16{,}000 - 70\frac{l}{r}$.

Maximum intensity of stress (f) in a column of **A** square inches area of cross-section, l inches length, **J** inches^4 moment of inertia about the axis about which bending occurs and **y** inches distance from that axis to the most stressed fiber, due to a direct load of **P** pounds and a bending moment of **M** inch-pounds.

542 $$f = \frac{P}{A} + \frac{My}{J - \frac{Pl^2}{cE}} \text{ pounds per sq. in. approx.}$$

NOTE. The constant **c** for the common case of pin-ended columns subject to bending due to a uniformly distributed load may be taken as 10.

Maximum intensity of stress (f) in a short column of **A** square inches area of cross-section, due to a load of **P** pounds applied **a** inches distant from the X axis of symmetry and **b** inches distant from the Y axis of symmetry, if J_x inches[4] is the moment of inertia about the X axis, **y** inches the distance from the X axis to the most stressed fiber, J_y inches[4] the moment of inertia about the Y axis and **x** inches the distance from Y axis to the most stressed fiber.

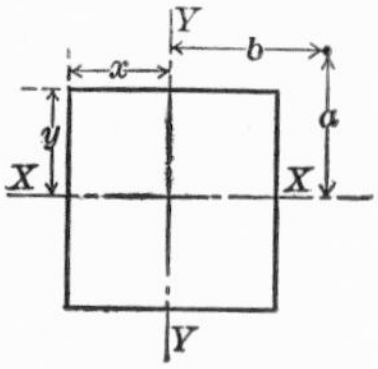

FIG. 543

543 $$f = \frac{P}{A} + \frac{Pay}{J_x} + \frac{Pbx}{J_y} \text{ pounds per sq. in.}$$

SHAFTS

Maximum intensity of shear (s) in a shaft of **r** inches radius and of J_0 inches[4] polar moment of inertia due to a torque (twisting moment) of **M** inch-pounds.

544 $$s = \frac{Mr}{J_0} \text{ pounds per sq. in.}$$

NOTE. For a solid round shaft $s = \frac{2M}{\pi r^3}$.

Angle (θ) of twist in a solid circular shaft, of **r** inches radius, **l** inches in length and with E_s pounds per square inch modulus of elasticity in shear, due to a torque of **M** inch-pounds.

545 $$\theta = \frac{2Ml}{\pi r^4 E_s} \text{ radians.}$$

NOTE. E_s for steel is commonly taken as 12,000,000.

Horse-power (**P**) transmitted by a shaft making **n** revolutions per minute under a torque of **M** inch-pounds.

546 $$P = \frac{2\,\pi n M}{33{,}000 \times 12} \text{ horse-power.}$$

Diameter (**d**) of a solid circular shaft to transmit H.P. horse-power at **n** revolutions per minute with a fiber stress in shear of **s** pounds per square inch.

547 $$d = \sqrt[3]{\frac{321{,}000 \text{ H.P.}}{ns}} \text{ inches.}$$

Maximum intensity (s′) **of shearing stress and** (f′) **of tensile or compression stress** due to combined twisting and bending in a shaft where **s** is the maximum intensity of shear due to the torque and **f** is the maximum intensity of tension or compression due to the bending.

548 $$s' = \frac{1}{2}\sqrt{4\,s^2 + f^2} \text{ pounds per sq. in.}$$

549 $$f' = \frac{1}{2}f + \frac{1}{2}\sqrt{4\,s^2 + f^2} \text{ pounds per sq. in.}$$

HYDRAULICS

HYDROSTATICS

Intensity of pressure ($\mathbf{p}$) due to a head of $\mathbf{h}$ feet in a liquid weighing $\mathbf{w}$ pounds per cubic foot.

550 $$\mathbf{p} = \mathbf{wh} \text{ pounds per sq. ft.}$$

NOTE. In water, the intensity of pressure corresponding to a head of $\mathbf{h}$ feet is 0.434 $\mathbf{h}$ pounds per square inch.

Pressure head ($\mathbf{h}$) corresponding to a pressure of $\mathbf{p}$ pounds per square foot in a liquid weighing $\mathbf{w}$ pounds per cubic foot.

551 $$\mathbf{h} = \frac{\mathbf{p}}{\mathbf{w}} \text{ feet.}$$

NOTE. In water, the pressure head corresponding to an intensity of pressure of $\mathbf{p}$ pounds per square inch is 2.3 $\mathbf{p}$ feet.

Total normal pressure ($\mathbf{P}$) on a plane or curved surface $\mathbf{A}$ square feet in area immersed in a liquid weighing $\mathbf{w}$ pounds per cubic foot with a head of $\mathbf{h_0}$ feet on its center of gravity.

552 $$\mathbf{P} = \mathbf{wAh_0} \text{ pounds.}$$

NOTE. The total pressure on a plane surface may be represented by a resultant force of $\mathbf{P}$ pounds acting normally to the area at its center of pressure.

Distance ($\mathbf{x_c}$) **to the center of pressure** of a plane area, measured from the surface of the liquid along the plane of the area, if $\mathbf{S}$ is the statical moment in feet^3 about the axis formed by the intersection of the plane of the area and the surface of the liquid and $\mathbf{J}$ is the moment of inertia in feet^4 about the same axis.

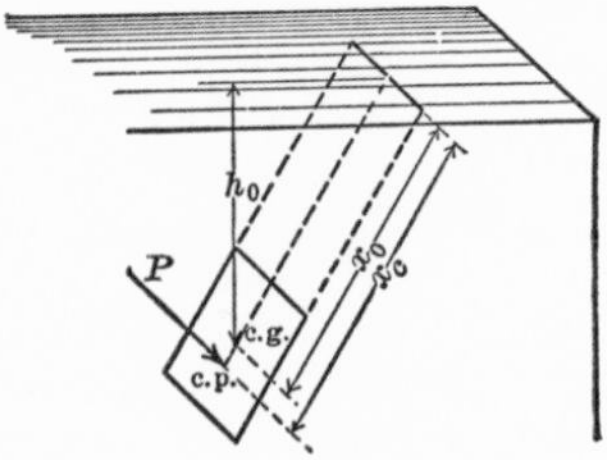

FIG. 553.

553 $$\mathbf{x_c} = \frac{\mathbf{J}}{\mathbf{S}} \text{ feet.}$$

NOTE. If $\mathbf{x_0}$ is the distance in feet from the center of gravity of the area to the surface axis, $\mathbf{K_0}$ the radius of gyration in feet about the center of

gravity axis, A the area in square feet and J_0 the moment of inertia in feet[4] about the center of gravity axis,

$$x_c = \frac{J_0 + Ax_0^2}{Ax_0} \text{ feet} \quad \text{and} \quad x_c - x_0 = \frac{K_0^2}{x_0} \text{ feet.}$$

Special cases of 553. **Rectangle** of altitude **d** feet and base **b** feet parallel to the surface: $x_c - x_0 = \dfrac{d^2}{12\, x_0}$ feet.

Rectangle with one base coinciding with the surface of the liquid: $x_c = \frac{2}{3}\, d$ ft.

Triangle of altitude **d** feet and base, **b** feet, parallel to the surface of the liquid with its vertex upward: $x_c - x_0 = \dfrac{d^2}{18\, x_0}$ feet.

Triangle of altitude **d** feet with its vertex at the surface: $x_c = \frac{3}{4}\, d$ feet.

Triangle of altitude **d** feet with its base in the surface and its vertex down: $x_c = \frac{1}{2}\, d$ feet.

Circle of radius **r** feet: $x_c - x_0 = \dfrac{r^2}{4\, x_0}$ feet.

Circle of radius **r** feet with a point in the surface: $x_c = \frac{5}{4}\, r$ feet.

Component of normal pressure (P_c) on a plane area of **A** square feet with h_0 feet head on its center of gravity and a projection of A_c square feet on a plane perpendicular to the component of pressure.

554 $$P_c = wA_ch_0 \text{ pounds.}$$

Vertical component of pressure (P_v) on a plane area of **A** square feet with h_0 feet head on its center of gravity and A_h square feet horizontal projection of area.

555 $$P_v = wA_hh_0 \text{ pounds.}$$

Horizontal component of pressure (P_h) on any area of **A** square feet with A_v square feet vertical projection of area and h_0 feet head on the center of gravity of the projected area.

556 $$P_h = wA_vh_0 \text{ pounds.}$$

Resultant pressure (P_{bc}) on an area bc of A_{bc} square feet with a head above its base of h_1 feet on one side and h_2 feet on the other side, or a difference of head of **h** feet.

557 $$P_{bc} = wA_{bc}(h_1 - h_2) = wA_{bc}h \text{ pounds.}$$

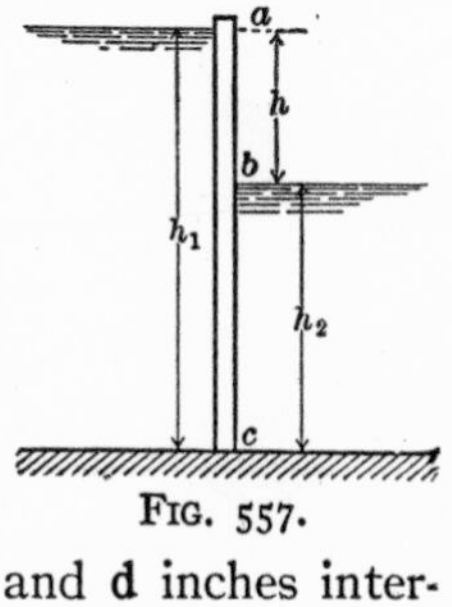

FIG. 557.

Stress (f) **in a pipe** of **t** inches thickness and **d** inches internal diameter due to a pressure of **p** pounds per square inch.

558 $$f = \frac{pd}{2t} \text{ pounds per sq. in.}$$

Thickness (t) of a pipe of **d** inches internal diameter to withstand a pressure of **p** pounds per square inch with a fiber stress of **f** pounds per square inch.

559 $$t = \frac{pd}{2f} \text{ inches.}$$

Practical formula for thickness (**t**) recommended by the New England Water Works Association.

560 For cast-iron pipes $$t = \frac{(p + p')d}{6600} + \frac{1}{4} \text{ inches.}$$

561 For riveted steel pipes $$t = \frac{(p + p')d}{2f} \text{ inches.}$$

NOTE. p' is an additional pressure in pounds per square inch which allows for water hammer and the following arbitrary values are recommended for various diameters **d** of the pipe in inches:

d	p′	d	p′
4 to 10	120	24	85
12 to 14	110	30	80
16 to 18	100	36	75
20	90	42 to 60	70

Difference in water pressure ($p_1 - p_2$) in two pipes as indicated by a differential gage with an oil of specific gravity **s**, when the difference in level of the surfaces of separation of the oil and water is **z** feet and the difference in level of the two pipes is **h** feet.

(*a*) When the oil has a specific gravity less than 1. (See Fig. 562.)

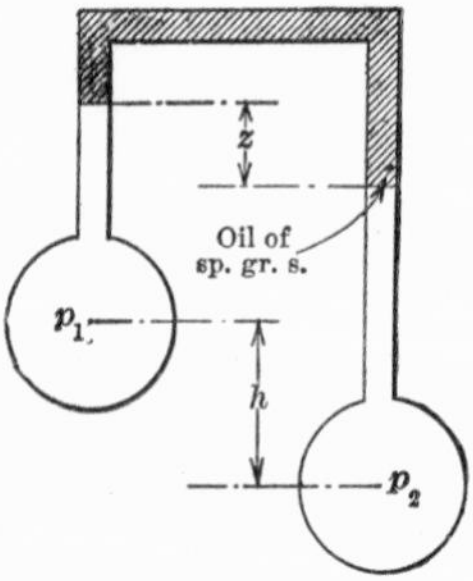

FIG. 562.

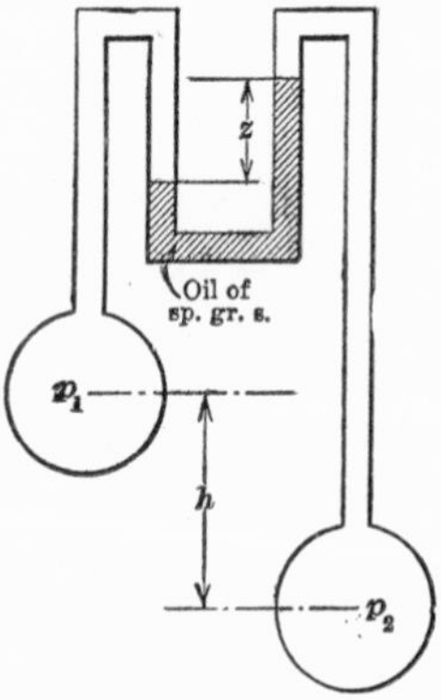

FIG. 563.

562 $$p_1 - p_2 = 0.434\,[z(1 - s) - h] \text{ pounds per sq. in.}$$

(*b*) When the oil has a specific gravity greater than 1. (See Fig. 563.)

563 $p_1 - p_2 = 0.434\,[z\,(s - 1) - h]$ pounds per sq. in.

HYDRODYNAMICS

Conservation of Energy. In steady flow the total energy at any section is equal to the total energy at any further section in the direction of flow, plus the loss of energy due to friction in the distance between the two sections.

Pressure Energy (W_{pr}) per pound of water weighing **w** pounds per cubic foot due to a pressure of **p** pounds per square foot.

564 $$W_{pr} = \frac{p}{w} = 0.016\,p \text{ foot-pounds.}$$

Potential Energy (W_p) per pound of water due to a height of **z** feet of the center of gravity of the section above the datum level.

565 $$W_p = z \text{ foot-pounds.}$$

Kinetic Energy (W) per pound of water due to a velocity of **v** feet per second, the acceleration due to gravity being **g** feet per second per second.

566 $$W = \frac{v^2}{2g} \text{ foot-pounds.}$$

Bernouilli's Theorem. In steady flow the total head (pressure head plus potential head plus velocity head) at any section is equal to the total head at any further section in the direction of flow, plus the lost head due to friction between these two sections

567 $$\frac{p_1}{w} + z_1 + \frac{v_1^2}{2g} = \frac{p_2}{w} + z_2 + \frac{v_2^2}{2g} + \text{lost head.}$$

NOTE. This is also known as the conservation of energy equation.

Power (P) available at a section of **A** square feet area in a moving stream of water, due to a pressure of **p** pounds per square foot, a velocity of **v** feet per second and a height of **z** feet above the datum level.

568 $$P = wvA\left(\frac{p}{w} + z + \frac{v^2}{2g}\right) \text{ foot-pounds per sec.}$$

Horse-power (H.P.) available at any section of a stream.

569 $$\text{H.P.} = \frac{wvA\left(\frac{p}{w} + z + \frac{v^2}{2g}\right)}{550} \text{ horse-power.}$$

Power (P) available in a jet **A** square feet in area discharging with a velocity of **v** feet per second.

570 $$P = \frac{wv^3A}{2g} \text{ foot-pounds per second.}$$

ORIFICES

Theoretical velocity of discharge (v) through an orifice due to a head of **h** feet over the center of gravity of the orifice.

571 $$v = \sqrt{2gh} \text{ feet per second.}$$

Actual velocity of discharge (v) if the coefficient of velocity for the orifice is c_v.

572 $$v = c_v\sqrt{2gh} \text{ feet per second.}$$

Quantity of discharge (Q) through an orifice **A** square feet in area due to a head of **h** feet over the center of gravity of the orifice if the coefficient of discharge is **c**.

NOTE. Orifice coefficients are given on page 302.

573 $$Q = cA\sqrt{2gh} \text{ cubic feet per second.}$$

Coefficient of discharge (c) in terms of the coefficient of velocity c_v and the coefficient of contraction c_c.

574 $$c = c_vc_c.$$

Quantity of discharge (Q) through a submerged orifice **A** square feet in area due to a head of h_1 feet on one side of the orifice and h_2 feet on the other side, the coefficient of discharge being **c**.

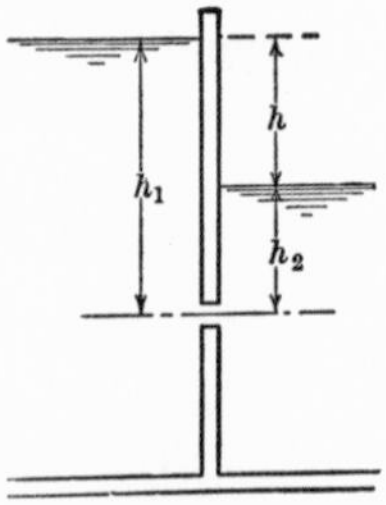

FIG. 575.

575 $$Q = cA\sqrt{2g(h_1 - h_2)} \text{ cubic feet per second.}$$

NOTE. If $h = h_1 - h_2$, $Q = cA\sqrt{2gh}$ cubic feet per sec.

Quantity of discharge (Q) through a large rectangular orifice **b** feet in width with a small head of h_1 feet above the top of the orifice and a head of h_2 feet above the bottom of the orifice, the coefficient of discharge being **c**.

576 $$Q = \frac{2}{3} cb \sqrt{2g}\,(h_2^{\frac{3}{2}} - h_1^{\frac{3}{2}}) \text{ cubic feet per second.}$$

Velocity of discharge (v) **and quantity of discharge** (Q) through an orifice A_1 square feet in area, considering the velocity of approach in the approach channel of A_2 square feet area, due to a pressure head of **h** feet, if the coefficient of discharge is **c** and the coefficient of velocity is c_v.

577 $$v = c_v \sqrt{\frac{2gh}{1 - \left(\frac{A_1 c}{A_2}\right)^2}} \text{ feet per second.}$$

578 $$Q = A_1 \sqrt{\frac{2gh}{\left(\frac{1}{c}\right)^2 - \left(\frac{A_1}{A_2}\right)^2}} \text{ cubic feet per second.}$$

Time (t) to lower the water in a vessel of A_1 square feet constant cross-section through an orifice A_2 square feet in area, from an original head of h_1 feet over the orifice to a final head of h_2 feet.

579 $$t = \frac{2A_1}{cA_2\sqrt{2g}}(\sqrt{h_1} - \sqrt{h_2}) \text{ seconds.}$$

NOTE. In general, problems involving the time required to lower the water in a reservoir of any cross-section may be solved thus: Let **A** = cross-sectional area of the reservoir (this may be a variable in terms of **h**), Q = the rate of discharge through an orifice (or weir) as given by the ordinary formula and h_1 and h_2 the initial and final heads.

$$t = \int_{h_2}^{h_1} \frac{A\,dh}{Q} \text{ seconds.}$$

For a suppressed weir this would be

$$t = \int_{h_2}^{h_1} \frac{A\,dh}{3.33\,bh^{\frac{3}{2}}} \text{ seconds.}$$

Mean velocity of discharge (v_m) in lowering water in a vessel of constant cross-section, if the initial velocity of discharge is v_1 feet per second and the final velocity is v_2 feet per second.

580 $$v_m = \frac{v_1 + v_2}{2} \text{ feet per second.}$$

Constant head (h_m) which will produce the same mean velocity of discharge as is produced in lowering the water in a vessel of constant cross-section from an initial head of h_1 feet over the orifice to a final head of h_2 feet.

581 $$h_m = \left(\frac{\sqrt{h_1} + \sqrt{h_2}}{2}\right)^2 \text{ feet.}$$

WEIRS

Theoretical discharge (**Q**) over a rectangular **weir b** feet in width due to a head of **H** feet over the crest.

582 $$Q = \tfrac{2}{3}\, b \sqrt{2g}\, H^{\frac{3}{2}} \text{ cubic feet per second.}$$

NOTE. If the velocity head due to the velocity of approach **v** feet per second in the channel back of the weir is **h** feet: $Q = \frac{2}{3} b \sqrt{2g}\,[(H + h)^{\frac{3}{2}} - h^{\frac{3}{2}}]$ cubic feet per second. The actual discharge may be obtained by multiplying the theoretical discharge by a coefficient **c** which varies from 0.60 to 0.63 for contracted weirs and from 0.62 to 0.65 for suppressed weirs.

Francis Formula for discharge (**Q**) over a rectangular weir **b** feet in width due to a head of **H** feet over the crest.

For a suppressed weir.

583 $$Q = 3.33\, bH^{\frac{3}{2}} \text{ cubic feet per second.}$$

For a suppressed weir considering the velocity head **h** due to the velocity of approach.

584 $$Q = 3.33\, b\,[(H + h)^{\frac{3}{2}} - h^{\frac{3}{2}}] \text{ cubic feet per second.}$$

For a contracted weir.

585 $$Q = 3.33\,(b - 0.2\,H)\, H^{\frac{3}{2}} \text{ cubic feet per second.}$$

For a contracted weir considering the velocity head **h** due to the velocity of approach.

586 $$Q = 3.33\,(b - 0.2\,H)\,[(H + h)^{\frac{3}{2}} - h^{\frac{3}{2}}] \text{ cubic feet per second.}$$

NOTE. In case contraction occurs on only one side of the weir the term for width becomes $(b - 0.1\,H)$.

Bazin Formula for discharge (**Q**) over a rectangular suppressed weir **b** feet in width due to a lead of **H** feet over the crest and a height **p** feet of the crest above the bottom of the channel.

587 $$Q = \left[0.405 + \frac{0.00984}{H}\right]\left[1 + 0.55\left(\frac{H}{p + H}\right)^2\right] b \sqrt{2g}\, H^{\frac{3}{2}}$$

cubic feet per second.

Fteley and Stearns' Formula for discharge (Q) over a suppressed weir **b** feet in width due to a head of **H** feet over the crest.

588 $$Q = 3.31\,b\,H^{\frac{3}{2}} + 0.007\,b \text{ cubic feet per second.}$$

NOTE. Considering the velocity head **h** due to the velocity of approach. $Q = 3.31\,b\,(H + 1.5\,h)^{\frac{3}{2}} + 0.007\,b$ cubic feet per second.

Hamilton Smith Formula for discharge (Q) over a rectangular weir **b** feet in width due to a head of **H** feet over the crest, if the coefficient of discharge is **c**.

NOTE. A table on page **303** gives values of **c** for both suppressed and contracted weirs.

For a contracted or a suppressed weir (**c** to be properly chosen for the type of weir).

589 $$Q = c\,\tfrac{2}{3}\,b\,\sqrt{2\,g}\,H^{\frac{3}{2}} \text{ cubic feet per second.}$$

NOTE. For a suppressed weir considering the velocity head **h** due to the velocity of approach, $Q = c\frac{2}{3}\,b\,\sqrt{2\,g}\,(H + \frac{4}{3}\,h)^{\frac{3}{2}}$ cubic feet per second. For a contracted weir considering the velocity head **h** due to the velocity of approach, $Q = c\frac{2}{3}\,b\,\sqrt{2\,g}\,(H + 1.4\,h)^{\frac{3}{2}}$ cubic feet per second.

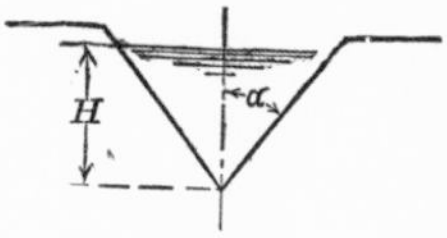

FIG. 590.

Discharge (Q) over a triangular weir, with the sides making an angle of α degrees with the vertical, due to a head of **H** feet over the crest.

590 $$Q = c\,\tfrac{8}{15}\,\tan\alpha\,\sqrt{2\,g}\,H^{\frac{5}{2}} \text{ cubic feet per second.}$$

NOTE. If $\alpha = 45°$ (90° notch), $Q = 2.53\,H^{\frac{5}{2}}$ cubic feet per second.

Discharge (Q) over a trapezoidal weir. Compute by adding the discharge over a suppressed weir **b** feet in width to that over

$H^{5/2} = \sqrt{H^5}$

$(H^5)^{1/2} = H^{\frac{5}{2}}$

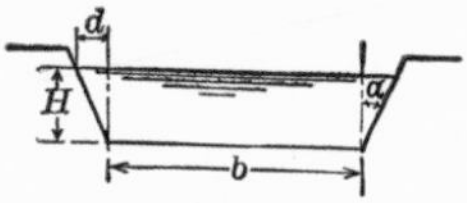

FIG. 591a.

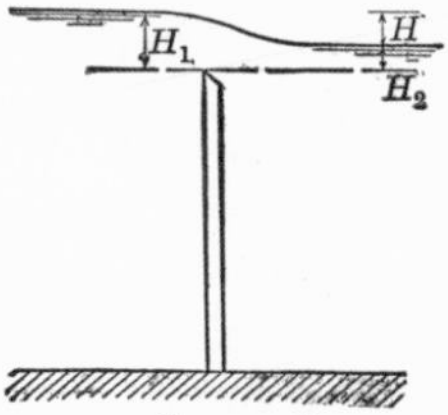

FIG. 591b.

the triangular weir formed by the sloping sides. A general solution is obtained by summing up the discharges through a series of differential orifices, giving: $Q = c \int_0^H b' \sqrt{2g}\, h^{\frac{1}{2}}\, dh$ cubic feet per second. (**b′**, the width of the differential orifice, varies with **h**.)

NOTE. In the Cippoletti weir **d** is made equal to $\frac{H}{4}$ and the formula becomes $Q = 3.37\, bH^{\frac{3}{2}}$ cubic feet per second.

Discharge (Q) over a submerged weir b feet in width due to a head of H_1 feet over the crest on the upstream side and a head of H_2 feet on the downstream side.

591 $$Q = \tfrac{2}{3} cb \sqrt{2g}\,(H_1 - H_2)^{\frac{3}{2}} + cbH_2 \sqrt{2g\,(H_1 - H_2)}$$ cubic feet per second.

Time (t) to lower the surface of a prismatic reservoir of **A** square feet superficial area by means of a suppressed weir **b** feet in width, from an initial head of H_1 feet over the crest to a final head of H_2 feet over the crest.

592 $$t = 0.6 \frac{A}{b} \left(\frac{1}{\sqrt{H_2}} - \frac{1}{\sqrt{H_1}} \right) \text{ seconds.}$$

NOTE. This value is based on formula 583.

VENTURI METER

Quantity of water (Q) flowing through a Venturi Meter with an area of A_1 square feet in the main pipe and an area A_2 square feet in the throat and a pressure head of h_1 feet in the main pipe and of h_2 feet in the throat, if the coefficient of the meter is **c**.

593 $$Q = c \frac{A_1 A_2}{\sqrt{A_1^2 - A_2^2}} \sqrt{2g\,(h_1 - h_2)} \text{ cubic feet per second.}$$

FLOW THROUGH PIPES*

Solution by Bernouilli's Theorem. If the total head at any point in the pipe-system (preferably at the source) is known, the velocity of discharge at the end can be computed by applying Bernouilli's Theorem between these two points, provided the losses of head can be determined. Following are expressions for the important losses of head which may occur.

* These formulas apply to pipes flowing full under pressure, otherwise the pipe may be treated as an open channel.

Friction loss (h_f) in a pipe of **d** feet internal diameter and **l** feet length with a velocity of **v** feet per second and a friction factor **f**.

594
$$h_f = f \frac{l}{d} \frac{v^2}{2g} \text{ feet.}$$

NOTE. A mean value for the friction factor for clean cast-iron pipes is 0.02. A table on page 304 gives values for various sizes of pipes and different velocities. In long pipe-lines it is accurate enough to consider that the total head **H** is used up in overcoming friction in the pipe. Then $H = f \frac{l}{d} \frac{v^2}{2g}$ feet and $Q = Av$ cubic feet per second.

Loss at entrance to a pipe (h_e) if the velocity of flow in the pipe is **v** feet per second.

595
$$h_e = 0.5 \frac{v^2}{2g} \text{ feet.}$$

Loss due to sudden expansion (h_x) where one pipe is abruptly followed by a second pipe of larger diameter, if the velocity in the smaller pipe is v_1 feet per second and that in the larger pipe is v_2 feet per second.

596
$$h_x = \frac{(v_1 - v_2)^2}{2g} \text{ feet.}$$

Loss due to sudden contraction (h_c) where one pipe is abruptly followed by a second pipe of smaller diameter, if the velocity in the smaller pipe is **v** feet per second and c_c is a coefficient.

597
$$h_c = c_c \frac{v^2}{2g} \text{ feet.}$$

NOTE. Values of c_c:

Ratio of areas	0.1	0.2	0.3	0.4	0.5	0.8	1.00
c_c	0.362	0.338	0.308	0.267	0.221	0.053	0.00

Loss due to bends (h_b).

598
$$h_b = c_b \frac{v^2}{2g} \text{ feet.}$$

NOTE. Values of c_b: (**d** is the diameter of the pipe in feet and **r** is the radius of the bend in feet).

$\frac{d}{r}$	0.2	0.4	0.6	0.8	1.00
c_b	0.131	0.138	0.158	0.206	0.294

Nozzle loss (h_n) if the velocity of discharge is **v** feet per second and the velocity coefficient of the nozzle is c_v.

599
$$h_n = \left(\frac{1}{c_v^2} - 1\right) \frac{v^2}{2g} \text{ feet.}$$

Quantity of discharge (**Q**) in a pipe **A** square feet in **area** where the velocity is **v** feet per second.

600 $$Q = Av \text{ cubic feet per second.}$$

Diameter of pipe (**d**) required to deliver **Q** cubic feet of water per second under a head of **h** feet if the friction factor is **f**.

601 $$d = \sqrt[5]{\frac{f\,l}{2\,gh}\left(\frac{4\,Q}{\pi}\right)^2} \text{ feet.}$$

Hydraulic Gradient is a line the ordinates to which show the pressure heads at the different points in the pipe system. It

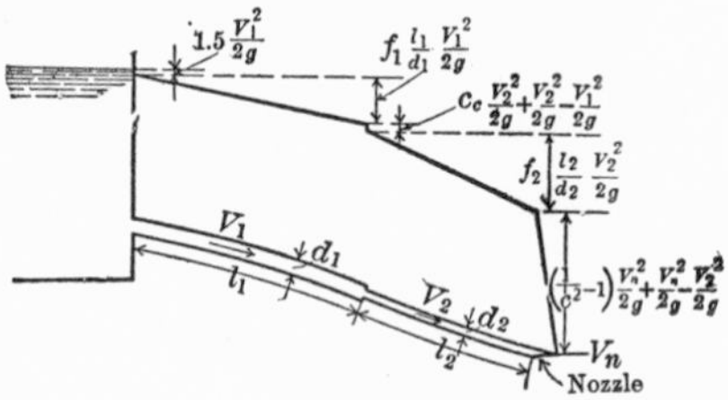

FIG. 602.

may also be defined as the line to which water would rise in piezometer tubes placed at intervals along the pipe.

Solution by Chezy Formula. Quantity (**Q**) and velocity (**v**) of flow through a pipe when the hydraulic radius is **r** feet, and the slope of the hydraulic gradient is **s** and the coefficient for the Chezy Formula is **c**. (See page 304.)

602 $$v = c\sqrt{rs} \text{ feet per second.}$$

603 $$Q = Av \text{ cubic feet per second.}$$

NOTE. **r** equals the area in square feet divided by the wetted perimeter in feet and **s** equals the head in feet divided by the length of the pipe in feet or the slope of the hydraulic gradient. Try **c** = 124, compute **v**, look up new value of **c**, etc.

FLOW IN OPEN CHANNELS

Chezy Formula for quantity (**Q**) **and velocity** (**v**) of flow in an open channel **A** square feet in sectional area with **p** feet wetted perimeter, **r** feet hydraulic radius, **h** feet drop of water surface **in distance l** feet and slope **s** of water surface. $\left(s = \frac{h}{l}\right)$

604 $$r = \frac{A}{p} \text{ feet.} \quad v = c\sqrt{rs} \text{ feet per second.}$$

$$Q = Av \text{ cubic feet per second.}$$

NOTE. c is the coefficient and is usually found either by the Kutter Formula or by the Bazin Formula.

Kutter Formula.

605 $$v = c\sqrt{rs} \text{ feet per second,}$$

where
$$c = \frac{41.6 + \frac{1.811}{n} + \frac{0.00281}{s}}{1 + \left(41.6 + \frac{0.00281}{s}\right)\frac{n}{\sqrt{r}}}.$$

NOTE. Specific values of c are given in a table on page 305. n is the coefficient of roughness and has the following values:

Channel Lining	*n*
Smooth wooden flume	0.009
Neat cement and glazed pipe	0.010
Unplaned timber	0.012
Ashlar and brick work	0.013
Rubble masonry	0.017
Very firm gravel	0.020
Earth free from stone and weeds	0.025
Earth with stone and weeds	0.030
Earth in bad condition	0.035

Bazin Formula.

606 $$v = c\sqrt{rs} \text{ feet per second,}$$

where
$$c = \frac{87}{0.552 + \frac{m}{\sqrt{r}}}.$$

NOTE. Specific values of c are given in a table on page 305. m is the coefficient of roughness and has the following values:

Channel Lining	*m*
Smooth cement or matched boards	0.06
Planks and bricks	0.16
Masonry	0.46
Regular earth beds	0.85
Canals in good order	1.30
Canals in bad order	1.75

DYNAMIC ACTION OF JETS

Reaction of a Jet (P) A square feet in area, the head on the orifice being **h** feet and the weight of the liquid **w** pounds per cubic foot.

607 $$P = 2\ Awh \text{ pounds (theoretical).}$$

NOTE. P equals about 1.2 Awh pounds (actual).

Power of a Jet (P) discharging with a velocity of **v** feet per second.

608 $$P = \frac{wv^3A}{2g} \text{ foot-pounds per second.}$$

NOTE. If h_v is the velocity head and Q (= Av) the quantity of flow in cubic feet per second, $W = wQh_v$ foot-pounds.

Force (F) exerted on a fixed curve vane by a jet **A** square feet in area and **v** feet per second velocity.

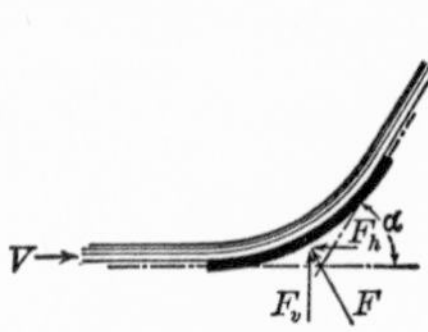

FIG. 609.

609 $$F = \frac{Awv^2}{g}\sqrt{2(1 - \cos\alpha)} \text{ pounds.}$$

Vertical component of force (F_v) exerted by a jet on a fixed curved vane.

610 $$F_v = \frac{Awv^2}{g}\sin\alpha \text{ pounds.}$$

Horizontal component of force (F_h) exerted by a jet on a fixed curved vane.

611 $$F_h = \frac{Awv^2}{g}(1 - \cos\alpha) \text{ pounds.}$$

Force (F) exerted by a jet on a flat fixed plate perpendicular to the jet.

612 $$F = \frac{Awv^2}{g} \text{ pounds.}$$

Force (F) exerted on a moving curved vane by a jet **A** square feet in area with a velocity of **v** feet per second, the vane moving in the direction of the flow of the jet with a velocity of v_0 feet per second.

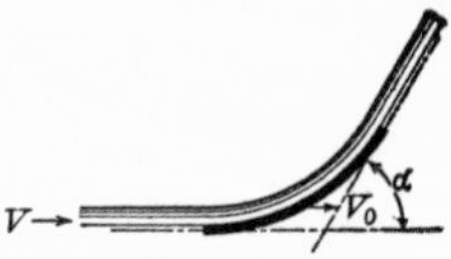

FIG. 613.

613 $$F = \frac{wA(v - v_0)^2}{g}\sqrt{2(1 - \cos\alpha)} \text{ pounds.}$$

Vertical component of force (F_v) exerted by a jet on a moving curved vane.

614 $$F_v = \frac{wA(v - v_0)^2}{g} \sin \alpha \text{ pounds.}$$

Horizontal component of force (F_h) exerted by a jet on a moving curved vane.

615 $$F_h = \frac{wA(v - v_0)^2}{g} (1 - \cos \alpha) \text{ pounds.}$$

NOTE. If there is a series of vanes. $F_h = \frac{wAv}{g}(v - v_0)(1 - \cos \alpha)$ pounds.

$$F_v = \frac{wAv}{g}(v - v_0) \sin \alpha \text{ pounds.}$$

Power (P) exerted on a (moving) vane.

616 $$P_h = F_h v_0 \text{ foot-pounds per second.}$$

NOTE. Maximum efficiency for a series of vanes occurs where $v_0 = \frac{v}{2}$ if there is no friction loss; then, $P = \frac{wAv^3}{4g}(1 - \cos \alpha)$ foot-pounds per second.

HEAT

In the following formulas, when specific units are not stated, any units may be used provided identical properties are expressed in the same units. Absolute pressure is indicated by **p**, total volume by **V**, specific volume by **v**, absolute temperature by **T**, and thermometer temperature by **t**. In all formulas containing indicated units, the temperature is measured in Fahrenheit degrees.

Measurement of Heat. Heat is the transient form of energy transmitted from one body to another when the two bodies are not at the same temperature. The ratio of the quantity of heat required to increase the temperature of a body in a specified state to that required to increase the temperature of an equal mass of water through the same temperature is called the specific heat of the body.

The unit of energy commonly used in the measurement of heat is the British thermal unit (B.t.u.) which equals 2.930×10^{-4} kilowatt-hours or substantially $\frac{1}{180}$th of the quantity of heat required to raise one pound of water from the ice-point to the steam-point at standard atmospheric pressure.

The quantity of heat (**Q**) added to **M** pounds of a substance having a constant specific heat (**c**) and causing the temperature to increase from t_1 to t_2 is

617 $$Q = Mkc\,(t_2 - t_1) \text{ units.}$$

NOTE. See page 314 for definite values of **c**. The constant **k** depends on the units of measurement, as follows:

Q	M	$t_2 - t_1$	k
gram-calories	grams	Cent.	1
kilogram-calories	kilograms	Cent.	1
British thermal units	pounds	Cent.	1.8
British thermal units	pounds	Fahr.	1
joules	grams	Cent.	4.18
joules	pounds	Fahr.	1054
kilowatt-hours	kilograms	Cent.	1.16×10^{-3}
kilowatt-hours	pounds	Fahr.	2.93×10^{-4}

If the specific heat varies with the temperature (as it usually does) according to the relation

$$c = a + bt + ft^2$$

where **a**, **b** and **f** are constants which are determined by experiment, then

618 $$Q = Mk\left[a\,(t_2 - t_1) + \frac{b}{2}\,(t_2^2 - t_1^2) + \frac{f}{3}\,(t_2^3 - t_1^3)\right] \text{units.}$$

The quantity of heat added to a substance can also be determined by applying the First Law of Thermodynamics which states that energy can be neither created nor destroyed. This principle may be expressed in an equation as follows:

619 $$Q = (W + \Delta E)/778 \text{ B.t.u.}$$

where **Q** is the heat interchange in B.t.u., **W** is the external work done in foot-pounds, and **ΔE** is the change in the internal energy in foot-pounds.

NOTE. Although the accepted value for the mechanical equivalent of heat is 778.26 foot-pounds, it is sufficient to use 778 foot-pounds in the solution of most engineering problems.

Influence of Heat on the Length of a Solid Body. If heat is applied to a solid body which has a length l_0 at a temperature of **0** degrees Centigrade, the length l_t at a temperature of **t** degrees Centigrade is

620 $$l_t = l_0\,(1 + at)$$

NOTE. See page 315 for definite values of **a** (the Centigrade mean coefficient of linear expansion). The mean coefficient of cubical expansion equals 3 **a**, approximately. When the temperature is expressed in Fahrenheit degrees, **620** becomes $l_t = l_{32}\left[1 + a\,\frac{(t - 32)}{1.8}\right]$.

Measurement of External Work. External work is the result of a force acting through a distance to overcome external resistances. For mechanical processes this work may be expressed by

621 $$W = 144\int_{V_1}^{V_2} p\, dV \text{ foot-pounds}$$

where **p** is the intensity of pressure in pounds absolute per square inch and **V** is the total volume expressed in cubic feet.

If the relation which exists between pressure and volume is known, the external work can be determined, as follows:

622 $p = \text{constant}$ $\quad W = 144\, p\,(V_2 - V_1)$ foot-pounds

623 $V = \text{constant}$ $\quad W = 0$ foot-pounds

624 $pV = \text{constant}$ $\quad W = 144\, p_1V_1 \ln \frac{V_2}{V_1}$ foot-pounds

625 $pV^n = \text{constant}$ $\quad W = \frac{144\,(p_1V_1 - p_2V_2)}{n - 1}$ foot-pounds

Measurement of the Change in the Internal Energy. Internal energy is the form of energy which is stored up within the body. For engineering purposes the internal energy of a substance may be assumed to include the internal vibration energy (due to change in temperature) and internal potential energy (due to change in volume). The internal energy is a function of the temperature and volume and is a property of the substance. Although the absolute amount of internal energy in a substance cannot be measured, the change in the internal energy is independent of the path and can be determined by means of the First Law of Thermodynamics.

Measurement of the Change in Entropy. The Second Law of Thermodynamics states that heat will not, of its own accord, flow from a cold to a relatively hotter body. It can be shown by means of this law that heat can never be completely converted into mechanical energy, not even under ideal conditions. The fractional part of the heat which is wasted is termed the unavailable energy. In order to measure this unavailable energy the term "change in entropy" (Δs) has been introduced and is now used extensively in order to simplify the solution of certain thermodynamic problems. The change in entropy for reversible processes between the absolute temperatures T_1 and T_2 may be determined by:

626 $$\Delta s = M\int \frac{dQ}{T} = M\int_{T_1}^{T_2} \frac{c\,dT}{T} \text{ units of entropy.}$$

If the specific heat remains constant, then

627 $$\Delta s = Mc\int_{T_1}^{T_2} \frac{dT}{T} = Mc \ln \frac{T_2}{T_1} \text{ units of entropy.}$$

If the specific heat varies with the temperature according to the relation

$$c = a + bt + ft^2$$

then

628 $$\Delta s = M\int_{T_1}^{T_2} \frac{c\,dT}{T} = M\left[(a - 460\,b + 211{,}600\,f)\ln\frac{T_2}{T_1} + (b - 920\,f)(T_2 - T_1) + \frac{f}{2}(T_2^2 - T_1^2)\right] \text{ units of entropy.}$$

If the temperature remains constant, then

629 $$\Delta s = M\left(\frac{Q}{T}\right) \text{ units of entropy.}$$

If no heat is added or rejected to or from the substance and the expansion or compression is frictionless, then

630 $$\Delta s = 0 \text{ units of entropy.}$$

Steady Flow. When the same quantity of working fluid progresses continuously and uniformly in one direction, the process is termed the steady flow condition. If the conservation of energy principle is applied to such a process between two sections 1 and 2, the equation for each pound of working fluid in its simplest form for engineering applications is

631 $$E_1 + 144\,p_1v_1 + \frac{U_1^2}{64.4} = E_2 + 144\,p_2v_2 + \frac{U_2^2}{64.4} + W + 778\,Q_{loss} \text{ foot-pounds.}$$

where **E** is the internal energy in foot-pounds, 144 **pv** is the flow work in foot-pounds, $\frac{U^2}{64.4}$ is the kinetic energy in foot-pounds, **W** is the external work done in foot-pounds, and Q_{loss} is the heat lost to or from the surroundings in B.t.u.

NOTE. In the expression $\frac{U^2}{64.4}$, **U** is the velocity of the fluid flowing in feet per second and 64.4 is equal to 2 g (where **g** is the acceleration due to gravity and is assumed equal 32.2 feet per second per second).

Enthalpy. The combination **E** + 144 **pv** occurs so often that the special term "enthalpy" has been universally adopted. This property is represented by the symbol **h** when the combination is expressed in B.t.u.; that is, (**E** + 144 **pv**)/778.

PERFECT GASES

Characteristic Equation. The relation between pressure, volume and absolute temperature can be determined by combining the two experimental gas laws, those of Boyle and Charles, or Gay-Lussac. This relation for any two conditions 1 and 2 may be expressed by

632 $$\frac{p_1V_1}{T_1} = \frac{p_2V_2}{T_2} = MR.$$

Values of the gas constant (R) for various gases are given, as follows: air, 53.3; carbon dioxide, 34.9; carbon monoxide, 55.1; helium, 386; hydrogen, 767; nitrogen, 55.1; oxygen, 48.3.

Fundamental Equations. The dual relation between pressure and volume, temperature and pressure, or temperature and volume for many changes met in practice may be represented by exponential equations. These equations are

633 $$pV^n = \text{constant} \qquad Tp^{\frac{1-n}{n}} = \text{constant} \qquad TV^{n-1} = \text{constant}.$$

The exponent (n) may be determined from the following relations

634 $$n = \frac{\log p_1 - \log p_2}{\log V_2 - \log V_1},$$

or

635 $$n = \frac{c - c_p}{c - c_v} = \frac{c - kc_v}{c - c_v}$$

where **c** is the specific heat for a polytropic change (pV^n = constant), c_p is the specific heat at constant pressure and c_v is the specific heat at constant volume and **k** is $\frac{c_p}{c_v}$.

NOTE. Another useful relation between c_p and c_v is $778\,(c_p - c_v) = R$.

The change in the internal energy (ΔE) is independent of the path and may be determined by the following equations:

636 $$\Delta E = 778\,Mc_v\,(T_2 - T_1) \text{ foot-pounds}$$

637 $$= M\,(778\,c_p - R)\,(T_2 - T_1) \text{ foot-pounds}$$

638 $$= \frac{MR\,(T_2 - T_1)}{k - 1} \text{ foot-pounds}$$

639 $$= \frac{144\,(p_2V_2 - p_1V_1)}{k - 1} \text{ foot-pounds.}$$

The change of entropy (Δs) is also independent of the path and may be determined by the following equations:

640 $$\Delta s = M\left[c_v \ln\frac{T_2}{T_1} + (c_p - c_v)\ln\frac{V_2}{V_1}\right] \text{ units of entropy}$$

641 $$= M\left[c_p \ln\frac{T_2}{T_1} - (c_p - c_v)\ln\frac{P_2}{P_1}\right] \text{ units of entropy}$$

642 $$= M\left[c_v \ln\frac{P_2}{P_1} + c_p \ln\frac{V_2}{V_1}\right] \text{ units of entropy.}$$

The heat interchange and the external work done **are** dependent on the path or the character of the change which takes place between two conditions 1 and 2. Of the innumerable possible changes, the only ones of importance to the engineer are: constant pressure changes, during which the pressure remains constant; constant volume changes, during which the volume remains constant; isothermal changes, during which the temperature remains constant; adiabatic changes, during which no heat is received from or rejected to external bodies; polytropic changes, during which the heat supplied to or withdrawn from the gas by external bodies is directly proportional to the change in temperature.

A summary of the convenient formulas for these paths is given in Table I on page 162.

LIQUIDS AND VAPORS

Physical Conditions. A liquid and its vapor may exist in either of the following six conditions:

(1) Compressed or subcooled liquid is a liquid at a temperature less than the saturation temperature corresponding to the pressure.

(2) Saturated liquid is a liquid which under its given pressure will begin to vaporize when heat is added to it.

(3) Saturated vapor is a vapor which under its given pressure will start changing to the liquid form when heat is removed.

(4) Wet vapor is a physical mixture of saturated liquid and saturated vapor. In each pound of mixture, the fractional part by weight which is saturated vapor is designated by the symbol **x**.

(5) Superheated vapor is a vapor the temperature of which is greater than the saturation temperature corresponding to the pressure imposed on it.

TABLE I

Summary of Convenient Formulas for Perfect Gases between Conditions 1 and 2

Path	pV-Relation	Heat Interchange (B.t.u.)	External Work Done (ft.-lbs.)	Change of Entropy (units of entropy)	Specific Heat
Constant Pressure	$p_1V_1^0 = p_2V_2^0$ $p = \text{const.}$	$Mc_p\,(T_2 - T_1)$ $\frac{MR}{778}\left(\frac{k}{k-1}\right)(T_2 - T_1)$	$144\,p\,(V_2 - V_1)$ $MR\,(T_2 - T_1)$	$Mc_p \ln \frac{T_2}{T_1}$ $Mc_p \ln \frac{V_2}{V_1}$	c_p
Constant Volume	$p_1V_1^\infty = p_2V_2^\infty$ $V = \text{const.}$	$Mc_v\,(T_2 - T_1)$ $\frac{MR}{778}\left(\frac{1}{k-1}\right)(T_2 - T_1)$	0	$Mc_v \ln \frac{T_2}{T_1}$ $Mc_v \ln \frac{p_2}{p_1}$	c_v
Isothermal or Isodynamic	$p_1V_1 = p_2V_2$ $pV = \text{const.}$	$MT\,(s_2 - s_1)$ $0.1851\,p_1V_1 \ln \frac{V_2}{V_1}$ $\frac{MRT}{778} \ln \frac{V_2}{V_1}$	$144\,p_1V_1 \ln \frac{V_2}{V_1}$ $MRT \ln \frac{V_2}{V_1}$	$\frac{Q}{T}$	∞
Reversible Adiabatic or Isentropic	$p_1V_1^k = p_2V_2^k$ $pV^k = \text{const.}$	0	$\frac{144\,(p_1V_1 - p_2V_2)}{k-1}$ $\frac{MR\,(T_1 - T_2)}{k-1}$	0	0
Polytropic	$p_1V_1^n = p_2V_2^n$ $pV^n = \text{const.}$	$Mc_v\left(\frac{n-k}{n-1}\right)(T_2 - T_1)$	$\frac{144\,(P_1V_1 - p_2V_2)}{n-1}$ $\frac{MR\,(T_1 - T_2)}{n-1}$	$Mc_v\left(\frac{n-k}{n-1}\right) \ln \frac{T_2}{T_1}$	$c_v\left(\frac{n-k}{n-1}\right)$

(6) Supersaturated vapor is vapor the temperature and specific volume of which are less than those corresponding to the saturated condition for the pressure imposed on it. This condition occurs only during rapid expansion as in nozzles and is of special importance to turbine designers.

Properties of Liquids and Vapors. The various properties of the more important liquids and vapors are available in tabulated form. In general the properties given are pressure (**p**), temperature (**t**), specific volume (**v**), enthalpy (**h**), internal energy (**E**) and entropy (**s**).

Properties of Steam. The properties of saturated liquid, saturated vapor and superheated vapor are given on pages 306 to 313.

The properties of compressed liquid can be computed from table 4 in Keenan and Keyes "Thermodynamic Properties of Steam."

The properties of wet vapor can be computed from the saturated properties as follows:

643 $$v = v_f + xv_{fg}, \quad h = h_f + xh_{fg}, \quad s = s_f + xs_{fg}$$
$$E = u_f + xu_{fg} = h - 0.1851\,pv.$$

Thermodynamic Processes. The heat interchange (**Q**), work done (**W**), change in the internal energy (**ΔE**) and change of entropy (**Δs**) for processes most frequently met in engineering practice are given in Table II on page 164.

FLOW OF GASES AND VAPORS THROUGH NOZZLES AND ORIFICES

Steady Flow Equation. Equation 631 for the steady flow of the working fluid when applied to nozzles and orifices, provided there is no loss or gain of heat, no friction, and no external work is done, reduces to

644 $$E_1 + 144\,p_1v_1 + \frac{U_1^2}{64.4} = E_2 + 144\,p_2v_2 + \frac{U_2^2}{64.4}\ \text{foot-pounds.}$$

Velocity. Since **h** may be substituted for **E + 144 pv** divided by 778, the change in kinetic energy is

645 $$\frac{U_2^2 - U_1^2}{64.4} = 778\,(h_1 - h_2)\ \text{foot-pounds.}$$

If the initial velocity U_1 is small, it may be neglected giving

646 $$\frac{U_2^2}{64.4} = 778\,(h_1 - h_2)\ \text{foot-pounds, or}$$

TABLE II

Summary of Convenient Formulas for Steam between States 1 and 2

Path	Heat Interchange (B.t.u.)	Work Done (B.t.u.)	Change in the Internal Energy (B.t.u.)	Change of Entropy (units of entropy)
Constant Pressure $p = \text{const.}$	$M(h_2 - h_1)$	$0.1851\, p(V_2 - V_1)$	$M(E_2 - E_1)$	$M(s_2 - s_1)$
Constant Volume $V = \text{const.}$	$M(E_2 - E_1)$	0	$M(E_2 - E_1)$	$M(s_2 - s_1)$
Reversible Adiabatic $s = \text{const.}$	0	$M(E_1 - E_2)$	$M(E_2 - E_1)$	0
Isothermal $T = \text{const.}$	$MT(s_2 - s_1)$	$M[T(s_2 - s_1) - (E_2 - E_1)]$	$M(E_2 - E_1)$	$M(s_2 - s_1)$
Isodynamic $E = \text{const.}$	$\frac{M(T_1 + T_2)}{2}(s_2 - s_1)$	$\frac{M(T_1 + T_2)}{2}(s_2 - s_1)$	0	$M(s_2 - s_1)$
Exponential $pV^n = \text{const.}$	$W + \Delta E$	$\frac{0.1851\,(p_1V_1 - p_2V_2)}{n - 1}$	$M(E_2 - E_1)$	$M(s_2 - s_1)$

647 $$U_2 = 223.8\sqrt{h_1 - h_2} \text{ feet per second.}$$

Weight. The weight of working fluid flowing must be constant throughout the process. If **G** represents this weight in pounds per second, **a** the area in square feet, **U** the velocity in feet per second and **v** the specific volume in cubic feet per pound at any pressure, then

648 $$G = \frac{a_1U_1}{v_1} = \frac{a_2U_2}{v_2} \text{ pounds per second.}$$

This weight is a maximum when the absolute pressure P_t at the throat is

649 $$p_t = p_1\left(\frac{2}{n+1}\right)^{\frac{n}{n-1}} \text{ pounds per square inch.}$$

For dry saturated steam $n = 1.135$, then

650 $$p_t = 0.58\, p_1 \text{ pounds per square inch.}$$

For superheated steam $n = 1.30$, then

651 $$p_t = 0.55\, p_1 \text{ pounds per square inch.}$$

For diatomic (two atoms per molecule) gases $n = 1.40$, then

652 $$p_t = 0.53\, p_1 \text{ pounds per square inch.}$$

The pressure (p_t) that makes the weight of working fluid flowing a maximum is called Critical Pressure. When the final absolute pressure p_2 is less than the critical pressure, then the weight discharged remains constant and the term applied for such a condition is unretarded flow. When the final absolute pressure is greater than the critical pressure, then the weight discharged decreases as p_2 increases and the flow is said to be retarded.

FLOW OF GASES

Velocity. For a perfect gas c_pT may be substituted for **h** and **k** for **n**. Equation 647 may be modified to give

653 $$U_2 = 223.8\sqrt{c_p(T_1 - T_2)} \text{ feet per second}$$

or

654 $$U_2 = \sqrt{\frac{2\,gkp_1v_1}{k-1}\left[1 - \left(\frac{p_2}{p_1}\right)^{\frac{k-1}{k}}\right]} \text{ feet per second.}$$

Weight. Substituting equation 654 in equation 648 gives

655 $$G = a_2\sqrt{\frac{2\,gk}{k-1}\left(\frac{p_1}{v_1}\right)\left[\left(\frac{p_2}{p_1}\right)^{\frac{2}{k}} - \left(\frac{p_2}{p_1}\right)^{\frac{k+1}{k}}\right]} \text{ pounds per second.}$$

It is often more convenient to use the throat area a_t. The equations used must be classified depending on the type of flow, retarded or unretarded. All units in feet.

For retarded flow of diatomic gases ($k = 1.40$) equation 655 reduces to

656 $$G = \frac{15.03\,a_t p_1}{\sqrt{RT_1}}\sqrt{\left(\frac{p_2}{p_1}\right)^{1.43} - \left(\frac{p_2}{p_1}\right)^{1.71}} \text{ pounds per second}$$

and for air ($R = 53.34$)

657 $$G = \frac{2.056\,a_t p_1}{\sqrt{T_1}}\sqrt{\left(\frac{p_2}{p_1}\right)^{1.43} - \left(\frac{p_2}{p_1}\right)^{1.71}} \text{ pounds per second.}$$

Fliegner's empirical formula for (inch units) retarded flow of air is

658 $$G = \frac{1.06\,a_t}{\sqrt{T_1}}\sqrt{p_2\,(p_1 - p_2)} \text{ pounds per second.}$$

For unretarded flow of diatomic gases ($k = 1.40$) equation 655 reduces (inch or foot units) to

659 $$G = \frac{3.885\,a_t p_1}{\sqrt{RT_1}} \text{ pounds per second}$$

and for air ($R = 53.34$)

660 $$G = \frac{0.532\,a_t p_1}{\sqrt{T_1}} \text{ pounds per second.}$$

Fliegner's empirical formula for unretarded flow of air is

661 $$G = \frac{0.53\,a_t p_1}{\sqrt{T_1}} \text{ pounds per second.}$$

FLOW OF STEAM

Velocity. For steam, since the enthalpy (h) may be determined from the steam tables, the velocity is

662 $$U_2 = 223.8\,\sqrt{h_1 - h_2} \text{ feet per second.}$$

Weight. Although it is possible to deduce equations involving the exponent **n**, the most convenient form (foot units) is

663 $$G = \frac{a_2 U_2}{v_2} \text{ pounds per second.}$$

If the throat area a_t is used,

664 $$G = \frac{a_t U_t}{v_t} \text{ pounds per second.}$$

The throat pressure equals p_2 for retarded flow and equals the critical pressure ($0.58\ p_1$ for wet or dry saturated vapor and $0.55\ p_1$ for superheated vapor) for unretarded flow.

Rankine's empirical formula for retarded flow of dry saturated steam (inch units) is

665 $$G = 0.0292\ a_t \sqrt{p_2\ (p_1 - p_2)} \text{ pounds per second.}$$

Rankine's empirical formula for unretarded flow of dry saturated vapor (inch or foot units) is

666 $$G = \frac{a_t p_1}{70} \text{ pounds per second.}$$

Grashof's empirical formula for unretarded flow of dry saturated vapor (inch units) is

667 $$G = \frac{a_t p_1^{0.97}}{60} \text{ pounds per second.}$$

Equations 665, 666 and 667 should be divided by $\sqrt{x_1}$ for wet steam and $1 + 0.00065\ \Delta t$ for superheated steam.

NOTE. Δt is the number of Fahrenheit degrees of superheat.

STEAM CALORIMETERS

Throttling Calorimeter. Equation 631 for the steady flow of the working fluid when applied to a throttling calorimeter, provided there is no loss or gain of heat, no external work is done and the initial and final velocities are equal or negligible, reduces to

668 $$h_1 = h_2 \text{ B.t.u.}$$

This calorimeter is limited in its use since the steam in the calorimeter must be superheated. For reliable results at least 10° of superheat must be available. The quality (**x**) can be determined from

669 $$x = \frac{h_2 - h_{f_1}}{h_{fg_1}}.$$

NOTE. The percentage priming equals $(1 - x)\,100$.

Separating Calorimeters. This type of calorimeter is designed to separate the moisture from the steam. The drip (G_m) is collected and weighed. The saturated steam (G_s), for the same time interval, may be condensed and weighed or discharged through an orifice. The priming $(1 - x)$ can be determined from

670 $$1 - x = \frac{G_m}{G_s + G_m}.$$

STEAM ENGINES

Mean Effective Pressure. The indicator card shows the pressure distribution in the cylinder of a steam engine at every point of the working and exhaust strokes. The mean effective pressure (**M.E.P.** or **P**) in pounds per square inch is

671 $$\text{M.E.P.} = \frac{aS}{l} \text{ pounds per square inch}$$

where **a** is the area of the card in square inches, **l** is the length of the card in inches and **S** is the scale of the indicator spring in pounds per square inch per inch of height, or

672 $$\text{M.E.P.} = \left[p_1 \times C + p_1 (C + Cl) \ln \frac{R + Cl}{C + Cl} - p_2 (1 - K) - p_2 (K + Cl) \ln \frac{K + Cl}{Cl}\right] \times \text{D.F. pounds per square inch,}$$

where p_1 is the admission pressure in pounds absolute per square inch, p_2 is the exhaust pressure in pounds absolute per square inch, **C** is the per cent cut-off, **R** is the per cent release, **K** is the per cent compression and **Cl** is the per cent clearance. The diagram factor (**D.F.**) is usually between 0.85 and 0.95.

NOTE. The per cent events are expressed as decimal fractions.

In compound engines it is customary to neglect the clearance in the conventional card. If **E** represents the expansion ratio, then

673 $$\text{M.E.P.} = \left[p_1 \times \frac{1}{E} + p_1 \times \frac{1}{E} \ln E - p_2\right] \times \text{D.F. pounds per square inch.}$$

The diagram factor (**D.F.**) is usually between 0.65 and 0.75. The number of expansions are

674 $$E = \frac{\text{Volume of low-pressure cylinder}}{\text{Volume of high-pressure cylinder at cut-off}}$$

675 $$= \frac{D^2}{d^2 x C_h}$$

where **D** and **d** are the diameters of the low and high pressure cylinders, respectively, and C_h is the per cent cut-off in the high pressure cylinder expressed as a decimal fraction.

Indicated Horsepower. Having determined the mean effective pressure (**P**) from the indicator card of an engine with a stroke of **L** feet, a piston area exposed to the steam pressure of **A** square inches and a speed of **N** revolutions per minute, the indicated horsepower (**I.H.P**) is

676 $$\text{I.H.P.} = \frac{PLAN}{33{,}000} \text{ horsepower.}$$

NOTE. The indicated horsepower should be figured for the head end and crank end separately as the mean effective pressure and exposed piston area are not the same for both ends.

Brake Horsepower. The output of a steam engine may be determined by means of a friction brake. If **F** represents the net force in pounds acting at a distance **R** feet from the center of the shaft rotating at **N** revolutions per minute, then the brake horsepower (**B.H.P.**) is

677 $$\text{B.H.P.} = \frac{2\pi RNF}{33{,}000} \text{ horsepower.}$$

Mechanical Efficiency. The ratio of the brake horsepower (**B.H.P.**) to the indicated horsepower (**I.H.P.**) is termed mechanical efficiency (ϵ_M), hence

678 $$\epsilon_M = \frac{\text{B.H.P}}{\text{I.H.P.}}.$$

Carnot Efficiency. The Carnot cycle consists of the reception and rejection of heat energy, each at constant temperature together with frictionless adiabatic expansion and compression. The efficiency for such a cycle (ϵ_c) is

679 $$\epsilon_c = \frac{Q_1 - Q_2}{Q_1} = \frac{T_1 - T_2}{T_1}.$$

Rankine Efficiency. The Rankine cycle for a steam engine consists of the reception of heat energy at constant pressure, a frictionless adiabatic expansion to the exhaust pressure and the rejection of heat energy at constant pressure. The approximate efficiency for such a cycle (ϵ_R) is

680 $$\epsilon_R = \frac{Q_1 - Q_2}{Q_1} = \frac{h_1 - h_2}{h_1 - h_{f_2}} = \frac{2545}{w_R (h_1 - h_{f_2})}$$

where h_1 is the enthalpy at the initial conditions, h_2 is the enthalpy after isentropic expansion to the exhaust pressure, h_{f_2} is the enthalpy of the saturated liquid at the exhaust pressure and w_R is the ideal steam consumption in pounds per horsepower-hour.

Thermal Efficiency. The actual cycle for a steam engine takes into account the fact that heat losses occur in the actual steam engine. The efficiency for such a cycle (ϵ_T) is

681 $$\epsilon_T = \frac{Q_1 - Q_2 - Q_{losses}}{Q_1} = \frac{2545}{w_A (h_1 - h_{f_2})}$$

where w_A is the actual steam consumption in pounds per horse-power-hour.

Engine Efficiency or Rankine Cycle Ratio. The ratio of the ideal steam consumption per horsepower-hour (w_R) to the actual steam consumption in pounds per horsepower-hour (w_A) is termed engine efficiency (ϵ_E), cylinder efficiency, or Rankine cycle ratio, hence

682 $$E = \frac{w_R}{w_A} = \frac{2545}{w_A (h_1 - h_2)} = \frac{\epsilon_T}{\epsilon_R}.$$

INTERNAL COMBUSTION ENGINES

Compression Ratio. The ratio of the total volume at the beginning of compression (V_1) and the volume at the end of compression or clearance volume (V_c) is a significant ratio for the various internal combustion engine cycles. This ratio is known as compression ratio (r_k) and may also be expressed in terms of the piston displacement (P.D.), hence

683 $$r_k = \frac{V_1}{V_c} = \frac{P.D. + V_c}{V_c}.$$

Otto Efficiency. The Otto cycle consists of a frictionless adiabatic compression, constant volume burning, frictionless adiabatic expansion and rejection of heat energy at constant volume. If 1 and 2 are used to designate the states at the beginning and end of compression, respectively, then the efficiency for such a cycle (ϵ_o) is

684 $$\epsilon_o = 1 - \frac{T_1}{T_2} = 1 - \left(\frac{p_1}{p_2}\right)^{\frac{k-1}{k}} = 1 - \left(\frac{V_2}{V_1}\right)^{k-1}$$
$$= 1 - \left(\frac{V_c}{P.D. + V_c}\right)^{k-1} = 1 - \left(\frac{1}{r_k}\right)^{k-1}$$

NOTE. For explanation of the exponent **k** see section on perfect gases.

Joule or Brayton Efficiency. The Joule cycle consists of a frictionless adiabatic compression, constant pressure burning, frictionless adiabatic expansion and rejection of heat energy at constant pressure. If 1 and 2 are used to designate the states at the beginning and end of compression, respectively, then the efficiency for such a cycle (ϵ_J) is

685 $$\epsilon_J = 1 - \frac{T_1}{T_2} = 1 - \left(\frac{p_1}{p_2}\right)^{\frac{k-1}{k}} = 1 - \left(\frac{V_2}{V_1}\right)^{k-1} = 1 - \left(\frac{1}{r_k}\right)^{k-1}.$$

Diesel Efficiency. The Diesel cycle consists of a frictionless adiabatic compression, constant pressure burning, frictionless adiabatic expansion and the rejection of heat energy at constant volume. If 1 and 2 are used to designate the states at the beginning and end of compression, respectively, and 3 and 4 are used to designate the states at the beginning and end of expansion, respectively, then the efficiency for such a cycle (ϵ_D) is

686 $$\epsilon_D = 1 - \frac{1}{k}\left(\frac{T_4 - T_1}{T_3 - T_2}\right) = 1 - \left(\frac{1}{r_k}\right)^{k-1}\left[\frac{r_c^k - 1}{k(r_c - 1)}\right]$$

where r_c represents the ratio of the total volume at the end of burning (V_3) to the volume at the start of burning (V_2). This ratio is termed cut-off ratio.

Thermal Efficiency. The actual cycle for an internal combustion engine takes into account the fact that heat losses occur in the actual engine. The efficiency for such a cycle (ϵ_T) is

687 $$\epsilon_T = \frac{Q_1 - Q_2 - Q_{losses}}{Q_1} = \frac{2545}{w_A(H_1)}$$

where w_A is the actual fuel consumption in pounds per horsepower-hour and H_1 is the calorific heating value of the fuel per pound.

Engine Efficiency. The ratio of the ideal fuel consumption per horsepower-hour (w_I) to the actual fuel consumption per horsepower-hour (w_A) is termed engine efficiency (ϵ_E), hence

688 $$\epsilon_E = \frac{w_I}{w_A} = \frac{\epsilon_T}{\epsilon_I}.$$

NOTE. The ideal efficiency (ϵ_I) can be either the Otto, Joule or Diesel cycle, depending on which one of these cycles the actual engine is operating.

Brake Horsepower. The output of an internal combustion engine may be determined by means of a friction brake. Equation 677 is applicable in this case.

The empirical equation for determining the brake horsepower (**B.H.P.**) for an engine with **n** cylinders of **d** inches diameter and **L** inches stroke at **N** revolutions per minute, the clearance being **m** per cent of the stroke, is

689 $$\text{B.H.P.} = \frac{d^2LnN}{14{,}000}\left(0.48 - \frac{1}{10\ m}\right) \text{ horsepower.}$$

Diameter. If an engine cylinder is designed for maximum obtainable indicated horsepower (**I.H.P.**) with a mean effective pressure (**M.E.P.**) pounds per square inch, the number of explosions per minute at full load being **y** and the stroke **L** in feet being **x** times the diameter (**d**) in feet, then

690 $$d = \sqrt[3]{\frac{300\ (\text{I.H.P.})}{(\text{M.E.P.})\ xy}} \text{ feet.}$$

STEAM BOILERS

Maximum Allowable Working Pressure. For a steam boiler drum with a shell of **r** inches radius, **t** inches thick, an ultimate tensile strength of **f** pounds per square inch, factor of safety **F.S.** and an efficiency of the longitudinal joint ϵ per cent the maximum allowable working pressure (**p**) in pounds per square inch gage is

691 $$p = \frac{ft\,\epsilon}{\text{F.S.}\,r} \text{ pounds per square inch.}$$

Thickness of Bumped Head. For a bumped head of bumped radius **r** inches, working pressure of **p** pounds per square inch gage, an ultimate tensile strength of **f** pounds per square inch and a factor of safety **F.S.**, the thickness (**t**) in inches is

692
$$t = \frac{F.S.rp}{Kf} \text{ inches.}$$

NOTE. **K** = 1 for convex heads and **K** = 0.6 for concave heads. The factor of safety (**F.S.**) is usually taken as 5.

Boiler Horsepower. One boiler horsepower is the evaporation of 34.5 pounds of water per hour at 212° F. and atmospheric pressure. If G_a pounds of water per hour enter the boiler with an enthalpy h_{f2} and leaves as steam with an enthalpy h_1 then the boiler horsepower (P_B) is

693
$$P_B = \frac{G_a\,(h_1 - h_{f2})}{33{,}475} \text{ horsepower.}$$

Equivalent Evaporation. The numerator of equation 693 represents the actual heat absorbed per hour. Since 970.3 B.t.u. are required to evaporate one pound of water "from and at 212° F.," the equivalent evaporation (G_e) in pounds per hour is

694
$$G_e = \frac{G_a\,(h_1 - h_{f2})}{970.3} \text{ pounds per hour.}$$

Factor of Evaporation. In order to determine the equivalent evaporation per hour (G_e) it is necessary to multiply the actual evaporation per hour (G_a) by a factor. This factor is termed factor of evaporation (**F**) and is

695
$$F = \frac{h_1 - h_{f2}}{970.3}.$$

Boiler Efficiency. The ratio of the heat absorbed in the boiler to the heat supplied by the fuel is termed boiler efficiency ϵ_B and is

696
$$\epsilon_B = \frac{G_a\,(h_1 - h_{f2})}{G_f\,(H)}$$

where G_f is the weight of fuel in pounds per hour and **H** is the calorific heating value in B.t.u. per pound of fuel.

CHIMNEYS AND DRAFT

Intensity of Draft. For a chimney **H** feet high whose gases have an absolute temperature of T_1 degrees and an outside absolute temperature of T_2 degrees, the intensity of the draft (**D**) in inches of water is

697 $$D = 7.64\ H\left(\frac{1}{T_2} - \frac{1}{T_1}\right) \text{ inches of water.}$$

NOTE. This formula neglects the effect of friction. For a chimney with a friction factor **f**, **H** feet high, **C** feet circumference, **A** square feet passage area and discharging **G** pounds of gases per second, the draft loss (**d**) in inches of water is

698 $$d = \frac{fG^2CH}{A^3} \text{ inches of water}$$

where **f** = 0.0015 for steel stacks with gases at 600° F. absolute (0.0011 at 810° F. absolute) and 0.0020 for brick or brick-lined stacks with gases at 650° F. absolute (0.0015 at 810° F. absolute).

Effective Area of Chimney. The retardation of ascending gases by friction within the stack has the effect of decreasing the inside cross-sectional area, or of lining the chimney with a layer of gas with no velocity. If the thickness of this lining is assumed to be 2 inches for all chimneys, then the effective area (**E**) for square or round chimneys with **A** square feet of passage area is approximately

699 $$E = A - 0.6\sqrt{A} \text{ square feet.}$$

Boiler Horsepower. For a chimney **H** feet high and **A** square feet of passage area, the Kent's empirical formula for the boiler horsepower (P_B) is

700 $$P_B = 3.33\ (A - 0.6\sqrt{A})\ \sqrt{H} \text{ horsepower.}$$

NOTE. This formula is based on the assumptions that the boiler horsepower capacity varies as the effective area (**E**) and the available draft is sufficient to effect combustion of 5 pounds of coal per hour per rated boiler horsepower (the water heating surface divided by 10).

FUELS AND COMBUSTION

Heating Value for a Solid Fuel. In the case of a solid fuel which contains **C** per cent* of fixed and volatile carbon, **H** per cent of hydrogen, **O** per cent of oxygen and **S** per cent of sulphur, the heating value (**Q**) per pound of fuel-as-fired is

701 $$Q = 14{,}500\ C + 62{,}000\left[H - \frac{O}{8}\right] + 4{,}000\ S \text{ B.t.u.}$$

* Percentages by weight are expressed as a decimal fraction.

Heating Value for Liquid Fuel. In the case of a liquid fuel which contains **C** per cent* of carbon and **H** per cent of hydrogen, the heating value **Q** per pound of fuel is

702 $$Q = 13{,}500\ C + 60{,}890\ H \text{ B.t.u.}$$

If the Baumé reading is known, the heating value **Q** per pound of fuel is

703 $$Q = 18{,}650 + 40\ (\text{Baumé reading} - 10) \text{ B.t.u.}$$

Weight of Dry Flue Gases per Pound of Carbon. The flue gas analysis indicates the percentage CO_2, CO, O_2 and N_2 by volume. The weight (G_1) of dry flue gas per pound of carbon is given by

704 $$G_1 = \frac{11\ CO_2 + 8\ O_2 + 7\ (CO + N_2)}{3\ (CO_2 + CO)} \text{ pounds}$$

or

705 $$= \frac{4\ CO_2 + O_2 + 700}{3\ (CO_2 + CO)} \text{ pounds.}$$

Weight of Dry Flue Gases per Pound of Coal-as-fired. The percentage by weight of carbon in the coal (C_c) and in the ash (C_a) can be determined from the coal and ash analyses. If the pounds of coal-as-fired (M_c), the pounds of ash (M_a) and the gas analysis are known, then the weight (G_2) of dry flue gas per pound of coal-as-fired is given by

706 $$G_2 = \frac{M_c C_c - M_a C_a}{M_c}\left[\frac{4\ CO_2 + O_2 + 700}{3\ (CO_2 + CO)}\right] \text{ pounds.}$$

Actual Weight of Dry Air per Pound of Coal-as-fired. If air is assumed to be 77 per cent nitrogen (N_2) by weight, the weight (G_3) of dry air per pound of coal-as-fired is given by

707 $$G_3 = \frac{M_c C_c - M_a C_a}{M_c}\left[\frac{3.032\ N_2}{CO_2 + CO}\right] \text{ pounds.}$$

Theoretical Weight of Dry Air per Pound of Coal-as-fired. The carbon, hydrogen and sulphur enter into combination with the oxygen of the air to produce combustion reactions. In the case of a coal-as-fired which contains **C** per cent by weight of fixed and volatile carbon, **H** per cent of hydrogen, **O** per cent of oxygen and **S** per cent of sulphur, the theoretical weight (G_4) of dry air per pound of coal-as-fired are given by

708 $$G_4 = 11.57\ C + 34.8\left(H - \frac{O}{8}\right) + 4.35\ S \text{ pounds.}$$

Percentage of Excess Air. Since G_3 gives the actual weight and G_4 the theoretical weight of air required for combustion, it follows that

709 $$\text{Excess air} = \left(\frac{G_3 - G_4}{G_4}\right) 100 \text{ per cent.}$$

PUMPS

Capacity. The theoretical cubic feet per minute displacement (V) of a pump which makes **N** pumping strokes of **L** feet forward per minute and has a piston of **A** square inches effective area is

710 $$V = \frac{ALN}{144} \text{ cubic feet per minute.}$$

NOTE. If the pump is double-acting the total displacement is the sum of the displacements on the forward and return strokes, the effective area varies for the two sides of the piston. Due to clearance, slip, imperfect valve action, etc., the actual displacement is reduced as much as 50 per cent in some cases.

Water Horsepower. For a pump which discharges **G** pounds of water per minute through a total head **H** feet, the water horsepower (**W.H.P.**) is

711 $$W.H.P. = \frac{GH}{33{,}000} \text{ horsepower.}$$

NOTE. The total head must include the suction lift, the discharge lift, friction and velocity heads.

Overall Thermal Efficiency. For steam-driven pumping units it is customary to express the ratio of the heat actually converted into work in lifting the water to the heat supplied as overall thermal efficiency (ϵ_t), hence

712 $$\epsilon_t = \frac{2545}{w_a\,(h_1 - h_{f2})}$$

where w_a is the actual steam consumption in pounds per water horsepower-hour.

Duty. The term " duty " is applied to steam-driven pumping units to indicate the foot-pounds of work done for every million B.t.u. supplied. For a pump which discharges **G** pounds of water per minute through a total head of **H** feet while using **M**

pounds of steam per minute with **Q** B.t.u. available per pound, the duty (**D**) is

713 $$D = \frac{GH}{MQ} \times 10^6 \text{ foot-pounds per million B.t.u.}$$

714 $$= \frac{W.H.P.}{MQ} \times 33 \times 10^9 \text{ foot-pounds per million B.t.u.}$$

715 $$= \epsilon_t \times 778 \times 10^6 \text{ foot-pounds per million B.t.u.}$$

AIR COMPRESSORS

Capacity. The capacity of an air compressor is usually measured in terms of the cubic feet of "free air" handled per minute (V_a), which is air at atmospheric pressure (p_a) and atmospheric temperature (t_a). If **P.D.** represents the piston displacement in cubic feet per minute of a compressor operating with suction pressure p_a pounds absolute per square inch, a discharge pressure p_2 pounds absolute per square inch, compression according to the law pV^n = a constant, and **m** per cent clearance (expressed as a decimal), then

716 $$V_a = P.D.\left[1 + m - m\left(\frac{p_2}{p_a}\right)^{\frac{1}{n}}\right] \text{ cubic feet per minute.}$$

NOTE. For general expressions between pressure, volume and temperature see section on perfect gases. For air compressors $n = 1.20$ to 1.35.

Volumetric Efficiency. The ratio of the volume of "free air" (V_a) to the piston displacement (**P.D.**) is termed displacement or volumetric efficiency (ϵ_v).

For single-stage compression

717 $$\epsilon_v = \frac{V_a}{P.D.} = 1 + m - m\left(\frac{p_2}{p_a}\right)^{\frac{1}{n}}.$$

For multi-stage air compressors

718 $$\epsilon_v = \frac{V_a}{P.D.} = 1 + m - m\left(\frac{p_x}{p_a}\right)^{\frac{1}{n}}$$

where p_x is the discharge pressure in pounds absolute per square inch leaving the first stage cylinder.

For two-stage compression

719 $$p_x = \sqrt{p_a p_2} \text{ pounds per square inch.}$$

For three-stage compression

720 $p_x = \sqrt[3]{p_a^2 p_2}$ and $p_y = \sqrt[3]{p_a p_2^2}$ pounds per square inch where p_y is the discharge pressure in pounds absolute per square inch leaving the second stage cylinder.

Power. The power (P) required in foot-pounds per minute to compress V_a cubic feet of "free air" per minute polytropically according to the law, pv^n = a constant, from atmospheric pressure p_a to a discharge pressure p_2 for single-stage compression is

721 $$P = 144\, p_a V_a \frac{n}{n-1}\left[\left(\frac{p_2}{p_a}\right)^{\frac{n-1}{n}} - 1\right] \text{ foot-pounds per minute.}$$

For two-stage compression

722 $$P = 288\, p_a V_a \frac{n}{n-1}\left[\left(\frac{p_2}{p_a}\right)^{\frac{n-1}{2n}} - 1\right] \text{ foot-pounds per minute.}$$

For three-stage compression

723 $$P = 432\, p_a V_a \frac{n}{n-1}\left[\left(\frac{p_2}{p_a}\right)^{\frac{n-1}{3n}} - 1\right] \text{ foot-pounds per minute.}$$

The power (P) required in foot-pounds per minute to compress V_a cubic feet of "free air" per minute isothermally according to the law pV^n = a constant from atmospheric pressure (p_a) to a discharge pressure (p_2) is

724 $$P = 144\, p_a V_a \ln \frac{p_2}{p_a} \text{ foot-pounds per minute.}$$

Efficiency of Compression. The ratio of the isothermal power to the polytropic power is termed efficiency of compression (ϵ_c).

For single-stage compression

725 $$\epsilon_c = \frac{\ln \frac{p_2}{p_a}}{\frac{n}{n-1}\left[\left(\frac{p_2}{p_a}\right)^{\frac{n-1}{n}} - 1\right]}.$$

For two-stage compression

726 $$\epsilon_c = \frac{\ln \frac{p_2}{p_a}}{\frac{2n}{n-1}\left[\left(\frac{p_2}{p_a}\right)^{\frac{n-1}{2n}} - 1\right]}.$$

For three-stage compression

727 $$\epsilon_c = \frac{\ln \frac{p_2}{p_a}}{\frac{3n}{n-1}\left[\left(\frac{p_2}{p_a}\right)^{\frac{n-1}{3n}} - 1\right]}.$$

COMPRESSION REFRIGERATION

Ideal Compression Refrigeration Cycle. In order to simplify the references to conditions in the ideal compression refrigeration cycle, a complete description of the cycle is given. The compressor draws the vapor (usually saturated or slightly superheated) from the evaporator at condition 1, compresses it adiabatically and without friction to condition 2 in the superheated region and then discharges the vapor to the condenser. The cooling water condenses the vapor to a saturated liquid at condition 3. The liquid is drawn off and then passes through an expansion valve to condition 4. This partially vaporized liquid now enters the evaporator where further evaporation takes place before entering the compressor.

Refrigerating Effect. The refrigerant enters the evaporator with an enthalpy of h_{f_3} B.t.u. per pound and leaves with an enthalpy of h_1 B.t.u. per pound. If G_r pounds of refrigerant are circulated per minute and the refrigerating effect per minute is represented by **R,** then

728 $$R = G_r\,(h_1 - h_{f_3}) \text{ B.t.u. per minute.}$$

Capacity. The cubic feet per minute (V_1) handled by a compressor operating at **N** revolutions per minute with a piston displacement (**P.D.**) per revolution and drawing in G_r pounds of refrigerant per minute, each pound having a specific volume v_1 cubic feet is

729 $$V_1 = N \times P.D. = G_r v_1 \text{ cubic feet per minute.}$$

NOTE. This formula assumes no clearance. If the refrigerant is compressed from a suction pressure p_1 pounds absolute per square inch to a discharge pressure p_2 pounds absolute per square inch according to the law pV^n = a constant and a clearance of **m** per cent expressed as a decimal, then

730 $$V_1 = N \times P.D.\left[1 + m - m\left(\frac{p_2}{p_1}\right)^{\frac{1}{n}}\right] \text{ cubic feet per minute.}$$

Tonnage. One ton of refrigeration is the heat equivalent to the melting of one ton (2000 pounds) of ice at 32° F. in 24 hours. Since one pound of ice melting at 32° F. will absorb approximately 144 B.t.u., then a ton of refrigeration will absorb 288,000 B.t.u. per day or 200 B.t.u. per minute, hence

731 $$\text{Tonnage} = \frac{R}{200} = \frac{G_r\,(h_1 - h_{f_3})}{200}\ \text{tons.}$$

Power. The refrigerant enters the compressor with an enthalpy h_1 and leaves with an enthalpy h_2 B.t.u. per pound. If G_r pounds of refrigerant are circulated per minute, the power (P) expressed in foot-pounds per minute for adiabatic compression is

732 $$P = 778\ G_r\,(h_2 - h_1)\ \text{foot-pounds per minute.}$$

If V_1 cubic feet of refrigerant per minute enter the compressor with a suction pressure p_1 pounds absolute per square inch and is compressed to a discharge pressure p_2 pounds absolute per square inch according to the law pV^n = a constant, then the power (P) is

733 $$P = 144\ p_1V_1\frac{n}{n-1}\left[\left(\frac{p_2}{p_1}\right)^{\frac{n-1}{n}} - 1\right]\text{foot-pounds per minute.}$$

Coefficient of Performance. The ratio of the refrigerating effect R to the power $\left(\frac{P}{778}\right)$ is called the coefficient of performance (c. of p.). Hence, for adiabatic compression,

734 $$\text{c. of p.} = \frac{778\ R}{P} = \frac{h_1 - h_{f_3}}{h_2 - h_1}$$

and, for polytropic compression,

735 $$\text{c. of p.} = \frac{5.40\ G_r\,(h_1 - h_{f_3})}{p_1V_1\frac{n}{n-1}\left[\left(\frac{p_2}{p_1}\right)^{\frac{n-1}{n}} - 1\right]}.$$

Heat Removed in the Condenser. If G pounds of refrigerant per minute enter the condenser with an enthalpy h_2 and leave with an enthalpy h_{f_3} B.t.u. per pound, then the heat removed per minute in the condenser (Q_c) is.

736 $$Q_c = \frac{W}{778} + R = G\,(h_2 - h_{f_3})\ \text{B.t.u. per minute.}$$

Weight of Cooling Water Required. If Q_c B.t.u. per minute are to be removed in the condenser and the temperatures of the cooling water entering and leaving the condenser are t_c and t_h, respectively, then the pounds of cooling water per minute (G_w) required are

737 $$G_w = \frac{Q_c}{h_{f_h} - h_{f_c}} = \frac{G\,(h_2 - h_{f_3})}{h_{f_h} - h_{f_c}} \text{ pounds per minute.}$$

HEAT TRANSMISSION

Fundamental Equation. The heat transmitted in engineering apparatus is affected by a combination of the heat transferred by conduction, convection and radiation. If the temperature is low and the rate of flow of the fluid over the surface is high, the radiation factor is ignored. The fundamental equation for the heat (Q) conducted in time (t), through a material having a thermal conductivity (k) and a surface area (S) which is normal to the flow of heat and of thickness (x) in the direction of the flow of heat with a temperature difference (θ) between its surfaces, is

738 $$Q = \frac{ckS\theta t}{x} \text{ units in time (t).}$$

NOTE. Average values of **k** for various engineering materials expressed in gram-calories per second per square centimeter per centimeter per degree Centigrade, are given on page 317. The constant **c** depends on the units of measurement as follows:

Q	S	x	θ	t	c	*c
gram-calories	sq. cms.	cms.	Cent.	seconds	1	0.000344
kilogram-calories	sq. meters	cms.	Cent.	hours	36,000	12.4
British thermal units	sq. feet	inches	Fahr.	hours	2903	1
joules	sq. cms.	cms.	Cent.	seconds	4.18	0.00144
joules	sq. feet	inches	Fahr.	seconds	851	0.293
kilowatt-hours	sq. meters	cms.	Cent.	hours	41.8	0.0144
kilowatt-hours	sq. feet	inches	Fahr.	hours	0.851	0.000293

* Values of **c** if **k** is expressed in B.t.u. per hour per square foot per inch per degree Fahrenheit.

If the heat is transmitted through a body composed of an

inside film, two materials and an outside film, equation **738**, when expressed in English units, becomes

739 $$Q = \frac{\theta_m}{\frac{1}{a_1 S_{mf1}} + \frac{x_1}{k_1 S_{m1}} + \frac{x_2}{k_2 S_{m2}} + \frac{1}{a_2 S_{mf2}}} \text{ B.t.u. per hour}$$

where θ_m is the mean temperature difference in degrees Fahrenheit between the two fluids while passing over the body, a_1 and a_2 are the inside and outside film coefficients in B.t.u. per hour per square foot per degree Fahrenheit, S_{mf1} and S_{mf2} are the areas of the inside and outside films in square feet, x_1 and x_2 are the thicknesses of the materials in inches, k_1 and k_2 are the conductivities of the materials in B.t.u. per hour per square foot per inch of thickness per degree Fahrenheit, and S_{m1} and S_{m2} are the mean surface areas of the materials.

Mean Temperature Difference. For building walls, roofs, partitions, etc., steam and refrigerating pipes carrying wet or saturated vapor and surrounded by atmospheric air, the heat is assumed to be transmitted from the hot fluid at a uniform temperature (t_1) to a cold fluid at a uniform temperature (t_2). For these cases the mean temperature difference (θ_m) is

740 $$\theta_m = t_1 - t_2 \text{ degrees Fahrenheit.}$$

In heat exchangers such as boilers, superheaters, condensers, economizers, liquid and gas heaters or coolers, the temperature of either one or both fluids changes. If the hot fluid enters the apparatus at a temperature t_1 and leaves at t_2 and the contiguous cold fluid temperatures are t_a and t_b, respectively, then

741 $$\theta_m = \frac{(t_1 - t_a) - (t_2 - t_b)}{\ln \frac{t_1 - t_a}{t_2 - t_b}} \text{ degrees Fahrenheit.}$$

NOTE. Equation **741** gives the logarithmic mean temperature difference and is applicable only when the overall coefficient of heat transfer (K), the weight (W) of the hot fluid and the weight (w) of the cold fluid and their specific heat (C) and (c), respectively, are approximately constant during the transfer of heat.

Mean Surface Area. The most important surfaces encountered in engineering practice are the plane or uniform cross-sectional surface, cylindrical and spherical surfaces. If S_1 and

S_2 represent the inside and outside surface areas, respectively, then the mean surface area (S_m) for the plane surface is

742 $$S_m = \frac{S_2 + S_1}{2} = S \text{ square feet.}$$

For the cylindrical surface the mean surface area (S_m) is

743 $$S_m = \frac{S_2 - S_1}{\ln \frac{S_2}{S_1}} = \frac{2\pi L (r_2 - r_1)}{\ln \frac{r_2}{r_1}} \text{ square feet}$$

where **L** is the length in feet and r_1 and r_2 are the inside and outside radii in feet, respectively.

For the spherical surface the mean surface area (S_m) is

744 $$S_m = \sqrt{S_2 S_1} = 4\pi r_1 r_2 \text{ square feet.}$$

Overall Coefficient of Heat Transfer. It is desirable in the solution of engineering problems involving the transfer of heat through typical walls, roofs, partitions, floors, pipes, heat exchangers, etc., to use a coefficient of heat transmission which will take into account the effects of conduction, convection and radiation, together with the type, thickness and position of the materials, and which may be used with the difference of the temperatures of the fluid temperatures on each side of the composite section. This quantity is termed "over-all coefficient of heat transfer" (**K**) and is expressed in B.t.u. per hour per square foot of surface area per degree Fahrenheit. The heat transmitted per hour (**Q**) becomes

745 $$Q = KS\theta_m \text{ B.t.u. per hour.}$$

NOTE. Average values of **K** for the usual building structures are given below.

Overall Coefficients of Heat Transfer (K) for Building Structures*
Expressed in B.t.u. per hour per square foot per degree Fahrenheit

Walls. Thickness in inches.	8	12	16
Brick, without interior plaster	0.50	0.36	0.28
Brick, with interior plaster	0.46	0.34	0.27
Concrete, without interior plaster	0.69	0.54	0.48

* Correction for exposure:

	North	East	South	West
Multiply K by	1.3	1.1	1.0	1.2

	8	12	16
Concrete, with interior plaster	0.62	0.49	0.44
Haydite, without interior plaster	0.36	0.26	0.21
Haydite, with interior plaster	0.34	0.24	0.20
Hollow tile, without interior plaster	0.40	0.30	0.25
Hollow tile, with interior plaster	0.38	0.29	0.24
Limestone, without interior plaster	0.71	0.49	0.37
Limestone, with interior plaster	0.64	0.45	0.35

Wood, shingled or clapboarded, with interior plaster	0.25
Stucco, with interior plaster	0.30
Brick veneer, with interior plaster	0.27

Partitions

4-inch hollow clay tile, plaster both sides	0.40
4-inch common brick, plaster both sides	0.43
4-inch hollow gypsum tile, plaster both sides	0.27
Wood lath and plaster on one side of studding	0.62
Wood lath and plaster on both sides of studding	0.34
Metal lath and plaster on one side of studding	0.69
Metal lath and plaster on both sides of studding	0.39
Plaster board and plaster on one side of studding	0.61
Plaster board and plaster on both sides of studding	0.34
2-inch corkboard and plaster on one side of studding	0.12
2-inch corkboard and plaster on both sides of studding	0.063

Floors. Thickness in inches.	4	6	8	10
Concrete, no ceiling and no flooring	0.65	0.59	0.53	0.49
Concrete, plastered ceiling and no flooring	0.59	0.54	0.50	0.45
Concrete, no ceiling and terrazzo flooring	0.61	0.56	0.51	0.47
Concrete, plastered ceiling and terrazzo flooring	0.56	0.52	0.47	0.44
Concrete, on ground and no flooring	1.07	0.90	0.79	0.70
Concrete, on ground and terrazzo flooring	0.98	0.84	0.74	0.66

Frame construction, no ceiling, maple or oak flooring on yellow pine sub-flooring on joists	0.34
Frame construction, metal lath and plaster ceiling, maple or oak flooring on yellow pine sub-flooring on joists	0.35
Frame construction, wood lath and plaster ceiling, maple or oak flooring on yellow pine sub-flooring on joists	0.24
Frame construction, plaster board ceiling, maple or oak flooring on yellow pine sub-flooring on joists	0.24

Roofs, tar and gravel. Thickness in inches.	2	4	6
Concrete, no ceiling and no insulation	0.82	0.72	0.64
Concrete, no ceiling and 1 inch rigid insulation	0.24	0.23	0.22
Concrete, metal lath and plaster ceiling and no insulation	0.42	0.40	0.37
Concrete, metal lath and plaster ceiling and 1 inch rigid insulation	0.19	0.18	0.18
1 inch wood, no ceiling and no insulation			0.49
1 inch wood, no ceiling and 1 inch rigid insulation			0.20
1 inch wood, metal lath and plaster ceiling and no insulation			0.32
1 inch wood, metal lath and plaster ceiling and 1 inch rigid insulation			0.16
Metal, no ceiling and no insulation			0.95
Metal, no ceiling and 1 inch rigid insulation			0.25
Metal, metal lath and plaster ceiling and no insulation			0.46
Metal, metal lath and plaster ceiling and 1 inch rigid insulation			0.19
Wood shingles, rafters exposed			0.46
Wood shingles, metal lath and plaster			0.30
Wood shingles, wood lath and plaster			0.29
Wood shingles, plaster board and plaster			0.29
Asphalt shingles, rafters exposed			0.56
Asphalt shingles, metal lath and plaster			0.34
Asphalt shingles, wood lath and plaster			0.32
Asphalt shingles, plaster board and plaster			0.32

Glass

Single windows and skylights	1.13
Double windows and skylights	0.45
Triple windows and skylights	0.281
Hollow glass tile wall, 6 × 6 × 2 inches thick blocks	
Wind velocity 15 mph, outside surface; still air, inside surface	0.60
Still air outside and inside surfaces	0.48

Doors. Nominal thickness in inches.

	1	$1\frac{1}{4}$	$1\frac{1}{2}$	$1\frac{3}{4}$	2	$2\frac{1}{2}$	3
Wood	0.69	0.59	0.52	0.51	0.46	0.38	0.33

AIR AND VAPOR MIXTURES

Specific or Absolute Humidity. The weight of water vapor per unit volume of space occupied, expressed in grains or pounds per cubic foot, is termed absolute humidity. In order to simplify the solution of problems involving air and vapor mixtures, it is convenient to express the weight of water per cubic foot (d_s) in terms of the weight of dry air per cubic foot (d_a). This ratio has no specific name although the term absolute humidity

(ϕ) is more often applied to this ratio than to the one previously given.

746 $$\phi = \frac{d_s}{d_a} = \frac{v_a}{v_s} = 0.622\left(\frac{p_s}{B - p_s}\right) \text{ pounds.}$$

NOTE. In equation 746 the perfect gas laws are assumed to hold for both the water vapor and the dry air present in the moisture-laden air. The total pressure of the moisture-laden air (B) expressed in pounds absolute per square inch is assumed to be equal to the sum of the partial pressure exerted by the water vapor (p_s) and the partial pressure exerted by the dry air (p_a), both expressed in pounds absolute per square inch.

Relative Humidity. The ratio of the actual density of the water vapor in the moisture-laden air (d_s) to the density of saturated vapor (d_{sat}) at the same temperature is termed relative humidity (H). Assuming the perfect gas laws to satisfy this low pressure vapor, then

747 $$H = \frac{d_s}{d_{sat}} = \frac{v_{sat}}{v_s} = \frac{p_s}{p_{sat}}.$$

NOTE. Although p_s may be determined from Ferrell's or Carrier's equation in terms of the barometric reading, the wet-bulb and dry-bulb temperatures, it is customary to use equation 747 for determining the partial pressure (p_s) and the specific volume (v_s) of the water vapor. In engineering practice the psychrometric tables are used for determining the relative humidity (H).

Temperature, dry bulb	Difference between wet and dry bulb									
	2°	4°	6°	8°	10°	12°	14°	16°	18°	20°
32° F.	79	59	39	20	2	..	..	..	..	..
40° F.	84	68	52	37	22	8	..	..	..	..
45° F.	86	71	59	44	32	19	6	..	..	..
50° F.	87	74	62	50	38	26	16	5	..	..
55° F.	88	76	65	54	43	33	24	14	5	..
60° F.	89	78	68	58	48	39	30	22	13	5
65° F.	90	80	70	61	52	44	35	28	20	12
70° F.	90	81	72	64	56	48	40	32	26	19
75° F.	91	82	74	66	58	51	44	37	30	24
80° F.	92	83	75	68	61	54	47	41	34	29
85° F.	92	84	77	70	63	56	50	44	38	33
90° F.	92	85	78	71	65	58	52	47	41	36
95° F.	93	86	79	72	66	60	54	49	44	39
100° F.	93	86	80	74	68	62	57	52	46	42

NOTE. The relative humidity should range between 35 and 45.

Dew Point. When moisture-laden air is cooled until the temperature reaches that corresponding to the saturation temperature for the partial pressure of the water vapor, condensation or precipitation begins. This temperature is called the dew point.

Determination of Weight of Moisture Precipitated. In order to precipitate moisture from V cubic feet of moisture-laden air at condition 1 it is necessary to cool the air to the dew point temperature (t_3) for the final condition 2 desired. The pounds of moisture precipitated (M_p) is

748 $$M_p = \frac{V}{v_{a_1}}(\phi_1 - \phi_2) = \frac{V}{v_{a_1}}\left(\frac{v_{a_1}}{v_{s_1}} - \frac{v_{a_2}}{v_{s_2}}\right) \text{ pounds.}$$

NOTE. The specific volume of the dry air (v_a) may be determined from

749 $$v_a = \frac{53.34\ (t + 460)}{144\ (B - p_s)} \text{ cubic feet per pound}$$

and the specific volume of the water vapor (v_s) may be determined from

750 $$v_s = \frac{v_{sat}}{H} \text{ cubic feet per pound.}$$

Determination of the Quantity of Heat Removed from the Moisture-Laden Air. In order to remove the moisture (M_p), as given in equation 748, it is necessary to supply refrigeration. This refrigeration must cool the dry air and the water vapor in addition to precipitating the moisture, thus the total amount of heat removed (R) in B.t.u. is

751 $$R = \frac{V}{v_{a_1}}\left[0.241\ (t_1 - t_3) + \frac{v_{a_2}}{v_{s_2}}(h_{s_1} - h_{s_3}) + \left(\frac{v_{a_1}}{v_{s_1}} - \frac{v_{a_2}}{v_{s_2}}\right)(h_{s_1} - h_{f_3})\right] \text{B.t.u.}$$

NOTE. h_s may be assumed to be the same as the enthalpy (h) for the saturated vapor at the same temperature.

Determination of the Heat Added. In order to precipitate the required moisture from the air at condition 1, it was necessary to cool the air to the dew point temperature (t_3) for condition 2. The saturated air must now be heated in order that the temperature desired (t_2) may be obtained. This heat must be supplied to the dry air and the water vapor, thus the total amount of heat added (Q) in B.t.u. is

752 $$Q = \frac{V}{v_{a_1}}\left[0.241\ (t_3 - t_2) + \frac{v_{a_2}}{v_{s_a}}(h_{s_2} - h_{s_3})\right] \text{B.t.u.}$$

ELECTRICITY

MAGNETISM

Force (F) between two poles* of **m** and **m′** unit poles strength respectively separated by a distance of **d** centimeters.

753 $$F = \frac{mm'}{d^2} \text{ dynes.}$$

NOTE. Unlike poles attract and like poles repel.

Magnetic intensity (H) at a point distant **d** centimeters from a pole* of **m** unit poles strength.

754 $$H = \frac{m}{d^2} \text{ oersteds.}$$

NOTE. The magnetic intensity at a point is measured in magnitude and direction by the force acting on a unit **N** pole concentrated at the point and may be due to poles or electric current.

Force (F) acting upon a pole of **m** unit poles strength placed in a magnetic field of uniform magnetic intensity **H** oersteds.

755 $$F = mH \text{ dynes.}$$

Magnetic flux (Φ) due to a pole of **m** unit poles strength.

756 $$\Phi = 4\,\pi m \text{ maxwells.}$$

NOTE. The flux leaves a **N** pole and enters a **S** pole.

Intensity of magnetization (J) at any point in a magnet of constant section which has a pole strength per unit area of **σ** unit poles per square centimeter distributed uniformly over the end surfaces of the magnet.

757 $$J = \sigma \text{ unit poles per square centimeter.}$$

* The dimensions of the surface over which each pole is distributed are assumed to be negligible compared with all other dimensions and the permeability of the surrounding medium is unity.

Magnetic flux density (B) at a point in a magnet where the magnetic intensity is H oersteds and the intensity of magnetization is J unit poles per square centimeter.

758 $$B = H + 4\pi J \text{ gausses.}$$

NOTE. The addition is vectorial.

Magnetic flux density (B) produced by a magnetic intensity of H oersteds in a medium where the permeability corresponding to the stated magnetic intensity is μ.

759 $$B = \mu H \text{ gausses.}$$

Permeability (μ) of a medium in which a magnetic intensity of H oersteds produces a magnetic flux density of B gausses.

760 $$\mu = \frac{B}{H}.$$

Susceptibility (κ) of a medium in which a magnetic intensity of H oersteds produces an intensity of magnetization of J unit poles per square centimeter.

761 $$\kappa = \frac{J}{H}.$$

Permeability (μ) of a medium of susceptibility κ.

762 $$\mu = 1 + 4\pi\kappa.$$

Force (F) between two poles distributed over two plane surfaces A square centimeters in area, separated by an air gap in which the uniform flux density is B gausses, and surrounded by a medium of permeability μ.

763 $$F = \frac{B^2 A}{8\pi\mu} \text{ dynes.}$$

Energy of magnetic field per cubic centimeter (W) in a medium where the flux density is B gausses and the constant permeability is μ or where the magnetic intensity is H oersteds and the constant permeability is μ.

764 $$W = \frac{B^2}{8\pi\mu} = \frac{\mu H^2}{8\pi} \text{ ergs.}$$

ELECTROMAGNETISM

Magnetic intensity (dH) at a point distant **d** centimeters from a circuit element of length **dl** centimeters carrying a current of **I** amperes, **θ** being the angle between the circuit element and the line joining the element and the point.

765 $$dH = \frac{I \sin \theta \, dl}{10 \, d^2} \text{ oersteds.}$$

NOTE. The values of dH must be summed for all elements **dl** of a complete circuit. The summation must be made vectorially.

Magnetic intensity (H) at a point distant **d** centimeters on a normal from the axis of a cylindrical straight wire conducting a current of **I** amperes with uniform density throughout the wire.

Case I. Distance **d** negligible compared with length of wire and not less than radius of wire.

766 $$H = \frac{0.2 \, I}{d} \text{ oersteds.}$$

Case II. Distance **d** not negligible compared with length of wire and not less than radius of wire.

767 $$H = \frac{0.1 \, I}{d} (\sin \theta_1 + \sin \theta_2) \text{ oersteds.}$$

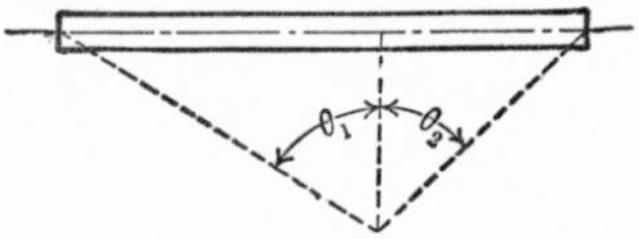

FIG. 767.

Case III. Distance **d** negligible compared with length of wire and not greater than the radius **R** of the wire.

768 $$H = \frac{0.2 \, Id}{R^2} \text{ oersteds.}$$

Case IV. A hollow cylindrical wire of internal radius **r** centimeters and external radius **R** centimeters. Distance **d** not greater than **R**, not less than **r** and negligible compared with the length of the wire.

769 $$H = \frac{0.2 \, I \, (d^2 - r^2)}{d \, (R^2 - r^2)} \text{ oersteds.}$$

NOTE. In each case the direction of the magnetic intensity at the point is normal to a plane including the point and the axis of the wire and is in a clockwise direction when viewing the wire from the end at which the current enters. The magnetic intensity at a point on the axis of the wire in each case is zero and in Case IV is zero throughout the air core.

Magnetic intensity (H) at a point on a line through the center and normal to the plane of a circular turn of wire of negligible section conducting a current of **I** amperes, the radius of the circular turn being **r** centimeters and the distance of the point from the wire being **d** centimeters.

770 $$H = \frac{0.628\ Ir^2}{d^3} \text{ oersteds.}$$

NOTE. The magnetic intensity at the center of the circular turn is $\frac{0.628\ I}{r}$ oersteds. At the center of curvature of an arc of length **l** centimeters and radius of curvature **r** centimeters the magnetic intensity is $\frac{0.1\ Il}{r^2}$ oersteds. When its section is negligible the above formulas also apply to a compact coil of **N** turns each conducting a current of **I** amperes if the current is taken as **NI** amperes. The direction of the magnetic intensity in each case is along a line through the center of curvature and normal to the plane of the wire and away from a viewing point at which the current is seen to flow in a clockwise direction.

Magnetic intensity (H) at a point inside a long coil of constant section wound uniformly with **n** turns of wire per centimeter of axial length, each turn conducting a current of **I** amperes.

Case I. Magnetic intensity at any point in the plane of the central turn when the section of the coil is of any shape and its dimensions are negligible compared with the axial length of the coil.

771 $$H = 1.26\ nI \text{ oersteds.}$$

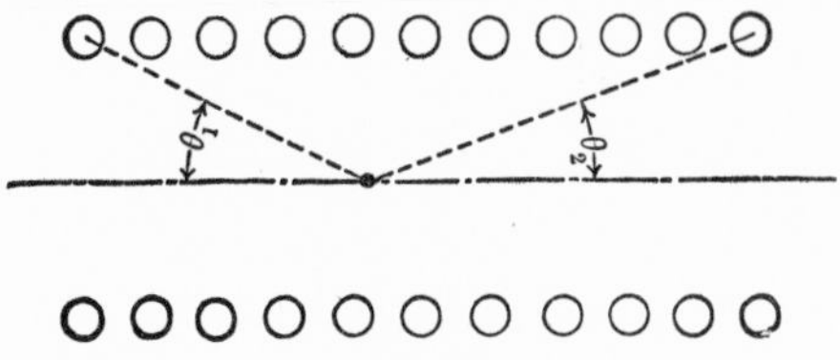

FIG. 772.

Case II. Magnetic intensity at any point on the axis of a cylindrical helix wound with wire of negligible section.

772 $$H = 0.628\ nI\ (\cos\theta_1 + \cos\theta_2)\ \text{oersteds.}$$

NOTE. The direction of the magnetic intensity in either case is determined as in 770.

Magnetic intensity (H) at a point distant **d** centimeters from the axis of and within a toroidal coil of **N** turns conducting a current of **I** amperes and wound uniformly on a surface generated by the revolution of a circle **r** centimeters in radius about an axis **R** centimeters from the center of the circle.

773 $$H = \frac{0.2\ NI}{d}\ \text{oersteds.}$$

NOTE. The average magnetic intensity within the coil is $\frac{0.4\ NI\ (R - \sqrt{R^2 - r^2})}{r^2}$ oersteds. Formula 773 also applies to a coil wound on a surface generated by the revolution of a rectangle, with sides **a** centimeters and **b** centimeters in length respectively, about an axis **R** centimeters from the center of the rectangle. The average magnetic intensity within this coil, taking **b** parallel and **a** perpendicular to the axis, is $\frac{0.2\ NI}{a}\ \ln \frac{R + \frac{a}{2}}{R - \frac{a}{2}}$ oersteds. ln equals $\log_e$.

Magnetomotive force ($\mathfrak{F}$) due to **N** turns of wire each conducting a current of **I** amperes in the same direction of rotation.

774 $$\mathfrak{F} = 1.26\ NI\ \text{gilberts.}$$

Reluctance ($\mathcal{R}$) **and Permeance** ($\mathcal{P}$) between the bases of a right prism or cylinder of permeability μ in which the direction of the flux density at all points is normal to the bases, the area of each base being **A** square centimeters and the perpendicular distance between the bases **l** centimeters.

775 $$\mathcal{R} = \frac{l}{\mu A}\ \text{gilberts per maxwell, or rels.}$$

776 $$\mathcal{P} = \frac{\mu A}{l}\ \text{maxwells per gilbert, or perms.}$$

NOTE. The total reluctance of several reluctances connected in series without abrupt change of section at any point equals the sum of the several reluctances. The total permeance of several permeances connected in parallel equals the sum of the separate permeances.

Magnetic flux (Φ) established by a magnetomotive force of $\mathfrak{F}$ gilberts in a magnetic circuit of $\mathcal{R}$ gilberts per maxwell, or rels reluctance.

777 $$\Phi = \frac{\mathfrak{F}}{\mathfrak{R}} \text{ maxwells.}$$

NOTE. See 775, 776, and note to 853.

Force (F) on a portion of a conductor of effective length l centimeters carrying a current **I** amperes and placed in a magnetic field of uniform flux density **B** gausses.

778 $$F = 0.1 \text{ BlI dynes.}$$

NOTE. The effective length of a portion of the conductor is the shortest distance between the ends of a projection of the portion of the conductor on a plane normal to the flux density. The respective directions of the force, flux density, and current in the effective length are represented by the directions in which the thumb, index, and middle fingers of the left hand point when held in positions respectively perpendicular to each other.

Force (F) per centimeter length between two parallel straight wires d centimeters apart and conducting currents of I_1 and I_2 amperes respectively. [distance between wires negligible compared with their lengths but the cross section of each wire may be finite]

779 $$F = \frac{0.02\, I_1 I_2}{d} \text{ dynes.}$$

NOTE. The force is an attraction if the currents flow in the same direction and is a repulsion if the currents flow in opposite directions.

Force (F) per centimeter length between two circuits each composed of two straight wires of negligible section, located in parallel planes as shown in Fig. 780 and conducting currents of I_1 and I_2 amperes respectively. [distance between planes negligible compared with length of wires and all dimensions in centimeters]

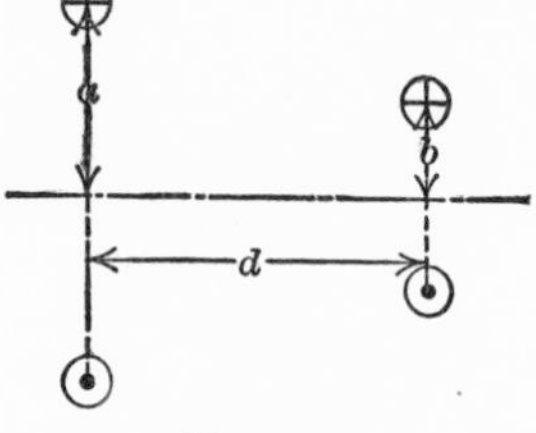

FIG. 780.

780 $$F = 0.04\ I_1I_2d\left(\frac{1}{(a-b)^2+d^2} - \frac{1}{(a+b)^2+d^2}\right)\text{dynes.}$$

NOTE. With the current directions as shown in Fig. 780 the force is an attraction and if the current in either circuit is reversed the force is a repulsion.

Force (F) between two parallel circular coaxial turns of negligible section located as in Fig. 780 and conducting currents of I_1 and I_2 amperes respectively.

Case I. Radii **a** and **b** nearly equal and **d** very small compared with either **a** or **b**.

781 $$F = \frac{0.126\ adI_1I_2}{(a-b)^2+d^2}\text{dynes.}$$

Case II. Radius **b** small compared with **a**.

782 $$F = \frac{0.592\ I_1I_2a^2b^2d}{(a^2+d^2)^{\frac{5}{2}}}\text{dynes.}$$

NOTE. The direction of the force is determined as in **780**.

Torque (T) acting on a circuit conducting a current of **I** amperes and enclosing an effective area **A** square centimeters, placed in a magnetic field in which the uniform flux density is **B** gausses.

783 $$T = 0.1\ BAI\ \text{cm.-dynes.}$$

NOTE. The effective area of a closed circuit is the maximum area obtained by projecting the closed circuit on a plane parallel to the flux density. The closed circuit will turn in such a direction that the summation of the fluxes enclosed by the circuit due to itself and the external field respectively will be a maximum.

Torque (T) acting upon a small circular turn enclosing an area of **A** square centimeters, conducting a current of I_2 amperes and with its center coinciding with the center of a large circular turn **r** centimeters in radius and conducting a current of I_1 amperes, the angle between the planes of the two coils being θ.

784 $$T = \frac{0.0628\ AI_1I_2 \sin\theta}{r}\text{cm.-dynes.}$$

NOTE. The direction of the torque is determined as in **783**.

Self-inductance (L) of a coil of **N** turns of wire of negligible section through which a current of **I** amperes establishes a magnetic flux of Φ maxwells.

785 $$L = \frac{N\Phi 10^{-8}}{I} \text{ henrys.}$$

NOTE. If the flux does not link all of the turns the self-inductance is given by $\frac{N_1\phi_1 + N_2\phi_2, \text{ etc.}}{I} \times 10^{-8}$ henrys, where ϕ_1 represents the flux linking N_1 turns, ϕ_2 the flux linking N_2 turns, etc. The self-inductance of a wire of appreciable section conducting a current of **I** amperes is given by $\frac{I_1\phi_1 + I_2\phi_2, \text{ etc.}}{I^2} \times 10^{-8}$ henrys, where ϕ_1 represents the flux linking the current I_1, ϕ_2 the flux linking the current I_2, etc., the summation of I_1, I_2, etc., being equal to **I**. If the conductors and the medium surrounding any circuit are of constant permeability, the self-inductance is independent of the current and may also be determined by **800** or **836**. When the conductors or the surrounding medium are not of constant permeability the self-inductance of a circuit varies with the current and has no definite meaning since its values determined by **785**, **800** or **836** do not agree. In the following cases when no mention is made of the dimensions of a conductor section it is assumed that they are negligible and when such dimensions are given it is assumed that the current density throughout the section is uniform.

Self-inductance (L) per centimeter axial length of the turns near the center of an air solenoid, **A** square centimeters in sectional area, wound uniformly with **n** turns per centimeter length. [dimensions of sectional area negligible compared with the axial length]

786 $$L = 12.6\, n^2 A \times 10^{-9} \text{ henrys.}$$

NOTE. If the solenoid is filled completely with a medium of constant permeability μ the self-inductance per centimeter length is $12.6\, n^2 \mu A \times 10^{-9}$ henrys and if filled partially throughout its length with a medium of constant permeability μ and **B** square centimeters in constant sectional area the self-inductance per centimeter length is $12.6\, n^2(\mu B + A - B) \times 10^{-9}$ henrys.

Self-inductance (L) of a single-layer short solenoid of **N** turns, **l** centimeters in axial length and **r** centimeters in radius. [l small compared with r]

787 $$L = 12.6\, rN^2 \left\{ \ln \frac{8r}{l} - \frac{1}{2} + \frac{l^2}{32 r^2}\left(\ln \frac{8r}{l} + \frac{1}{4} \right) \right\} \times 10^{-9} \text{ henrys.}$$

Self-inductance (L) of a multiple-layer short solenoid of **N** turns, **l** centimeters in axial length, **R** centimeters in external radius and **r** centimeters in internal radius. [l small compared with R or r]

$$788 \quad L = 12.6\,aN^2\left\{\ln\frac{8\,a}{b}\left(1+\frac{3\,b^2}{16\,a^2}\right)-\left(2+\frac{b^2}{16\,a^2}\right)\right\}\times 10^{-9}\,\text{henrys.}$$

NOTE. $a = \frac{R+r}{2}$ and $b = 0.2235\,(l + R - r)$. ln equals $\log_e$.

Self-inductance (L) of a toroidal coil wound uniformly with a single layer of **N** turns on a surface generated by the revolution of a circle **r** centimeters in radius about an axis **R** centimeters from the center of the circle.

789 $$L = 12.6\,N^2\,(R - \sqrt{R^2 - r^2}) \times 10^{-9}\ \text{henrys.}$$

Self-inductance (L) of a toroidal coil of rectangular section, **r** and **R** centimeters in internal and external radius respectively, sides of section (**R** − **r**) centimeters and **l** centimeters respectively and wound uniformly with a single layer of **N** turns.

790 $$L = 2\,N^2 l\,\ln\frac{R}{r} \times 10^{-9}\ \text{henrys.}$$

Self-inductance (L) per centimeter length of one of two parallel straight cylindrical wires each **r** centimeters in radius, their axes **d** centimeters apart and conducting the same current in opposite directions. [distance **d** small compared with the length of the wires]

791 $$L = \left(2\,\ln\frac{d}{r} + 0.5\right) \times 10^{-9}\ \text{henrys.}$$

NOTE. The self-inductance of each wire per mile is $0.08047 + 0.7411\,\log\frac{d}{r}$ millihenrys and for two wires is twice as great. Formula 791 also gives the self-inductance per centimeter length of one of three wires, located at the vertices of an equilateral triangle (d is the distance between the axes of any two wires), provided the algebraic sum of the instantaneous currents conducted respectively by the three wires in the same direction equals zero.

Self-inductance (L) per mile length of one of three unsymmetrically spaced but completely transposed wires each **r** inches in radius, with axial spacings of d_{12}, d_{23}, and d_{13} inches, and with the algebraic sum of the instantaneous currents conducted respectively by the three wires in the same direction equal to zero.

792 $$L = 0.08047 + 0.7411\,\log\frac{\sqrt[3]{d_{12}d_{23}d_{13}}}{r}\ \text{millihenrys.}$$

NOTE. A completely transposed three-wire circuit is one in which each wire occupies each position for one-third of the distance.

Self-inductance (L) per centimeter length of two straight cylindrical concentric wires of equal section conducting the same current in opposite directions; the inner radius of the outer conductor being **b** centimeters and the radius of the solid inner conductor being **c** centimeters.

793 $$L = \left(2 \ln \frac{b}{c} + \frac{1}{2} + \frac{c^2}{3b^2} - \frac{c^4}{12b^4} + \frac{c^6}{30b^6} - \ldots\right) \times 10^{-9} \text{ henrys.}$$

Self-inductance (L) of a single circular turn of wire of circular section, the mean radius of the turn being **R** centimeters and the radius of the section **r** centimeters.

794 $$L = 12.6R\left\{\left(1 + \frac{r^2}{8R^2}\right)\ln \frac{8R}{r} + \frac{r^2}{24R^2} - 1.75\right\} \times 10^{-9} \text{ henrys.}$$

Mutual-inductance (M) of two coils in which a current of I_1 amperes in one establishes a flux of Φ_2 maxwells through the N_2 turns of the other.

795 $$M = \frac{N_2\Phi_2}{I_1} \times 10^{-8} \text{ henrys.}$$

Mutual-inductance (M) of two parallel circular coaxial turns each **r** centimeters in radius and their planes **d** centimeters apart. [**d** small compared with **r**]

796 $$M = 12.6r\left\{\ln \frac{8r}{d}\left(1 + \frac{3d^2}{16r^2}\right) - \left(2 + \frac{d^2}{16r^2}\right)\right\} \times 10^{-9} \text{ henrys.}$$

Mutual-inductance (M) of two concentric solenoids, the exterior of N_1 turns and length **l** centimeters and the interior of N_2 turns and sectional area A_2 square centimeters. [the axial length of the interior solenoid small compared with the axial length of the exterior solenoid]

797 $$M = \frac{12.6\, N_1N_2A_2}{l} \times 10^{-9} \text{ henrys.}$$

Self-inductance (L) of two series connections of self-inductance L_1 and L_2 henrys respectively and mutual-inductance M_{12} henrys.

798 $$L = L_1 + L_2 \pm 2\,M_{12} \text{ henrys.}$$

NOTE. The sign is + when the mutual fluxes are in conjunction and is − when the mutual fluxes are in opposition. The mutual-inductance (**M**) of two series connections of self-inductance L_1 and L_2 henrys respectively with p* per cent coupling is given by $M = p\sqrt{L_1L_2}$ henrys.

Self-inductance (**L**) of several coils of self-inductances L_1, L_2, L_3, etc., wound on the same core with 100 per cent coupling in conjunction (+ sign) or opposition (− sign).

799 $$L = (\sqrt{L_1} \pm \sqrt{L_2} \pm \sqrt{L_3} \pm \text{etc.})^2 \text{ henrys.}$$

Energy of magnetic field (**W**) established by a circuit of constant self-inductance **L** henrys and conducting a current of **I** amperes.

800 $$W = \tfrac{1}{2}\,LI^2 \text{ joules.}$$

Energy (**W**) required to change the magnetic flux linking **N** turns of wire conducting a current of **i** amperes.

801 $$W = \frac{N}{10}\int i\,d\phi \text{ ergs.}$$

NOTE. When ϕ is caused by **i**, an increase of **i** converts electric to magnetic energy, and a decrease of **i** converts magnetic to electric energy.

Hysteresis loss per cubic centimeter per second (P_h) in a medium in which a variable magnetic flux of maximum density B_m gausses changes from positive to negative to positive maximum **f** times per second.

802 $$P_h = \eta f B_m^{1.6} \times 10^{-7} \text{ watts per cubic centimeter.}$$

NOTE. This is an empirical equation and in some cases the exponent of B_m may differ appreciably from 1.6. The hysteresis loop must be symmetrical with no re-entrant loops. The hysteresis coefficient (η) varies in different materials as follows: cast iron, 0.012; cast steel, 0.005; hipernik (50 Ni) 0.00015; low-carbon sheet steel, 0.003; permalloy (78 Ni), 0.0001; pure Norway iron, 0.002; silicon sheet steel, 0.00046 to 0.001. Formula 802 does not apply to the hysteresis loss in iron rotated in a magnetic field. In the latter case at low flux densities the loss may be twice as much as that due to an alternating flux but declines in value as the flux density increases. For soft iron the loss by either process will be about the same at 15,000 gausses and at 20,000 gausses the loss due to rotation is practically zero.

* p is expressed as a decimal fraction and represents the per cent of the flux caused by one coil which links the other.

Eddy-current loss (P_e) in thin laminations placed in a sinusoidally varying magnetic flux.

803 $$P_e = \frac{1.64\,(tfB_m)^2}{\rho \times 10^{16}} \text{ watts per cubic centimeter.}$$

NOTE. t is the thickness of the laminations in centimeters, f is the frequency of flux variation in cycles per second, B_m is the maximum flux density in gausses, and ρ is the resistivity of the laminations in ohms per centimeter cube.

ELECTROSTATICS

Charge per unit area (σ) on a body charged uniformly with q statcoulombs over a surface area of **A** square centimeters.

804 $$\sigma = \frac{q}{A} \text{ statcoulombs per square centimeter.}$$

Force (f) between two bodies* charged with q and q′ statcoulombs respectively, and separated by a distance of **d** centimeters.

805 $$f = \frac{qq'}{kd^2} \text{ dynes.}$$

NOTE. Unlike charges attract and like charges repel.

Dielectric intensity (F) at a point distant **d** centimeters from a body* charged with q statcoulombs.

806 $$F = \frac{q}{kd^2} \text{ statvolts per centimeter or dynes per statcoulomb.}$$

NOTE. The dielectric intensity at a point is measured in magnitude and direction by the force acting on a positive charge of one statcoulomb concentrated at the point and may be due to charges, changing magnetic flux or contact emf. The dielectric intensity within a conducting body is zero if it conducts no current.

Dielectric intensity (F) at a point where the magnetic flux density changes at a rate of $\frac{dB}{dt}$ gausses per second.

807 $$F = \frac{dB}{dt} \times \frac{10^{-10}}{3} \text{ statvolts per cm.}$$

* The dimensions of the surface over which each charge is distributed are assumed to be negligible compared with all other dimensions and the dielectric constant of the surrounding medium is **k**.

NOTE. If the point moves through a magnetic field of **B** gausses magnetic flux density at a velocity perpendicular to the flux density of **v** centimeters per second, $\mathbf{F} = \frac{\mathbf{Bv}}{3} \times 10^{-10}$ statvolts per cm. Dielectric intensity may also be due to contact emf; a contact emf of **E** statvolts produced uniformly in a distance of **d** centimeters establishes a dielectric intensity of $\frac{\mathbf{E}}{\mathbf{d}}$ statvolts per cm. throughout that distance.

Potential (**V**) between a point distant **d** centimeters from a body* charged with **q** statcoulombs and a point at an infinite distance from the charged body.

808 $$\mathbf{V} = \frac{\mathbf{q}}{\mathbf{kd}} \text{ statvolts.}$$

NOTE. The potential between two points in any medium is measured by the work done in moving a positive charge of one statcoulomb from one point to the other against the force due to all existing charges and is independent of the path.

Dielectric intensity (**F**) at a point on a normal through the center of a circular disc uniformly charged on one side with **σ** statcoulombs per square centimeter, the angle between the normal and a line from the point to the edge of the disc being **θ**. [the dielectric constant of the surrounding medium is **k**]

809 $$\mathbf{F} = \frac{2\pi\sigma}{\mathbf{k}}(1 - \cos\theta) \text{ statvolts per cm.}$$

Dielectric intensity (**F**) at a point opposite the centers and between two plane parallel surfaces† each charged uniformly and oppositely with **σ** statcoulombs per square centimeter.

810 $$\mathbf{F} = \frac{4\pi\sigma}{\mathbf{k}} \text{ statvolts per cm.}$$

Force (**f**) acting on a body charged with **q** statcoulombs placed in a field of uniform dielectric intensity **F** statvolts per cm.

811 $$\mathbf{f} = \mathbf{qF} \text{ dynes.}$$

* The dimensions of the surface over which each charge is distributed are assumed to be negligible compared with all other dimensions and the dielectric constant of the surrounding medium is **k**.

† The distance between the surfaces is assumed to be negligible compared with all other dimensions and the dielectric constant of the medium between the surfaces is **k**.

Force (f) acting between two parallel surfaces† each **A** square centimeters in area, and charged uniformly and oppositely with σ statcoulombs per square centimeter.

812 $$f = \frac{2\pi\sigma^2 A}{k} \text{ dynes.}$$

Charge (Q) per surface required to produce a force of **f** dynes between two parallel surfaces† each uniformly and equally charged over an area of **A** square centimeters.

813 $$Q = \sqrt{\frac{kAf}{2\pi}} \text{ statcoulombs.}$$

Potential (V) between two parallel surfaces† each uniformly, equally and oppositely charged over an area of **A** square centimeters, spaced **d** centimeters apart and acted upon by a force of **f** dynes.

814 $$V = d\sqrt{\frac{8\pi f}{kA}} \text{ statvolts.}$$

Potential (V) between two parallel surfaces† charged uniformly and oppositely with σ statcoulombs per sq. cm., the dielectric constant of the medium between the surfaces being k_1 for a distance of d_1 centimeters and k_2 for a distance of d_2 centimeters.

815 $$V = 12.6\,\sigma\left(\frac{d_1}{k_1} + \frac{d_2}{k_2}\right) \text{ statvolts.}$$

Dielectric flux (ψ) due to a body charged with **q** statcoulombs.

816 $$\psi = 4\pi q \text{ lines.}$$

Intensity of electrisation (J) in a nonconducting plate charged uniformly and oppositely over two of its parallel surfaces with σ statcoulombs per square centimeter.

817 $$J = \sigma \text{ statcoulombs per sq. cm.}$$

NOTE. The intensity of electrisation within a conducting body is zero.

† The distance between the surfaces is assumed to be negligible compared with all other dimensions and the dielectric constant of the medium between the surfaces is **k**.

Dielectric flux density (**D**) at a point in a nonconducting body where the field intensity is **F** statvolts per centimeter and the intensity of electrisation is **J** statcoulombs per square centimeter.

818 $$\mathbf{D} = \mathbf{F} + 4\,\pi\mathbf{J} \text{ lines per sq. cm.}$$

NOTE. The addition is vectorial. The dielectric flux density within a conducting body is zero if it conducts no current.

Dielectric constant (**k**) of a medium in which a dielectric intensity of **F** statvolts per centimeter produces a dielectric flux density of **D** lines per square centimeter.

819 $$\mathbf{k} = \frac{\mathbf{D}}{\mathbf{F}}.$$

NOTE. The dielectric constant of various substances is given on page 321.

Capacitance (**C**) of a condenser which is charged with **Q** coulombs when the potential between its terminals is **V** volts.

820 $$\mathbf{C} = \frac{\mathbf{Q}}{\mathbf{V}} \text{ farads.}$$

Capacitance (**C**) **of a parallel plate condenser** in which the positive and negative charges are each distributed uniformly over a surface area of **A** square centimeters, the uniform distance between the oppositely charged surfaces is **d** centimeters and the medium between the oppositely charged surfaces is of dielectric constant **k**. [**d** is assumed to be small compared with all other dimensions]

821 $$\mathbf{C} = \frac{\mathbf{kA}}{36\,\pi\mathbf{d} \times 10^5} \text{ microfarads.}$$

Capacitance (**C**) **of two concentric spheres**; the inner $\mathbf{r}_1$ centimeters in external radius, the outer $\mathbf{r}_2$ centimeters in internal radius and separated by a medium of dielectric constant **k**.

822 $$\mathbf{C} = \frac{\mathbf{r}_1\mathbf{r}_2\mathbf{k}}{9\,(\mathbf{r}_2 - \mathbf{r}_1) \times 10^5} \text{ microfarads.}$$

Capacitance (**C**) **of two coaxial cylinders** per centimeter axial length; the inner $\mathbf{r}_1$ centimeters in external radius, the outer $\mathbf{r}_2$ centimeters in internal radius and separated by a medium of dielectric constant **k**. [**ln** equals $\log_e$]

823 $$C = \frac{k}{18 \ln \frac{r_2}{r_1} \times 10^5} \text{ microfarads.}$$

NOTE. The capacitance per mile is $\frac{0.03882\ k}{\log \frac{r_2}{r_1}}$ microfarads.

Capacitance (C) of two parallel cylinders per centimeter length; each cylinder **r** centimeters in radius, their centers separated by a distance of **d** centimeters and immersed in a medium of dielectric constant **k**. [**r** small compared with **d** and all dimensions small compared with distance to surrounding objects]

824 $$C = \frac{k}{36 \ln \frac{d}{r} \times 10^5} \text{ microfarads.}$$

NOTE. The capacitance per mile is $\frac{1.941\ k \times 10^{-2}}{\log \frac{d}{r}}$ microfarads. The capacitance per conductor (to neutral) of a balanced 3-phase transmission line with conductors located at the vertices of an equilateral triangle equals $\frac{3.882\ k \times 10^{-2}}{\log \frac{d}{r}}$ microfarads per mile.

Capacitance (C) to neutral per mile of one conductor of a balanced 3-phase transmission line with unsymmetrical spacing but completely transposed, d_{12}, d_{23} and d_{13} being the axial spacings in inches and **r** being the conductor radius in inches.

825 $$C = \frac{3.882 \times 10^{-2}}{\log \frac{\sqrt[3]{d_{12} d_{23} d_{13}}}{r}} \text{ microfarads.}$$

NOTE. See note following 792.

Total capacitance (C_0) of several series condensers of capacitance C_1, C_2 and C_3 farads respectively.

826 $$C_0 = \frac{1}{\frac{1}{C_1} + \frac{1}{C_2} + \frac{1}{C_3}} \text{ farads.}$$

Total charge (Q_0) on several series condensers charged with Q_1, Q_2 and Q_3 coulombs respectively.

827 $$Q_0 = Q_1 = Q_2 = Q_3 \text{ coulombs.}$$

Potential (V_0) between the end terminals of several series condensers when the potential between the terminals of each condenser is V_1, V_2 and V_3 volts respectively.

828 $$V_0 = V_1 + V_2 + V_3 \text{ volts.}$$

Total capacitance (C_0) of several parallel condensers of capacitance C_1, C_2 and C_3 farads respectively.

829 $$C_0 = C_1 + C_2 + C_3 \text{ farads.}$$

Total charge (Q_0) on several parallel condensers charged with Q_1, Q_2 and Q_3 coulombs respectively.

830 $$Q_0 = Q_1 + Q_2 + Q_3 \text{ coulombs.}$$

Potential (V_0) between the common terminals of several parallel condensers when the potential between the terminals of each condenser is V_1, V_2 and V_3 volts respectively.

831 $$V_0 = V_1 = V_2 = V_3 \text{ volts.}$$

Energy of electrostatic field (W) per cubic centimeter in a medium of dielectric constant **k** where the dielectric flux density is **D** lines per square centimeter or the dielectric intensity is **F** statvolts per centimeter.

832 $$W = \frac{D^2}{8\pi k} = \frac{kF^2}{8\pi} \text{ ergs.}$$

Energy (W) stored in a condenser of **C** farads capacitance charged with **Q** coulombs, the potential between its terminals being **V** volts.

833 $$W = \frac{1}{2}CV^2 = \frac{1}{2}\frac{Q^2}{C} = \frac{1}{2}QV \text{ joules.}$$

DIRECT CURRENTS

Electromotive force (E) induced in a coil of **N** turns linked by a magnetic flux which changes at a rate of $\frac{d\phi}{dt}$ maxwells per second.

834 $$E = N\frac{d\phi}{dt} \times 10^{-8} \text{ volts.}$$

NOTE. The direction of the emf is such that any current produced by it would establish a magnetic flux through the coil opposing the change in flux to which the emf is due.

Electromotive force (E) induced in a conductor **l** centimeters in effective length all points of which move in parallel straight lines with an effective velocity of **v** centimeters per second through a magnetic field of uniform flux density **B** gausses.

835 $$E = Blv \times 10^{-8} \text{ volts.}$$

NOTE. The effective length of the conductor is the shortest distance between the ends of a projection of the conductor on a plane normal to the flux density. The effective velocity of the conductor is the component velocity of its projection normal to the effective length and in a plane normal to the flux density. The respective directions of the effective velocity, the flux density and the induced emf in the effective length are represented by the directions in which the thumb, index and middle fingers of the right hand point when held in positions respectively perpendicular to each other.

Electromotive force (E) induced in a circuit of **L** henrys self-inductance in which the current is changing at a rate of $\frac{di}{dt}$ amperes per second.

836 $$E = L\frac{di}{dt} \text{ volts.}$$

NOTE. An increasing current induces an emf opposite in direction to the current and a decreasing current induces an emf in the same direction as the current.

Total electromotive force (E_0) of several sources of emf, E_1, E_2, E_3, etc., connected in series, each emf being measured in volts.

837 $$E_0 = E_1 + E_2 + E_3, \text{ etc., volts.}$$

NOTE. The addition is algebraic.

Resistance (R_1) between the ends of a conductor l_1 in length and A_1 in uniform section made of a material a specimen of which l_2 in length and A_2 in uniform section has a resistance of R_2 ohms.

838 $$R_1 = \frac{l_1 A_2 R_2}{l_2 A_1} \text{ ohms.}$$

NOTE. The temperature and the respective units of length and area in each case must be the same. When the length l_2 and the area A_2 of the specimen are each unity the resistance R_2 is called the resistivity (ρ) of the material per unit length and area specifying the units of length, area and resistance and the temperature. The resistivity of various materials is given on page 321. The resistance obtained by 838 or 839 applies rigorously only to conductors in which the current is constant.

Resistance (R_1) between the ends of a conductor l_1 in length and m_1 in mass made of a material a specimen of which l_2 in length and m_2 in mass has a resistance of R_2 ohms.

839 $$R_1 = \frac{l_1^2 m_2 R_2}{l_2^2 m_1} \text{ ohms.}$$

NOTE. Read note to 838 substituting " mass " for " area " throughout.

Conductance (G) of a conductor of R ohms resistance.

840 $$G = \frac{1}{R} \text{ mhos.}$$

Resistance (R_2) of a conductor at t_2 degrees Cent. which has a resistance of R_1 ohms at t_1 degrees Cent. and is made of a material which has a resistance-temperature coefficient of α at t_1 degrees Cent.

841 $$R_2 = R_1 [1 + \alpha (t_2 - t_1)] \text{ ohms.}$$

NOTE. Specific values of α for various materials are given on page 321.

Temperature (t_2) of a conductor when its resistance is R_2 ohms and which has a resistance of R_1 ohms at a temperature of t_1 degrees Cent., the resistance-temperature coefficient of the material at t_1 degrees Cent. being α.

842 $$t_2 = \frac{R_2 - R_1}{\alpha R_1} + t_1 \text{ degrees Cent.}$$

Resistance (R) between the bases of the frustum of a cone, l centimeters in height with bases of r_1 and r_2 centimeters radius respectively, made of a material of ρ ohms per centimeter-cube resistivity. [r_1 and r_2 small compared with l]

843 $$R = \frac{\rho l}{\pi r_1 r_2} \text{ ohms.}$$

Resistance (R) between two concentric cylindrical surfaces l centimeters in axial length, the exterior r_2 centimeters and the interior r_1 centimeters in radius, the resistivity of the medium between the cylindrical surfaces being ρ ohms per centimeter-cube. [ln equals $\log_e$]

844 $$R = \frac{\rho}{2\pi l} \ln \frac{r_2}{r_1} \text{ ohms.}$$

Total resistance (R_s) of a series circuit the respective parts of which have resistances of R_1, R_2, R_3, etc., each resistance being measured in ohms.

845 $$R_s = R_1 + R_2 + R_3, \text{ etc., ohms.}$$

Equivalent resistance (R_p) of a parallel circuit the respective branches of which have resistances of R_1, R_2, R_3, etc., and contain no emf, each resistance being measured in ohms.

846 $$\frac{1}{R_p} = \frac{1}{R_1} + \frac{1}{R_2} + \frac{1}{R_3}, \text{ etc., mhos.}$$

NOTE. When there are only two branches 846 reduces to $R_p = \frac{R_1 R_2}{R_1 + R_2}$ ohms and when there are n branches each of equal resistance, R_1 ohms, $R_p = \frac{R_1}{n}$ ohms.

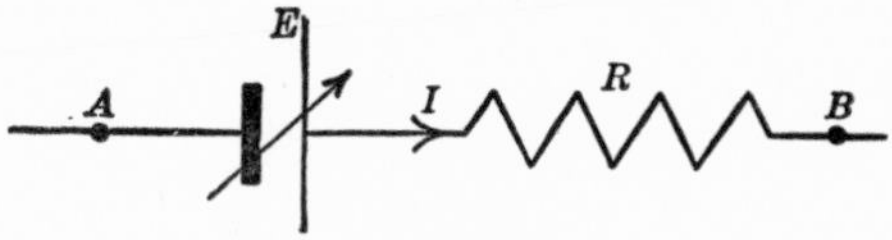

FIG. 847.

Potential (V_{AB}) between the ends A and B of a part-circuit in which the current flowing from A to B is I_{AB} amperes, the resistance from A to B is R_{AB} ohms and the emf in the part-circuit acting from A to B is E_{AB} volts.

847 $$V_{AB} = +E_{AB} - I_{AB}R_{AB} \text{ volts.}$$

NOTE. The sign of the emf or current is positive when acting in the direction shown in Fig. 847 and is negative when acting in the opposite direction. When V_{AB} is positive it is called a potential rise from A to B and when V_{AB} is negative it is called a potential drop from A to B. If E_{AB} is zero, $V_{AB} = -I_{AB}R_{AB}$ volts and if either I_{AB} or R_{AB} is zero, $V_{AB} = +E_{AB}$ volts.

Potential (V_{AD}) between the ends A and D of a part-circuit, the potentials between the ends of its constituent parts being V_{AB}, V_{BC} and V_{CD} measured in volts.

848 $$V_{AD} = V_{AB} + V_{BC} + V_{CD} \text{ volts.}$$

NOTE. The addition is algebraic.

Current (I_{AB}) flowing from the end A to the end B in a part-circuit under the conditions indicated in Fig. 847.

849 $$I_{AB} = \frac{+E_{AB} - V_{AB}}{R_{AB}} \text{ amperes.}$$

NOTE. The direction of the current is determined by its sign, a positive sign indicating a flow from A to B and a negative sign a flow from B to A. When V_{AB} equals zero, $I_{AB} = \frac{+E_{AB}}{R_{AB}}$ amperes and when E_{AB} equals zero, $I_{AB} = \frac{-V_{AB}}{R_{AB}}$ amperes, these simple forms of 849 being known as Ohm's Law. When V_{AB} is a function of the current the value of V_{AB} substituted in 849 must be known for the particular current I_{AB}.

Total current (I_0) flowing toward a junction from which the currents I_1, I_2, I_3, etc., flow away, all currents being measured in amperes.

850 $$I_0 = I_1 + I_2 + I_3, \text{ etc., amperes.}$$

Current (I_1) flowing in a branch of R_1 ohms resistance connected in parallel with a branch of R_2 ohms resistance conducting a current of I_2 amperes, the emf within either branch being zero.

851 $$I_1 = \frac{I_2R_2}{R_1} \text{ amperes.}$$

Current (I_1) flowing in a branch of R_1 ohms resistance connected in parallel with a branch of R_2 ohms resistance, the sum of the currents in the two branches being I_0 and the emf within either branch being zero.

852 $$I_1 = \frac{I_0 R_2}{R_1 + R_2} \text{ amperes.}$$

Current (**I**) flowing in any branch of a network (Fig. 853). The magnitudes of the current, total emf and total resistance respectively in any branch are indicated (as in the branch ACB) by the symbols $\mathbf{I_1}$, $\mathbf{E_1}$ and $\mathbf{R_1}$, and the respective directions of the current and emf are indicated by arrows, any unknown direction being assumed arbitrarily. Since the potential between any two points is independent of the path (for example, $V_{ACB} = V_{ADB} = V_{AFB}$) we may write from **847**

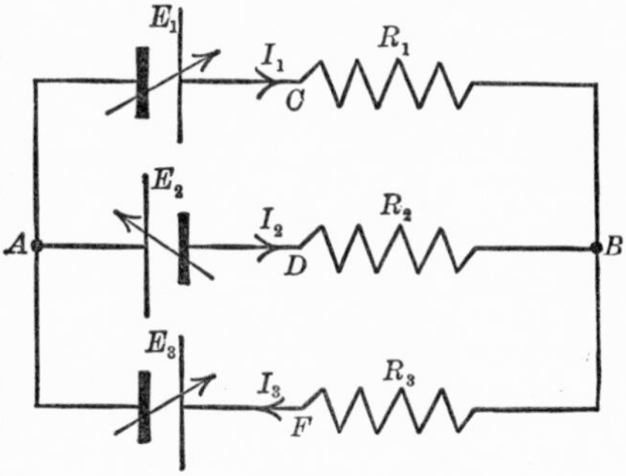

FIG. 853.

853 (1) $+E_1 - I_1R_1 = -E_2 - I_2R_2,$

(2) $+E_1 - I_1R_1 = +E_3 + I_3R_3,$

and from **850**

(3) $I_3 = I_1 + I_2.$

NOTE. The magnitude and direction of each current may be determined by solving the simultaneous equations, a positive value of the current indicating the same direction and a negative value indicating the opposite direction to that assumed in the figure. The equations written under **853** state the principles known as Kirchhoff's Laws. In a magnetic circuit containing several branches of known permeability similar equations may be written substituting magnetomotive force for electromotive force, flux for current and reluctance for resistance.

Power (**P**) delivered to or from a part-circuit conducting a current of **I** amperes and across which the potential is **V** volts.

854 $$P = VI \text{ watts.}$$

NOTE. A potential rise in the direction of the current indicates power delivered from, and a potential drop in the direction of the current indicates power delivered to the part-circuit. Multiply **854** or **855** by seconds to obtain joules or by hours divided by 1000 to obtain kilowatt-hours.

Power (**P**) delivered to a part-circuit of **R** ohms resistance containing no emf and conducting a current of **I** amperes.

855 $$P = I^2R \text{ watts.}$$

Quantity of electricity (**Q**) transmitted in **t** seconds through a circuit conducting a current of **I** amperes.

856 $$Q = It \text{ coulombs.}$$

Quantity of electricity (**Q**) transmitted through a circuit of **R** ohms resistance when the flux linking **N** turns of the circuit is changed from ϕ_1 to ϕ_2 maxwells.

857 $$Q = \frac{(\phi_1 - \phi_2)N}{R \times 10^8} \text{ coulombs.}$$

Voltage (V_L) at the load end of a transmission line of R_l ohms total resistance and conducting a current of I_l amperes, the voltage at the generator end being V_G volts.

858 $$V_L = V_G - I_lR_l \text{ volts.}$$

Power (P_L) received at the load end of a transmission line under the conditions stated in **858**.

859 $$P_L = V_GI_l - I_l^2R_l \text{ watts.}$$

Energy (W_L) received at the load end of a transmission line in **h** hours under the constant conditions stated in **858**.

860 $$W_L = [V_GI_l - I_l^2R_l]\frac{h}{1000} \text{ kilowatt-hours.}$$

Efficiency (η) of a transmission line under the conditions stated in **858**.

861 $$\eta = \frac{V_GI_l - I_l^2R_l}{V_GI_l} = \frac{V_LI_l}{V_GI_l} = \frac{V_L}{V_G}.$$

Current (I_l) conducted by a transmission line of R_l ohms total resistance when the power delivered at the load end is P_L watts and the voltage at the generator end is V_G volts.

862 $$I_l = \frac{V_G \pm \sqrt{(V_G)^2 - 4\,R_lP_L}}{2\,R_l} \text{ amperes.}$$

Area (**A**) of **l** feet of copper wire at 25° C. conducting **a current** of **I** amperes and in which the potential drop is **V** volts.

863 $$A = \frac{10.6\ Il}{V} \text{ circular mils.}$$

NOTE. For stranded wire the constant is 10.8.

Weight (G) of copper wire required to transmit energy a distance of l feet at a rate of P_L watts when the load end and generator end voltages are V_L and V_G volts respectively.

864 $$G = 0.000128 \frac{P_L l^2}{(V_G - V_L)V_L} \text{ pounds.}$$

Sectional area (A) of copper wire required to transmit energy under the conditions stated in 864.

865 $$A = \frac{21.2\ lP_L}{(V_G - V_L)V_L} \text{ circular mils.}$$

Sectional area (A) of a copper wire for which the total annual cost of transmitting energy over a line conducting a constant current of I amperes will be a minimum.

866 $$A = 593\ I\sqrt{\frac{ch}{c'p}} \text{ circular mils.}$$

NOTE. **c** is the cost of the generated energy in dollars per kilowatt-hour, **c′** is the cost of the bare copper wire in dollars per pound, **h** is the number of hours per year that the line is in use and **p** is the annual percentage rate of interest on the capital invested in copper which will pay the annual capital interest, taxes and depreciation of the copper. Equation **866** states the principle known as Kelvin's Law.

TRANSIENT CURRENTS*

Current (i) flowing in a series circuit of **R** ohms resistance and **L** henrys self-inductance **t** seconds after a constant emf of **E** volts is impressed upon the circuit.

867 $$i = \frac{E}{R}\left(1 - \epsilon^{-\frac{Rt}{L}}\right) + I\epsilon^{-\frac{Rt}{L}} \text{ amperes.}$$

NOTE. **I** is the current in amperes flowing in the circuit at the instant before the emf is impressed. It is a positive quantity if flowing in conjunction and is a negative quantity if flowing in opposition to the emf.

* The value of ϵ throughout is 2.718

Current (i) flowing in a series circuit of **R** ohms resistance and **L** henrys self-inductance **t** seconds after its source of emf is short-circuited, the current flowing in the circuit at the instant before the short-circuit being **I** amperes.

868 $$i = I\epsilon^{-\frac{Rt}{L}} \text{ amperes.}$$

Current (i) flowing in a circuit of **R** ohms resistance and **C** farads series capacitance **t** seconds after a constant emf of **E** volts is impressed upon the circuit.

869 $$i = \left(\frac{E - V}{R}\right)\epsilon^{-\frac{t}{RC}} \text{ amperes.}$$

NOTE. **V** is the potential across the condenser at the instant before the emf is impressed. It is a positive quantity if acting in opposition and a negative quantity if acting in conjunction with the impressed emf.

Charge (q) on the condenser at any time **t** under the conditions stated in 869.

870 $$q = CE\left(1 - \epsilon^{-\frac{t}{RC}}\right) + CV\epsilon^{-\frac{t}{RC}} \text{ coulombs.}$$

Potential (v) across the condenser at any time **t** under the conditions stated in 869.

871 $$v = E\left(1 - \epsilon^{-\frac{t}{RC}}\right) + V\epsilon^{-\frac{t}{RC}} \text{ volts.}$$

Current (i) flowing in a circuit of **R** ohms resistance and **C** farads series capacitance **t** seconds after its source of emf is short-circuited, the potential across the condenser at the instant before the short-circuit being **V** volts.

872 $$i = \frac{V}{R}\epsilon^{-\frac{t}{RC}} \text{ amperes.}$$

Charge (q) on the condenser at any time **t** under the conditions stated in 872.

873 $$q = CV\epsilon^{-\frac{t}{RC}} \text{ coulombs.}$$

Potential (v) across the condenser at any time **t** under the conditions stated in 872.

874 $$v = V\epsilon^{-\frac{t}{RC}} \text{ volts.}$$

Current (i) flowing in a circuit of **R** ohms resistance, **L** henrys self-inductance and **C** farads series capacitance **t** seconds after a constant emf of **E** volts is impressed upon the circuit, the potential across the condenser and the current flowing in the circuit at the instant before the emf is impressed being **V** volts and **I** amperes respectively.

Case I. $R^2C > 4\ L$.

875 $$i = \left\{\frac{E - V - aLI}{(b - a)L}\right\}\epsilon^{-at} - \left\{\frac{E - V - bLI}{(b - a)L}\right\}\epsilon^{-bt} \text{ amperes.}$$

Case II. $R^2C = 4\ L$.

876 $$i = \left\{I + \left(\frac{2\ (E - V) - RI}{2\ L}\right)t\right\}\epsilon^{-\frac{Rt}{2L}} \text{ amperes.}$$

Case III. $R^2C < 4\ L$.

877 $$i = \left\{\left(\frac{2\ (E - V) - RI}{2\ \omega_1 L}\right)\sin \omega_1 t + I \cos \omega_1 t\right\}\epsilon^{-\frac{Rt}{2L}} \text{ amperes.}$$

NOTE. $a = \dfrac{RC - \sqrt{R^2C^2 - 4\ LC}}{2\ LC}$, $b = \dfrac{RC + \sqrt{R^2C^2 - 4\ LC}}{2\ LC}$ and $\omega_1 = \dfrac{\sqrt{4\ LC - R^2C^2}}{2\ LC}$. The current (I) is positive when flowing in the same direction as the impressed emf and the sign of the potential (V) is obtained as in 869.

Current (i) flowing in a circuit of **R** ohms resistance, **L** henrys self-inductance and **C** farads series capacitance **t** seconds after its source of emf is short-circuited, the potential across the condenser and the current flowing in the circuit at the instant before the short-circuit being **V** volts and **I** amperes respectively.

NOTE. Write 875, 876, or 877 making E zero in each case.

Current (i) flowing in a circuit of **R** ohms resistance, **L** henrys self-inductance and **C** farads series capacitance **t** seconds after a harmonic emf, $e = E_m \sin(\omega t + \alpha)$ volts, is impressed upon the circuit, the potential across the condenser and the current flowing in the circuit at the instant before the emf is impressed being **V** volts and **I** amperes respectively.

Case I. $R^2C > 4L$.

878 $$i = G\epsilon^{-at} - H\epsilon^{-bt} + \frac{E_m}{Z}\sin(\omega t + \alpha - \theta) \text{ amperes.}$$

Case II. $R^2C = 4L$.

879 $$i = (J + Kt)\epsilon^{-\frac{Rt}{2L}} + \frac{E_m}{Z}\sin(\omega t + \alpha - \theta) \text{ amperes.}$$

Case III. $R^2C < 4L$.

880 $$i = \left\{M \sin \omega_1 t + N \cos \omega_1 t\right\}\epsilon^{-\frac{Rt}{2L}} + \frac{E_m}{Z}\sin(\omega t + \alpha - \theta) \text{ amperes.}$$

NOTE. $$G = \frac{E_m \sin \alpha - V - aLI - \frac{E_m L}{Z}\left\{b \sin(\alpha - \theta) + \omega \cos(\alpha - \theta)\right\}}{(b - a)L}$$

$$H = \frac{E_m \sin \alpha - V - bLI - \frac{E_m L}{Z}\left\{a \sin(\alpha - \theta) + \omega \cos(\alpha - \theta)\right\}}{(b - a)L}$$

$$J = I - \frac{E_m}{Z}\sin(\alpha - \theta) \qquad \omega_1, \text{ a and b as in 877.}$$

$$K = \frac{1}{L}\left\{E_m \sin \alpha - V - \frac{RI}{2} - \frac{E_m}{Z}\left(\frac{R}{2}\sin(\alpha - \theta) + L\omega \cos(\alpha - \theta)\right)\right\}$$

$$M = \frac{K}{\omega_1} \qquad N = J \qquad \omega = 2\pi f \qquad \theta = \cos^{-1}\frac{R}{Z}$$

$$Z = \sqrt{R^2 + \left(\omega L - \frac{1}{\omega C}\right)^2}$$

$\alpha = \sin^{-1}\frac{e}{E_m}$, where e equals the algebraic value of the harmonic emf at the instant that it is impressed on the circuit. The current (I) is positive if flowing in the same direction as the impressed emf and the potential (V) is positive if acting in opposition to the impressed emf, both at time (t) equals zero. If the circuit contains no condenser the series capacitance is infinite and $a = 0$, $b = \frac{R}{L}$ and $Z = \sqrt{R^2 + \omega^2 L^2}$.

HARMONIC ALTERNATING CURRENTS

Electromotive force (e) of a harmonic emf of maximum value E_m volts and angular velocity ω radians per second at any harmonic time **t** seconds.

881 $$e = E_m \sin \omega t \text{ volts.}$$

NOTE. A harmonic cycle is a single sequence of harmonic values from zero to positive maximum to zero to negative maximum to zero. The harmonic frequency (f) is the sequence rate in cycles per second. The angular velocity (ω) in radians per second equals 2π times the frequency (f). The harmonic time (t) is the time in seconds measured from the instant when the harmonic value is zero and is increasing to a positive maximum. When a harmonic emf is indicated by the expression, $e = E_m \sin(\omega t + \alpha)$, harmonic time is measured from the instant when $e = E_m \sin \alpha$.

Current (i) flowing at any harmonic emf time **t** seconds in a circuit of **R** ohms resistance, **L** henrys self-inductance and C farads series capacitance upon which a harmonic emf, $e = E_m \sin \omega t$, is impressed.

882 $$i = \frac{E_m}{\sqrt{R^2 + \left(L\omega - \frac{1}{C\omega}\right)}} \sin\left\{\omega t - \tan^{-1}\left(\frac{L\omega - \frac{1}{C\omega}}{R}\right)\right\} \text{ amperes.}$$

NOTE. It is assumed that the emf has been impressed upon the circuit long enough to produce a harmonic current. The early transient current is given by 878, 879 or 880.

Maximum current (I_m) flowing in a circuit under the conditions stated in 882.

883 $$I_m = \frac{E_m}{\sqrt{R^2 + \left(L\omega - \frac{1}{C\omega}\right)^2}} \text{ amperes.}$$

Effective or root-mean-square emf (E) of a harmonic emf, $e = E_m \sin \omega t$ volts.

884 $$E = \frac{E_m}{\sqrt{2}} \text{ volts.}$$

NOTE. The effective current (I) of a harmonic current equals $\frac{I_m}{\sqrt{2}}$ amperes.

Average emf (E_a) of a harmonic emf, $e = E_m \sin \omega t$ volts.

885 $$E_a = \frac{2\,E_m}{\pi} \text{ volts.}$$

NOTE. The average current (I_a) of a harmonic current equals $\frac{2\,I_m}{\pi}$ amperes.

Form factor (f.f.) and amplitude factor (**a.f.**) respectively of a harmonic emf, $e = E_m \sin \omega t$ volts.

886 $$\text{f.f.} = \frac{E}{E_a} = 1.11.$$

$$\text{a.f.} = \frac{E_m}{E} = 1.414.$$

NOTE. The form factor (**f.f.**) and amplitude factor (**a.f.**) respectively of a harmonic current are $\frac{I}{I_a} = 1.11$ and $\frac{I_m}{I} = 1.414$.

Reactance (X) of a circuit of **L** henrys self-inductance and **C** farads series capacitance when conducting a harmonic current of **ω** radians per second angular velocity.

887 $$X = L\omega - \frac{1}{C\omega} \text{ ohms.}$$

NOTE. $L\omega$ is called the inductive reactance and $\frac{1}{C\omega}$ the capacitive reactance of the circuit, each measured in ohms.

Impedance (Z) of a circuit of **R** ohms resistance and **X** ohms series reactance.

888 $$Z = \sqrt{R^2 + X^2} \text{ ohms.}$$

Phase angle (θ) of a circuit of **R** ohms resistance and **X** ohms series reactance.

889 $$\theta = \tan^{-1}\frac{X}{R}.$$

NOTE. The phase angle (θ) of a circuit in radians divided by the angular velocity (ω) of the conducted current in radians per second equals the time **t** in seconds by which the harmonic current lags or leads the harmonic emf. A positive value of $\frac{X}{R}$ indicates a lagging current and a negative value a leading current.

Power factor (p.f.) of a part-circuit of **R** ohms resistance, **Z** ohms impedance and phase angle **θ** containing no generated emf.

890 $$\text{p.f.} = \frac{R}{Z} = \cos\theta.$$

Total resistance (R_s) of a series circuit. See 845.

Total reactance (X_s) of a series circuit the respective parts of which have reactances of X_1, X_2, X_3, etc., ohms.

891 $$X_s = X_1 + X_2 + X_3, \text{ etc., ohms.}$$

NOTE. The addition is algebraic, inductive reactance being positive and capacitive reactance negative.

Total impedance (Z_s) of a series circuit of R_s ohms total resistance and X_s ohms total reactance. See 888.

NOTE. The total impedance of a series circuit does not equal the sum of the impedances of its respective parts unless the ratio of reactance to resistance in each part is the same and the net reactances are of the same sign.

Power (P) delivered to or from a part-circuit conducting an effective current of I amperes across which the effective potential rise in the direction of the current is V volts, the phase angle between the current and the potential rise being θ.

892 $$P = VI \cos \theta \text{ watts.}$$

NOTE. Positive power indicates net power delivered from, and negative power indicates net power delivered to the part-circuit. Multiply 892 or 895 by seconds to obtain energy in joules and by hours divided by 1000 to obtain energy in kilowatt-hours.

Reactive power (Q) under the conditions stated in 892.

893 $$Q = VI \sin \theta \text{ vars.}$$

NOTE. Leading reactive power is considered by convention to be positive. Lagging reactive power is considered negative.

Volt-amperes (V-A) **or apparent power** under the conditions stated in 892.

894 $$V\text{-}A = VI \text{ volt-amperes.}$$

NOTE. No significance is ascribed to the sign of apparent power.

Power (P) delivered to a part-circuit of R ohms resistance conducting an effective current of I amperes and containing no generated emf.

895 $$P = I^2R \text{ watts.}$$

NOTE. The net power delivered to a reactance is zero.

Reactive power (**Q**) delivered to a part-circuit of **X** ohms reactance conducting an effective current of **I** amperes and containing no generated emf.

896 $$Q = I^2X \text{ vars.}$$

NOTE. Q is positive for a capacitive reactance and negative for an inductive reactance. The reactive power delivered to a resistance is zero.

Volt-amperes (**V-A**) delivered to a part-circuit of **Z** ohms impedance conducting an effective current of **I** amperes and containing no generated emf.

897 $$V\text{-}A = I^2Z \text{ volt-amperes.}$$

Volt-amperes (**V-A**) corresponding to a power of **P** watts and a reactive power of **Q** vars.

898 $$V\text{-}A = \sqrt{P^2 + Q^2}.$$

Effective vector expression ($\dot{E}$) and ($\dot{I}$) for an emf, $e = E_m \sin(\omega t + \alpha)$ volts, and a current, $i = I_m \sin(\omega t - \beta)$ amperes.

899 $$\dot{E} = \left(\frac{E_m}{\sqrt{2}}\cos\alpha + j\frac{E_m}{\sqrt{2}}\sin\alpha\right) = \frac{E_m}{\sqrt{2}}\underline{/\alpha} \text{ volts.}$$

$$\dot{I} = \left(\frac{I_m}{\sqrt{2}}\cos\beta - j\frac{I_m}{\sqrt{2}}\sin\beta\right) = \frac{I_m}{\sqrt{2}}\underline{/-\beta} \text{ amperes.}$$

NOTE. In symbolic notation the horizontal component of a vector is without prefix and its sign is + to the right and − to the left of the **Y** axis; the vertical component is designated by the prefix **j** and its sign is + above and − below the **X** axis. In some mathematical operations the symbol **j** has the value $\sqrt{-1}$. The symbols $\underline{/\alpha}$ and $\underline{/-\beta}$ indicate vectors making the angles $+\alpha$ and $-\beta$, respectively, with the **X** axis and having magnitudes given by the quantity preceding the symbols. This is known as the polar form of the vector expression.

Vector electromotive force ($\dot{E}_{AD}$) in a circuit the constituent parts of which contain the vector emf's $\dot{E}_{AB}$, $\dot{E}_{BC}$ and $\dot{E}_{CD}$ volts.

900 $$\dot{E}_{AD} = \dot{E}_{AB} + \dot{E}_{BC} + \dot{E}_{CD} \text{ volts.}$$

NOTE. Each vector emf must be referred to the same axis of reference. The subscripts in each case indicate the direction of emf rise.

Vector current ($\dot{I}_{BA}$) flowing from B toward a junction A from which the vector currents $\dot{I}_{AC}$, $\dot{I}_{AD}$ and $\dot{I}_{AF}$ amperes flow away.

901 $$\dot{I}_{BA} = \dot{I}_{AC} + \dot{I}_{AD} + \dot{I}_{AF} \text{ amperes.}$$

Electromotive force equivalent (E) of a vector emf, $\dot{E} = (a + jb)$ volts.

902 $$E = \sqrt{a^2 + b^2} \text{ volts.}$$

NOTE. In polar form $\dot{E} = \sqrt{a^2 + b^2} \angle \tan^{-1}\frac{b}{a}$. The current equivalent (I) of a vector current $\dot{I} = (c + jd)$ amperes is $\sqrt{c^2 + d^2}$ amperes, and in polar form $I = \sqrt{c^2 + d^2} \angle \tan^{-1}\frac{d}{c}$.

Symbolic expression ($\dot{Z}$) for the impedance of a circuit of R ohms resistance and X ohms reactance.

903 $$\dot{Z} = (R + jX) \text{ ohms.}$$

NOTE. The resistance component has no prefix and is always +; the reactance component has the prefix j, a + sign indicating net inductive reactance and a − sign net capacitive reactance. In polar form, $\dot{Z} = \sqrt{R^2 + X^2} \angle \tan^{-1}\frac{X}{R}$.

Symbolic impedance ($\dot{Z}_{AD}$) between the ends A and D of a part-circuit containing several series parts of symbolic impedance $\dot{Z}_{AB}$, $\dot{Z}_{BC}$ and $\dot{Z}_{CD}$ ohms respectively.

904 $$\dot{Z}_{AD} = \dot{Z}_{AB} + \dot{Z}_{BC} + \dot{Z}_{CD} \text{ ohms}$$
$$= (R_{AB} + R_{BC} + R_{CD}) + j(X_{AB} + X_{BC} + X_{CD}) \text{ ohms.}$$

Vector current ($\dot{I}$) flowing in the direction of an emf rise, $\dot{E} = (a + jb)$ volts acting in a circuit of symbolic impedance $\dot{Z} = (r + jx)$ ohms.

905 $$\dot{I} = \left(\frac{a + jb}{r + jx}\right) \text{ amperes.}$$

NOTE. To rationalize 905 multiply both numerator and denominator by the denominator with the sign of its j term reversed. We then have

$$\dot{I} = \frac{(a + jb)\ (r - jx)}{(r + jx)\ (r - jx)} = \frac{ar - j^2bx + jbr - jax}{r^2 - j^2x^2}.$$

In this operation $j = \sqrt{-1}$ or $j^2 = -1$. Hence

$$\dot{I} = \frac{(ar + bx) + j\,(br - ax)}{r^2 + x^2} = \left(\frac{ar + bx}{r^2 + x^2}\right) + j\left(\frac{br - ax}{r^2 + x^2}\right).$$

Alternately, in polar form

$$\dot{I} = \frac{\sqrt{a^2 + b^2} \angle \tan^{-1}\frac{b}{a}}{\sqrt{r^2 + x^2} \angle \tan^{-1}\frac{x}{r}} = \frac{\sqrt{a^2 + b^2}}{\sqrt{r^2 + x^2}} \angle \tan^{-1}\frac{b}{a} - \tan^{-1}\frac{x}{r}.$$

Vector potential rise ($\dot{V}_{AB}$) between the ends A and B of a part-circuit of symbolic impedance $\dot{Z}_{AB}$ ohms conducting a current of vector value $\dot{I}_{AB}$ amperes and containing an emf rise of vector value $\dot{E}_{AB}$ volts.

906 $$\dot{V}_{AB} = +\dot{E}_{AB} - \dot{I}_{AB}\dot{Z}_{AB} \text{ volts.}$$

NOTE. If $\dot{E}_{AB} = \mathbf{a} + \mathbf{jb}$, $\dot{I}_{AB} = \mathbf{c} + \mathbf{jd}$ and $\dot{Z}_{AB} = \mathbf{r} + \mathbf{jx}$,

$$\dot{V}_{AB} = a + jb - (c + jd)(r + jx)$$
$$= a + jb - cr - j^2dx - jcx - jdr$$

and since $j^2 = -1$,

$$\dot{V}_{AB} = (a - cr + dx) + j(b - cx - dr).$$

Vector potential rise ($\dot{V}_{AD}$) between the ends A and D of a part-circuit containing several series parts across which the respective vector potential rises are $\dot{V}_{AB}$, $\dot{V}_{BC}$ and $\dot{V}_{CD}$ volts.

907 $$\dot{V}_{AD} = \dot{V}_{AB} + \dot{V}_{BC} + \dot{V}_{CD} \text{ volts.}$$

Power (**P**) delivered to or from a part-circuit conducting a vector current $\dot{I} = (\mathbf{c} + \mathbf{jd})$ amperes and across which the vector potential rise in the direction of the current is $\dot{V} = (\mathbf{a} + \mathbf{jb})$ volts.

908 $$P = (ac + bd) \text{ watts.}$$

NOTE. The signs of **a, b, c** and **d** are preserved in **908.** Positive power indicates power delivered from, and negative power indicates power delivered to the part-circuit. The power does not equal $(a + jb)(c + jd)$.

Reactive power (**Q**) under the conditions stated in **908.**

909 $$Q = (ad - bc) \text{ vars.}$$

NOTE. The signs of **a, b, c** and **d** should be preserved in **909.** Leading reactive power is positive; lagging reactive power is negative.

Conductance (**G**) **and susceptance** (**B**) of a branch of **R** ohms resistance, **X** ohms series reactance and **Z** ohms impedance.

910 $$G = \frac{R}{R^2 + X^2} = \frac{R}{Z^2} \text{ mhos.}$$

$$B = \frac{X}{R^2 + X^2} = \frac{X}{Z^2} \text{ mhos.}$$

Admittance (**Y**) of a branch of **Z** ohms impedance, **G** mhos conductance and **B** mhos susceptance.

911 $$Y = \frac{1}{Z} = \sqrt{G^2 + B^2} \text{ mhos.}$$

Total conductance (G_0) of several parallel branches of G_1, G_2 and G_3 mhos conductance respectively.

912 $$G_0 = G_1 + G_2 + G_3 \text{ mhos.}$$

Total susceptance (B_0) of several parallel branches of B_1, B_2 and B_3 mhos susceptance respectively.

913 $$B_0 = B_1 + B_2 + B_3 \text{ mhos.}$$

NOTE. The addition is algebraic, inductive susceptance being positive and capacitive susceptance negative.

Total admittance (Y_0) of several parallel branches of total conductance G_0 mhos and total susceptance B_0 mhos. See 911.

NOTE. The total admittance of a parallel circuit does not equal the sum of the admittances of the respective branches unless the ratio of susceptance to conductance in each branch is the same and the net susceptances are of the same sign.

Phase-angle (θ) of a circuit of G mhos conductance and B mhos susceptance.

914 $$\theta = \tan^{-1}\frac{B}{G}.$$

Power factor (p.f.) of a part-circuit of G mhos conductance and Y mhos admittance, containing no generated emf.

915 $$\text{p.f.} = \frac{G}{Y}.$$

Resistance (R) **and reactance** (X) of a circuit of G mhos conductance, B mhos susceptance and Y mhos admittance.

916 $$R = \frac{G}{G^2 + B^2} = \frac{G}{Y^2} \text{ ohms.}$$

$$X = \frac{B}{G^2 + B^2} = \frac{B}{Y^2} \text{ ohms.}$$

Impedance (Z) of a circuit of Y mhos admittance.

917 $$Z = \frac{1}{Y} \text{ ohms.}$$

Symbolic expression ($\underset{\cdot}{Y}$) for the admittance of a **circuit of** G mhos conductance and B mhos susceptance.

918 $$\underset{\cdot}{Y} = (G - jB) \text{ mhos.}$$

NOTE. In polar form, $\underset{\cdot}{Y} = \sqrt{G^2 + B^2} \,\big/ \tan^{-1}\frac{B}{G}$.

Symbolic admittance ($\underset{\cdot}{Y}_0$) of a parallel circuit containing several branches of symbolic admittance $\underset{\cdot}{Y}_1$, $\underset{\cdot}{Y}_2$ and $\underset{\cdot}{Y}_3$ mhos respectively.

919 $$\underset{\cdot}{Y}_0 = \underset{\cdot}{Y}_1 + \underset{\cdot}{Y}_2 + \underset{\cdot}{Y}_3 \text{ mhos.}$$

Vector current ($\underset{\cdot}{I}$) flowing in the direction of an emf rise $\underset{\cdot}{E}$ acting in a circuit of $\underset{\cdot}{Y}$ mhos symbolic admittance.

920 $$\underset{\cdot}{I} = \underset{\cdot}{E}\underset{\cdot}{Y} \text{ amperes.}$$

Nomenclature of 3-phase circuits. Line emf or voltage, E_l volts; phase emf or voltage, E_p volts; line current, I_l amperes; phase current, I_p amperes; phase angle between phase voltage and phase current, θ_p. In 925 and 927 $\underset{\cdot}{E}_a$, $\underset{\cdot}{E}_b$ and $\underset{\cdot}{E}_c$ are any three voltage vectors which may exist in a three-phase system, such as voltages to neutral or to ground, line-to-line voltages, induced voltages, etc. Likewise $\underset{\cdot}{I}_a$, $\underset{\cdot}{I}_b$ and $\underset{\cdot}{I}_c$ in 926 and 928 may be line currents, phase currents, the currents in a Δ-connected winding, etc. The subscripts 1, 2 and 0 in 925 to 928 denote, respectively, positive-, negative-, and zero-sequence components. Positive phase rotation, ABC in counter-clockwise direction.

Conditions for balanced 3-phase circuit: all phase currents, phase emf's, and phase voltages, respectively, equal and differing in phase by 120 degrees. Conditions for unbalanced 3-phase circuit: phase currents, phase emf's, or phase voltages, respectively, unequal or not differing in phase by 120 degrees.

Balanced Y-connected branches (Fig. 921).

921
$$E_l = \sqrt{3}\, E_p.$$
$$I_l = I_p.$$
$$\underset{\cdot}{E}_{OA} + \underset{\cdot}{E}_{OB} + \underset{\cdot}{E}_{OC} = 0.$$
$$\underset{\cdot}{E}_{AB} + \underset{\cdot}{E}_{BC} + \underset{\cdot}{E}_{CA} = 0.$$
$$\underset{\cdot}{E}_{AB} = \underset{\cdot}{E}_{OB} - \underset{\cdot}{E}_{OA} = \sqrt{3}\,\underset{\cdot}{E}_{OB}\big/\underline{30^\circ} = \sqrt{3}\,\underset{\cdot}{E}_{OC}\big/\underline{90^\circ}.$$
$$\underset{\cdot}{I}_{OA} + \underset{\cdot}{I}_{OB} + \underset{\cdot}{I}_{OC} = 0.$$

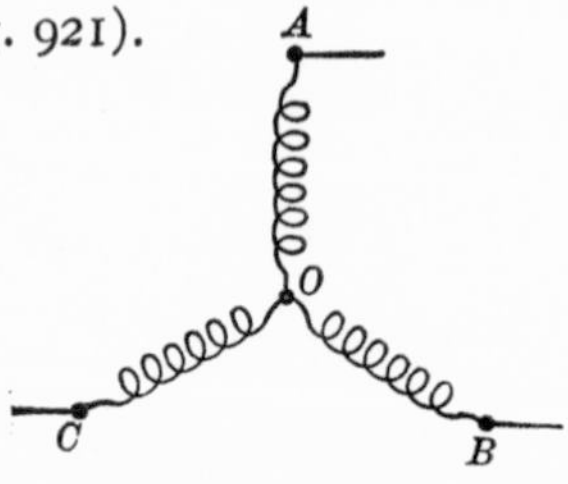

FIG. 921.

Balanced Δ-connected branches (Fig. 922).

922 $E_l = E_p.$

$I_l = \sqrt{3}\, I_p.$

$\underset{\cdot}{E}_{AB} + \underset{\cdot}{E}_{BC} + \underset{\cdot}{E}_{CA} = 0.$

$\underset{\cdot}{I}_{AB} + \underset{\cdot}{I}_{BC} + \underset{\cdot}{I}_{CA} = 0.$

$\underset{\cdot}{I}_{AA'} = \underset{\cdot}{I}_{CA} - \underset{\cdot}{I}_{AB} =$

$\sqrt{3}\, \underset{\cdot}{I}_{CA} \underline{/30^\circ} = \sqrt{3}\, \underset{\cdot}{I}_{BC} \overline{/90^\circ}$

$\underset{\cdot}{I}_{AA'} + \underset{\cdot}{I}_{BB'} + \underset{\cdot}{I}_{CC'} = 0.$

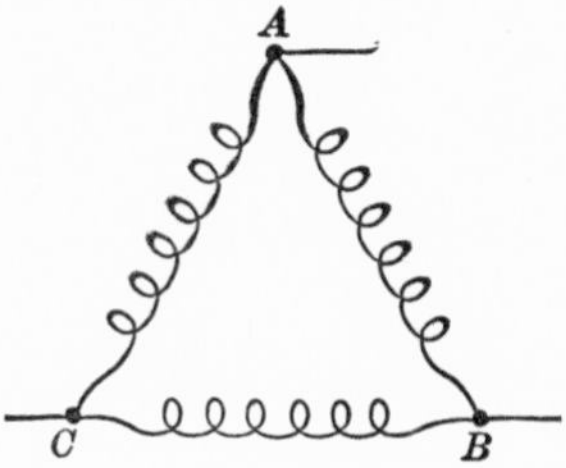

FIG. 922.

Unbalanced Y-connected branches (Fig. 921).

923
$$\underset{\cdot}{E}_{AB} + \underset{\cdot}{E}_{BC} + \underset{\cdot}{E}_{CA} = 0.$$
$$\underset{\cdot}{E}_{AB} = \underset{\cdot}{E}_{OB} - \underset{\cdot}{E}_{OA}$$
$$\underset{\cdot}{I}_{OA} + \underset{\cdot}{I}_{OB} + \underset{\cdot}{I}_{OC} = 0.$$

NOTE. If there is a current flowing out of the point o in a neutral connection, it must be added vectorially to the left-hand side of the last equation.

Unbalanced Δ-connected branches (Fig. 922).

924
$$\underset{\cdot}{E}_{AB} + \underset{\cdot}{E}_{BC} + \underset{\cdot}{E}_{CA} = 0.$$
$$\underset{\cdot}{I}_{AA'} = \underset{\cdot}{I}_{CA} - \underset{\cdot}{I}_{AB}$$

Symmetrical components of voltage in an unbalanced 3-phase system.

925
$$\underset{\cdot}{E}_{a1} = \tfrac{1}{3}\left(\underset{\cdot}{E}_a + \underset{\cdot}{E}_b \underline{/120^\circ} + \underset{\cdot}{E}_c \overline{/120^\circ}\right)$$
$$\underset{\cdot}{E}_{a2} = \tfrac{1}{3}\left(\underset{\cdot}{E}_a + \underset{\cdot}{E}_b \overline{/120^\circ} + \underset{\cdot}{E}_c \underline{/120^\circ}\right)$$
$$\underset{\cdot}{E}_0 = \tfrac{1}{3}\left(\underset{\cdot}{E}_a + \underset{\cdot}{E}_b + \underset{\cdot}{E}_c\right)$$

Symmetrical components of current in an unbalanced 3-phase system.

926 Replace $\underset{\cdot}{E}$ in 925 by $\underset{\cdot}{I}$.

Three-phase voltages in terms of the symmetrical components of voltage.

927
$$\underset{\cdot}{E}_a = \underset{\cdot}{E}_{a1} + \underset{\cdot}{E}_{a2} + \underset{\cdot}{E}_0$$
$$\underset{\cdot}{E}_b = \underset{\cdot}{E}_{a1} \overline{/120^\circ} + \underset{\cdot}{E}_{a2} \underline{/120^\circ} + \underset{\cdot}{E}_0$$
$$\underset{\cdot}{E}_c = \underset{\cdot}{E}_{a1} \underline{/120^\circ} + \underset{\cdot}{E}_{a2} \overline{/120^\circ} + \underset{\cdot}{E}_0$$

Three-phase currents in terms of the symmetrical components of current.

928 Replace E in 927 by I.

Y-connected impedances which are equivalent to a given set of Δ-connected impedances so far as conditions at the terminals are concerned. See Fig. 929.

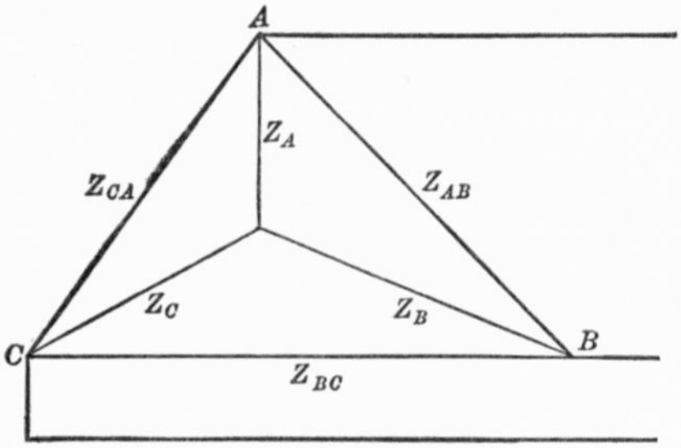

FIG. 929.

929

$$Z_A = \frac{Z_{CA}Z_{AB}}{Z_{AB} + Z_{BC} + Z_{CA}}$$

$$Z_B = \frac{Z_{AB}Z_{BC}}{Z_{AB} + Z_{BC} + Z_{CA}}$$

$$Z_C = \frac{Z_{BC}Z_{CA}}{Z_{AB} + Z_{BC} + Z_{CA}}$$

NOTE. If the impedances are balanced, the Y-connected impedances are $\frac{1}{3}$ of the Δ-connected impedances.

Δ-connected impedances which are equivalent to a given set of Y-connected impedances so far as conditions at the terminals are concerned. See Fig. 929.

930

$$Z_{AB} = \frac{Z_AZ_B + Z_BZ_C + Z_CZ_A}{Z_C}$$

$$Z_{BC} = \frac{Z_AZ_B + Z_BZ_C + Z_CZ_A}{Z_A}$$

$$Z_{CA} = \frac{Z_AZ_B + Z_BZ_C + Z_CZ_A}{Z_B}$$

NOTE. If the impedances are balanced, the Δ-connected impedances are 3 times the Y-connected impedances.

Power (**P**) delivered to or from a balanced 3-phase line.

931 $$P = \sqrt{3}\, E_l I_l \cos \theta_p \text{ watts.}$$

Power factor (**p.f.**) of a balanced 3-phase load.

932 $$\text{p.f.} = \cos \theta_p.$$

Power (**P**) delivered to or from an unbalanced 3-phase line.

933 $$P = E_{p_1} I_{p_1} \cos \theta_{p_1} + E_{p_2} I_{p_2} \cos \theta_{p_2} + E_{p_3} I_{p_3} \cos \theta_{p_3} \text{ watts.}$$

NOTE. The subscripts 1, 2 and 3 here denote the three phases.

Power factor (**p.f.**) of an unbalanced 3-phase load.

934 $$\text{p.f.} = \frac{P}{E_{p_1} I_{p_1} + E_{p_2} I_{p_2} + E_{p_3} I_p}.$$

NOTE. The subscripts 1, 2 and 3 here denote the three phases.

Power (**P**) measured by two wattmeters connected in a 3-phase line as shown in Fig. 935.

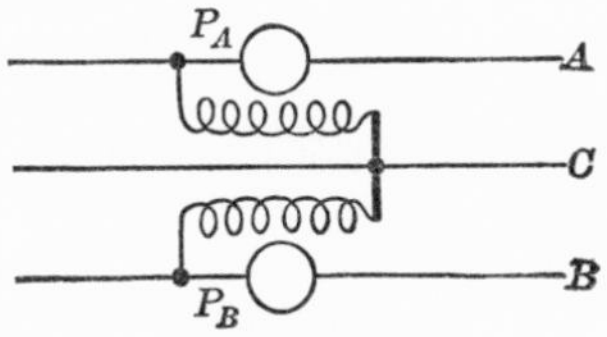

FIG. 935.

935 $$P = P_A \pm P_B \text{ watts.}$$

NOTE. To determine use of + or − sign, break connection of potential coil of wattmeter A at line C and connect to line B. A wattmeter deflection on scale indicates the use of the + sign and a deflection off scale indicates the use of the − sign.

Power (**P_A**) **and** (**P_B**) respectively measured by two wattmeters connected in a 3-phase balanced line as shown in Fig. 935.

936 $$P_A = E_l I_l \cos (30^\circ - \theta_p) \text{ watts.}$$
$$P_B = E_l I_l \cos (30^\circ + \theta_p) \text{ watts.}$$

Phase angle (**θ**) of a balanced 3-phase load when two wattmeters connected as shown in Fig. 935 measure **P_A** and **P_B** watts, respectively.

937 $$\theta = \tan^{-1} \sqrt{3}\, \frac{P_A - P_B}{P_A + P_B}.$$

NOTE. The additions and subtractions are algebraic.

NON-HARMONIC ALTERNATING CURRENTS

Effective or root-mean-square value (E) of a periodically time-varying emf or voltage **e** = f(t) volts having a period of **T** seconds.

938 $$E = \sqrt{\frac{1}{T}\int_0^T e^2\,dt} \text{ volts.}$$

Effective or root-mean-square value (I) of a periodically time-varying current **i** = f(t) amperes having a period of **T** seconds.

939 $$I = \sqrt{\frac{1}{T}\int_0^T i^2\,dt} \text{ amperes.}$$

Electromotive force (e) of a non-harmonic emf or voltage at any harmonic time **t** seconds.

940 $$e = E_{m_1}\sin(\omega t + \theta_1) + E_{m_3}\sin(3\,\omega t + \theta_3) + E_{m_5}\sin(5\,\omega t + \theta_5) + \ldots \text{ volts.}$$

NOTE. E_{m_1}, E_{m_3}, E_{m_5}, etc., represent the maximum values of the first, third, fifth, etc., harmonics, and θ_1, θ_2, θ_3, etc., their respective phase angles with a common axis of reference. The angular velocity ω is that of the fundamental or first harmonic. Alternators do not normally generate even harmonics of voltage. Some non-sinusoidal voltages, such as the outputs of rectifiers, however, may contain odd or even harmonics and average or d-c components. In such cases the terms $E_0 + E_{m_2}\sin(2\,\omega t + \theta_2) + E_{m_4}\sin(4\,\omega t + \theta_4) + \ldots$ should be added to the right-hand member of 940.

Current (i) at any harmonic time **t** seconds flowing in a circuit of **R** ohms resistance, **L** henrys self-inductance and **C** farads series capacitance upon which a non-harmonic emf of the form stated in 940 is impressed.

941 $$i = I_{m_1}\sin(\omega t + \theta_1') + I_{m_3}\sin(3\,\omega t + \theta_3') + I_{m_5}\sin(5\,\omega t + \theta_5') + \text{, etc., amperes.}$$

NOTE.

$$I_{m_1} = \frac{E_{m_1}}{\sqrt{R^2 + \left(L\omega - \frac{1}{C\omega}\right)^2}}, \qquad \theta_1' = \theta_1 - \tan^{-1}\left(\frac{L\omega - \frac{1}{C\omega}}{R}\right),$$

$$I_{m_3} = \frac{E_{m_3}}{\sqrt{R^2 + \left(3\,L\omega - \frac{1}{3\,C\omega}\right)^2}}, \qquad \theta_3' = \theta_3 - \tan^{-1}\left(\frac{3\,L\omega - \frac{1}{3\,C\omega}}{R}\right),$$

$$I_{m_5} = \frac{E_{m_5}}{\sqrt{R^2 + \left(5\,L\omega - \frac{1}{5\,C\omega}\right)^2}}, \quad \theta_5' = \theta_5 - \tan^{-1}\left(\frac{5\,L\omega - \frac{1}{5\,C\omega}}{R}\right).$$

Effective emf (E) of a non-harmonic emf of the form stated in 940.

942 $$E = \sqrt{\frac{(E_{m_1})^2 + (E_{m_3})^2 + (E_{m_5})^2}{2}} \text{ volts.}$$

NOTE. The effective value of a non-harmonic potential or current is obtained in the same manner.

Power (P) delivered to a part-circuit conducting a non-harmonic current of the form stated in 941 and upon which is impressed a non-harmonic emf of the form stated in 940.

943 $$P = \frac{E_{m_1} I_{m_1}}{2} \cos(\theta_1 - \theta_1') + \frac{E_{m_3} I_{m_3}}{2} \cos(\theta_3 - \theta_3') + \frac{E_{m_5} I_{m_5}}{2} \cos(\theta_5 - \theta_5') +, \text{ etc., watts.}$$

Power factor (p.f.) of a part-circuit conducting a non-harmonic current of effective value **I** amperes and absorbing energy at a rate of **P** watts, the effective value of the non-harmonic potential between its ends being **V** volts.

944 $$\text{p.f.} = \frac{P}{VI}.$$

Harmonic emf and current equivalent to the non-harmonic forms stated in 940 and 941.

945 $$e = \sqrt{2}\,E \sin \omega t \text{ volts;}$$
$$i = \sqrt{2}\,I \sin(\ t \pm \cos^{-1} \text{p.f.}) \text{ amperes.}$$

NOTE. **p.f.** is the power factor of the circuit upon which the non-harmonic emf is impressed.

Resistance (R) of a part-circuit containing no source of generated emf, conducting an effective current of **I** amperes and absorbing energy at a rate of **P** watts.

946 $$R = \frac{P}{I^2} \text{ ohms.}$$

Impedance (Z) of a part-circuit containing no source of generated emf, conducting an effective current of **I** amperes and across which the potential is **V** volts.

947 $$Z = \frac{V}{I} \text{ ohms.}$$

Reactance (X) of a part-circuit of **R** ohms resistance and **Z** ohms impedance.

948 $$X = \sqrt{Z^2 - R^2} \text{ ohms.}$$

NOTE. The reactance of a part-circuit to a non-harmonic current does not equal $\left(L\omega - \frac{1}{C\omega}\right)$ ohms.

DIRECT CURRENT MACHINERY

Dynamos

NOTE. Unless indicated otherwise each formula applies to a generator or a motor.

Nomenclature and units of measurement. Emf generated in armature, **E** volts; terminal potential or voltage, **V** volts; armature current, **I** amperes; line current, I_l amperes; shunt field current, I_f amperes; series field current, I_s amperes; armature resistance between brushes, **R** ohms; shunt field resistance including rheostat, R_f ohms; series field resistance including shunt, R_s ohms; number of poles, **p**; shunt field turns per pole, N_f; series field turns per pole, N_s; number of armature paths between terminals, **m**; number of armature conductors, **Z**; magnetic flux per pole, **Φ** maxwells; armature speed, **n** revolutions per minute; armature torque, **T** pound-feet.

Electromotive force (E) generated in the armature of a dynamo.

949 $$E = \frac{p\Phi Zn}{6\,m \times 10^9} \text{ volts.}$$

Shunt field current (I_{fd}) equivalent to the demagnetizing magnetomotive force of the armature of a dynamo per pole when the armature current is **I** amperes and the brushes are shifted through an angle of θ space degrees from the neutral plane to improve commutation.

950 $$I_{fd} = \frac{ZI\theta}{360\ N_f m} \text{ amperes.}$$

Shunt field current (I_{fs}) equivalent to the magnetomotive force of the series turns of a dynamo per pole.

951 $$I_{fs} = \frac{N_s}{N_f} I_s \text{ amperes.}$$

Net field current (I_{fn}) of a dynamo at any load.

952 $$I_{fn} = I_f - I_{fd} \pm I_{fs} \text{ amperes.}$$

NOTE. The sign before I_{fs} is + for a cumulative and − for a differential compound dynamo.

Terminal voltage (V) of a shunt dynamo when the armature current is **I** amperes and the generated emf is **E** volts.

953 $$V = E \pm IR \text{ volts.}$$

NOTE. The sign before **IR** is + for a motor and − for a generator. In a series or long-shunt compound dynamo, $V = E \pm I\ (R + R_s)$ volts and in a short-shunt compound dynamo, $V = E \pm IR \pm I_sR_s$ volts.

Armature speed (n) of a dynamo when the generated emf is **E** volts.

954 $$n = \frac{6\ Em \times 10^9}{p\Phi Z} \text{ revs. per minute.}$$

Armature torque (T) of a dynamo when the armature current is **I** amperes.

955 $$T = \frac{0.1175\ Z\Phi Ip}{m \times 10^8} \text{ pound-feet.}$$

Rotational losses (P_r) of a dynamo which, operated as a shunt motor at no load with a voltage between brushes of **V** volts, takes an armature current of **I** amperes.

956 $$P_r = VI - I^2R \text{ watts.}$$

NOTE. To determine the rotational losses corresponding to a definite load the dynamo, operated as a shunt motor at no load, must be run at the same speed and with the same generated emf as when running at the definite load.

Copper losses (P_k) of a dynamo at any load.

957 Shunt field, $P_f = I_f^2 R_f = VI_f$ watts.
Series field, $P_s = I_s^2 R_s$ watts.
Armature, $P_a = I^2 R$ watts.

Power input (P_i) **to a generator** at any load.

958 $P_i = EI + P_r$ watts.
$= P_o + P_k + P_r$ watts.
$= 0.1420\ nT$ watts.
$= 1.903\ nT \times 10^{-4}$ horse-power.

Power output (P_o) **of a generator** at any load.

959 $P_o = VI_l$ watts.

Power input (P_i) **to a motor** at any load.

960 $P_i = VI_l$ watts.

Power output (P_o) **of a motor** at any load.

961 $P_o = EI - P_r$ watts.
$= P_i - P_k - P_r$ watts.
$= 0.1420\ nT$ watts.
$= 1.903\ nT \times 10^{-4}$ horse-power.

Efficiency (η) of a dynamo at any load.

962 $$\eta = \frac{P_o}{P_i} = \frac{P_o}{P_o + P_k + P_r} = \frac{P_i - P_k - P_r}{P_i}.$$

ALTERNATING CURRENT MACHINERY

NOTE. Sinusoidal voltages and currents are assumed throughout and their magnitudes are expressed by effective values.

Synchronous Machines

NOTE. Three-phase machines with Y-connected windings are assumed throughout. All machine impedances, voltages, and currents are phase values for a Y-connection. Unless indicated otherwise each formula applies to a synchronous generator, motor, or condenser.

Frequency (f) of the voltage generated in a synchronous machine having **p** poles, the speed of the armature or field being **n** revolutions per minute.

963 $$f = \frac{pn}{120} \text{ cycles per second.}$$

Synchronous internal voltage or excitation voltage (E_i) generated in the armature of a synchronous machine.

964
$$E_i = 4.44\, fN\Phi \cos\frac{\beta}{2}\left(\frac{\sin\frac{m\alpha}{2}}{m\sin\frac{\alpha}{2}}\right) \times 10^{-8} \text{ volts.}$$

NOTE. N is the number of armature series turns per phase or one-half the number of series conductors on the armature divided by the number of phases, Φ is the main field flux per pole in maxwells, β is the pitch deficiency or the difference in electrical degrees between the pole pitch (180°) and the coil pitch, m is the number of slots per pole per phase and α is the angle between adjacent slot centers in electrical degrees. Electrical degrees equal space degrees multiplied by $\frac{p}{2}$. This equation assumes that the arrangement of the winding before each pole is the same and hence that there are an integral number of slots per pole per phase. If ϕ_r, the resultant air-gap flux per pole corresponding to the mmf R in 967, be used instead of ϕ, the voltage given by 964 is the air-gap voltage E_a.

Field magnetomotive force (F) in a cylindrical-rotor machine.

965
$$F = \frac{4}{\pi} N_f I_f \left(\frac{\sin\frac{n_f \alpha_f}{2}}{n_f \sin\frac{\alpha_f}{2}}\right) \text{ ampere-turns per pole.}$$

NOTE. F is the maximum value of the fundamental field mmf. N_f is the number of field turns per pole carrying the current I_f amperes per turn. n_f is the number of rotor slots per pole, and α_f is the electrical angle between adjacent slots in a belt. 965 assumes that the field winding is a regular, distributed, full-pitch winding.

Armature magnetomotive force (A) in a cylindrical-rotor machine.

966
$$A = 0.90\, KN_a I \text{ ampere-turns per pole.}$$

NOTE. K equals $\cos\frac{\beta}{2}\left(\frac{\sin\frac{m\alpha}{2}}{m\sin\frac{\alpha}{2}}\right)$ as explained in 964. N_a is the number of armature series turns per pole. I is the armature current. A may be obtained in equivalent field amperes by dividing by N_f.

Resultant magnetomotive force (R) in a machine with a field magnetomotive force F ampere-turns per pole and an armature magnetomotive force A ampere-turns per pole.

967 $\underset{\cdot}{R} = \underset{\cdot}{F} + \underset{\cdot}{A}$ ampere-turns per pole.

NOTE. The addition must be made vectorially. See 970, 971, and Fig. 970.

Characteristic curves (Fig. 968) of a synchronous machine. Data are plotted as follows: **OCC** (open-circuit characteristic), terminal voltage at no-load versus field current; the air-gap line is drawn tangent to the lower part of **OCC**; **SCC** (short-circuit characteristic), armature current with the armature terminals short-circuited versus field current; **ZPF** (zero-power-factor characteristic), terminal voltage versus field current with the armature supplying a constant current I_0 to a zero-power-factor lagging load at its terminals, I_0 usually being taken equal to rated armature current.

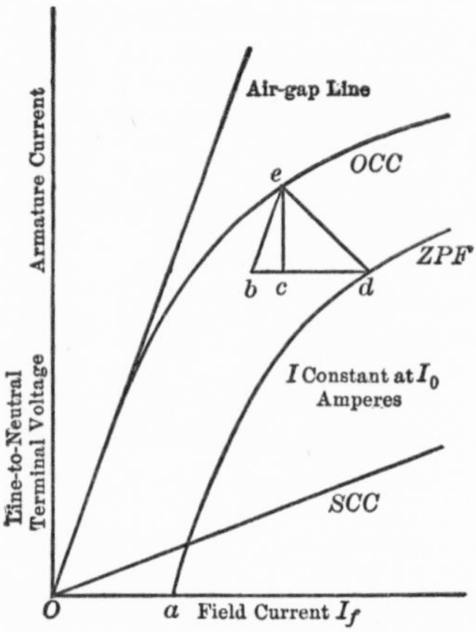

FIG. 968.

Potier triangle cde (Fig. 968). Procedure in obtaining Potier triangle: Choose a point **d** well up on the curved part of **ZPF**; draw **db** parallel to and equal to **ao**; draw **be** parallel to the air-gap line; draw **ec** perpendicular to **db**; **cde** is then the Potier triangle.

Potier reactance (X_p) from Potier triangle (see Fig. 968).

968 $$X_p = \frac{\text{ec in volts}}{I_0 \text{ in amperes}} \text{ ohms.}$$

NOTE. I_0 is the constant armature current for which **ZPF** is drawn. X_p is very nearly equal to and is frequently used for the armature leakage reactance X_a.

Armature magnetomotive force (A) corresponding to the armature current $I = I_0$ from Potier triangle (see Fig. 968).

969 **A** = **cd** equivalent field amperes.

NOTE. To obtain **A** in ampere-turns per pole as in 966, substitute **Cd** for I_f in 965. **A** for any other value of **I** may be obtained by direct proportion.

General-method vector diagram for a cylindrical-rotor machine with armature current **I**, terminal voltage **V**, and phase angle **θ** between **I** and **V**, Fig. 970. Figure 970 is drawn for generator operation with a lagging power-factor load. All voltage vectors are voltage rises. For motor operation, **I** and the voltage drop vector **−V** are **θ** degrees out of phase.

Air-gap voltage $\mathbf{E_a}$ is given by

970 $$\dot{\mathbf{E}}_a = \dot{\mathbf{V}} + \dot{\mathbf{I}}\,(\mathbf{r} + \mathbf{j}\mathbf{X_a}),$$

in which the voltages $\dot{\mathbf{E}}_a$ and $\dot{\mathbf{V}}$ are rises and **r** is the armature resistance.

A is obtained from **969** or **966**.

R is obtained in equivalent field amperes by entering **OCC** (Fig. 968) with $\mathbf{E_a}$ volts and reading the corresponding field current.

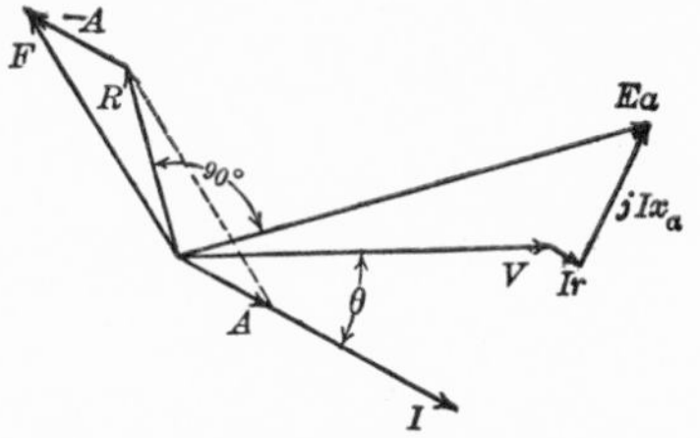

FIG. 970.

Field current ($\mathbf{I_f}$) required in a cylindrical-rotor machine with armature current **I**, terminal voltage **V**, and phase angle **θ** between **I** and **V**.

971 $$\mathbf{I_f} = \text{magnitude of } (\dot{\mathbf{R}} - \dot{\mathbf{A}}) \text{ amperes.}$$

NOTE. **R** and **A** must be in equivalent field amperes in **971**; if they are in ampere-turns per pole, use equation **965** to convert ampere-turns per pole into equivalent field amperes. See Fig. 970. **R** and **A** obtained as in **970**.

Excitation voltage ($\mathbf{E_i}$) under conditions in 971.

972 Enter **OCC** (Fig. 968) with $\mathbf{I_f}$ from **971** and read the corresponding voltage.

For salient-pole machines the methods of **965** to **972** will give approximately correct results. In **965**, **F** should be taken as $\mathbf{N_f I_f}$, and in **966** the factor 0.75 should be used instead of 0.9. For more accurate results the Blondel two-reaction method should be used.

Total power output ($\mathbf{P_o}$) of a cylindrical-rotor synchronous machine with excitation voltage $\mathbf{E_i}$ volts, terminal voltage **V**

volts, angle δ electrical degrees between the vectors $\dot{E}_i$ and $\dot{V}$, and synchronous reactance X_s ohms.

973 $$P_o = \frac{3\ VE_i}{X_s} \sin \delta \text{ watts.}$$

NOTE. Losses are neglected. X_s should be properly adjusted for saturation. The maximum power output is $\frac{3\ VE_i}{X_s}$ watts; it may not be attainable without loss of synchronism. $\frac{3\ VE_i}{X_s}$ watts gives the breakdown or pull-out power for a cylindrical-rotor motor connected directly to a power system of capacity large compared to that of the motor.

Total power output (P_o) of a salient-pole synchronous machine with excitation voltage E_d volts, terminal voltage V volts, angle δ electrical degrees between the vectors $\dot{E}_d$ and $\dot{V}$, direct-axis synchronous reactance X_d ohms, and quadrature-axis synchronous reactance X_q ohms.

974 $$P_o = 3\left(\frac{VE_d}{X_d} \sin \delta + \frac{V^2 (X_d - X_q)}{2\ X_d X_q} \sin 2\ \delta\right) \text{watts.}$$

NOTE. Losses are neglected. The reactances should be properly adjusted for saturation. The maximum value of P_o, indicated by 974 as δ varies, may not be attainable without loss of synchronism. This maximum value gives the breakdown or pull-out power for a salient-pole motor connected directly to a power system of capacity large compared to that of the motor.

Efficiency (η) of a synchronous machine when the output is P_o watts.

975 $$\eta = \frac{P_o}{P_o + P_a + P_c + P_s + P_{fw} + P_f}.$$

NOTE. All powers are total 3-phase powers. P_a, the armature copper loss, equals three times the armature current squared times the ohmic resistance per phase. P_c is the open-circuit core loss (hysteresis and eddy-current losses). To determine P_c, enter a curve of open-circuit core loss versus open-circuit voltage with a voltage equal to the vector sum of the terminal voltage plus the armature resistance drop, and read the corresponding value of P_c. P_s is the stray-load loss (skin effect and eddy-current losses in the armature conductors plus local core losses due to armature leakage flux) and may be found from a curve of stray-load loss versus armature current. P_{fw} is the friction and windage loss, and P_f is the field copper loss.

Synchronous Converters

Effective alternating voltage (V_{ac}) between slip-rings of a synchronous converter when the direct voltage between brushes is V_{dc} volts.

976 $$V_{ac} = 0.707\, V_{dc} \sin\frac{\pi}{n} \text{ volts.}$$

NOTE. n, the number of slip-rings, equals two for a single-phase machine and for a polyphase machine equals the number of phases.

Alternating line current (I_{ac}) of a synchronous converter when the direct armature current is I_{dc} amperes.

977 $$I_{ac} = \frac{2.83\, I_{dc}}{\eta n\,(\text{p.f.})} \text{ amperes.}$$

NOTE. The efficiency (η) is approximately 0.95.

Armature copper loss (P_a) of a synchronous converter when the direct armature current is I_{dc} amperes.

978 $$P_a = \left[\frac{8}{\left(\eta n\,(\text{p.f.}) \sin\frac{\pi}{n}\right)^2} - \frac{1.62}{\eta} + 1\right] I_{dc}^2 R_{dc} \text{ watts.}$$

NOTE. R_{dc} is the resistance of the armature between the direct current brushes.

Field current (I_{fa}) equivalent to the armature magnetomotive force of a synchronous converter when the power output is P_o watts.

979 $$I_{fa} = \frac{1.5\, K N_a P_o \tan\theta}{p \eta n V_{ac} N_f} \text{ amperes.}$$

NOTE. K and N_a as in 966. N_f as in 965. p is the number of poles and θ the power factor angle.

Net field current (I_{fn}) of a synchronous converter at any stated load.

980 $$I_{fn} = I_f + I_{fs} \pm I_{fa} \text{ amperes.}$$

NOTE. I_f is the actual shunt field current, I_{fa} is determined by 979 and I_{fs}, the equivalent shunt field current for the series field equals $\frac{I_{dc} N_s}{N_f}$ where I_{dc} is the direct line current of a short-shunt compound machine, N_s and N_f are respectively the number of series field turns and shunt field turns per pole.

Conversion	a-c to d-c		d-c to a-c	
Current phase	lag	lead	lag	lead
Sign before I_{fa}	+	−	−	+

Efficiency (η) of a synchronous converter when operating **a-c** to **d-c** and delivering P_o watts.

981 $$\eta = \frac{P_o}{P_o + P_a + P_c + P_{fw} + P_f + P_s}.$$

NOTE. P_a is determined by 978, P_c as in 975 where the curve is entered with the terminal voltage. P_{fw} is found by test, P_f and P_s by 957. This is the efficiency of the converter alone and does not include the losses in any associated transformers.

Transformers

Electromotive force (E) induced in N turns linked by a flux, $\phi = \Phi_m \sin 2\pi ft$ maxwells.

982 $$E = 4.44\, Nf\Phi_m\, 10^{-8} \text{ volts.}$$

Core loss (P_c) of a transformer at any load.

983 $$P_c = P_h + P_e \text{ watts.}$$

NOTE. P_h is the hysteresis loss and P_e the eddy-current loss in the magnetic circuit of the transformer. See 802 and 803.

Ratio of transformation (T_1) from primary to secondary of a transformer wound with two coils of N_1 (primary) and N_2 (secondary) turns respectively.

984 $$T_1 = \frac{N_1}{N_2}.$$

NOTE. T_1 equals the ratio $\left(\frac{E_1}{E_2}\right)$ of the emf's induced respectively in the primary and secondary coils and equals approximately the ratio $\left(\frac{V_1}{V_2}\right)$ of terminal voltages of the primary and secondary coils or the ratio $\left(\frac{I_2}{I_1}\right)$ of the secondary and primary currents. The ratio of transformation (T_2) from secondary to primary equals $\frac{1}{T_1}$.

Magnetizing current (I_m) in a coil of N turns wound on a magnetic circuit of uniform maximum permeability (μ), l centimeters in mean length, **A** square centimeters in mean section and conducting a flux, $\phi = \Phi_m \sin 2\pi ft$ maxwells.

985 $$I_m = \frac{10\,\Phi_m l}{4\,\pi N \mu A \sqrt{2}} \text{ amperes (approx.)}$$

Core loss current (I_c) in a coil containing an induced emf of **E** volts and wound on a magnetic circuit in which the core loss is P_c watts.

986 $$I_c = \frac{P_c}{E} \text{ amperes.}$$

No load current (I_n) taken by a transformer which requires a magnetizing current of I_m amperes and a core loss current of I_c amperes.

987 $$I_n = \sqrt{I_m^2 + I_c^2} \text{ amperes.}$$

Equivalent resistance (R_1) and **equivalent reactance** (X_1) between the primary terminals of a transformer which has a primary resistance of r_1 ohms, a primary leakage reactance of x_1 ohms, a secondary resistance of r_2 ohms, a secondary leakage reactance of x_2 ohms and primary to secondary ratio of transformation of T_1.

988 $$R_1 = r_1 + T_1^2 r_2 \text{ ohms.}$$
$$X_1 = x_1 + T_1^2 x_2 \text{ ohms.}$$

NOTE. The equivalent resistance and reactance respectively between the secondary terminals is given by $R_2 = r_2 + T_2^2 r_1$ ohms and $X_2 = x_2 + T_2^2 x_1$ ohms. The equivalent impedance in each case equals $\sqrt{R^2 + X^2}$ ohms and $Z_1 = T_1^2 Z_2$ ohms.

Equivalent resistance (R_1) between the primary terminals of a transformer which, with short-circuited secondary, absorbs P_i watts with a primary current of I_1 amperes.

989 $$R_1 = \frac{P_i}{I_1^2} \text{ ohms.}$$

Equivalent impedance (Z_1) between the primary terminals of a transformer which, with secondary short-circuited and with

V_1 volts between the primary terminals, takes a primary current of I_1 amperes.

990 $$Z_1 = \frac{V_1}{I_1} \text{ ohms.}$$

Equivalent reactance (X_1) between the primary terminals of a transformer of equivalent resistance (R_1) ohms and equivalent impedance (Z_1) ohms between the primary terminals.

991 $$X_1 = \sqrt{Z_1^2 - R_1^2} \text{ ohms.}$$

Primary voltage (V_1) of a transformer of ratio of transformation (T_1), equivalent resistance and reactance respectively between secondary terminals (R_2) and (X_2) ohms, secondary terminal voltage (V_2) volts, secondary current (I_2) amperes and power factor of the load on the secondary ($\cos \theta_2$).

992 $$V_1 = T_1 \sqrt{(V_2 \cos \theta_2 + I_2R_2)^2 + (V_2 \sin \theta_2 \pm I_2X_2)^2} \text{ volts.}$$

NOTE. The sign before I_2X_2 is + for zero or lagging current phase and − for leading current phase.

Voltage regulation (v.r.) of a transformer at any load; V_1, V_2 and T_1 as in 992.

993 $$\text{v.r.} = \frac{V_1 - T_1V_2}{T_1V_2}.$$

Efficiency (η) of a transformer at any load.

994 $$\eta = \frac{I_2V_2 \cos \theta_2}{I_2V_2 \cos \theta_2 + I_2^2R_2 + P_c}.$$

Induction Machines

NOTE. Three-phase Y-wound machines are assumed throughout and unless indicated otherwise each formula applies to a generator or a motor. All rotor resistances and reactances are referred to the stator. All impedances, voltages, and currents are phase values for a Y connection. All values of power **P** are total 3-phase powers.

Equivalent effective resistance (R_1) of an induction machine.

995 $$R_1 = \frac{P_i}{3\, I_1^2} \text{ ohms.}$$

NOTE. P_i is the power input on blocked run and I_1 is the stator blocked-run current.

Equivalent impedance (Z_1) of an induction machine.

996 $$Z_1 = \frac{V_1}{I_1} \text{ ohms.}$$

NOTE. V_1 is the stator terminal voltage during blocked run and I_1 is the stator blocked-run current.

Equivalent reactance (X_1) of an induction machine.

997 $$X_1 = \sqrt{Z_1^2 - R_1^2} \text{ ohms.}$$

NOTE. Z_1 and R_1 are determined as in 996 and 995.

Rotor resistance (r_2) of an induction machine referred to the stator.

998 $$r_2 = T_1^2 r_2' \text{ ohms.}$$

NOTE. r_2' is the actual rotor resistance and T_1 is the ratio of transformation from stator to rotor or the ratio of the emf's induced in the stator and rotor respectively during blocked-run.

Rotor leakage reactance (x_2) of an induction machine referred to the stator.

999 $$x_2 = T_1^2 x_2' \text{ ohms.}$$

NOTE. Read note to 998, substituting x_2' for r_2' and reactance for resistance.

Equivalent effective resistance (R_1) of an induction machine of r_{1e} ohms effective stator resistance and r_{2e} ohms effective rotor resistance referred to the stator.

1000 $$R_1 = r_{1e} + r_{2e} \text{ ohms.}$$

Equivalent reactance (X_1) of an induction machine of x_1 ohms stator leakage reactance and x_2 ohms rotor leakage reactance referred to the stator.

1001 $$X_1 = x_1 + x_2 \text{ ohms.}$$

Synchronous speed (n_1) of an induction machine having **p** poles, the frequency of the impressed voltage being **f** cycles per second.

1002 $$n_1 = \frac{120\,f}{p} \text{ revolutions per minute.}$$

Slip (s) of an induction machine of synchronous speed (n_1) revolutions per minute when the rotor speed is n_2 revolutions per minute.

1003 $$s = 1 - \frac{n_2}{n_1}.$$

Equivalent circuit of an induction motor, Fig. 1004. r_{1e} and r_{2e} as in 1000. x_1 and x_2 as in 1001. Z_n, the exciting impedance, and its components r_n and x_n may be determined from a no-load run. The determination is similar to that for R_1, Z_1 and X_1 in 995, 996 and 997 except that the no-load voltage and current and the no-load power input minus friction and windage losses must be used. $r_{2e}\left(\frac{1-s}{s}\right)$ is the resistance equivalent for the rotational power, which is the shaft power output plus friction and windage and a small core loss.

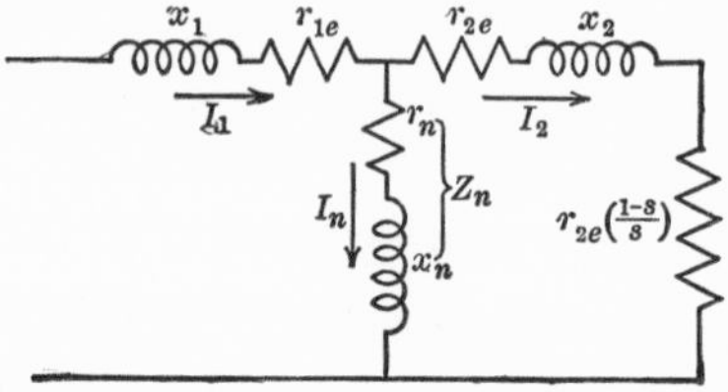

FIG. 1004.

Induced stator emf (E_n) of an induction machine at no load.

1004 $$E_n = V_1 - I_n \sqrt{r_{1e}^2 + x_1^2} \text{ volts.}$$

NOTE. V_1 is the stator terminal voltage, I_n the no-load line current, r_{1e} and x_1 as in 1000 and 1001.

Rotor current (I_2) referred to stator of an induction machine at slip (s).

1005 $$I_2 = \frac{E_n}{\sqrt{\left(r_{1e} + \frac{r_{20}}{s}\right)^2 + (X_1)^2}} \text{ amperes.}$$

NOTE. For a wound-rotor induction motor with an effective external resistance R_e ohms per phase referred to the stator, substitute ($r_{2e} + R_e$) for r_{20}. To find the starting rotor current of an induction motor make s equal one.

Stator current (I_1) of an induction machine at slip (s).

1006 $$I_1 = \sqrt{I_2^2 + I_n^2 + 2\,I_2 I_n \sin\alpha} \text{ amperes.}$$

NOTE. I_2 is determined by 1005, I_n is the no-load current and $\alpha = \sin^{-1}$ (p.f. at no load) $+ \tan^{-1}\left(\frac{sx_2}{r_{20}}\right)$. For a wound-rotor motor, substitute $(r_{2e} + R_e)$ for r_{20} in the expression for α. The starting stator current is given by making s equal one, the starting rotor current being determined as indicated in 1005.

Power output (P_o) of an induction machine at slip (s).

1007 $$P_o = 3\,I_2^2 r_{20}\left(\frac{1 - s}{s}\right) - P_{fw} \text{ watts.}$$

NOTE. For a wound-rotor motor r_{20} should be increased as indicated in 1005. When the slip is negative P_o is negative and gives the power input to an induction generator. P_{fw} is the friction and windage loss.

Power input (P_i) to an induction machine at slip (s).

1008 $$P_i = 3\frac{I_2^2 r_{20}}{s} + 3\,I_1^2 r_{1e} + P_n - 3\,I_n^2 r_{1e} \text{ watts.}$$

NOTE. P_n is the power in watts taken at no load. For a wound-rotor motor r_{20} should be increased as indicated in 1005. When the slip is negative P_i is negative and gives the power output of an induction generator.

Output torque (T) of an induction motor at slip (s).

1009 $$T = 0.1761\frac{PI_2^2 r_{20}}{fs} - T_{fw} \text{ pound-feet.}$$

NOTE. Read comment on r_{20} for a wound-rotor motor and s under starting conditions in 1005. T_{fw} is the friction and windage torque. If P_{fw} is known in watts, T_{fw} equals $\frac{0.0587\,p\,P_{fw}}{f\,(1 - s)}$ pound-feet.

Slip (s) of an induction motor at any stated load.

1010 $$s = \frac{r_{20}\left(\frac{3\,E_n^2}{P_o} - 2\,r_{1e} - 2\,r_{20}\right)}{\left(\frac{3\,E_n^2}{P_o} - 2\,r_{1e} - 2\,r_{20}\right)^2 - Z_1^2} \text{ approximately.}$$

NOTE. Read comment on r_{20} for a wound-rotor motor in 1005.

Slip (s) of an induction generator at any stated load.

1011 $$s = \frac{r_{20}\dfrac{3\,E_n^2}{P_o}}{\left(\dfrac{3\,E_n^2}{P_o}\right) - Z_1^2 + r_{1e}\left(\dfrac{3\,E_n^2}{P_o}\right)} \text{ approximately.}$$

Efficiency (η) of an induction machine.

1012 $$\eta = \frac{P_o}{P_i}.$$

A-C POWER TRANSMISSION

NOTE. Three-phase power networks are assumed throughout. All apparatus is assumed to be Y-connected, and all impedances and admittances are for one phase or line on this basis. All currents are line currents (phase currents for a Y-connection). Unless otherwise stated, all voltages except in 1013 and 1014 are line-to-neutral voltages (phase voltages for a Y-connection). In 1013 and 1014 line-to-line voltages are used.

Per unit impedance (Z_{pu}) in terms of the impedance **Z** in ohms in one phase of a 3-phase system, expressed on a 3-phase base of **kva** kilovolt-amperes and a line-to-line base voltage of V_b volts.

1013 $$Z_{pu} = \frac{Z \times kva}{V_b^2} \times 10^3 \text{ per unit.}$$

NOTE. Per cent impedance is 100 Z_{pu}. Per unit or per cent impedances of electrical machinery are customarily expressed with the rated kilovolt-amperes and voltage as base quantities. In computations on any one power network, however, the same kilovolt-ampere base and voltage bases consistent with the transformer ratios must be used throughout. For changing from one base quantity to another, per cent and per unit impedances are directly proportional to the **kva** base and inversely proportional to the square of the voltage base.

Per unit admittance (Y_{pu}) in terms of the admittance **Y** in mhos in one phase of a 3-phase system, expressed on a 3-phase base of **kva** kilovolt-amperes and a line-to-line voltage base of V_b volts.

1014 $$Y_{pu} = \frac{YV_b^2}{kva} \times 10^{-3}.$$

NOTE. Per cent admittance is 100 Y_{pu}. In computations on any one power network the same kilovolt-ampere base and voltage bases consistent with the transformer ratios must be used. For changing from one base quantity to another, per cent and per unit admittances are inversely proportional to the **kva** base and directly proportional to the square of the voltage base.

Voltage ($\mathbf{E_s}$) at the sending end of a transmission line of **R** ohms resistance per conductor and **X** ohms reactance per conductor conducting a current of **I** amperes, the voltage at the receiving end being $\mathbf{E_r}$ volts and the phase angle between the receiving-end voltage and the line current being θ_r.

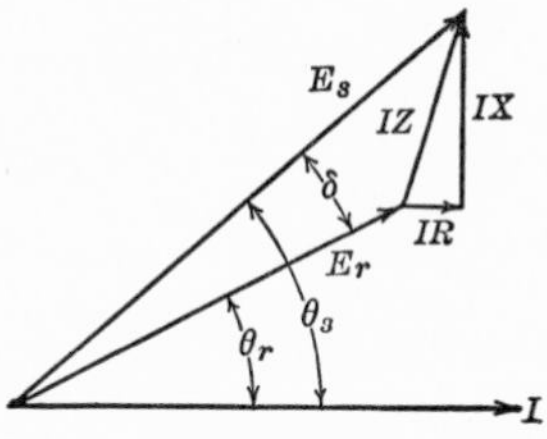

FIG. 1015.

1015 $$E_s = \sqrt{(E_r \cos\theta_r + IR)^2 + (E_r \sin\theta_r \pm IX)^2} \text{ volts.}$$

NOTE. See vector diagram, Fig. 1015. When **I** lags or is in phase with $\mathbf{E_r}$, the sign before **IX** is +, and when **I** leads $\mathbf{E_r}$, the sign before **IX** is −. 1015 neglects capacitance. For transmission lines longer than 40 miles, the capacitance should be included. For cables longer than about 2 miles the capacitance should be included. See **1020, 1026** and **1027.**

For application to single-phase lines, use 2 **R** and 2 **X** in place of **R** and **X.**

Phase-angle (θ_s) between the sending-end voltage and the line current under the conditions stated in 1015.

1016 $$\theta_s = \tan^{-1}\frac{E_r \sin\theta_r \pm IX}{E_r \cos\theta_r + IR}$$

NOTE. The power factor at the sending end is $\cos\theta_s$.

Power input ($\mathbf{P_s}$) to the sending end of a transmission line with δ degrees angular displacement between sending- and receiving-end voltages, the line having an impedance **Z** ohms and an impedance angle $\theta = \tan^{-1}\frac{X}{R}$.

1017 $$P_s = \frac{E_s^2}{Z}\cos\theta - \frac{E_s E_r}{Z}\cos(\delta + \theta) \text{ watts.}$$

NOTE. δ is positive if E_s leads E_r. Capacitance is neglected. $\mathbf{P_s}$ will be total 3-phase power if line-to-line voltages are used and power per phase if line-to-neutral voltages are used.

The maximum value is $\frac{E_s E_r}{Z} - \frac{E_s^2}{Z}\cos\theta$; this maximum value may not be attainable without instability in the system.

Power output (P_r) at the receiving end of a transmission line under the conditions in **1017**.

1018 $$P_r = \frac{E_sE_r}{Z}\cos(\delta - \theta) - \frac{E_r^2}{Z}\cos\theta \text{ watts.}$$

NOTE. Read first paragraph of note to **1017**, substituting P_r for P_s. The maximum value of P_r is $\frac{E_sE_r}{Z} - \frac{E_r^2}{Z}\cos\theta$; this maximum value may not be attainable without instability in the system.

Efficiency (η) of a transmission line under the conditions stated in **1015**.

1019 $$\eta = \frac{E_rI\cos\theta_r}{E_rI\cos\theta_r + I^2R} = \frac{E_r\cos\theta_r}{E_s\cos\theta_s} = \frac{P_r}{P_s}.$$

Nominal π (Fig. **1020a**) **and nominal T** (Fig. **1020b**) equivalent circuits for a transmission line of length l miles having an impedance **z** ohms per mile composed of resistance **r** ohms per mile and reactance **x** ohms per mile, and an admittance **y** mhos per mile composed of a leakage conductance **g** mhos per mile and a capacitive susceptance of **b** mhos per mile.

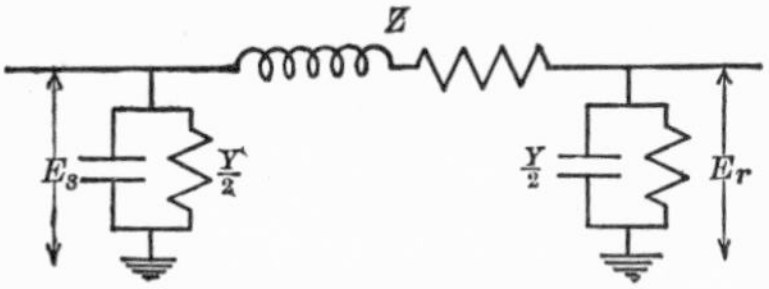

FIG. 1020*a*.

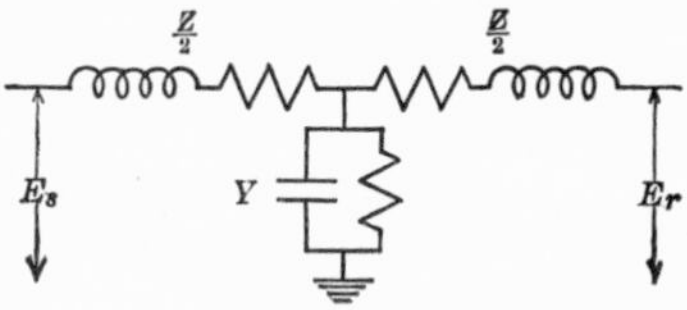

FIG. 1020*b*.

1020 $$\begin{aligned} Z &= zl \text{ ohms} \\ &= (r + jx)l \text{ vector ohms} \\ Y &= yl \text{ mhos} \\ &= (g + jb) \text{ vector mhos.} \end{aligned}$$

NOTE. The admittance **Y** is almost always purely capacitive, leakage usually being negligible. Nominal π and **T** circuits are usually used for transmission lines of lengths between about 40 and 100 miles. Below 40 miles the capacitance may be neglected; above 100 miles the equivalent π or **T** should be used (see **1026** and **1027**). For cables the nominal π and **T** circuits are usually used for lengths between about 2 and 5 miles. Below 2 miles the capacitance may be neglected; above 5 miles the equivalent π or **T** should be used.

The long transmission line. Nomenclature: length of line, l miles. Resistance, inductive reactance, capacitance, and leakage conductance, respectively, **r** ohms per mile, **x** ohms per mile, **c** farads per mile, and **g** mhos per mile; all per conductor. Impedance, **z** vector ohms per mile. Admittance, **y** vector mhos per mile. In the following formulas, **z** and **y** are vectors equal, respectively, to $\mathbf{r} + \mathbf{jx}$ and $\mathbf{g} + \mathbf{j\omega c}$. Sending-end voltage and current, respectively, $\mathbf{E_s}$ vector volts and $\mathbf{I_s}$ vector amperes. Receiving-end voltage and current, respectively, $\mathbf{E_r}$ vector volts and $\mathbf{I_r}$ vector amperes. In the following formulas, $\mathbf{E_s}$, $\mathbf{I_s}$, $\mathbf{E_r}$ and $\mathbf{I_r}$ are vector quantities with a common, arbitrary reference axis.

Propagation constant (α) of a transmission line.

1021 $\alpha = \sqrt{\mathbf{zy}} = \sqrt{(\mathbf{r} + \mathbf{jx})\,(\mathbf{g} + \mathbf{j\omega c})}$ hyps per mile.

Note. The real part of α is called the attenuation constant. The imaginary part is called the wave-length constant, phase constant, or velocity constant.

Surge impedance or characteristic impedance ($\mathbf{Z_0}$) of a transmission line.

1022 $\mathbf{Z_0} = \sqrt{\dfrac{\mathbf{z}}{\mathbf{y}}} = \sqrt{\dfrac{\mathbf{r} + \mathbf{jx}}{\mathbf{g} + \mathbf{j\omega c}}}$ vector ohms.

Hyperbolic angle (θ) of a transmission line.

1023 $\theta = \alpha \mathbf{l}$ numerics.

Current (i) **and voltage** (e) at a point on the line distant **x** miles from the receiving end in terms of receiving-end voltage and current.

1024 $\mathbf{i} = \mathbf{I_r} \cosh \alpha \mathbf{x} + \dfrac{\mathbf{E_r}}{\mathbf{Z_0}} \sinh \alpha \mathbf{x}$ vector amperes.

$\mathbf{e} = \mathbf{E_r} \cosh \alpha \mathbf{x} + \mathbf{I_r Z_0} \sinh \alpha \mathbf{x}$ vector volts.

Note. To obtain $\mathbf{I_s}$ and $\mathbf{E_s}$, substitute θ for $\alpha \mathbf{x}$.

Current (i) **and voltage** (e) at a point on the line distant **x** miles from the receiving end in terms of sending-end voltage and current.

1025 $\mathbf{i} = \mathbf{I_s} \cosh (1 - \mathbf{x})\alpha - \dfrac{\mathbf{E_s}}{\mathbf{Z_0}} \sinh (1 - \mathbf{x})\alpha$ vector amperes.

$\mathbf{e} = \mathbf{E_s} \cosh (1 - \mathbf{x})\alpha - \mathbf{I_s Z_0} \sinh (1 - \mathbf{x})\alpha$ vector volts.

Note. To obtain $\mathbf{I_r}$ and $\mathbf{E_r}$, substitute θ for $(1 - \mathbf{x})\alpha$.

Equivalent π circuit for a long transmission line. See Fig. 1026.

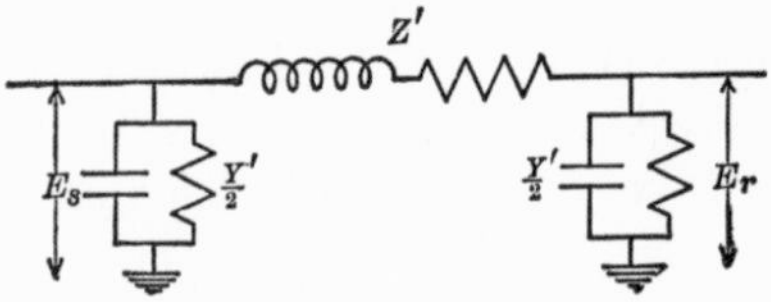

FIG. 1026.

1026
$$\mathbf{Z'} = zl\frac{\sinh \theta}{\theta} \text{ vector ohms.}$$

$$\mathbf{Y'} = yl\frac{\tanh \frac{\theta}{2}}{\frac{\theta}{2}} \text{ vector mhos.}$$

NOTE. Equivalent π and **T** circuits will represent exactly the performance of smooth transmission lines at their terminals under steady-state conditions. For transmission lines of lengths below about 100 miles, the nominal π and **T** circuits (see 1020) may be used unless very precise results are desired.

Equivalent T circuit for a long transmission line. See Fig. 1027.

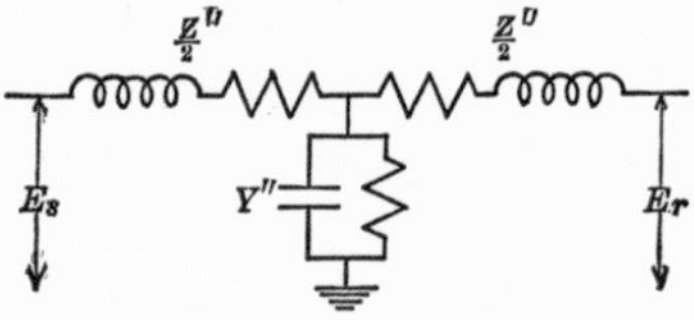

FIG. 1027.

1027
$$\mathbf{Z''} = zl\frac{\tanh \frac{\theta}{2}}{\frac{\theta}{2}} \text{ vector ohms.}$$

$$\mathbf{Y''} = yl\frac{\sinh \theta}{\theta}$$

NOTE. See note to 1026.

General circuit constants (A, B, C and D) are often used to express the steady-state performance of a network consisting of any combination of constant impedances to which power is supplied at one point and received at another. See Fig. 1028 and note to **1028**.

Sending-end voltage (E_s) and current (I_s) in terms of the receiving-end voltage E_r volts and current I_r amperes for the network of Fig. 1028.

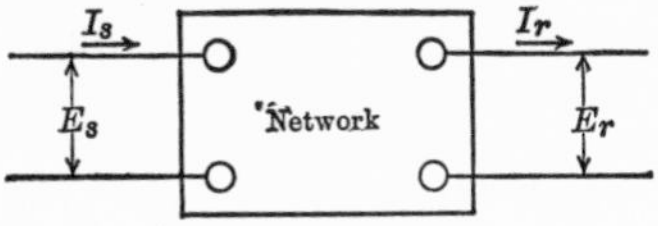

FIG. 1028.

1028 $E_s = AE_r + BI_r$ vector volts.
$I_s = CE_r + DI_r$ vector amperes.

NOTE. All quantities are vector quantities. These are the defining equations for general circuit constants. A and B are found for a given network by obtaining from circuit equations an expression for E_s in terms of E_r and I_r; A is then the coefficient of E_r in this expression, and B is the coefficient of I_r. C and D are found similarly from the expression for I_s in terms of E_r and I_r. Thus, for a network consisting of a series impedance Z, $E_s = E_r + ZI_r$, and $I_s = I_r$; hence for this network, $A = 1$, $B = Z$, $C = 0$, and $D = 1$.

Receiving-end voltage (E_r) and current (I_r) in terms of the sending-end voltage E_s volts and current I_s amperes for the network of Fig. 1028.

1029 $E_r = DE_s - BI_s$ vector volts.
$I_r = -CE_s + AI_s$ vector amperes.

NOTE. All quantities are vector quantities.

$+$	Plus Positive
$-$	Minus Negative
$\pm$	Plus or minus Positive or negative
$\mp$	Minus or plus Negative or positive
$\times$ or $\cdot$	Multiplied by
$\div$ or $:$	Divided by
$=$ or $::$	Equals, as
$\neq$	Does not equal
$\approx$	Equals approximately
$>$	Greater than
$<$	Less than
$\geqq$	Greater than or equal to
$\leqq$	Less than or equal to
$\equiv$	Is identical to
$\rightarrow$ or $\doteq$	Approaches as a limit
$\propto$	Varies directly as
$\therefore$	Therefore
$\sqrt{\ }$	Square root
$\sqrt[n]{\ }$	nth root
a^n	nth power of a
$n!$	$1 \cdot 2 \cdot 3 \cdots n$
log	Common logarithm Briggsian "
ln or $\log_e$	Natural logarithm Hyberbolic " Napierian "
e or ϵ	Base (2.718) of natural system of logarithms
π	Pi (3.1416)
$\angle$	Angle
$\perp$	Perpendicular to
$\parallel$	Parallel to
( )	parentheses
[ ]	brackets
{ }	braces
—	vinculum
a°	a degrees (angle)
a'	a minutes (angle) a prime
a''	a seconds (angle) a second a double-prime
a'''	a third a triple-prime

a_n	a sub n
sin	Sine
cos	Cosine
tan	Tangent
cot	Cotangent
sec	Secant
csc	Cosecant
vers	Versed sine
covers	Coversed sine
exsec	Exsecant
$\sin^{-1} a$	Anti-sine a Angle whose sine is a Inverse sine a
sinh	Hyperbolic sine
cosh	Hyperbolic cosine
tanh	Hyperbolic tangent
$\sinh^{-1} a$	Anti-hyperbolic sine a Angle whose hyperbolic sine is a
$P(x, y)$	Rect. coörd. of point P
$P(r, \theta)$	Polar coörd. of point P
$f(x)$, $F(x)$ or $\phi(x)$	Function of x
$\triangle y$	Increment of y
Σ	Summation of
∞	Infinity
dy	Differential of y
$\frac{dy}{dx}$ or $f'(x)$	Derivative of $y = f(x)$ with respect to x
$\frac{d^2y}{dx^2}$ or $f''(x)$	Second deriv. of $y = f(x)$ with respect to x
$\frac{d^ny}{dx^n}$ or $f^{(n)}(x)$	nth deriv. of $y = f(x)$ with respect to x
$\frac{\partial z}{\partial x}$	Partial derivative of z with respect to x
$\frac{\partial^2 z}{\partial x \, \partial y}$	Second partial deriv. of z with respect to y and x
$\int$	Integral of
$\int_a^b$	Integral between the limits a and b
j	Imaginary quantity ($\sqrt{-1}$)
$\underset{\bullet}{x} = a + jb$	Symbolic vector notation

MATHEMATICAL TABLES

PROPERTIES OF MATERIALS AND CONVERSION FACTORS

N	0	1	2	3	4	5	6	7	8	9
0	....	0000	3010	4771	6021	6990	7782	8451	9031	9542
1	0000	0414	0792	1139	1461	1761	2041	2304	2553	2788
2	3010	3222	3424	3617	3802	3979	4150	4314	4472	4624
3	4771	4914	5051	5185	5315	5441	5563	5682	5798	5911
4	6021	6128	6232	6335	6435	6532	6628	6721	6812	6902
5	6990	7076	7160	7243	7324	7404	7482	7559	7634	7709
6	7782	7853	7924	7993	8062	8129	8195	8261	8325	8388
7	8451	8513	8573	8633	8692	8751	8808	8865	8921	8976
8	9031	9085	9138	9191	9243	9294	9345	9395	9445	9494
9	9542	9590	9638	9685	9731	9777	9823	9868	9912	9956
10	0000	0043	0086	0128	0170	0212	0253	0294	0334	0374
11	0414	0453	0492	0531	0569	0607	0645	0682	0719	0755
12	0792	0828	0864	0899	0934	0969	1004	1038	1072	1106
13	1139	1173	1206	1239	1271	1303	1335	1367	1399	1430
14	1461	1492	1523	1553	1584	1614	1644	1673	1703	1732
15	1761	1790	1818	1847	1875	1903	1931	1959	1987	2014
16	2041	2068	2095	2122	2148	2175	2201	2227	2253	2279
17	2304	2330	2355	2380	2405	2430	2455	2480	2504	2529
18	2553	2577	2601	2625	2648	2672	2695	2718	2742	2765
19	2788	2810	2833	2856	2878	2900	2923	2945	2967	2989
20	3010	3032	3054	3075	3096	3118	3139	3160	3181	3201
21	3222	3243	3263	3284	3304	3324	3345	3365	3385	3404
22	3424	3444	3464	3483	3502	3522	3541	3560	3579	3598
23	3617	3636	3655	3674	3692	3711	3729	3747	3766	3784
24	3802	3820	3838	3856	3874	3892	3909	3927	3945	3962
25	3979	3997	4014	4031	4048	4065	4082	4099	4116	4133
26	4150	4166	4183	4200	4216	4232	4249	4265	4281	4298
27	4314	4330	4346	4362	4378	4393	4409	4425	4440	4456
28	4472	4487	4502	4518	4533	4548	4564	4579	4594	4609
29	4624	4639	4654	4669	4683	4698	4713	4728	4742	4757
30	4771	4786	4800	4814	4829	4843	4857	4871	4886	4900
31	4914	4928	4942	4955	4969	4983	4997	5011	5024	5038
32	5051	5065	5079	5092	5105	5119	5132	5145	5159	5172
33	5185	5198	5211	5224	5237	5250	5263	5276	5289	5302
34	5315	5328	5340	5353	5366	5378	5391	5403	5416	5428
35	5441	5453	5465	5478	5490	5502	5514	5527	5539	5551
36	5563	5575	5587	5599	5611	5623	5635	5647	5658	5670
37	5682	5694	5705	5717	5729	5740	5752	5763	5775	5786
38	5798	5809	5821	5832	5843	5855	5866	5877	5888	5899
39	5911	5922	5933	5944	5955	5966	5977	5988	5999	6010
40	6021	6031	6042	6053	6064	6075	6085	6096	6107	6117
41	6128	6138	6149	6160	6170	6180	6191	6201	6212	6222
42	6232	6243	6253	6263	6274	6284	6294	6304	6314	6325
43	6335	6345	6355	6365	6375	6385	6395	6405	6415	6425
44	6435	6444	6454	6464	6474	6484	6493	6503	6513	6522
45	6532	6542	6551	6561	6571	6580	6590	6599	6609	6618
46	6628	6637	6646	6656	6665	6675	6684	6693	6702	6712
47	6721	6730	6739	6749	6758	6767	6776	6785	6794	6803
48	6812	6821	6830	6839	6848	6857	6866	6875	6884	6893
49	6902	6911	6920	6928	6937	6946	6955	6964	6972	6981
50	6990	6998	7007	7016	7024	7033	7042	7050	7059	7067
N	0	1	2	3	4	5	6	7	8	9

N	0	1	2	3	4	5	6	7	8	9
50	6990	6998	7007	7016	7024	7033	7042	7050	7059	7067
51	7076	7084	7093	7101	7110	7118	7126	7135	7143	7152
52	7160	7168	7177	7185	7193	7202	7210	7218	7226	7235
53	7243	7251	7259	7267	7275	7284	7292	7300	7308	7316
54	7324	7332	7340	7348	7356	7364	7372	7380	7388	7396
55	7404	7412	7419	7427	7435	7443	7451	7459	7466	7474
56	7482	7490	7497	7505	7513	7520	7528	7536	7543	7551
57	7559	7566	7574	7582	7589	7597	7604	7612	7619	7627
58	7634	7642	7649	7657	7664	7672	7679	7686	7694	7701
59	7709	7716	7723	7731	7738	7745	7752	7760	7767	7774
60	7782	7789	7796	7803	7810	7818	7825	7832	7839	7846
61	7853	7860	7868	7875	7882	7889	7896	7903	7910	7917
62	7924	7931	7938	7945	7952	7959	7966	7973	7980	7987
63	7993	8000	8007	8014	8021	8028	8035	8041	8048	8055
64	8062	8069	8075	8082	8089	8096	8102	8109	8116	8122
65	8129	8136	8142	8149	8156	8162	8169	8176	8182	8189
66	8195	8202	8209	8215	8222	8228	8235	8241	8248	8254
67	8261	8267	8274	8280	8287	8293	8299	8306	8312	8319
68	8325	8331	8338	8344	8351	8357	8363	8370	8376	8382
69	8388	8395	8401	8407	8414	8420	8426	8432	8439	8445
70	8451	8457	8463	8470	8476	8482	8488	8494	8500	8506
71	8513	8519	8525	8531	8537	8543	8549	8555	8561	8567
72	8573	8579	8585	8591	8597	8603	8609	8615	8621	8627
73	8633	8639	8645	8651	8657	8663	8669	8675	8681	8686
74	8692	8698	8704	8710	8716	8722	8727	8733	8739	8745
75	8751	8756	8762	8768	8774	8779	8785	8791	8797	8802
76	8808	8814	8820	8825	8831	8837	8842	8848	8854	8859
77	8865	8871	8876	8882	8887	8893	8899	8904	8910	8915
78	8921	8927	8932	8938	8943	8949	8954	8960	8965	8971
79	8976	8982	8987	8993	8998	9004	9009	9015	9020	9025
80	9031	9036	9042	9047	9053	9058	9063	9069	9074	9079
81	9085	9090	9096	9101	9106	9112	9117	9122	9128	9133
82	9138	9143	9149	9154	9159	9165	9170	9175	9180	9186
83	9191	9196	9201	9206	9212	9217	9222	9227	9232	9238
84	9243	9248	9253	9258	9263	9269	9274	9279	9284	9289
85	9294	9299	9304	9309	9315	9320	9325	9330	9335	9340
86	9345	9350	9355	9360	9365	9370	9375	9380	9385	9390
87	9395	9400	9405	9410	9415	9420	9425	9430	9435	9440
88	9445	9450	9455	9460	9465	9469	9474	9479	9484	9489
89	9494	9499	9504	9509	9513	9518	9523	9528	9533	9538
90	9542	9547	9552	9557	9562	9566	9571	9576	9581	9586
91	9590	9595	9600	9605	9609	9614	9619	9624	9628	9633
92	9638	9643	9647	9652	9657	9661	9666	9671	9675	9680
93	9685	9689	9694	9699	9703	9708	9713	9717	9722	9727
94	9731	9736	9741	9745	9750	9754	9759	9763	9768	9773
95	9777	9782	9786	9791	9795	9800	9805	9809	9814	9818
96	9823	9827	9832	9836	9841	9845	9850	9854	9859	9863
97	9868	9872	9877	9881	9886	9890	9894	9899	9903	9908
98	9912	9917	9921	9926	9930	9934	9939	9943	9948	9952
99	9956	9961	9965	9969	9974	9978	9983	9987	9991	9996
100	0000	0004	0009	0013	0017	0022	0026	0030	0035	0039
N	0	1	2	3	4	5	6	7	8	9

N	0.0	1.0	2.0	3.0	4.0	5.0	6.0	7.0	8.0	9.0
	....	0.0000	0.6931	1.0986	1.3863	1.6094	1.7918	1.9459	2.0794	2.1972
10	2.3026	2.3979	2.4849	2.5649	2.6391	2.7081	2.7726	2.8332	2.8904	2.9444
20	9957	3.0445	3.0910	3.1355	3.1781	3.2189	3.2581	3.2958	3.3322	3.3673
30	3.4012	4340	4657	4965	5264	5553	5835	6109	6376	6636
40	6889	7136	7377	7612	7842	8067	8286	8501	8712	8918
50	9120	9318	9512	9703	9890	4.0073	4.0254	4.0431	4.0604	4.0775
60	4.0943	4.1109	4.1271	4.1431	4.1589	1744	1897	2047	2195	2341
70	2485	2627	2767	2905	3041	3175	3307	3408	3567	3694
80	3820	3944	4067	4188	4308	4427	4543	4659	4773	4886
90	4998	5109	5218	5326	5433	5539	5643	5747	5850	5951
100	6052	6151	6250	6347	6444	6540	6634	6728	6821	6913
110	7005	7095	7185	7274	7362	7449	7536	7622	7707	7791
120	7875	7958	8040	8122	8203	8283	8363	8442	8520	8598
130	8675	8752	8828	8903	8978	9053	9127	9200	9273	9345
140	9416	9488	9558	9628	9698	9767	9836	9904	9972	5.0039
150	5.0106	5.0173	5.0239	5.0304	5.0370	5.0434	5.0499	5.0562	5.0626	0689
160	0752	0814	0876	0938	0999	1059	1120	1180	1240	1299
170	1358	1417	1475	1533	1591	1648	1705	1761	1818	1874
180	1930	1985	2040	2095	2149	2204	2257	2311	2364	2417
190	2470	2523	2575	2627	2679	2730	2781	2832	2883	2933
200	2983	3033	3083	3132	3181	3230	3279	3327	3375	3423
210	3471	3519	3566	3613	3660	3706	3753	3799	3845	3891
220	3936	3982	4027	4072	4116	4161	4205	4250	4293	4337
230	4381	4424	4467	4510	4553	4596	4638	4681	4723	4765
240	4806	4848	4889	4931	4972	5013	5053	5094	5134	5175
250	5215	5255	5294	5334	5373	5413	5452	5491	5530	5568
260	5607	5645	5683	5722	5759	5797	5835	5872	5910	5947
270	5984	6021	6058	6095	6131	6168	6204	6240	6276	6312
280	6348	6384	6419	6454	6490	6525	6560	6595	6630	6664
290	6699	6733	6768	6802	6836	6870	6904	6937	6971	7004
300	7038	7071	7104	7137	7170	7203	7236	7268	7301	7333
310	7366	7398	7430	7462	7494	7526	7557	7589	7621	7652
320	7683	7714	7746	7777	7807	7838	7869	7900	7930	7961
330	7991	8021	8051	8081	8111	8141	8171	8201	8230	8260
340	8289	8319	8348	8377	8406	8435	8464	8493	8522	8551
350	8579	8608	8636	8665	8693	8721	8749	8777	8805	8833
360	8861	8889	8916	8944	8972	8999	9026	9054	9081	9108
370	9135	9162	9189	9216	9243	9269	9296	9322	9349	9375
380	9402	9428	9454	9480	9506	9532	9558	9584	9610	9636
390	9661	9687	9713	9738	9764	9789	9814	9839	9865	9890
400	9915	9940	9965	9989	6.0014	6.0039	6.0064	6.0088	6.0113	6.0137
410	6.0162	6.0186	6.0210	6.0234	0259	0283	0307	0331	0355	0379
420	0403	0426	0450	0474	0497	0521	0544	0568	0591	0615
430	0638	0661	0684	0707	0730	0753	0776	0799	0822	0845
440	0868	0890	0913	0936	0958	0981	1003	1026	1048	1070
450	1092	1115	1137	1159	1181	1203	1225	1247	1269	1291
460	1312	1334	1356	1377	1399	1420	1442	1463	1485	1506
470	1527	1549	1570	1591	1612	1633	1654	1675	1696	1717
480	1738	1759	1779	1800	1821	1841	1862	1883	1903	1924
490	1944	1964	1985	2005	2025	2046	2066	2086	2106	2126
500	2146	2166	2186	2206	2226	2246	2265	2285	2305	2324
N	0.0	1.0	2.0	3.0	4.0	5.0	6.0	7.0	8.0	9.0

N	0.0	1.0	2.0	3.0	4.0	5.0	6.0	7.0	8.0	9.0
500	6.2146	6.2166	6.2186	6.2206	6.2226	6.2246	6.2265	6.2285	6.2305	6.2324
510	2344	2364	2383	2403	2422	2442	2461	2480	2500	2519
520	2538	2558	2577	2596	2615	2634	2653	2672	2691	2710
530	2729	2748	2766	2785	2804	2823	2841	2860	2879	2897
540	2916	2934	2953	2971	2989	3008	3026	3044	3063	3081
550	3099	3117	3135	3154	3172	3190	3208	3226	3244	3261
560	3279	3297	3315	3333	3351	3368	3386	3404	3421	3439
570	3456	3474	3491	3509	3526	3544	3561	3578	3596	3613
580	3630	3648	3665	3682	3699	3716	3733	3750	3767	3784
590	3801	3818	3835	3852	3869	3886	3902	3919	3936	3953
600	3969	3986	4003	4019	4036	4052	4069	4085	4102	4118
610	4135	4151	4167	4184	4200	4216	4232	4249	4265	4281
620	4297	4313	4329	4345	4362	4378	4394	4409	4425	4441
630	4457	4473	4489	4505	4520	4536	4552	4568	4583	4599
640	4615	4630	4646	4661	4677	4693	4708	4723	4739	4754
650	4770	4785	4800	4816	4831	4846	4862	4877	4892	4907
660	4922	4938	4953	4968	4983	4998	5013	5028	5043	5058
670	5073	5088	5103	5117	5132	5147	5162	5177	5191	5206
680	5221	5236	5250	5265	5280	5294	5309	5323	5338	5352
690	5367	5381	5396	5410	5425	5439	5453	5468	5482	5497
700	5511	5525	5539	5554	5568	5582	5596	5610	5624	5639
710	5653	5667	5681	5695	5709	5723	5737	5751	5765	5779
720	5793	5806	5820	5834	5848	5862	5876	5889	5903	5917
730	5930	5944	5958	5971	5985	5999	6012	6026	6039	6053
740	6067	6080	6093	6107	6120	6134	6147	6161	6174	6187
750	6201	6214	6227	6241	6254	6267	6280	6294	6307	6320
760	6333	6346	6359	6373	6386	6399	6412	6425	6438	6451
770	6464	6477	6490	6503	6516	6529	6542	6554	6567	6580
780	6593	6606	6619	6631	6644	6657	6670	6682	6695	6708
790	6720	6733	6746	6758	6771	6783	6796	6809	6821	6834
800	6846	6859	6871	6884	6896	6908	6921	6933	6946	6958
810	6970	6983	6995	7007	7020	7032	7044	7056	7069	7081
820	7093	7105	7117	7130	7142	7154	7166	7178	7190	7202
830	7214	7226	7238	7250	7262	7274	7286	7298	7310	7322
840	7334	7346	7358	7370	7382	7393	7405	7417	7429	7441
850	7452	7464	7476	7488	7499	7511	7523	7534	7546	7558
860	7569	7581	7593	7604	7616	7627	7639	7650	7662	7673
870	7685	7696	7708	7719	7731	7742	7754	7765	7776	7788
880	7799	7811	7822	7833	7845	7856	7867	7878	7890	7901
890	7912	7923	7935	7946	7957	7968	7979	7991	8002	8013
900	8024	8035	8046	8057	8068	8079	8090	8101	8112	8123
910	8134	8145	8156	8167	8178	8189	8200	8211	8222	8233
920	8244	8255	8265	8276	8287	8298	8309	8320	8330	8341
930	8352	8363	8373	8384	8395	8405	8416	8427	8437	8448
940	8459	8469	8480	8491	8501	8512	8522	8533	8544	8554
950	8565	8575	8586	8596	8607	8617	8628	8638	8648	8659
960	8669	8680	8690	8701	8711	8721	8732	8742	8752	8763
970	8773	8783	8794	8804	8814	8824	8835	8845	8855	8865
980	8876	8886	8896	8906	8916	8926	8937	8947	8957	8967
990	8977	8987	8997	9007	9017	9027	9037	9048	9058	9068
1000	9078	9088	9098	9108	9117	9127	9137	9147	9157	9167
N	0.0	1.0	2.0	3.0	4.0	5.0	6.0	7.0	8.0	9.0

Numbers (1 to 50), Squares, Cubes, Square Roots, Cube Roots, Reciprocals, Circumferences and Circular Areas

N	N^2	N^3	$\sqrt{N}$	$\sqrt[3]{N}$	$\frac{1000}{N}$	πN	$\frac{\pi N^2}{4}$
1	1	1	1.0000	1.0000	1000.000	3.142	0.7854
2	4	8	1.4142	1.2599	500.000	6.283	3.1416
3	9	27	1.7321	1.4422	333.333	9.425	7.0686
4	16	64	2.0000	1.5874	250.000	12.566	12.5664
5	25	125	2.2361	1.7100	200.000	15.708	19.6350
6	36	216	2.4495	1.8171	166.667	18.850	28.2743
7	49	343	2.6458	1.9129	142.857	21.991	38.4845
8	64	512	2.8284	2.0000	125.000	25.133	50.2655
9	81	729	3.0000	2.0801	111.111	28.274	63.6173
10	100	1000	3.1623	2.1544	100.000	31.416	78.5398
11	121	1331	3.3166	2.2240	90.9091	34.558	95.0332
12	144	1728	3.4641	2.2894	83.3333	37.699	113.097
13	169	2197	3.6056	2.3513	76.9231	40.841	132.732
14	196	2744	3.7417	2.4101	71.4286	43.982	153.938
15	225	3375	3.8730	2.4662	66.6667	47.124	176.715
16	256	4096	4.0000	2.5198	62.5000	50.265	201.062
17	289	4913	4.1231	2.5713	58.8235	53.407	226.980
18	324	5832	4.2426	2.6207	55.5556	56.549	254.469
19	361	6859	4.3589	2.6684	52.6316	59.690	283.529
20	400	8000	4.4721	2.7144	50.0000	62.832	314.159
21	441	9261	4.5826	2.7589	47.6190	65.973	346.361
22	484	10648	4.6904	2.8020	45.4545	69.115	380.133
23	529	12167	4.7958	2.8439	43.4783	72.257	415.476
24	576	13824	4.8990	2.8845	41.6667	75.398	452.389
25	625	15625	5.0000	2.9240	40.0000	78.540	490.874
26	676	17576	5.0990	2.9625	38.4615	81.681	530.929
27	729	19683	5.1962	3.0000	37.0370	84.823	572.555
28	784	21952	5.2915	3.0366	35.7143	87.965	615.752
29	841	24389	5.3852	3.0723	34.4828	91.106	660.520
30	900	27000	5.4772	3.1072	33.3333	94.248	706.858
31	961	29791	5.5678	3.1414	32.2581	97.389	754.768
32	1024	32768	5.6569	3.1748	31.2500	100.531	804.248
33	1089	35937	5.7446	3.2075	30.3030	103.673	855.299
34	1156	39304	5.8310	3.2396	29.4118	106.814	907.920
35	1225	42875	5.9161	3.2711	28.5714	109.956	962.113
36	1296	46656	6.0000	3.3019	27.7778	113.097	1017.88
37	1369	50653	6.0828	3.3322	27.0270	116.239	1075.21
38	1444	54872	6.1644	3.3620	26.3158	119.381	1134.11
39	1521	59319	6.2450	3.3912	25.6410	122.522	1194.59
40	1600	64000	6.3246	3.4200	25.0000	125.66	1256.64
41	1681	68921	6.4031	3.4482	24.3902	128.81	1320.25
42	1764	74088	6.4807	3.4760	23.8095	131.95	1385.44
43	1849	79507	6.5574	3.5034	23.2558	135.09	1452.20
44	1936	85184	6.6332	3.5303	22.7273	138.23	1520.53
45	2025	91125	6.7082	3.5569	22.2222	141.37	1590.43
46	2116	97336	6.7823	3.5830	21.7391	144.51	1661.90
47	2209	103823	6.8557	3.6088	21.2766	147.65	1734.94
48	2304	110592	6.9282	3.6342	20.8333	150.80	1809.56
49	2401	117649	7.0000	3.6593	20.4082	153.94	1885.74
50	2500	125000	7.0711	3.6840	20.0000	157.08	1963.50

N	N^2	N^3	$\sqrt{N}$	$\sqrt[3]{N}$	$\frac{1000}{N}$	πN	$\frac{\pi N^2}{4}$
51	2601	132651	7.1414	3.7084	19.6078	160.22	2042.82
52	2704	140608	7.2111	3.7325	19.2308	163.36	2123.72
53	2809	148877	7.2801	3.7563	18.8679	166.50	2206.18
54	2916	157464	7.3485	3.7798	18.5185	169.65	2290.22
55	3025	166375	7.4162	3.8030	18.1818	172.79	2375.83
56	3136	175616	7.4833	3.8259	17.8571	175.93	2463.01
57	3249	185193	7.5498	3.8485	17.5439	179.07	2551.76
58	3364	195112	7.6158	3.8709	17.2414	182.21	2642.08
59	3481	205379	7.6811	3.8930	16.9492	185.35	2733.97
60	3600	2 6000	7.7460	3.9149	16.6667	188.50	2827.43
61	3721	226981	7.8102	3.9365	16.3934	191.64	2922.47
62	3844	238328	7.8740	3.9579	16.1290	194.78	3019.07
63	3969	250047	7.9373	3.9791	15.8730	197.92	3117.25
64	4096	262144	8.0000	4.0000	15.6250	201.06	3216.99
65	4225	274625	8.0623	4.0207	15.3846	204.20	3318.31
66	4356	287496	8.1240	4.0412	15.1515	207.35	3421.19
67	4489	300763	8.1854	4.0615	14.9254	210.49	3525.65
68	4624	314432	8.2462	4.08 7	14.7059	213.63	3631.68
69	4761	328509	8.3066	4.1016	14.4928	216.77	3739.28
70	4900	343000	8.3666	4.1213	14.2857	219.91	3848.45
71	5041	357911	8 4261	4.1408	14.0845	223.05	3959.19
72	5184	373248	8.4853	4.1602	13.8889	226.19	4071.50
73	5329	389017	8.5440	4.1793	13.6986	229.34	4185.39
74	5476	405224	8.6023	4 1983	13 5135	232.48	4300.84
75	5625	421875	8.6603	4.2172	13.3333	235.62	4417.86
76	5776	438976	8.7178	4.2358	13.1579	238.76	4536.46
77	5929	456533	8.7750	4 2543	12.9870	241.90	4656.63
78	6084	474552	8.8318	4.2727	12.8205	245.04	4778.36
79	6241	493039	8.8882	4.2908	12.6582	248.19	4901.67
80	6400	512000	8.9443	4.3089	12.5000	251.33	5026.55
81	6561	531441	9.0000	4.3267	12.3457	254.47	5153.00
82	6724	551368	9.0554	4.3445	12.1951	257.61	5281.02
83	6889	571787	9 1104	4.3621	12.0482	260.75	5410.61
84	7056	592704	9.1652	4.3795	11.9048	263.89	5541.77
85	7225	614125	9.2195	4.3968	11.7647	267.04	5674.50
86	7396	636056	9.2736	4.4140	11.6279	270.18	5808.80
87	7569	658503	9.3274	4.4310	11.4943	273.32	5944.68
88	7744	681472	9.3808	4.4480	11.3636	276.46	6082.12
89	7921	704969	9.4340	4.4647	11.2360	279.60	6221.14
90	8100	729000	9.4868	4.4814	11.1111	282.74	6361.73
91	8281	753571	9.5394	4.4979	10.9890	285.88	6503.88
92	8464	778688	9.5917	4.5144	10.8696	289.03	6647.61
93	8649	804357	9.6437	4.5307	10.7527	292.17	6792.91
94	8836	830584	9.6954	4.5468	10.6383	295.31	6939.78
95	9025	857375	9.7468	4.5629	10.5263	298.45	7088.22
96	9216	884736	9.7980	4.5789	10.4167	301.59	7238.23
97	9409	912673	9.8489	4.5947	10.3093	304.73	7389.81
98	9604	941192	9.8995	4.6104	10.2041	307.88	7542.96
99	9801	970299	9.9499	4.6261	10.1010	311.02	7697.69
100	10000	1000000	10.0000	4.6416	10.0000	314.16	7853.98

Numbers (101 to 150), Squares, Cubes, Square Roots, Cube Roots, Reciprocals, Circumferences and Circular Areas

N	N^2	N^3	$\sqrt{N}$	$\sqrt[3]{N}$	$\frac{1000}{N}$	πN	$\frac{\pi N^2}{4}$
101	10201	1030301	10.0499	4.6570	9.90099	317.30	8011.85
102	10404	1061208	10.0995	4.6723	9.80392	320.44	8171.28
103	10609	1092727	10.1489	4.6875	9.70874	323.58	8332.29
104	10816	1124864	10.1980	4.7027	9.61538	326.73	8494.87
105	11025	1157625	10.2470	4.7177	9.52381	329.87	8659.01
106	11236	1191016	10.2956	4.7326	9.43396	333.01	8824.73
107	11449	1225043	10.3441	4.7475	9.34579	336.15	8992.02
108	11664	1259712	10.3923	4.7622	9.25926	339.29	9160.88
109	11881	1295029	10.4403	4.7769	9.17431	342.43	9331.32
110	12100	1331000	10.4881	4.7914	9.09091	345.58	9503.32
111	12321	1367631	10.5357	4.8059	9.00901	348.72	9676.89
112	12544	1404928	10.5830	4.8203	8.92857	351.86	9852.03
113	12769	1442897	10.6301	4.8346	8.84956	355.00	10028.7
114	12996	1481544	10.6771	4.8488	8.77193	358.14	10207.0
115	13225	1520875	10.7238	4.8629	8.69565	361.28	10386.9
116	13456	1560896	10.7703	4.8770	8.62069	364.42	10568.3
117	13689	1601613	10.8167	4.8910	8.54701	367.57	10751.3
118	13924	1643032	10.8628	4.9049	8.47458	370.71	10935.9
119	14161	1685159	10.9087	4.9187	8.40336	373.85	11122.0
120	14400	1728000	10.9545	4.9324	8.33333	376.99	11309.7
121	14641	1771561	11.0000	4.9461	8.26446	380.13	11499.0
122	14884	1815848	11.0454	4.9597	8.19672	383.27	11689.9
123	15129	1860867	11.0905	4.9732	8.13008	386.42	11882.3
124	15376	1906624	11.1355	4.9866	8.06452	389.56	12076.3
125	15625	1953125	11.1803	5.0000	8.00000	392.70	12271.8
126	15876	2000376	11.2250	5.0133	7.93651	395.84	12469.0
127	16129	2048383	11.2694	5.0265	7.87402	398.98	12667.7
128	16384	2097152	11.3137	5.0397	7.81250	402.12	12868.0
129	16641	2146689	11.3578	5.0528	7.75194	405.27	13069.8
130	16900	2197000	11.4018	5.0658	7.69231	408.41	13273.2
131	17161	2248091	11.4455	5.0788	7.63359	411.55	13478.2
132	17424	2299968	11.4891	5.0916	7.57576	414.69	13684.8
133	17689	2352637	11.5326	5.1045	7.51880	417.83	13892.9
134	17956	2406104	11.5758	5.1172	7.46269	420.97	14102.6
135	18225	2460375	11.6190	5.1299	7.40741	424.12	14313.9
136	18496	2515456	11.6619	5.1426	7.35294	427.26	14526.7
137	18769	2571353	11.7047	5.1551	7.29927	430.40	14741.1
138	19044	2628072	11.7473	5.1676	7.24638	433.54	14957.1
139	19321	2685619	11.7898	5.1801	7.19424	436.68	15174.7
140	19600	2744000	11.8322	5.1925	7.14286	439.82	15393.8
141	19881	2803221	11.8743	5.2048	7.09220	442.96	15614.5
142	20164	2863288	11.9164	5.2171	7.04255	446.11	15836.8
143	20449	2924207	11.9583	5.2293	6.99301	449.25	16060.6
144	20736	2985984	12.0000	5.2415	6.94444	452.39	16286.0
145	21025	3048625	12.0416	5.2536	6.89655	455.53	16513.0
146	21316	3112136	12.0830	5.2656	6.84932	458.67	16741.5
147	21609	3176523	12.1244	5.2776	6.80272	461.81	16971.7
148	21904	3241792	12.1655	5.2896	6.75676	464.96	17203.4
149	22201	3307949	12.2066	5.3015	6.71141	468.10	17436.6
150	22500	3375000	12.2474	5.3133	6.66667	471.24	17671.5

N	N^2	N^3	$\sqrt{N}$	$\sqrt[3]{N}$	$\frac{1000}{N}$	πN	$\frac{\pi N^2}{4}$
151	22801	3442951	12.2882	5.3251	6.62252	474.38	17907.9
152	23104	3511808	12.3288	5.3368	6.57895	477.52	18145.8
153	23409	3581577	12.3693	5.3485	6.53595	480.66	18385.4
154	23716	3652264	12.4097	5.3601	6.49351	483.81	18626.5
155	24025	3723875	12.4499	5.3717	6.45161	486.95	18869.2
156	24336	3796416	12.4900	5.3832	6.41026	490.09	19113.4
157	24649	3869893	12.5300	5.3947	6.36943	493.23	19359.3
158	24964	3944312	12.5698	5.4061	6.32911	496.37	19606.7
159	25281	4019679	12.6095	5.4175	6.28931	499.51	19855.7
160	25600	4096000	12.6491	5.4288	6.25000	502.65	20106.2
161	25921	4173281	12.6886	5.4401	6.21118	505.80	20358.3
162	26244	4251528	12.7279	5.4514	6.17284	508.94	20612.0
163	26569	4330747	12.7671	5.4626	6.13497	512.08	20867.2
164	26896	4410944	12.8062	5.4737	6.09756	515.22	21124.1
165	27225	4492125	12.8452	5.4848	6.06061	518.36	21382.5
166	27556	4574296	12.8841	5.4959	6.02410	521.50	21642.4
167	27889	4657463	12.9228	5.5069	5.98802	524.65	21904.0
168	28224	4741632	12.9615	5.5178	5.95238	527.79	22167.1
169	28561	4826809	13.0000	5.5288	5.91716	530.93	22431.8
170	28900	4913000	13.0384	5.5397	5.88235	534.07	22698.0
171	29241	5000211	13.0767	5.5505	5.84795	537.21	22965.8
172	29584	5088448	13.1149	5.5613	5.81395	540.35	23235.2
173	29929	5177717	13.1529	5.5721	5.78035	543.50	23506.2
174	30276	5268024	13.1909	5.5828	5.74713	546.64	23778.7
175	30625	5359375	13.2288	5.5934	5.71429	549.78	24052.8
176	30976	5451776	13.2665	5.6041	5.68182	552.92	24328.5
177	31329	5545233	13.3041	5.6147	5.64972	556.06	24605.7
178	31684	5639752	13.3417	5.6252	5.61798	559.20	24884.6
179	32041	5735339	13.3791	5.6357	5.58659	562.35	25164.9
180	32400	5832000	13.4164	5.6462	5.55556	565.49	25446.9
181	32761	5929741	13.4536	5.6567	5.52486	568.63	25730.4
182	33124	6028568	13.4907	5.6671	5.49451	571.77	26015.5
183	33489	6128487	13.5277	5.6774	5.46448	574.91	26302.2
184	33856	6229504	13.5647	5.6877	5.43478	578.05	26590.4
185	34225	6331625	13.6015	5.6980	5.40541	581.19	26880.3
186	34596	6434856	13.6382	5.7083	5.37634	584.34	27171.6
187	34969	6539203	13.6748	5.7185	5.34759	587.48	27464.6
188	35344	6644672	13.7113	5.7287	5.31915	590.62	27759.1
189	35721	6751269	13.7477	5.7388	5.29101	593.76	28055.2
190	36100	6859000	13.7840	5.7489	5.26316	596.90	28352.9
191	36481	6967871	13.8203	5.7590	5.23560	600.04	28652.1
192	36864	7077888	13.8564	5.7690	5.20833	603.19	28952.9
193	37249	7189057	13.8924	5.7790	5.18135	606.33	29255.3
194	37636	7301384	13.9284	5.7890	5.15464	609.47	29559.2
195	38025	7414875	13.9642	5.7989	5.12821	612.61	29864.8
196	38416	7529536	14.0000	5.8088	5.10204	615.75	30171.9
197	38809	7645373	14.0357	5.8186	5.07614	618.89	30480.5
198	39204	7762392	14.0712	5.8285	5.05051	622.04	30790.7
199	39601	7880599	14.1067	5.8383	5.02513	625.18	31102.6
200	40000	8000000	14.1421	5.8480	5.00000	628.32	31415.9

N	N^2	N^3	$\sqrt{N}$	$\sqrt[3]{N}$	$\frac{1000}{N}$	πN	$\frac{\pi N^2}{4}$
201	40401	8120601	14.1774	5.8578	4.97512	631.46	31730.9
202	40804	8242408	14.2127	5.8675	4.95050	634.60	32047.4
203	41209	8365427	14.2478	5.8771	4.92611	637.74	32365.5
204	41616	8489664	14.2829	5.8868	4.90196	640.89	32685.1
205	42025	8615125	14.3178	5.8964	4.87805	644.03	33006.4
206	42436	8741816	14.3527	5.9059	4.85437	647.17	33329.2
207	42849	8869743	14.3875	5.9155	4.83092	650.31	33653.5
208	43264	8998912	14.4222	5.9250	4.80769	653.45	33979.5
209	43681	9129329	14.4568	5.9345	4.78469	656.59	34307.0
210	44100	9261000	14.4914	5.9439	4.76190	659.73	34636.1
211	44521	9393931	14.5258	5.9533	4.73934	662.88	34966.7
212	44944	9528128	14.5602	5.9627	4.71698	666.02	35298.9
213	45369	9663597	14.5945	5.9721	4.69484	669.16	35632.7
214	45796	9800344	14.6287	5.9814	4.67290	672.30	35968.1
215	46225	9938375	14.6629	5.9907	4.65116	675.44	36305.0
216	46656	10077696	14.6969	6.0000	4.62963	678.58	36643.5
217	47089	10218313	14.7309	6.0092	4.60829	681.73	36983.6
218	47524	10360232	14.7648	6.0185	4.58716	684.87	37325.3
219	47961	10503459	14.7986	6.0277	4.56621	688.01	37668.5
220	48400	10648000	14.8324	6.0368	4.54545	691.15	38013.3
221	48841	10793861	14.8661	6.0459	4.52489	694.29	38359.6
222	49284	10941048	14.8997	6.0550	4.50450	697.43	38707.6
223	49729	11089567	14.9332	6.0641	4.48431	700.58	39057.1
224	50176	11239424	14.9666	6.0732	4.46429	703.72	39408.1
225	50625	11390625	15.0000	6.0822	4.44444	706.86	39760.8
226	51076	11543176	15.0333	6.0912	4.42478	710.00	40115.0
227	51529	11697083	15.0665	6.1002	4.40529	713.14	40470.8
228	51984	11852352	15.0997	6.1091	4.38596	716.28	40828.1
229	52441	12008989	15.1327	6.1180	4.36681	719.42	41187.1
230	52900	12167000	15.1658	6.1269	4.34783	722.57	41547.6
231	53361	12326391	15.1987	6.1358	4.32900	725.71	41909.6
232	53824	12487168	15.2315	6.1446	4.31034	728.85	42273.3
233	54289	12649337	15.2643	6.1534	4.29185	731.99	42638.5
234	54756	12812904	15.2971	6.1622	4.27350	735.13	43005.3
235	55225	12977875	15.3297	6.1710	4.25532	738.27	43373.6
236	55696	13144256	15.3623	6.1797	4.23729	741.42	43743.5
237	56169	13312053	15.3948	6.1885	4.21941	744.56	44115.0
238	56644	13481272	15.4272	6.1972	4.20168	747.70	44488.1
239	57121	13651919	15.4596	6.2058	4.18410	750.84	44862.7
240	57600	13824000	15.4919	6.2145	4.16667	753.98	45238.9
241	58081	13997521	15.5242	6.2231	4.14938	757.12	45616.7
242	58564	14172488	15.5563	6.2317	4.13223	760.27	45996.1
243	59049	14348907	15.5885	6.2403	4.11523	763.41	46377.0
244	59536	14526784	15.6205	6.2488	4.09836	766.55	46759.5
245	60025	14706125	15.6525	6.2573	4.08163	769.69	47143.5
246	60516	14886936	15.6844	6.2658	4.06504	772.83	47529.2
247	61009	15069223	15.7162	6.2743	4.04858	775.97	47916.4
248	61504	15252992	15.7480	6.2828	4.03226	779.12	48305.1
249	62001	15438249	15.7797	6.2912	4.01606	782.26	48695.5
250	62500	15625000	15.8114	6.2996	4.00000	785.40	49087.4

N	N^2	N^3	$\sqrt{N}$	$\sqrt[3]{N}$	$\frac{1000}{N}$	πN	$\frac{\pi N^2}{4}$
251	63001	15813251	15.8430	6.3080	3.98406	788.54	49480.9
252	63504	16003008	15.8745	6.3164	3.96825	791.68	49875.9
253	64009	16194277	15.9060	6.3247	3.95257	794.82	50272.6
254	64516	16387064	15.9374	6.3330	3.93701	797.96	50670.7
255	65025	16581375	15.9687	6.3413	3.92157	801.11	51070.5
256	65536	16777216	16.0000	6.3496	3.90625	804.25	51471.9
257	66049	16974593	16.0312	6.3579	3.89105	807.39	51874.8
258	66564	17173512	16.0624	6.3661	3.87597	810.53	52279.2
259	67081	17373979	16.0935	6.3743	3.86100	813.67	52685.3
260	67600	17576000	16.1245	6.3825	3.84615	816.81	53092.9
261	68121	17779581	16.1555	6.3907	3.83142	819.96	53502.1
262	68644	17984728	16.1864	6.3988	3.81679	823.10	53912.9
263	69169	18191447	16.2173	6.4070	3.80228	826.24	54325.2
264	69696	18399744	16.2481	6.4151	3.78788	829.38	54739.1
265	70225	18609625	16.2788	6.4232	3.77358	832.52	55154.6
266	70756	18821096	16.3095	6.4312	3.75940	835.66	55571.6
267	71289	19034163	16.3401	6.4393	3.74532	838.81	55990.3
268	71824	19248832	16.3707	6.4473	3.73134	841.95	56410.4
269	72361	19465109	16.4012	6.4553	3.71747	845.09	56832.2
270	72900	19683000	16.4317	6.4633	3.70370	848.23	57255.5
271	73441	19902511	16.4621	6.4713	3.69004	851.37	57680.4
272	73984	20123648	16.4924	6.4792	3.67647	854.51	58106.9
273	74529	20346417	16.5227	6.4872	3.66300	857.66	58534.9
274	75076	20570824	16.5529	6.4951	3.64964	860.80	58964.6
275	75625	20796875	16.5831	6.5030	3.63636	863.94	59395.7
276	76176	21024576	16.6132	6.5108	3.62319	867.08	59828.5
277	76729	21253933	16.6433	6.5187	3.61011	870.22	60262.8
278	77284	21484952	16.6733	6.5265	3.59712	873.36	60698.7
279	77841	21717639	16.7033	6.5343	3.58423	876.50	61136.2
280	78400	21952000	16.7332	6.5421	3.57143	879.65	61575.2
281	78961	22188041	16.7631	6.5499	3.55872	882.79	62015.8
282	79524	22425768	16.7929	6.5577	3.54610	885.93	62458.0
283	80089	22665187	16.8226	6.5654	3.53357	889.07	62901.8
284	80656	22906304	16.8523	6.5731	3.52113	892.21	63347.1
285	81225	23149125	16.8819	6.5808	3.50877	895.35	63794.0
286	81796	23393656	16.9115	6.5885	3.49650	898.50	64242.4
287	82369	23639903	16.9411	6.5962	3.48432	901.64	64692.5
288	82944	23887872	16.9706	6.6039	3.47222	904.78	65144.1
289	83521	24137569	17.0000	6.6115	3.46021	907.92	65597.2
290	84100	24389000	17.0294	6.6191	3.44828	911.06	66052.0
291	84681	24642171	17.0587	6.6267	3.43643	914.20	66508.3
292	85264	24897088	17.0880	6.6343	3.42466	917.35	66966.2
293	85849	25153757	17.1172	6.6419	3.41297	920.49	67425.6
294	86436	25412184	17.1464	6.6494	3.40136	923.63	67886.7
295	87025	25672375	17.1756	6.6569	3.38983	926.77	68349.3
296	87616	25934336	17.2047	6.6644	3.37838	929.91	68813.5
297	88209	26198073	17.2337	6.6719	3.36700	933.05	69279.2
298	88804	26463592	17.2627	6.6794	3.35570	936.19	69746.5
299	89401	26730899	17.2916	6.6869	3.34448	939.34	70215.4
300	90000	27000000	17.3205	6.6943	3.33333	942.48	70685.8

N	N^2	N^3	$\sqrt{N}$	$\sqrt[3]{N}$	$\frac{1000}{N}$	πN	$\frac{\pi N^2}{4}$
301	90601	27270901	17.3494	6.7018	3.32226	945.62	71157.9
302	91204	27543608	17.3781	6.7092	3.31126	948.76	71631.5
303	91809	27818127	17.4069	6.7166	3.30033	951.90	72106.6
304	92416	28094464	17.4356	6.7240	3.28947	955.04	72583.4
305	93025	28372625	17.4642	6.7313	3.27869	958.19	73061.7
306	93636	28652616	17.4929	6.7387	3.26797	961.33	73541.5
307	94249	28934443	17.5214	6.7460	3.25733	964.47	74023.0
308	94864	29218112	17.5499	6.7533	3.24675	967.61	74506.0
309	95481	29503629	17.5784	6.7606	3.23625	970.75	74990.6
310	96100	29791000	17.6068	6.7679	3.22581	973.89	75476.8
311	96721	30080231	17.6352	6.7752	3.21543	977.04	75964.5
312	97344	30371328	17.6635	6.7824	3.20513	980.18	76453.8
313	97969	30664297	17.6918	6.7897	3.19489	983.32	76944.7
314	98596	30959144	17.7200	6.7969	3.18471	986.46	77437.1
315	99225	31255875	17.7482	6.8041	3.17460	989.60	77931.1
316	99856	31554496	17.7764	6.8113	3.16456	992.74	78426.7
317	100489	31855013	17.8045	6.8185	3.15457	995.88	78923.9
318	101124	32157432	17.8326	6.8256	3.14465	999.03	79422.6
319	101761	32461759	17.8606	6.8328	3.13480	1002.2	79922.9
320	102400	32768000	17.8885	6.8399	3.12500	1005.3	80424.8
321	103041	33076161	17.9165	6.8470	3.11527	1008.5	80928.2
322	103684	33386248	17.9444	6.8541	3.10559	1011.6	81433.2
323	104329	33698267	17.9722	6.8612	3.09598	1014.7	81939.8
324	104976	34012224	18.0000	6.8683	3.08642	1017.9	82448.0
325	105625	34328125	18.0278	6.8753	3.07692	1021.0	82957.7
326	106276	34645976	18.0555	6.8824	3.06749	1024.2	83469.0
327	106929	34965783	18.0831	6.8894	3.05810	1027.3	83981.8
328	107584	35287552	18.1108	6.8964	3.04878	1030.4	84496.3
329	108241	35611289	18.1384	6.9034	3.03951	1033.6	85012.3
330	108900	35937000	18.1659	6.9104	3.03030	1036.7	85529.9
331	109561	36264691	18.1934	6.9174	3.02115	1039.9	86049.0
332	110224	36594368	18.2209	6.9244	3.01205	1043.0	86569.7
333	110889	36926037	18.2483	6.9313	3.00300	1046.2	87092.0
334	111556	37259704	18.2757	6.9382	2.99401	1049.3	87615.9
335	112225	37595375	18.3030	6.9451	2.98507	1052.4	88141.3
336	112896	37933056	18.3303	6.9521	2.97619	1055.6	88668.3
337	113569	38272753	18.3576	6.9589	2.96736	1058.7	89196.9
338	114244	38614472	18.3848	6.9658	2.95858	1061.9	89727.0
339	114921	38958219	18.4120	6.9727	2.94985	1065.0	90258.7
340	115600	39304000	18.4391	6.9795	2.94118	1068.1	90792.0
341	116281	39651821	18.4662	6.9864	2.93255	1071.3	91326.9
342	116964	40001688	18.4932	6.9932	2.92398	1074.4	91863.3
343	117649	40353607	18.5203	7.0000	2.91545	1077.6	92401.3
344	118336	40707584	18.5472	7.0068	2.90698	1080.7	92940.9
345	119025	41063625	18.5742	7.0136	2.89855	1083.8	93482.0
346	119716	41421736	18.6011	7.0203	2.89017	1087.0	94024.7
347	120409	41781923	18.6279	7.0271	2.88184	1090.1	94569.0
348	121104	42144192	18.6548	7.0338	2.87356	1093.3	95114.9
349	121801	42508549	18.6815	7.0406	2.86533	1096.4	95662.3
350	122500	42875000	18.7083	7.0473	2.85714	1099.6	96211.3

N	N^2	N^3	$\sqrt{N}$	$\sqrt[3]{N}$	$\frac{1000}{N}$	πN	$\frac{\pi N^2}{4}$
351	123201	43243551	18.7350	7.0540	2.84900	1102.7	96761.8
352	123904	43614208	18.7617	7.0607	2.84091	1105.8	97314.0
353	124609	43986977	18.7883	7.0674	2.83286	1109.0	97867.7
354	125316	44361864	18.8149	7.0740	2.82486	1112.1	98423.0
355	126025	44738875	18.8414	7.0807	2.81690	1115.3	98979.8
356	126736	45118016	18.8680	7.0873	2.80899	1118.4	99538.2
357	127449	45499293	18.8944	7.0940	2.80112	1121.5	100098
358	128164	45882712	18.9209	7.1006	2.79330	1124.7	100660
359	128881	46268279	18.9473	7.1072	2.78552	1127.8	101223
360	129600	46656000	18.9737	7.1138	2.77778	1131.0	101788
361	130321	47045881	19.0000	7.1204	2.77008	1134.1	102354
362	131044	47437928	19.0263	7.1269	2.76243	1137.3	102922
363	131769	47832147	19.0526	7.1335	2.75482	1140.4	103491
364	132496	48228544	19.0788	7.1400	2.74725	1143.5	104062
365	133225	48627125	19.1050	7.1466	2.73973	1146.7	104635
366	133956	49027896	19.1311	7.1531	2.73224	1149.8	105209
367	134689	49430863	19.1572	7.1596	2.72480	1153.0	105785
368	135424	49836032	19.1833	7.1661	2.71739	1156.1	106362
369	136161	50243409	19.2094	7.1726	2.71003	1159.2	106941
370	136900	50653000	19.2354	7.1791	2.70270	1162.4	107521
371	137641	51064811	19.2614	7.1855	2.69542	1165.5	108103
372	138384	51478848	19.2873	7.1920	2.68817	1168.7	108687
373	139129	51895117	19.3132	7.1984	2.68097	1171.8	109272
374	139876	52313624	19.3391	7.2048	2.67380	1175.0	109858
375	140625	52734375	19.3649	7.2112	2.66667	1178.1	110447
376	141376	53157376	19.3907	7.2177	2.65957	1181.2	111036
377	142129	53582633	19.4165	7.2240	2.65252	1184.4	111628
378	142884	54010152	19.4422	7.2304	2.64550	1187.5	112221
379	143641	54439939	19.4679	7.2368	2.63852	1190.7	112815
380	144400	54872000	19.4936	7.2432	2.63158	1193.8	113411
381	145161	55306341	19.5192	7.2495	2.62467	1196.9	114009
382	145924	55742968	19.5448	7.2558	2.61780	1200.1	114608
383	146689	56181887	19.5704	7.2622	2.61097	1203.2	115209
384	147456	56623104	19.5959	7.2685	2.60417	1206.4	115812
385	148225	57066625	19.6214	7.2748	2.59740	1209.5	116416
386	148996	57512456	19.6469	7.2811	2.59067	1212.7	117021
387	149769	57960603	19.6723	7.2874	2.58398	1215.8	117628
388	150544	58411072	19.6977	7.2936	2.57732	1218.9	118237
389	151321	58863869	19.7231	7.2999	2.57069	1221.1	118847
390	152100	59319000	19.7484	7.3061	2.56410	1225.2	119459
391	152881	59776471	19.7737	7.3124	2.55755	1228.4	120072
392	153664	60236288	19.7990	7.3186	2.55102	1231.5	120687
393	154449	60698457	19.8242	7.3248	2.54453	1234.6	121304
394	155236	61162984	19.8494	7.3310	2.53807	1237.8	121922
395	156025	61629875	19.8746	7.3372	2.53165	1240.9	122542
396	156816	62099136	19.8997	7.3434	2.52525	1244.1	123163
397	157609	62570773	19.9249	7.3496	2.51889	1247.2	123786
398	158404	63044792	19.9499	7.3558	2.51256	1250.4	124410
399	159201	63521199	19.9750	7.3619	2.50627	1253.5	125036
400	160000	64000000	20.0000	7.3681	2.50000	1256.6	125664

N	N^2	N^3	$\sqrt{N}$	$\sqrt[3]{N}$	$\frac{1000}{N}$	πN	$\frac{\pi N^2}{4}$
401	160801	64481201	20.0250	7.3742	2.49377	1259.8	126293
402	161604	64964808	20.0499	7.3803	2.48756	1262.9	126923
403	162409	65450827	20.0749	7.3864	2.48139	1266.1	127556
404	163216	65939264	20.0998	7.3925	2.47525	1269.2	128190
405	164025	66430125	20.1246	7.3986	2.46914	1272.3	128825
406	164836	66923416	20.1494	7.4047	2.46305	1275.5	129462
407	165649	67419143	20.1742	7.4108	2.45700	1278.6	130100
408	166464	67917312	20.1990	7.4169	2.45098	1281.8	130741
409	167281	68417929	20.2237	7.4229	2.44499	1284.9	131382
410	168100	68921000	20.2485	7.4290	2.43902	1288.1	132025
411	168921	69426531	20.2731	7.4350	2.43309	1291.2	132670
412	169744	69934528	20.2978	7.4410	2.42718	1294.3	133317
413	170569	70444997	20.3224	7.4470	2.42131	1297.5	133965
414	171396	70957944	20.3470	7.4530	2.41546	1300.6	134614
415	172225	71473375	20.3715	7.4590	2.40964	1303.8	135265
416	173056	71991296	20.3961	7.4650	2.40385	1306.9	135918
417	173889	72511713	20.4206	7.4710	2.39808	1310.0	136572
418	174724	73034632	20.4450	7.4770	2.39234	1313.2	137228
419	175561	73560059	20.4695	7.4829	2.38664	1316.3	137885
420	176400	74088000	20.4939	7.4889	2.38095	1319.5	138544
421	177241	74618461	20.5183	7.4948	2.37530	1322.6	139205
422	178084	75151448	20.5426	7.5007	2.36967	1325.8	139867
423	178929	75686967	20.5670	7.5067	2.36407	1328.9	140531
424	179776	76225024	20.5913	7.5126	2.35849	1332.0	141196
425	180625	76765625	20.6155	7.5185	2.35294	1335.2	141863
426	181476	77308776	20.6398	7.5244	2.34742	1338.3	142531
427	182329	77854483	20.6640	7.5302	2.34192	1341.5	143201
428	183184	78402752	20.6882	7.5361	2.33645	1344.6	143872
429	184041	78953589	20.7123	7.5420	2.33100	1347.7	144545
430	184900	79507000	20.7364	7.5478	2.32558	1350.9	145220
431	185761	80062991	20.7605	7.5537	2.32019	1354.0	145896
432	186624	80621568	20.7846	7.5595	2.31482	1357.2	146574
433	187489	81182737	20.8087	7.5654	2.30947	1360.3	147254
434	188356	81746504	20.8327	7.5712	2.30415	1363.5	147934
435	189225	82312875	20.8567	7.5770	2.29885	1366.6	148617
436	190096	82881856	20.8806	7.5828	2.29358	1369.7	149301
437	190969	83453453	20.9045	7.5886	2.28833	1372.9	149987
438	191844	84027672	20.9284	7.5944	2.28311	1376.0	150674
439	192721	84604519	20.9523	7.6001	2.27790	1379.2	151363
440	193600	85184000	20.9762	7.6059	2.27273	1382.3	152053
441	194481	85766121	21.0000	7.6117	2.26757	1385.4	152745
442	195364	86350888	21.0238	7.6174	2.26244	1388.6	153439
443	196249	86938307	21.0476	7.6232	2.25734	1391.7	154134
444	197136	87528384	21.0713	7.6289	2.25225	1394.9	154830
445	198025	88121125	21.0950	7.6346	2.24719	1398.0	155528
446	198916	88716536	21.1187	7.6403	2.24215	1401.2	156228
447	199809	89314623	21.1424	7.6460	2.23714	1404.3	156930
448	200704	89915392	21.1660	7.6517	2.23214	1407.4	157633
449	201601	90518849	21.1896	7.6574	2.22717	1410.6	158337
450	202500	91125000	21.2132	7.6631	2.22222	1413.7	159043

N	N^2	N^3	$\sqrt{N}$	$\sqrt[3]{N}$	$\frac{1000}{N}$	πN	$\frac{\pi N^2}{4}$
451	203401	91733851	21.2368	7.6688	2.21730	1416.9	159751
452	204304	92345408	21.2603	7.6744	2.21239	1420.0	160460
453	205209	92959677	21.2838	7.6801	2.20751	1423.1	161171
454	206116	93576664	21.3073	7.6857	2.20264	1426.3	161883
455	207025	94196375	21.3307	7.6914	2.19780	1429.4	162597
456	207936	94818816	21.3542	7.6970	2.19298	1432.6	163313
457	208849	95443993	21.3776	7.7026	2.18818	1435.7	164030
458	209764	96071912	21.4009	7.7082	2.18341	1438.9	164748
459	210681	96702579	21.4243	7.7138	2.17865	1442.0	165468
460	211600	97336000	21.4476	7.7194	2.17391	1445.1	166190
461	212521	97972181	21.4709	7.7250	2.16920	1448.3	166914
462	213444	98611128	21.4942	7.7306	2.16450	1451.4	167639
463	214369	99252847	21.5174	7.7362	2.15983	1454.6	168365
464	215296	99897344	21.5407	7.7418	2.15517	1457.7	169093
465	216225	100544625	21.5639	7.7473	2.15054	1460.8	169823
466	217156	101194696	21.5870	7.7529	2.14592	1464.0	170554
467	218089	101847563	21.6102	7.7584	2.14133	1467.1	171287
468	219024	102503232	21.6333	7.7639	2.13675	1470.3	172021
469	219961	103161709	21.6564	7.7695	2.13220	1473.4	172757
470	220900	103823000	21.6795	7.7750	2.12766	1476.5	173494
471	221841	104487111	21.7025	7.7805	2.12314	1479.7	174234
472	222784	105154048	21.7256	7.7860	2.11864	1482.8	174974
473	223729	105823817	21.7486	7.7915	2.11417	1486.0	175716
474	224676	106496424	21.7715	7.7970	2.10971	1489.1	176460
475	225625	107171875	21.7945	7.8025	2.10526	1492.3	177205
476	226576	107850176	21.8174	7.8079	2.10084	1495.4	177952
477	227529	108531333	21.8403	7.8134	2.09644	1498.5	178701
478	228484	109215352	21.8632	7.8188	2.09205	1501.7	179451
479	229441	109902239	21.8861	7.8243	2.08768	1504.8	180203
480	230400	110592000	21.9089	7.8297	2.08333	1508.0	180956
481	231361	111284641	21.9317	7.8352	2.07900	1511.1	181711
482	232324	111980168	21.9545	7.8406	2.07469	1514.3	182467
483	233289	112678587	21.9773	7.8460	2.07039	1517.4	183225
484	234256	113379904	22.0000	7.8514	2.06612	1520.5	183984
485	235225	114084125	22.0227	7.8568	2.06186	1523.7	184745
486	236196	114791256	22.0454	7.8622	2.05761	1526.8	185508
487	237169	115501303	22.0681	7.8676	2.05339	1530.0	186272
488	238144	116214272	22.0907	7.8730	2.04918	1533.1	187038
489	239121	116930169	22.1133	7.8784	2.04499	1536.2	187805
490	240100	117649000	22.1359	7.8837	2.04082	1539.4	188574
491	241081	118370771	22.1585	7.8891	2.03666	1542.5	189345
492	242064	119095488	22.1811	7.8944	2.03252	1545.7	190117
493	243049	119823157	22.2036	7.8998	2.02840	1548.8	190890
494	244036	120553784	22.2261	7.9051	2.02429	1551.9	191665
495	245025	121287375	22.2486	7.9105	2.02020	1555.1	192442
496	246016	122023936	22.2711	7.9158	2.01613	1558.2	193221
497	247009	122763473	22.2935	7.9211	2.01207	1561.4	194000
498	248004	123505992	22.3159	7.9264	2.00803	1564.5	194782
499	249001	124251499	22.3383	7.9317	2.00401	1567.7	195565
500	250000	125000000	22.3607	7.9370	2.00000	1570.8	196350

 Numbers (501 to 550), Squares, Cubes, Square Roots, Cube Roots, Reciprocals, Circumferences and Circular Areas

N	N^2	N^3	$\sqrt{N}$	$\sqrt[3]{N}$	$\frac{1000}{N}$	πN	$\frac{\pi N^2}{4}$
501	251001	125751501	22.3830	7.9423	1.99601	1573.9	197136
502	252004	126506008	22.4054	7.9476	1.99203	1577.1	197923
503	253009	127263527	22.4277	7.9528	1.98807	1580.2	198713
504	254016	128024064	22.4499	7.9581	1.98413	1583.4	199504
505	255025	128787625	22.4722	7.9634	1.98020	1586.5	200296
506	256036	129554216	22.4944	7.9686	1.97629	1589.7	201090
507	257049	130323843	22.5167	7.9739	1.97239	1592.8	201886
508	258064	131096512	22.5389	7.9791	1.96850	1595.9	202683
509	259081	131872229	22.5610	7.9843	1.96464	1599.1	203482
510	260100	132651000	22.5832	7.9896	1.96078	1602.2	204282
511	261121	133432831	22.6053	7.9948	1.95695	1605.4	205084
512	262144	134217728	22.6274	8.0000	1.95312	1608.5	205887
513	263169	135005697	22.6495	8.0052	1.94932	1611.6	206692
514	264196	135796744	22.6716	8.0104	1.94553	1614.8	207499
515	265225	136590875	22.6936	8.0156	1.94175	1617.9	208307
516	266256	137388096	22.7156	8.0208	1.93798	1621.1	209117
517	267289	138188413	22.7376	8.0260	1.93424	1624.2	209928
518	268324	138991832	22.7596	8.0311	1.93050	1627.3	210741
519	269361	139798359	22.7816	8.0363	1.92678	1630.5	211556
520	270400	140608000	22.8035	8.0415	1.92308	1633.6	212372
521	271441	141420761	22.8254	8.0466	1.91939	1636.8	213189
522	272484	142236648	22.8473	8.0517	1.91571	1639.9	214008
523	273529	143055667	22.8692	8.0569	1.91205	1643.1	214829
524	274576	143877824	22.8910	8.0620	1.90840	1646.2	215651
525	275625	144703125	22.9129	8.0671	1.90476	1649.3	216475
526	276676	145531576	22.9347	8.0723	1.90114	1652.5	217301
527	277729	146363183	22.9565	8.0774	1.89753	1655.6	218128
528	278784	147197952	22.9783	8.0825	1.89394	1658.8	218956
529	279841	148035889	23.0000	8.0876	1.89036	1661.9	219787
530	280900	148877000	23.0217	8.0927	1.88679	1665.0	220618
531	281961	149721291	23.0434	8.0978	1.88324	1668.2	221452
532	283024	150568768	23.0651	8.1028	1.87970	1671.3	222287
533	284089	151419437	23.0868	8.1079	1.87617	1674.5	223123
534	285156	152273304	23.1084	8.1130	1.87266	1677.6	223961
535	286225	153130375	23.1301	8.1180	1.86916	1680.8	224801
536	287296	153990656	23.1517	8.1231	1.86567	1683.9	225642
537	288369	154854153	23.1733	8.1281	1.86220	1687.0	226484
538	289444	155720872	23.1948	8.1332	1.85874	1690.2	227329
539	290521	156590819	23.2164	8.1382	1.85529	1693.3	228175
540	291600	157464000	23.2379	8.1433	1.85185	1696.5	229022
541	292681	158340421	23.2594	8.1483	1.84843	1699.6	229871
542	293764	159220088	23.2809	8.1533	1.84502	1702.7	230722
543	294849	160103007	23.3024	8.1583	1.84162	1705.9	231574
544	295936	160989184	23.3238	8.1633	1.83824	1709.0	232428
545	297025	161878625	23.3452	8.1683	1.83486	1712.2	233283
546	298116	162771336	23.3666	8.1733	1.83150	1715.3	234140
547	299209	163667323	23.3880	8.1783	1.82815	1718.5	234998
548	300304	164566592	23.4094	8.1833	1.82482	1721.6	235858
549	301401	165469149	23.4307	8.1882	1.82149	1724.7	236720
550	302500	166375000	23.4521	8.1932	1.81818	1727.9	237583

N	N^2	N^3	$\sqrt{N}$	$\sqrt[3]{N}$	$\frac{1000}{N}$	πN	$\frac{\pi N^2}{4}$
551	303601	167284151	23.4734	8.1982	1.81488	1731.0	238448
552	304704	168196608	23.4947	8.2031	1.81159	1734.2	239314
553	305809	169112377	23.5160	8.2081	1.80832	1737.3	240182
554	306916	170031464	23.5372	8.2130	1.80505	1740.4	241051
555	308025	170953875	23.5584	8.2180	1.80180	1743.6	241922
556	309136	171879616	23.5797	8.2229	1.79856	1746.7	242795
557	310249	172808693	23.6008	8.2278	1.79533	1749.9	243669
558	311364	173741112	23.6220	8.2327	1.79211	1753.0	244545
559	312481	174676879	23.6432	8.2377	1.78891	1756.2	245422
560	313600	175616000	23.6643	8.2426	1.78571	1759.3	246301
561	314721	176558481	23.6854	8.2475	1.78253	1762.4	247181
562	315844	177504328	23.7065	8.2524	1.77936	1765.6	248063
563	316969	178453547	23.7276	8.2573	1.77620	1768.7	248947
564	318096	179406144	23.7487	8.2621	1.77305	1771.9	249832
565	319225	180362125	23.7697	8.2670	1.76991	1775.0	250719
566	320356	181321496	23.7908	8.2719	1.76678	1778.1	251607
567	321489	182284263	23.8118	8.2768	1.76367	1781.3	252497
568	322624	183250432	23.8328	8.2816	1.76056	1784.4	253388
569	323761	184220009	23.8537	8.2865	1.75747	1787.6	254281
570	324900	185193000	23.8747	8.2913	1.75439	1790.7	255176
571	326041	186169411	23.8956	8.2962	1.75131	1793.9	256072
572	327184	187149248	23.9165	8.3010	1.74825	1797.0	256970
573	328329	188132517	23.9374	8.3059	1.74520	1800.1	257869
574	329476	189119224	23.9583	8.3107	1.74216	1803.3	258770
575	330625	190109375	23.9792	8.3155	1.73913	1806.4	259672
576	331776	191102976	24.0000	8.3203	1.73611	1809.6	260576
577	332929	192100033	24.0208	8.3251	1.73310	1812.7	261482
578	334084	193100552	24.0416	8.3300	1.73010	1815.8	262389
579	335241	194104539	24.0624	8.3348	1.72712	1819.0	263298
580	336400	195112000	24.0832	8.3396	1.72414	1822.1	264208
581	337561	196122941	24.1039	8.3443	1.72117	1825.3	265120
582	338724	197137368	24.1247	8.3491	1.71821	1828.4	266033
583	339889	198155287	24.1454	8.3539	1.71527	1831.6	266948
584	341056	199176704	24.1661	8.3587	1.71233	1834.7	267865
585	342225	200201625	24.1868	8.3634	1.70940	1837.8	268783
586	343396	201230056	24.2074	8.3682	1.70649	1841.0	269701
587	344569	202262003	24.2281	8.3730	1.70358	1844.1	270624
588	345744	203297472	24.2487	8.3777	1.70068	1847.3	271547
589	346921	204336469	24.2693	8.3825	1.69779	1850.4	272471
590	348100	205379000	24.2899	8.3872	1.69492	1853.5	273397
591	349281	206425071	24.3105	8.3919	1.69205	1856.7	274325
592	350464	207474688	24.3311	8.3967	1.68919	1859.8	275254
593	351649	208527857	24.3516	8.4014	1.68634	1863.0	276184
594	352836	209584584	24.3721	8.4061	1.68350	1866.1	277117
595	354025	210644875	24.3926	8.4108	1.68067	1869.3	278051
596	355216	211708736	24.4131	8.4155	1.67785	1872.4	278986
597	356409	212776173	24.4336	8.4202	1.67504	1875.5	279923
598	357604	213847192	24.4540	8.4249	1.67224	1878.7	280862
599	358801	214921799	24.4745	8.4296	1.66945	1881.8	281802
600	360000	216000000	24.4949	8.4343	1.66667	1885.0	282743

N	N^2	N^3	$\sqrt{N}$	$\sqrt[3]{N}$	$\frac{1000}{N}$	πN	$\frac{\pi N^2}{4}$
601	361201	217081801	24.5153	8.4390	1.66389	1888.1	283687
602	362404	218167208	24.5357	8.4437	1.66113	1891.2	284631
603	363609	219256227	24.5561	8.4484	1.65837	1894.4	285578
604	364816	220348864	24.5764	8.4530	1.65563	1897.5	286526
605	366025	221445125	24.5967	8.4577	1.65289	1900.7	287475
606	367236	222545016	24.6171	8.4623	1.65017	1903.8	288426
607	368449	223648543	24.6374	8.4670	1.64745	1907.0	289379
608	369664	224755712	24.6577	8.4716	1.64474	1910.1	290333
609	370881	225866529	24.6779	8.4763	1.64204	1913.2	291289
610	372100	226981000	24.6982	8.4809	1.63934	1916.4	292247
611	373321	228099131	24.7184	8.4856	1.63666	1919.5	293206
612	374544	229220928	24.7386	8.4902	1.63399	1922.7	294166
613	375769	230346397	24.7588	8.4948	1.63132	1925.8	295128
614	376996	231475544	24.7790	8.4994	1.62866	1928.9	296092
615	378225	232608375	24.7992	8.5040	1.62602	1932.1	297057
616	379456	233744896	24.8193	8.5086	1.62338	1935.2	298024
617	380689	234885113	24.8395	8.5132	1.62075	1938.4	298992
618	381924	236029032	24.8596	8.5178	1.61812	1941.5	299962
619	383161	237176659	24.8797	8.5224	1.61551	1944.7	300934
620	384400	238328000	24.8998	8.5270	1.61290	1947.8	301907
621	385641	239483061	24.9199	8.5316	1.61031	1950.9	302882
622	386884	240641848	24.9399	8.5362	1.60772	1954.1	303858
623	388129	241804367	24.9600	8.5408	1.60514	1957.2	304836
624	389376	242970624	24.9800	8.5453	1.60256	1960.4	305815
625	390625	244140625	25.0000	8.5499	1.60000	1963.5	306796
626	391876	245314376	25.0200	8.5544	1.59744	1966.6	307779
627	39[illegible]129	246491883	25.0400	8.5590	1.59490	1969.8	308763
628	394384	247673152	25.0599	8.5635	1.59236	1972.9	309748
629	395641	248858189	25.0799	8.5681	1.58983	1976.1	310736
630	396900	250047000	25.0998	8.5726	1.58730	1979.2	311725
6[illegible]1	398161	251239591	25.1197	8.5772	1.58479	1982.4	312715
6[illegible]	399424	252435968	25.1396	8.5817	1.58228	1985.5	313707
633	400689	253636137	25.1595	8.5862	1.57978	1988.6	314700
634	401956	254840104	25.1794	8.5907	1.57729	1991.8	315696
635	403225	256047875	25.1992	8.5952	1.57480	1994.9	316692
636	404496	257259456	25.2190	8.5997	1.57233	1998.1	317690
637	405769	258474853	25.2389	8.6043	1.56986	2001.2	318690
638	407044	259694072	25.2587	8.6088	1.56740	2004.3	319692
639	408321	260917119	25.2784	8.6132	1.56495	2007.5	320695
640	409600	26[illegible]144000	25.2982	8.6177	1.56250	2010.6	321699
641	410881	2633[illegible]4721	25.3180	8.6222	1.56006	2013.8	322705
642	412164	264609288	25.3377	8.6267	1.55763	2016.9	323713
643	413449	265047707	25.3574	8.6312	1.55521	2020.0	324722
644	414736	267089904	25.3772	8.6357	1.55280	2023.2	325733
645	416025	2683[illegible]6125	25.3969	8.6401	1.55039	2026.3	326745
646	417316	269586136	25.4165	8.6446	1.54799	2029.5	327759
647	418609	270840023	25.4362	8.6490	1.54560	2032.6	328775
648	419904	272097792	25.4558	8.6535	1.54321	2035.8	329792
649	421201	273359449	25.4755	8.6579	1.54083	2038.9	330810
650	422500	274625000	25.4951	8.6624	1.53846	2042.0	331831

N	N^2	N^3	$\sqrt{N}$	$\sqrt[3]{N}$	$\frac{1000}{N}$	πN	$\frac{\pi N^2}{4}$
651	423801	275894451	25.5147	8.6668	1.53610	2045.2	332853
652	425104	277167808	25.5343	8.6713	1.53374	2048.3	333876
653	426409	278445077	25.5539	8.6757	1.53139	2051.5	334901
654	427716	279726264	25.5734	8.6801	1.52905	2054.6	335927
655	429025	281011375	25.5930	8.6845	1.52672	2057.7	336955
656	430336	282300416	25.6125	8.6890	1.52439	2060.9	337985
657	431649	283593393	25.6320	8.6934	1.52207	2064.0	339016
658	432964	284890312	25.6515	8.6978	1.51976	2067.2	340049
659	434281	286191179	25.6710	8.7022	1.51745	2070.3	341084
660	435600	287496000	25.6905	8.7066	1.51515	2073.5	342119
661	436921	288804781	25.7099	8.7110	1.51286	2076.6	343157
662	438244	290117528	25.7294	8.7154	1.51057	2079.7	344196
663	439569	291434247	25.7488	8.7198	1.50830	2082.9	345237
664	440896	292754944	25.7682	8.7241	1.50602	2086.0	346279
665	442225	294079625	25.7876	8.7285	1.50376	2089.2	347323
666	443556	295408296	25.8070	8.7329	1.50150	2092.3	348368
667	444889	296740963	25.8263	8.7373	1.49925	2095.4	349415
668	446224	298077632	25.8457	8.7416	1.49701	2098.6	350464
669	447561	299418309	25.8650	8.7460	1.49477	2101.7	351514
670	448900	300763000	25.8844	8.7503	1.49254	2104.9	352565
671	450241	302111711	25.9037	8.7547	1.49031	2108.0	353618
672	451584	303464448	25.9230	8.7590	1.48810	2111.2	354673
673	452929	304821217	25.9422	8.7634	1.48588	2114.3	355730
674	454276	306182024	25.9615	8.7677	1.48368	2117.4	356788
675	455625	307546875	25.9808	8.7721	1.48148	2120.6	357847
676	456976	308915776	26.0000	8.7764	1.47929	2123.7	358908
677	458329	310288733	26.0192	8.7807	1.47711	2126.9	359971
678	459684	311665752	26.0384	8.7850	1.47493	2130.0	361035
679	461041	313046839	26.0576	8.7893	1.47275	2133.1	362101
680	462400	314432000	26.0768	8.7937	1.47059	2136.3	363168
681	463761	315821241	26.0960	8.7980	1.46843	2139.4	364237
682	465124	317214568	26.1151	8.8023	1.46628	2142.6	365308
683	466489	318611987	26.1343	8.8066	1.46413	2145.7	366380
684	467856	320013504	26.1534	8.8109	1.46199	2148.9	367453
685	469225	321419125	26.1725	8.8152	1.45985	2152.0	368528
686	470596	322828856	26.1916	8.8194	1.45773	2155.1	369605
687	471969	324242703	26.2107	8.8237	1.45560	2158.3	370684
688	473344	325660672	26.2298	8.8280	1.45349	2161.4	371764
689	474721	327082769	26.2488	8.8323	1.45138	2164.6	372845
690	476100	328509000	26.2679	8.8366	1.44928	2167.7	373928
691	477481	329939371	26.2869	8.8408	1.44718	2170.8	375013
692	478864	331373888	26.3059	8.8451	1.44509	2174.0	376099
693	480249	332812557	26.3249	8.8493	1.44300	2177.1	377187
694	481636	334255384	26.3439	8.8536	1.44092	2180.3	378276
695	483025	335702375	26.3629	8.8578	1.43885	2183.4	379367
696	484416	337153536	26.3818	8.8621	1.43678	2186.6	380459
697	485809	338608873	26.4008	8.8663	1.43472	2189.7	381554
698	487204	340068392	26.4197	8.8706	1.43267	2192.8	382649
699	488601	341532099	26.4386	8.8748	1.43062	2196.0	383746
700	490000	343000000	26.4575	8.8790	1.42857	2199.1	384845

N	N^2	N^3	$\sqrt{N}$	$\sqrt[3]{N}$	$\frac{1000}{N}$	πN	$\frac{\pi N^2}{4}$
701	491401	344472101	26.4764	8.8833	1.42653	2202.3	385945
702	492804	345948408	26.4953	8.8875	1.42450	2205.4	387047
703	494209	347428927	26.5141	8.8917	1.42248	2208.5	388151
704	495616	348913664	26.5330	8.8959	1.42046	2211.7	389256
705	497025	350402625	26.5518	8.9001	1.41844	2214.8	390363
706	498436	351895816	26.5707	8.9043	1.41643	2218.0	391471
707	499849	353393243	26.5895	8.9085	1.41443	2221.1	392580
708	501264	354894912	26.6083	8.9127	1.41243	2224.3	393692
709	502681	356400829	26.6271	8.9169	1.41044	2227.4	394805
710	504100	357911000	26.6458	8.9211	1.40845	2230.5	395919
711	505521	359425431	26.6646	8.9253	1.40647	2233.7	397035
712	506944	360944128	26.6833	8.9295	1.40449	2236.8	398153
713	508369	362467097	26.7021	8.9337	1.40253	2240.0	399272
714	509796	363994344	26.7208	8.9378	1.40056	2243.1	400393
715	511225	365525875	26.7395	8.9420	1.39860	2246.2	401515
716	512656	367061696	26.7582	8.9462	1.39665	2249.4	402639
717	514089	368601813	26.7769	8.9503	1.39470	2252.5	403765
718	515524	370146232	26.7955	8.9545	1.39276	2255.7	404892
719	516961	371694959	26.8142	8.9587	1.39082	2258.8	406020
720	518400	373248000	26.8328	8.9628	1.38889	2261.9	407150
721	519841	374805361	26.8514	8.9670	1.38696	2265.1	408282
722	521284	376367048	26.8701	8.9711	1.38504	2268.2	409416
723	522729	377933067	26.8887	8.9752	1.38313	2271.4	410550
724	524176	379503424	26.9072	8.9794	1.38122	2274.5	411687
725	525625	381078125	26.9258	8.9835	1.37931	2277.7	412825
726	527076	382657176	26.9444	8.9876	1.37741	2280.8	413965
727	528529	384240583	26.9629	8.9918	1.37552	2283.9	415106
728	529984	385828352	26.9815	8.9959	1.37363	2287.1	416248
729	531441	387420489	27.0000	9.0000	1.37174	2290.2	417393
730	532900	389017000	27.0185	9.0041	1.36986	2293.4	418539
731	534361	390617891	27.0370	9.0082	1.36799	2296.5	419686
732	535824	392223168	27.0555	9.0123	1.36612	2299.7	420835
733	537289	393832837	27.0740	9.0164	1.36426	2302.8	421986
734	538756	395446904	27.0924	9.0205	1.36240	2305.9	423138
735	540225	397065375	27.1109	9.0246	1.36054	2309.1	424293
736	541696	398688256	27.1293	9.0287	1.35870	2312.2	425448
737	543169	400315553	27.1477	9.0328	1.35685	2315.4	426604
738	544644	401947272	27.1662	9.0369	1.35501	2318.5	427762
739	546121	403583419	27.1846	9.0410	1.35318	2321.6	428922
740	547600	405224000	27.2029	9.0450	1.35135	2324.8	430084
741	549081	406869021	27.2213	9.0491	1.34953	2327.9	431247
742	550564	408518488	27.2397	9.0532	1.34771	2331.1	432412
743	552049	410172407	27.2580	9.0572	1.34590	2334.2	433578
744	553536	411830784	27.2764	9.0613	1.34409	2337.3	434746
745	555025	413493625	27.2947	9.0654	1.34228	2340.5	435916
746	556516	415160936	27.3130	9.0694	1.34048	2343.6	437087
747	558009	416832723	27.3313	9.0735	1.33869	2346.8	438259
748	559504	418508992	27.3496	9.0775	1.33690	2349.9	439433
749	561001	420189749	27.3679	9.0816	1.33511	2353.1	440609
750	562500	421875000	27.3861	9.0856	1.33333	2356.2	441786

N	N^2	N^3	$\sqrt{N}$	$\sqrt[3]{N}$	$\frac{1000}{N}$	πN	$\frac{\pi N^2}{4}$
751	564001	423564751	27.4044	9.0896	1.33156	2359.3	442965
752	565504	425259008	27.4226	9.0937	1.32979	2362.5	444146
753	567009	426957777	27.4408	9.0977	1.32802	2365.6	445328
754	568516	428661064	27.4591	9.1017	1.32626	2368.8	446511
755	570025	430368875	27.4773	9.1057	1.32450	2371.9	447697
756	571536	432081216	27.4955	9.1098	1.32275	2375.0	448883
757	573049	433798093	27.5136	9.1138	1.32100	2378.2	450072
758	574564	435519512	27.5318	9.1178	1.31926	2381.3	451262
759	576081	437245479	27.5500	9.1218	1.31752	2384.5	452453
760	577600	438976000	27.5681	9.1258	1.31579	2387.6	453646
761	579121	440711081	27.5862	9.1298	1.31406	2390.8	454841
762	580644	442450728	27.6043	9.1338	1.31234	2393.9	456037
763	582169	444194947	27.6225	9.1378	1.31062	2397.0	457234
764	583696	445943744	27.6405	9.1418	1.30890	2400.2	458434
765	585225	447697125	27.6586	9.1458	1.30719	2403.3	459635
766	586756	449455096	27.6767	9.1498	1.30548	2406.5	460837
767	588289	451217663	27.6948	9.1537	1.30378	2409.6	462042
768	589824	452984832	27.7128	9.1577	1.30208	2412.7	463247
769	591361	454756609	27.7308	9.1617	1.30039	2415.9	464454
770	592900	456533000	27.7489	9.1657	1.29870	2419.0	465663
771	594441	458314011	27.7669	9.1696	1.29702	2422.2	466873
772	595984	460099648	27.7849	9.1736	1.29534	2425.3	468085
773	597529	461889917	27.8029	9.1775	1.29366	2428.5	469298
774	599076	463684824	27.8209	9.1815	1.29199	2431.6	470513
775	600625	465484375	27.8388	9.1855	1.29032	2434.7	471730
776	602176	467288576	27.8568	9.1894	1.28866	2437.9	472948
777	603729	469097433	27.8747	9.1933	1.28700	2441.0	474168
778	605284	470910952	27.8927	9.1973	1.28535	2444.2	475389
779	606841	472729139	27.9106	9.2012	1.28370	2447.3	476612
780	608400	474552000	27.9285	9.2052	1.28205	2450.4	477836
781	609961	476379541	27.9464	9.2091	1.28041	2453.6	479062
782	611524	478211768	27.9643	9.2130	1.27877	2456.7	480290
783	613089	480048687	27.9821	9.2170	1.27714	2459.9	481519
784	614656	481890304	28.0000	9.2209	1.27551	2463.0	482750
785	616225	483736625	28.0179	9.2248	1.27389	2466.2	483982
786	617796	485587656	28.0357	9.2287	1.27226	2469.3	485216
787	619369	487443403	28.0535	9 2326	1.27065	2472.4	486451
788	620944	489303872	28.0713	9.2365	1.26904	2475.6	487688
789	622521	491169069	28.0891	9.2404	1.26743	2478.7	488927
790	624100	493039000	28.1069	9.2443	1.26582	2481.9	490167
791	625681	494913671	28.1247	9.2482	1.26422	2485.0	491409
792	627264	496793088	28.1425	9.2521	1.26263	2488.1	492652
793	628849	498677257	28.1603	9.2560	1.26103	2491.3	493897
794	630436	500566184	28.1780	9.2599	1.25945	2494.4	495143
795	632025	502459875	28.1957	9.2638	1.25786	2497.6	496391
796	633616	504358336	28.2135	9.2677	1.25628	2500.7	497641
797	635209	506261573	28.2312	9.2716	1.25471	2503.8	498892
798	636804	508169592	28.2489	9.2754	1.25313	2507.0	500145
799	638401	510082399	28.2666	9.2793	1.25156	2510.1	501399
800	640000	512000000	28.2843	9.2832	1.25000	2513.3	502655

 Numbers (801 to 850), Squares, Cubes, Square Roots, Cube Roots, Reciprocals, Circumferences and Circular Areas

N	N^2	N^3	$\sqrt{N}$	$\sqrt[3]{N}$	$\frac{1000}{N}$	πN	$\frac{\pi N^2}{4}$
801	641601	513922401	28.3019	9.2870	1.24844	2516.4	503912
802	643204	515849608	28.3196	9.2909	1.24688	2519.6	505171
803	644809	517781627	28.3373	9.2948	1.24533	2522.7	506432
804	646416	519718464	28.3549	9.2986	1.24378	2525.8	507694
805	648025	521660125	28.3725	9.3025	1.24224	2529.0	508958
806	649636	523606616	28.3901	9.3063	1.24069	2532.1	510223
807	651249	525557943	28.4077	9.3102	1.23916	2535.3	511490
808	652864	527514112	28.4253	9.3140	1.23762	2538.4	512758
809	654481	529475129	28.4429	9.3179	1.23609	2541.5	514028
810	656100	531441000	28.4605	9.3217	1.23457	2544.7	515300
811	657721	533411731	28.4781	9.3255	1.23305	2547.8	516573
812	659344	535387328	28.4956	9.3294	1.23153	2551.0	517848
813	660969	537367797	28.5132	9.3332	1.23001	2554.1	519124
814	662596	539353144	28.5307	9.3370	1.22850	2557.3	520402
815	664225	541343375	28.5482	9.3408	1.22699	2560.4	521681
816	665856	543338496	28.5657	9.3447	1.22549	2563.5	522962
817	667489	545338513	28.5832	9.3485	1.22399	2566.7	524245
818	669124	547343432	28.6007	9.3523	1.22249	2569.8	525529
819	670761	549353259	28.6182	9.3561	1.22100	2573.0	526814
820	672400	551368000	28.6356	9.3599	1.21951	2576.1	528102
821	674041	553387661	28.6531	9.3637	1.21803	2579.2	529391
822	675684	555412248	28.6705	9.3675	1.21655	2582.4	530681
823	677329	557441767	28.6880	9.3713	1.21507	2585.5	531973
824	678976	559476224	28.7054	9.3751	1.21359	2588.7	533267
825	680625	561515625	28.7228	9.3789	1.21212	2591.8	534562
826	682276	563559976	28.7402	9.3827	1.21065	2595.0	535858
827	683929	565609283	28.7576	9.3865	1.20919	2598.1	537157
828	685584	567663552	28.7750	9.3902	1.20773	2601.2	538456
829	687241	569722789	28.7924	9.3940	1.20627	2604.4	539758
830	688900	571787000	28.8097	9.3978	1.20482	2607.5	541061
831	690561	573856191	28.8271	9.4016	1.20337	2610.7	542365
832	692224	575930368	28.8444	9.4053	1.20192	2613.8	543671
833	693889	578009537	28.8617	9.4091	1.20048	2616.9	544979
834	695556	580093704	28.8791	9.4129	1.19904	2620.1	546288
835	697225	582182875	28.8964	9.4166	1.19760	2623.2	547599
836	698896	584277056	28.9137	9.4204	1.19617	2626.4	548912
837	700569	586376253	28.9310	9.4241	1.19474	2629.5	550226
838	702244	588480472	28.9482	9.4279	1.19332	2632.7	551541
839	703921	590589719	28.9655	9.4316	1.19189	2635.8	552858
840	705600	592704000	28.9828	9.4354	1.19048	2638.9	554177
841	707281	594823321	29.0000	9.4391	1.18906	2642.1	555497
842	708964	596947688	29.0172	9.4429	1.18765	2645.2	556819
843	710649	599077107	29.0345	9.4466	1.18624	2648.4	558142
844	712336	601211584	29.0517	9.4503	1.18483	2651.5	559467
845	714025	603351125	29.0689	9.4541	1.18343	2654.6	560794
846	715716	605495736	29.0861	9.4578	1.18203	2657.8	562122
847	717409	607645423	29.1033	9.4615	1.18064	2660.9	563452
848	719104	609800192	29.1204	9.4652	1.17925	2664.1	564783
849	720801	611960049	29.1376	9.4690	1.17786	2667.2	566116
850	722500	614125000	29.1548	9.4727	1.17647	2670.4	567450

N	N^2	N^3	$\sqrt{N}$	$\sqrt[3]{N}$	$\frac{1000}{N}$	πN	$\frac{\pi N^2}{4}$
851	724201	616295051	29.1719	9.4764	1.17509	2673.5	568786
852	725904	618470208	29.1890	9.4801	1.17371	2676.6	570124
853	727609	620650477	29.2062	9.4838	1.17233	2679.8	571463
854	729316	622835864	29.2233	9.4875	1.17096	2682.9	572803
855	731025	625026375	29.2404	9.4912	1.16959	2686.1	574146
856	732736	627222016	29.2575	9.4949	1.16822	2689.2	575490
857	734449	629422793	29.2746	9.4986	1.16686	2692.3	576835
858	736164	631628712	29.2916	9.5023	1.16550	2695.5	578182
859	737881	633839779	29.3087	9.5060	1.16414	2698.6	579530
860	739600	636056000	29.3258	9.5097	16279	2701.8	580880
861	741321	638277381	29.3428	9.5134	1.16144	2704.9	582232
862	743044	640503928	29.3598	9.5171	1.16009	2708.1	583585
863	744769	642735647	29.3769	9.5207	1.15875	2711.2	584940
864	746496	644972544	29.3939	9.5244	1.15741	2714.3	586297
865	748225	647214625	29.4109	9.5281	1.15607	2717.5	587655
866	749956	649461896	29.4279	9.5317	1.15473	2720.6	589014
867	751689	651714363	29.4449	9.5354	1.15340	2723.8	590375
868	753424	653972032	29.4618	9.5391	1.15207	2726.9	591738
869	755161	656234909	29.4788	9.5427	1.15075	2730.0	593102
870	756900	658503000	29.4958	9.5464	1.14943	2733.2	594468
871	758641	660776311	29.5127	9.5501	1.14811	2736.3	595835
872	760384	663054848	29.5296	9.5537	1.14679	2739.5	597204
873	762129	665338617	29.5466	9.5574	1.14548	2742.6	598575
874	763876	667627624	29.5635	9.5610	1.14416	2745.8	599947
875	765625	669921875	29.5804	9.5647	1.14286	2748.9	601320
876	767376	672221376	29.5973	9.5683	1.14155	2752.0	602696
877	769129	674526133	29.6142	9.5719	1.14025	2755.2	604073
878	770884	676836152	29.6311	9.5756	1.13895	2758.3	605451
879	772641	679151439	29.6479	9.5792	1.13766	2761.5	606831
880	774400	681472000	29.6648	9.5828	1.13636	2764.6	608212
881	776161	683797841	29.6816	9.5865	1.13507	2767.7	609595
882	777924	686128968	29.6985	9.5901	1.13379	2770.9	610980
883	779689	688465387	29.7153	9.5937	1.13250	2774.0	612366
884	781456	690807104	29.7321	9.5973	1.13122	2777.2	613754
885	783225	693154125	29.7489	9.6010	1.12994	2780.3	615143
886	784996	695506456	29.7658	9.6046	1.12867	2783.5	616534
887	786769	697864103	29.7825	9.6082	1.12740	2786.6	617927
888	788544	700227072	29.7993	9.6118	1.12613	2789.7	619321
889	790321	702595369	29.8161	9.6154	1.12486	2792.9	620717
890	792100	704969000	29.8329	9.6190	1.12360	2796.0	622114
891	793881	707347971	29.8496	9.6226	1.12233	2799.2	623513
892	795664	709732288	29.8664	9.6262	1.12108	2802.3	624913
893	797449	712121957	29.8831	9.6298	1.11982	2805.4	626315
894	799236	714516984	29.8998	9.6334	1.11857	2808.6	627718
895	801025	716917375	29.9166	9.6370	1.11732	2811.7	629124
896	802816	719323136	29.9333	9.6406	1.11607	2814.9	630530
897	804609	721734273	29.9500	9.6442	1.11483	2818.0	631938
898	806404	724150792	29.9666	9.6477	1.11359	2821.2	633348
899	808201	726572699	29.9833	9.6513	1.11235	2824.3	634760
900	810000	729000000	30.0000	9.6549	1.11111	2827.4	636173

 Numbers (901 to 950), Squares, Cubes, Square Roots, Cube Roots, Reciprocals, Circumferences and Circular Areas

N	N^2	N^3	$\sqrt{N}$	$\sqrt[3]{N}$	$\frac{1000}{N}$	πN	$\frac{\pi N^2}{4}$
901	811801	731432701	30.0167	9.6585	1.10988	2830.6	637587
902	813604	733870808	30.0333	9.6620	1.10865	2833.7	639003
903	815409	736314327	30.0500	9.6656	1.10742	2836.9	640421
904	817216	738763264	30.0666	9.6692	1.10619	2840.0	641840
905	819025	741217625	30.0832	9.6727	1.10497	2843.1	643261
906	820836	743677416	30.0998	9.6763	1.10375	2846.3	644683
907	822649	746142643	30.1164	9.6799	1.10254	2849.4	646107
908	824464	748613312	30.1330	9.6834	1.10132	2852.6	647533
909	826281	751089429	30.1496	9.6870	1.10011	2855.7	648960
910	828100	753571000	30.1662	9.6905	1.09890	2858.8	650388
911	829921	756058031	30.1828	9.6941	1.09769	2862.0	651818
912	831744	758550528	30.1993	9.6976	1.09649	2865.1	653250
913	833569	761048497	30.2159	9.7012	1.09529	2868.3	654684
914	835396	763551944	30.2324	9.7047	1.09409	2871.4	656118
915	837225	766060875	30.2490	9.7082	1.09290	2874.6	657555
916	839056	768575296	30.2655	9.7118	1.09170	2877.7	658993
917	840889	771095213	30.2820	9.7153	1.09051	2880.8	660433
918	842724	773620632	30.2985	9.7188	1.08932	2884.0	661874
919	844561	776151559	30.3150	9.7224	1.08814	2887.1	663317
920	846400	778688000	30.3315	9.7259	1.08696	2890.3	664761
921	848241	781229961	30.3480	9.7294	1.08578	2893.4	666207
922	850084	783777448	30.3645	9.7329	1.08460	2896.5	667654
923	851929	786330467	30.3809	9.7364	1.08342	2899.7	669103
924	853776	788889024	30.3974	9.7400	1.08225	2902.8	670554
925	855625	791453125	30.4138	9.7435	1.08108	2906.0	672006
926	857476	794022776	30.4302	9.7470	1.07991	2909.1	673460
927	859329	796597983	30.4467	9.7505	1.07875	2912.3	674915
928	861184	799178752	30.4631	9.7540	1.07759	2915.4	676372
929	863041	801765089	30.4795	9.7575	1.07643	2918.5	677831
930	864900	804357000	30.4959	9.7610	1.07527	2921.7	679291
931	866761	806954491	30.5123	9.7645	1.07411	2924.8	680752
932	868624	809557568	30.5287	9.7680	1.07296	2928.0	682216
933	870489	812166237	30.5450	9.7715	1.07181	2931.1	683680
934	872356	814780504	30.5614	9.7750	1.07066	2934.2	685147
935	874225	817400375	30.5778	9.7785	1.06952	2937.4	686615
936	876096	820025856	30.5941	9.7819	1.06838	2940.5	688084
937	877969	822656953	30.6105	9.7854	1.06724	2943.7	689555
938	879844	825293672	30.6268	9.7889	1.06610	2946.8	691028
939	881721	827936019	30.6431	9.7924	1.06496	2950.0	692502
940	883600	830584000	30.6594	9.7959	1.06383	2953.1	693978
941	885481	833237621	30.6757	9.7993	1.06270	2956.2	695455
942	887364	835896888	30.6920	9.8028	1.06157	2959.4	696934
943	889249	838561807	30.7083	9.8063	1.06045	2962.5	698415
944	891136	841232384	30.7246	9.8097	1.05932	2965.7	699897
945	893025	843908625	30.7409	9.8132	1.05820	2968.8	701380
946	894916	846590536	30.7571	9.8167	1.05708	2971.9	702865
947	896809	849278123	30.7734	9.8201	1.05597	2975.1	704352
948	898704	851971392	30.7896	9.8236	1.05485	2978.2	705840
949	900601	854670349	30.8058	9.8270	1.05374	2981.4	707330
950	902500	857375000	30.8221	9.8305	1.05263	2984.5	708822

N	N^2	N^3	$\sqrt{N}$	$\sqrt[3]{N}$	$\frac{1000}{N}$	πN	$\frac{\pi N^2}{4}$
951	904401	860085351	30.8383	9.8339	1.05152	2987.7	710315
952	906304	862801408	30.8545	9.8374	1.05042	2990.8	711809
953	908209	865523177	30.8707	9.8408	1.04932	2993.9	713306
954	910116	868250664	30.8869	9.8443	1.04822	2997.1	714803
955	912025	870983875	30.9031	9.8477	1.04712	3000.2	716303
956	913936	873722816	30.9192	9.8511	1.04603	3003.4	717804
957	915849	876467493	30.9354	9.8546	1.04493	3006.5	719306
958	917764	879217912	30.9516	9.8580	1.04384	3009.6	720810
959	919681	881974079	30.9677	9.8614	1.04275	3012.8	722316
960	921600	884736000	30.9839	9.8648	1.04167	3015.9	723823
961	923521	887503681	31.0000	9.8683	1.04058	3019.1	725332
962	925444	890277128	31.0161	9.8717	1.03950	3022.2	726842
963	927369	893056347	31.0322	9.8751	1.03842	3025 4	728354
964	929296	895841344	31.0483	9.8785	1.03734	3028.5	729867
965	931225	898632125	31.0644	9.8819	1.03627	3031.6	731382
966	933156	901428696	31.0805	9.8854	1.03520	3034.8	732899
967	935089	904231063	31.0966	9.8888	1.03413	3037.9	734417
968	937024	907039232	31.1127	9.8922	1.03306	3041.1	735937
969	938961	909853209	31.1288	9.8956	1.03199	3044.2	737458
970	940900	912673000	31.1448	9.8990	1.03093	3047.3	738981
971	942841	915498611	31.1609	9.9024	1.02987	3050.5	740506
972	944784	918330048	31.1769	9.9058	1.02881	3053.6	742032
973	946729	921167317	31.1929	9.9092	1.02775	3056.8	743559
974	948676	924010424	31.2090	9.9126	1.02669	3059.9	745088
975	950625	926859375	31.2250	9.9160	1.02564	3063.1	746619
976	952576	929714176	31.2410	9.9194	1.02459	3066 2	748151
977	954529	932574833	31.2570	9.9227	1.02354	3069.3	749685
978	956484	935441352	31.2730	9.9261	1.02249	3072.5	751221
979	958441	938313739	31.2890	9.9295	1.02145	3075.6	752758
980	960400	941192000	31.3050	9.9329	1.02041	3078.8	754296
981	962361	944076141	31.3209	9.9363	1.01937	3081.9	755837
982	964324	946966168	31.3369	9.9396	1.01833	3085.0	757378
983	966289	949862087	31.3528	9.9430	1.01729	3088.2	758922
984	968256	952763904	31.3688	9.9464	1.01626	3091.3	760466
985	970225	955671625	31.3847	9.9497	1.01523	3094.5	762013
986	972196	958585256	31.4006	9.9531	1.01420	3097.6	763561
987	974169	961504803	31.4166	9.9565	1.01317	3100.8	765111
988	976144	964430272	31.4325	9.9598	1.01215	3103.9	766662
989	978121	967361669	31.4484	9.9632	1.01112	3107.0	768214
990	980100	970299000	31.4643	9.9666	1.01010	3110.2	769769
991	982081	973242271	31.4802	9.9699	1.00908	3113.3	771325
992	984064	976191488	31.4960	9.9733	1.00806	3116.5	772882
993	986049	979146657	31.5119	9.9766	1.00705	3119.6	774441
994	988036	982107784	31.5278	9.9800	1.00604	3122.7	776002
995	990025	985074875	31.5436	9.9833	1.00503	3125.9	777564
996	992016	988047936	31.5595	9.9866	1.00402	3129.0	779128
997	994009	991026973	31.5753	9.9900	1.00301	3132.2	780693
998	996004	994011992	31.5911	9.9933	1.00200	3135.3	782260
999	998001	997002999	31.6070	9.9967	1.00100	3138.5	783828
1000	1000000	1000000000	31.6228	10.0000	1.00000	3141.6	785398

Divide the smaller number b by a, and find c. Then $\sqrt{a^2 + b^2} = a + bc$

Values of c

b/a	0.000	0.001	0.002	0.003	0.004	0.005	0.006	0.007	0.008	0.009
0.00	0.0000	0.0005	0.0010	0.0015	0.0020	0.0025	0.0030	0.0035	0.0040	0.0045
0.01	0.0050	0.0055	0.0060	0.0065	0.0070	0.0075	0.0080	0.0085	0.0090	0.0095
0.02	0.0100	0.0105	0.0110	0.0115	0.0120	0.0125	0.0130	0.0135	0.0140	0.0145
0.03	0.0150	0.0155	0.0160	0.0165	0.0170	0.0175	0.0180	0.0185	0.0190	0.0195
0.04	0.0200	0.0205	0.0210	0.0215	0.0220	0.0225	0.0230	0.0235	0.0240	0.0245
0.05	0.0250	0.0255	0.0260	0.0265	0.0270	0.0275	0.0280	0.0285	0.0290	0.0295
0.06	0.0300	0.0305	0.0310	0.0315	0.0320	0.0325	0.0330	0.0335	0.0340	0.0345
0.07	0.0350	0.0355	0.0360	0.0365	0.0370	0.0375	0.0380	0.0384	0.0389	0.0394
0.08	0.0399	0.0404	0.0409	0.0414	0.0419	0.0424	0.0429	0.0434	0.0439	0.0444
0.09	0.0449	0.0454	0.0459	0.0464	0.0469	0.0474	0.0479	0.0484	0.0489	0.0494
0.10	0.0499	0.0504	0.0509	0.0514	0.0519	0.0524	0.0529	0.0534	0.0538	0.0543
0.11	0.0548	0.0553	0.0558	0.0563	0.0568	0.0573	0.0578	0.0583	0.0588	0.0593
0.12	0.0598	0.0603	0.0608	0.0613	0.0618	0.0623	0.0628	0.0633	0.0637	0.0642
0.13	0.0647	0.0652	0.0657	0.0662	0.0667	0.0672	0.0677	0.0682	0.0687	0.0692
0.14	0.0697	0.0702	0.0707	0.0711	0.0716	0.0721	0.0726	0.0731	0.0736	0.0741
0.15	0.0746	0.0751	0.0756	0.0761	0.0766	0.0770	0.0775	0.0780	0.0785	0.0790
0.16	0.0795	0.0800	0.0805	0.0810	0.0815	0.0820	0.0824	0.0829	0.0834	0.0839
0.17	0.0844	0.0849	0.0854	0.0859	0.0864	0.0868	0.0873	0.0878	0.0883	0.0888
0.18	0.0893	0.0898	0.0903	0.0908	0.0912	0.0917	0.0922	0.0927	0.0932	0.0937
0.19	0.0942	0.0946	0.0951	0.0956	0.0961	0.0966	0.0971	0.0976	0.0981	0.0985
0.20	0.0990	0.0995	0.1000	0.1005	0.1010	0.1015	0.1020	0.1024	0.1029	0.1034
0.21	0.1039	0.1044	0.1048	0.1053	0.1058	0.1063	0.1068	0.1073	0.1077	0.1082
0.22	0.1087	0.1092	0.1097	0.1102	0.1106	0.1111	0.1116	0.1121	0.1126	0.1130
0.23	0.1135	0.1140	0.1145	0.1150	0.1154	0.1159	0.1164	0.1169	0.1174	0.1178
0.24	0.1183	0.1188	0.1193	0.1198	0.1202	0.1207	0.1212	0.1217	0.1222	0.1226
0.25	0.1231	0.1236	0.1241	0.1245	0.1250	0.1255	0.1260	0.1265	0.1269	0.1274
0.26	0.1279	0.1284	0.1288	0.1293	0.1298	0.1303	0.1307	0.1312	0.1317	0.1322
0.27	0.1326	0.1331	0.1336	0.1341	0.1345	0.1350	0.1355	0.1359	0.1364	0.1369
0.28	0.1374	0.1378	0.1383	0.1388	0.1393	0.1397	0.1402	0.1407	0.1411	0.1416
0.29	0.1421	0.1425	0.1430	0.1435	0.1440	0.1444	0.1449	0.1454	0.1458	0.1463
0.30	0.1468	0.1472	0.1477	0.1482	0.1486	0.1491	0.1496	0.1500	0.1505	0.1510
0.31	0.1515	0.1519	0.1524	0.1528	0.1533	0.1538	0.1542	0.1547	0.1552	0.1556
0.32	0.1561	0.1566	0.1570	0.1575	0.1580	0.1584	0.1589	0.1594	0.1598	0.1603
0.33	0.1607	0.1612	0.1617	0.1621	0.1626	0.1631	0.1635	0.1640	0.1644	0.1649
0.34	0.1654	0.1658	0.1663	0.1667	0.1672	0.1677	0.1681	0.1686	0.1690	0.1695
0.35	0.1700	0.1704	0.1709	0.1713	0.1718	0.1722	0.1727	0.1732	0.1736	0.1741
0.36	0.1745	0.1750	0.1754	0.1759	0.1763	0.1768	0.1773	0.1777	0.1782	0.1786
0.37	0.1791	0.1795	0.1800	0.1804	0.1809	0.1813	0.1818	0.1822	0.1827	0.1832
0.38	0.1836	0.1841	0.1845	0.1850	0.1854	0.1859	0.1863	0.1868	0.1872	0.1877
0.39	0.1881	0.1886	0.1890	0.1895	0.1899	0.1903	0.1908	0.1912	0.1917	0.1921
0.40	0.1926	0.1930	0.1935	0.1939	0.1944	0.1948	0.1953	0.1957	0.1962	0.1966
0.41	0.1970	0.1975	0.1979	0.1984	0.1988	0.1993	0.1997	0.2002	0.2006	0.2010
0.42	0.2015	0.2019	0.2024	0.2028	0.2032	0.2037	0.2041	0.2046	0.2050	0.2055
0.43	0.2059	0.2063	0.2068	0.2072	0.2076	0.2081	0.2085	0.2090	0.2094	0.2098
0.44	0.2103	0.2107	0.2112	0.2116	0.2120	0.2125	0.2129	0.2133	0.2138	0.2142
0.45	0.2146	0.2151	0.2155	0.2159	0.2164	0.2168	0.2173	0.2177	0.2181	0.2185
0.46	0.2190	0.2194	0.2198	0.2203	0.2207	0.2211	0.2216	0.2220	0.2224	0.2229
0.47	0.2233	0.2237	0.2241	0.2246	0.2250	0.2254	0.2259	0.2263	0.2267	0.2271
0.48	0.2276	0.2280	0.2284	0.2289	0.2293	0.2297	0.2301	0.2306	0.2310	0.2314
0.49	0.2318	0.2323	0.2327	0.2331	0.2335	0.2340	0.2344	0.2348	0.2352	0.2357
0.50	0.2361	0.2365	0.2369	0.2373	0.2378	0.2382	0.2386	0.2390	0.2394	0.2399

Another method. $\sqrt{a^2 + b^2} = a\,[1 + (b/a)c]$

Divide the smaller number b by a, and find c. Then $\sqrt{a^2 + b^2} = a + bc$

Values of c

b/a	0.000	0.001	0.002	0.003	0.004	0.005	0.006	0.007	0.008	0.009
0.50	0.2361	0.2365	0.2369	0.2373	0.2378	0.2382	0.2386	0.2390	0.2394	0.2399
0.51	0.2403	0.2407	0.2411	0.2415	0.2420	0.2424	0.2428	0.2432	0.2436	0.2440
0.52	0.2445	0.2449	0.2453	0.2457	0.2461	0.2465	0.2470	0.2474	0.2478	0.2482
0.53	0.2486	0.2490	0.2495	0.2499	0.2503	0.2507	0.2511	0.2515	0.2519	0.2523
0.54	0.2528	0.2532	0.2536	0.2540	0.2544	0.2548	0.2552	0.2556	0.2560	0.2565
0.55	0.2569	0.2573	0.2577	0.2581	0.2585	0.2589	0.2593	0.2597	0.2601	0.2605
0.56	0.2609	0.2613	0.2618	0.2622	0.2626	0.2630	0.2634	0.2638	0.2642	0.2646
0.57	0.2650	0.2654	0.2658	0.2662	0.2666	0.2670	0.2674	0.2678	0.2682	0.2686
0.58	0.2690	0.2694	0.2698	0.2702	0.2706	0.2710	0.2714	0.2718	0.2722	0.2726
0.59	0.2730	0.2734	0.2738	0.2742	0.2746	0.2750	0.2754	0.2758	0.2762	0.2766
0.60	0.2770	0.2774	0.2778	0.2782	0.2786	0.2790	0.2794	0.2798	0.2801	0.2805
0.61	0.2809	0.2813	0.2817	0.2821	0.2825	0.2829	0.2833	0.2837	0.2841	0.2845
0.62	0.2849	0.2852	0.2856	0.2860	0.2864	0.2868	0.2872	0.2876	0.2880	0.2884
0.63	0.2888	0.2891	0.2895	0.2899	0.2903	0.2907	0.2911	0.2915	0.2918	0.2922
0.64	0.2926	0.2930	0.2934	0.2938	0.2941	0.2945	0.2949	0.2953	0.2957	0.2961
0.65	0.2964	0.2968	0.2972	0.2976	0.2980	0.2984	0.2987	0.2991	0.2995	0.2999
0.66	0.3003	0.3006	0.3010	0.3014	0.3018	0.3022	0.3025	0.3029	0.3033	0.3037
0.67	0.3040	0.3044	0.3048	0.3052	0.3055	0.3059	0.3063	0.3067	0.3070	0.3074
0.68	0.3078	0.3082	0.3085	0.3089	0.3093	0.3097	0.3100	0.3104	0.3108	0.3112
0.69	0.3115	0.3119	0.3122	0.3126	0.3130	0.3134	0.3137	0.3141	0.3145	0.3149
0.70	0.3152	0.3156	0.3160	0.3163	0.3167	0.3171	0.3174	0.3178	0.3182	0.3185
0.71	0.3189	0.3193	0.3196	0.3200	0.3204	0.3207	0.3211	0.3215	0.3218	0.3222
0.72	0.3226	0.3229	0.3233	0.3236	0.3240	0.3244	0.3247	0.3251	0.3255	0.3258
0.73	0.3262	0.3265	0.3269	0.3273	0.3276	0.3280	0.3283	0.3287	0.3291	0.3294
0.74	0.3298	0.3301	0.3305	0.3308	0.3312	0.3316	0.3319	0.3323	0.3326	0.3330
0.75	0.3333	0.3337	0.3340	0.3344	0.3348	0.3351	0.3355	0.3358	0.3362	0.3365
0.76	0.3369	0.3372	0.3376	0.3379	0.3383	0.3386	0.3390	0.3393	0.3397	0.3400
0.77	0.3404	0.3407	0.3411	0.3414	0.3418	0.3421	0.3425	0.3428	0.3432	0.3435
0.78	0.3439	0.3442	0.3446	0.3449	0.3453	0.3456	0.3460	0.3463	0.3467	0.3470
0.79	0.3473	0.3477	0.3480	0.3484	0.3487	0.3491	0.3494	0.3498	0.3501	0.3504
0.80	0.3508	0.3511	0.3515	0.3518	0.3522	0.3525	0.3528	0.3532	0.3535	0.3539
0.81	0.3542	0.3545	0.3549	0.3552	0.3556	0.3559	0.3562	0.3566	0.3569	0.3572
0.82	0.3576	0.3579	0.3583	0.3586	0.3589	0.3593	0.3596	0.3599	0.3603	0.3606
0.83	0.3609	0.3613	0.3616	0.3619	0.3623	0.3626	0.3629	0.3633	0.3636	0.3639
0.84	0.3643	0.3646	0.3649	0.3653	0.3656	0.3659	0.3663	0.3666	0.3669	0.3673
0.85	0.3676	0.3679	0.3682	0.3686	0.3689	0.3692	0.3696	0.3699	0.3702	0.3705
0.86	0.3709	0.3712	0.3715	0.3718	0.3722	0.3725	0.3728	0.3731	0.3735	0.3738
0.87	0.3741	0.3744	0.3748	0.3751	0.3754	0.3757	0.3761	0.3764	0.3767	0.3770
0.88	0.3774	0.3777	0.3780	0.3783	0.3786	0.3790	0.3793	0.3796	0.3799	0.3802
0.89	0.3806	0.3809	0.3812	0.3815	0.3818	0.3822	0.3825	0.3828	0.3831	0.3834
0.90	0.3837	0.3841	0.3844	0.3847	0.3850	0.3853	0.3856	0.3860	0.3863	0.3866
0.91	0.3869	0.3872	0.3875	0.3878	0.3882	0.3885	0.3888	0.3891	0.3894	0.3897
0.92	0.3900	0.3903	0.3907	0.3910	0.3913	0.3916	0.3919	0.3922	0.3925	0.3928
0.93	0.3931	0.3934	0.3938	0.3941	0.3944	0.3947	0.3950	0.3953	0.3956	0.3959
0.94	0.3962	0.3965	0.3968	0.3971	0.3974	0.3978	0.3981	0.3984	0.3987	0.3990
0.95	0.3993	0.3996	0.3999	0.4002	0.4005	0.4008	0.4011	0.4014	0.4017	0.4020
0.96	0.4023	0.4026	0.4029	0.4032	0.4035	0.4038	0.4041	0.4044	0.4047	0.4050
0.97	0.4053	0.4056	0.4059	0.4062	0.4065	0.4068	0.4071	0.4074	0.4077	0.4080
0.98	0.4083	0.4086	0.4089	0.4092	0.4095	0.4098	0.4101	0.4104	0.4107	0.4110
0.99	0.4113	0.4116	0.4119	0.4122	0.4125	0.4128	0.4130	0.4133	0.4136	0.4139
1.00	0.4142	0.4145	0.4148	0.4151	0.4154	0.4157	0.4160	0.4163	0.4166	0.4168

Another method. $\sqrt{a^2 + b^2} = a\,[1 + (b/a)c]$

Degs.	0.0	0.1	0.2	0.3	0.4	0.5	0.6	0.7	0.8	0.9
0	0.0000	0.0017	0.0035	0.0052	0.0070	0.0087	0.0105	0.0122	0.0140	0.0157
1	0.0175	0.0192	0.0209	0.0227	0.0244	0.0262	0.0279	0.0297	0.0314	0.0332
2	0.0349	0.0367	0.0384	0.0401	0.0419	0.0436	0.0454	0.0471	0.0489	0.0506
3	0.0524	0.0541	0.0559	0.0576	0.0593	0.0611	0.0628	0.0646	0.0663	0.0681
4	0.0698	0.0716	0.0733	0.0750	0.0768	0.0785	0.0803	0.0820	0.0838	0.0855
5	0.0873	0.0890	0.0908	0.0925	0.0942	0.0960	0.0977	0.0995	0.1012	0.1030
6	0.1047	0.1065	0.1082	0.1100	0.1117	0.1134	0.1152	0.1169	0.1187	0.1204
7	0.1222	0.1239	0.1257	0.1274	0.1292	0.1309	0.1326	0.1344	0.1361	0.1379
8	0.1396	0.1414	0.1431	0.1449	0.1466	0.1484	0.1501	0.1518	0.1536	0.1553
9	0.1571	0.1588	0.1606	0.1623	0.1641	0.1658	0.1676	0.1693	0.1710	0.1728
10	0.1745	0.1763	0.1780	0.1798	0.1815	0.1833	0.1850	0.1868	0.1885	0.1902
11	0.1920	0.1937	0.1955	0.1972	0.1990	0.2007	0.2025	0.2042	0.2059	0.2077
12	0.2094	0.2112	0.2129	0.2147	0.2164	0.2182	0.2199	0.2217	0.2234	0.2251
13	0.2269	0.2286	0.2304	0.2321	0.2339	0.2356	0.2374	0.2391	0.2409	0.2426
14	0.2443	0.2461	0.2478	0.2496	0.2513	0.2531	0.2548	0.2566	0.2583	0.2601
15	0.2618	0.2635	0.2653	0.2670	0.2688	0.2705	0.2723	0.2740	0.2758	0.2775
16	0.2793	0.2810	0.2827	0.2845	0.2862	0.2880	0.2897	0.2915	0.2932	0.2950
17	0.2967	0.2985	0.3002	0.3019	0.3037	0.3054	0.3072	0.3089	0.3107	0.3124
18	0.3142	0.3159	0.3176	0.3194	0.3211	0.3229	0.3246	0.3264	0.3281	0.3299
19	0.3316	0.3334	0.3351	0.3368	0.3386	0.3403	0.3421	0.3438	0.3456	0.3473
20	0.3491	0.3508	0.3526	0.3543	0.3560	0.3578	0.3595	0.3613	0.3630	0.3648
21	0.3665	0.3683	0.3700	0.3718	0.3735	0.3752	0.3770	0.3787	0.3805	0.3822
22	0.3840	0.3857	0.3875	0.3892	0.3910	0.3927	0.3944	0.3962	0.3979	0.3997
23	0.4014	0.4032	0.4049	0.4067	0.4084	0.4102	0.4119	0.4136	0.4154	0.4171
24	0.4189	0.4206	0.4224	0.4241	0.4259	0.4276	0.4294	0.4311	0.4328	0.4346
25	0.4363	0.4381	0.4398	0.4416	0.4433	0.4451	0.4468	0.4485	0.4503	0.4520
26	0.4538	0.4555	0.4573	0.4590	0.4608	0.4625	0.4643	0.4660	0.4677	0.4695
27	0.4712	0.4730	0.4747	0.4765	0.4782	0.4800	0.4817	0.4835	0.4852	0.4869
28	0.4887	0.4904	0.4922	0.4939	0.4957	0.4974	0.4992	0.5009	0.5027	0.5044
29	0.5061	0.5079	0.5096	0.5114	0.5131	0.5149	0.5166	0.5184	0.5201	0.5219
30	0.5236	0.5253	0.5271	0.5288	0.5306	0.5323	0.5341	0.5358	0.5376	0.5393
31	0.5411	0.5428	0.5445	0.5463	0.5480	0.5498	0.5515	0.5533	0.5550	0.5568
32	0.5585	0.5603	0.5620	0.5637	0.5655	0.5672	0.5690	0.5707	0.5725	0.5742
33	0.5760	0.5777	0.5794	0.5812	0.5829	0.5847	0.5864	0.5882	0.5899	0.5917
34	0.5934	0.5952	0.5969	0.5986	0.6004	0.6021	0.6039	0.6056	0.6074	0.6091
35	0.6109	0.6126	0.6144	0.6161	0.6178	0.6196	0.6213	0.6231	0.6248	0.6266
36	0.6283	0.6301	0.6318	0.6336	0.6353	0.6370	0.6388	0.6405	0.6423	0.6440
37	0.6458	0.6475	0.6493	0.6510	0.6528	0.6545	0.6562	0.6580	0.6597	0.6615
38	0.6632	0.6650	0.6667	0.6685	0.6702	0.6720	0.6737	0.6754	0.6772	0.6789
39	0.6807	0.6824	0.6842	0.6859	0.6877	0.6894	0.6912	0.6929	0.6946	0.6964
40	0.6981	0.6999	0.7016	0.7034	0.7051	0.7069	0.7086	0.7103	0.7121	0.7138
41	0.7156	0.7173	0.7191	0.7208	0.7226	0.7243	0.7261	0.7278	0.7295	0.7313
42	0.7330	0.7348	0.7365	0.7383	0.7400	0.7418	0.7435	0.7453	0.7470	0.7487
43	0.7505	0.7522	0.7540	0.7557	0.7575	0.7592	0.7610	0.7627	0.7645	0.7662
44	0.7679	0.7697	0.7714	0.7732	0.7749	0.7767	0.7784	0.7802	0.7819	0.7837
45	0.7854	0.7871	0.7889	0.7906	0.7924	0.7941	0.7959	0.7976	0.7994	0.8011
	0′	6′	12′	18′	24′	30′	36′	42′	48′	54′

90° = 1.5708 radians	$30° = \frac{\pi}{6}$, $45° = \frac{\pi}{4}$, $60° = \frac{\pi}{3}$, $90° = \frac{\pi}{2}$ radians
180° = 3.1416 radians	$120° = \frac{2\pi}{3}$, $135° = \frac{3\pi}{4}$, $150° = \frac{5\pi}{6}$, $180° = \pi$ radians
270° = 4.7124 radians	$210° = \frac{7\pi}{6}$, $225° = \frac{5\pi}{4}$, $240° = \frac{4\pi}{3}$, $270° = \frac{3\pi}{2}$ radians
360° = 6.2832 radians	$300° = \frac{5\pi}{3}$, $315° = \frac{7\pi}{4}$, $330° = \frac{11\pi}{6}$, $360° = 2\pi$ radians

Degs.	0.0	0.1	0.2	0.3	0.4	0.5	0.6	0.7	0.8	0.9
45	0.7854	0.7871	0.7889	0.7906	0.7924	0.7941	0.7959	0.7976	0.7994	0.8011
46	0.8029	0.8046	0.8063	0.8081	0.8098	0.8116	0.8133	0.8151	0.8168	0.8186
47	0.8203	0.8221	0.8238	0.8255	0.8273	0.8290	0.8308	0.8325	0.8343	0.8360
48	0.8378	0.8395	0.8412	0.8430	0.8447	0.8465	0.8482	0.8500	0.8517	0.8535
49	0.8552	0.8570	0.8587	0.8604	0.8622	0.8639	0.8657	0.8674	0.8692	0.8709
50	0.8727	0.8744	0.8762	0.8779	0.8796	0.8814	0.8831	0.8849	0.8866	0.8884
51	0.8901	0.8919	0.8936	0.8954	0.8971	0.8988	0.9006	0.9023	0.9041	0.9058
52	0.9076	0.9093	0.9111	0.9128	0.9146	0.9163	0.9180	0.9198	0.9215	0.9233
53	0.9250	0.9268	0.9285	0.9303	0.9320	0.9338	0.9355	0.9372	0.9390	0.9407
54	0.9425	0.9442	0.9460	0.9477	0.9495	0.9512	0.9529	0.9547	0.9564	0.9582
55	0.9599	0.9617	0.9634	0.9652	0.9669	0.9687	0.9704	0.9721	0.9739	0.9756
56	0.9774	0.9791	0.9809	0.9826	0.9844	0.9861	0.9879	0.9896	0.9913	0.9931
57	0.9948	0.9966	0.9983	1.0001	1.0018	1.0036	1.0053	1.0071	1.0088	1.0105
58	1.0123	1.0140	1.0158	1.0175	1.0193	1.0210	1.0228	1.0245	1.0263	1.0280
59	1.0297	1.0315	1.0332	1.0350	1.0367	1.0385	1.0402	1.0420	1.0437	1.0455
60	1.0472	1.0489	1.0507	1.0524	1.0542	1.0559	1.0577	1.0594	1.0612	1.0629
61	1.0647	1.0664	1.0681	1.0699	1.0716	1.0734	1.0751	1.0769	1.0786	1.0804
62	1.0821	1.0838	1.0856	1.0873	1.0891	1.0908	1.0926	1.0943	1.0961	1.0978
63	1.0996	1.1013	1.1030	1.1048	1.1065	1.1083	1.1100	1.1118	1.1135	1.1153
64	1.1170	1.1188	1.1205	1.1222	1.1240	1.1257	1.1275	1.1292	1.1310	1.1327
65	1.1345	1.1362	1.1380	1.1397	1.1414	1.1432	1.1449	1.1467	1.1484	1.1502
66	1.1519	1.1537	1.1554	1.1572	1.1589	1.1606	1.1624	1.1641	1.1659	1.1676
67	1.1694	1.1711	1.1729	1.1746	1.1764	1.1781	1.1798	1.1816	1.1833	1.1851
68	1.1868	1.1886	1.1903	1.1921	1.1938	1.1956	1.1973	1.1990	1.2008	1.2025
69	1.2043	1.2060	1.2078	1.2095	1.2113	1.2130	1.2147	1.2165	1.2182	1.2200
70	1.2217	1.2235	1.2252	1.2270	1.2287	1.2305	1.2322	1.2339	1.2357	1.2374
71	1.2392	1.2409	1.2427	1.2444	1.2462	1.2479	1.2497	1.2514	1.2531	1.2549
72	1.2566	1.2584	1.2601	1.2619	1.2636	1.2654	1.2671	1.2689	1.2706	1.2723
73	1.2741	1.2758	1.2776	1.2793	1.2811	1.2828	1.2846	1.2863	1.2881	1.2898
74	1.2915	1.2933	1.2950	1.2968	1.2985	1.3003	1.3020	1.3038	1.3055	1.3073
75	1.3090	1.3107	1.3125	1.3142	1.3160	1.3177	1.3195	1.3212	1.3230	1.3247
76	1.3265	1.3282	1.3299	1.3317	1.3334	1.3352	1.3369	1.3387	1.3404	1.3422
77	1.3439	1.3456	1.3474	1.3491	1.3509	1.3526	1.3544	1.3561	1.3579	1.3596
78	1.3614	1.3631	1.3648	1.3666	1.3683	1.3701	1.3718	1.3736	1.3753	1.3771
79	1.3788	1.3806	1.3823	1.3840	1.3858	1.3875	1.3893	1.3910	1.3928	1.3945
80	1.3963	1.3980	1.3998	1.4015	1.4032	1.4050	1.4067	1.4085	1.4102	1.4120
81	1.4137	1.4155	1.4172	1.4190	1.4207	1.4224	1.4242	1.4259	1.4277	1.4294
82	1.4312	1.4329	1.4347	1.4364	1.4382	1.4399	1.4416	1.4434	1.4451	1.4469
83	1.4486	1.4504	1.4521	1.4539	1.4556	1.4573	1.4591	1.4608	1.4626	1.4643
84	1.4661	1.4678	1.4696	1.4713	1.4731	1.4748	1.4765	1.4783	1.4800	1.4818
85	1.4835	1.4853	1.4870	1.4888	1.4905	1.4923	1.4940	1.4957	1.4975	1.4992
86	1.5010	1.5027	1.5045	1.5062	1.5080	1.5097	1.5115	1.5132	1.5149	1.5167
87	1.5184	1.5202	1.5219	1.5237	1.5254	1.5272	1.5289	1.5307	1.5324	1.5341
88	1.5359	1.5376	1.5394	1.5411	1.5429	1.5446	1.5464	1.5481	1.5499	1.5516
89	1.5533	1.5551	1.5568	1.5586	1.5603	1.5621	1.5638	1.5656	1.5673	1.5691
90	1.5708	1.5725	1.5743	1.5760	1.5778	1.5795	1.5813	1.5830	1.5848	1.5865
	0′	6′	12′	18′	24′	30′	36′	42′	48′	54′

$90° = 1.5708$ radians	$30° = \frac{\pi}{6}$, $45° = \frac{\pi}{4}$, $60° = \frac{\pi}{3}$, $90° = \frac{\pi}{2}$ radians
$180° = 3.1416$ radians	$120° = \frac{2\pi}{3}$, $135° = \frac{3\pi}{4}$, $150° = \frac{5\pi}{6}$, $180° = \pi$ radians
$270° = 4.7124$ radians	$210° = \frac{7\pi}{6}$, $225° = \frac{5\pi}{4}$, $240° = \frac{4\pi}{3}$, $270° = \frac{3\pi}{2}$ radians
$360° = 6.2832$ radians	$300° = \frac{5\pi}{3}$, $315° = \frac{7\pi}{4}$, $330° = \frac{11\pi}{6}$, $360° = 2\pi$ radians

Degs.	Function	0.0°	0.1°	0.2°	0.3°	0.4°	0.5°	0.6°	0.7°	0.8°	0.9°
	sin	0.0000	0.0017	0.0035	0.0052	0.0070	0.0087	0.0105	0.0122	0.0140	0.0157
0	cos	1.0000	1.0000	1.0000	1.0000	1.0000	1.0000	0.9999	0.9999	0.9999	0.9999
	tan	0.0000	0.0017	0.0035	0.0052	0.0070	0.0087	0.0105	0.0122	0.0140	0.0157
	sin	0.0175	0.0192	0.0209	0.0227	0.0244	0.0262	0.0279	0.0297	0.0314	0.0332
1	cos	0.9998	0.9998	0.9998	0.9997	0.9997	0.9997	0.9996	0.9996	0.9995	0.9995
	tan	0.0175	0.0192	0.0209	0.0227	0.0244	0.0262	0.0279	0.0297	0.0314	0.0332
	sin	0.0349	0.0366	0.0384	0.0401	0.0419	0.0436	0.0454	0.0471	0.0488	0.0506
2	cos	0.9994	0.9993	0.9993	0.9992	0.9991	0.9990	0.9990	0.9989	0.9988	0.9987
	tan	0.0349	0.0367	0.0384	0.0402	0.0419	0.0437	0.0454	0.0472	0.0489	0.0507
	sin	0.0523	0.0541	0.0558	0.0576	0.0593	0.0610	0.0628	0.0645	0.0663	0.0680
3	cos	0.9986	0.9985	0.9984	0.9983	0.9982	0.9981	0.9980	0.9979	0.9978	0.9977
	tan	0.0524	0.0542	0.0559	0.0577	0.0594	0.0612	0.0629	0.0647	0.0664	0.0682
	sin	0.0698	0.0715	0.0732	0.0750	0.0767	0.0785	0.0802	0.0819	0.0837	0.0854
4	cos	0.9976	0.9974	0.9973	0.9972	0.9971	0.9969	0.9968	0.9966	0.9965	0.9963
	tan	0.0699	0.0717	0.0734	0.0752	0.0769	0.0787	0.0805	0.0822	0.0840	0.0857
	sin	0.0872	0.0889	0.0906	0.0924	0.0941	0.0958	0.0976	0.0993	0.1011	0.1028
5	cos	0.9962	0.9960	0.9959	0.9957	0.9956	0.9954	0.9952	0.9951	0.9949	0.9947
	tan	0.0875	0.0892	0.0910	0.0928	0.0945	0.0963	0.0981	0.0998	0.1016	0.1033
	sin	0.1045	0.1063	0.1080	0.1097	0.1115	0.1132	0.1149	0.1167	0.1184	0.1201
6	cos	0.9945	0.9943	0.9942	0.9940	0.9938	0.9936	0.9934	0.9932	0.9930	0.9928
	tan	0.1051	0.1069	0.1086	0.1104	0.1122	0.1139	0.1157	0.1175	0.1192	0.1210
	sin	0.1219	0.1236	0.1253	0.1271	0.1288	0.1305	0.1323	0.1340	0.1357	0.1374
7	cos	0.9925	0.9923	0.9921	0.9919	0.9917	0.9914	0.9912	0.9910	0.9907	0.9905
	tan	0.1228	0.1246	0.1263	0.1281	0.1299	0.1317	0.1334	0.1352	0.1370	0.1388
	sin	0.1392	0.1409	0.1426	0.1444	0.1461	0.1478	0.1495	0.1513	0.1530	0.1547
8	cos	0.9903	0.9900	0.9898	0.9895	0.9893	0.9890	0.9888	0.9885	0.9882	0.9880
	tan	0.1405	0.1423	0.1441	0.1459	0.1477	0.1495	0.1512	0.1530	0.1548	0.1566
	sin	0.1564	0.1582	0.1599	0.1616	0.1633	0.1650	0.1668	0.1685	0.1702	0.1719
9	cos	0.9877	0.9874	0.9871	0.9869	0.9866	0.9863	0.9860	0.9857	0.9854	0.9851
	tan	0.1584	0.1602	0.1620	0.1638	0.1655	0.1673	0.1691	0.1709	0.1727	0.1745
	sin	0.1736	0.1754	0.1771	0.1788	0.1805	0.1822	0.1840	0.1857	0.1874	0.1891
10	cos	0.9848	0.9845	0.9842	0.9839	0.9836	0.9833	0.9829	0.9826	0.9823	0.9820
	tan	0.1763	0.1781	0.1799	0.1817	0.1835	0.1853	0.1871	0.1890	0.1908	0.1926
	sin	0.1908	0.1925	0.1942	0.1959	0.1977	0.1994	0.2011	0.2028	0.2045	0.2062
11	cos	0.9816	0.9813	0.9810	0.9806	0.9803	0.9799	0.9796	0.9792	0.9789	0.9785
	tan	0.1944	0.1962	0.1980	0.1998	0.2016	0.2035	0.2053	0.2071	0.2089	0.2107
	sin	0.2079	0.2096	0.2113	0.2130	0.2147	0.2164	0.2181	0.2198	0.2215	0.2232
12	cos	0.9781	0.9778	0.9774	0.9770	0.9767	0.9763	0.9759	0.9755	0.9751	0.9748
	tan	0.2126	0.2144	0.2162	0.2180	0.2199	0.2217	0.2235	0.2254	0.2272	0.2290
	sin	0.2250	0.2267	0.2284	0.2300	0.2318	0.2334	0.2351	0.2368	0.2385	0.2402
13	cos	0.9744	0.9740	0.9736	0.9732	0.9728	0.9724	0.9720	0.9715	0.9711	0.9707
	tan	0.2309	0.2327	0.2345	0.2364	0.2382	0.2401	0.2419	0.2438	0.2456	0.2475
	sin	0.2419	0.2436	0.2453	0.2470	0.2487	0.2504	0.2521	0.2538	0.2554	0.2571
14	cos	0.9703	0.9699	0.9694	0.9690	0.9686	0.9681	0.9677	0.9673	0.9668	0.9664
	tan	0.2493	0.2512	0.2530	0.2549	0.2568	0.2586	0.2605	0.2623	0.2642	0.2661
Degs.	Function	0′	6′	12′	18′	24′	30′	36′	42′	48′	54′

Degs.	Function	0.0°	0.1°	0.2°	0.3°	0.4°	0.5°	0.6°	0.7°	0.8°	0.9°
15	sin	0.2588	0.2605	0.2622	0.2639	0.2656	0.2672	0.2689	0.2706	0.2723	0.2740
	cos	0.9659	0.9655	0.9650	0.9646	0.9641	0.9636	0.9632	0.9627	0.9622	0.9617
	tan	0.2679	0.2698	0.2717	0.2736	0.2754	0.2773	0.2792	0.2811	0.2830	0.2849
16	sin	0.2756	0.2773	0.2790	0.2807	0.2823	0.2840	0.2857	0.2874	0.2890	0.2907
	cos	0.9613	0.9608	0.9603	0.9598	0.9593	0.9588	0.9583	0.9578	0.9573	0.9568
	tan	0.2867	0.2886	0.2905	0.2924	0.2943	0.2962	0.2981	0.3000	0.3019	0.3038
17	sin	0.2924	0.2940	0.2957	0.2974	0.2990	0.3007	0.3024	0.3040	0.3057	0.3074
	cos	0.9563	0.9558	0.9553	0.9548	0.9542	0.9537	0.9532	0.9527	0.9521	0.9516
	tan	0.3057	0.3076	0.3096	0.3115	0.3134	0.3153	0.3172	0.3191	0.3211	0.3230
18	sin	0.3090	0.3107	0.3123	0.3140	0.3156	0.3173	0.3190	0.3206	0.3223	0.3239
	cos	0.9511	0.9505	0.9500	0.9494	0.9489	0.9483	0.9478	0.9472	0.9466	0.9461
	tan	0.3249	0.3269	0.3288	0.3307	0.3327	0.3346	0.3365	0.3385	0.3404	0.3424
19	sin	0.3256	0.3272	0.3289	0.3305	0.3322	0.3338	0.3355	0.3371	0.3387	0.3404
	cos	0.9455	0.9449	0.9444	0.9438	0.9432	0.9426	0.9421	0.9415	0.9409	0.9403
	tan	0.3443	0.3463	0.3482	0.3502	0.3522	0.3541	0.3561	0.3581	0.3600	0.3620
20	sin	0.3420	0.3437	0.3453	0.3469	0.3486	0.3502	0.3518	0.3535	0.3551	0.3567
	cos	0.9397	0.9391	0.9385	0.9379	0.9373	0.9367	0.9361	0.9354	0.9348	0.9342
	tan	0.3640	0.3659	0.3679	0.3699	0.3719	0.3739	0.3759	0.3779	0.3799	0.3819
21	sin	0.3584	0.3600	0.3616	0.3633	0.3649	0.3665	0.3681	0.3697	0.3714	0.3730
	cos	0.9336	0.9330	0.9323	0.9317	0.9311	0.9304	0.9298	0.9291	0.9285	0.9278
	tan	0.3839	0.3859	0.3879	0.3899	0.3919	0.3939	0.3959	0.3979	0.4000	0.4020
22	sin	0.3746	0.3762	0.3778	0.3795	0.3811	0.3827	0.3843	0.3859	0.3875	0.3891
	cos	0.9272	0.9265	0.9259	0.9252	0.9245	0.9239	0.9232	0.9225	0.9219	0.9212
	tan	0.4040	0.4061	0.4081	0.4101	0.4122	0.4142	0.4163	0.4183	0.4204	0.4224
23	sin	0.3907	0.3923	0.3939	0.3955	0.3971	0.3987	0.4003	0.4019	0.4035	0.4051
	cos	0.9205	0.9198	0.9191	0.9184	0.9178	0.9171	0.9164	0.9157	0.9150	0.9143
	tan	0.4245	0.4265	0.4286	0.4307	0.4327	0.4348	0.4369	0.4390	0.4411	0.4431
24	sin	0.4067	0.4083	0.4099	0.4115	0.4131	0.4147	0.4163	0.4179	0.4195	0.4210
	cos	0.9135	0.9128	0.9121	0.9114	0.9107	0.9100	0.9092	0.9085	0.9078	0.9070
	tan	0.4452	0.4473	0.4494	0.4515	0.4536	0.4557	0.4578	0.4599	0.4621	0.4642
25	sin	0.4226	0.4242	0.4258	0.4274	0.4289	0.4305	0.4321	0.4337	0.4352	0.4368
	cos	0.9063	0.9056	0.9048	0.9041	0.9033	0.9026	0.9018	0.9011	0.9003	0.8996
	tan	0.4663	0.4684	0.4706	0.4727	0.4748	0.4770	0.4791	0.4813	0.4834	0.4856
26	sin	0.4384	0.4399	0.4415	0.4431	0.4446	0.4462	0.4478	0.4493	0.4509	0.4524
	cos	0.8988	0.8980	0.8973	0.8965	0.8957	0.8949	0.8942	0.8934	0.8926	0.8918
	tan	0.4877	0.4899	0.4921	0.4942	0.4964	0.4986	0.5008	0.5029	0.5051	0.5073
27	sin	0.4540	0.4555	0.4571	0.4586	0.4602	0.4617	0.4633	0.4648	0.4664	0.4679
	cos	0.8910	0.8902	0.8894	0.8886	0.8878	0.8870	0.8862	0.8854	0.8846	0.8838
	tan	0.5095	0.5117	0.5139	0.5161	0.5184	0.5206	0.5228	0.5250	0.5272	0.5295
28	sin	0.4695	0.4710	0.4726	0.4741	0.4756	0.4772	0.4787	0.4802	0.4818	0.4833
	cos	0.8829	0.8821	0.8813	0.8805	0.8796	0.8788	0.8780	0.8771	0.8763	0.8755
	tan	0.5317	0.5340	0.5362	0.5384	0.5407	0.5430	0.5452	0.5475	0.5498	0.5520
29	sin	0.4848	0.4863	0.4879	0.4894	0.4909	0.4924	0.4939	0.4955	0.4970	0.4985
	cos	0.8746	0.8738	0.8729	0.8721	0.8712	0.8704	0.8695	0.8686	0.8678	0.8669
	tan	0.5543	0.5566	0.5589	0.5612	0.5635	0.5658	0.5681	0.5704	0.5727	0.5750
Degs.	Function	0′	6′	12′	18′	24′	30′	36′	42′	48′	54′

Degs.	Function	0.0°	0.1°	0.2°	0.3°	0.4°	0.5°	0.6°	0.7°	0.8°	0.9°
30	sin	0.5000	0.5015	0.5030	0.5045	0.5060	0.5075	0.5090	0.5105	0.5120	0.5135
	cos	0.8660	0.8652	0.8643	0.8634	0.8625	0.8616	0.8607	0.8599	0.8590	0.8581
	tan	0.5774	0.5797	0.5820	0.5844	0.5867	0.5890	0.5914	0.5938	0.5961	0.5985
31	sin	0.5150	0.5165	0.5180	0.5195	0.5210	0.5225	0.5240	0.5255	0.5270	0.5284
	cos	0.8572	0.8563	0.8554	0.8545	0.8536	0.8526	0.8517	0.8508	0.8499	0.8490
	tan	0.6009	0.6032	0.6056	0.6080	0.6104	0.6128	0.6152	0.6176	0.6200	0.6224
32	sin	0.5299	0.5314	0.5329	0.5344	0.5358	0.5373	0.5388	0.5402	0.5417	0.5432
	cos	0.8480	0.8471	0.8462	0.8453	0.8443	0.8434	0.8425	0.8415	0.8406	0.8396
	tan	0.6249	0.6273	0.6297	0.6322	0.6346	0.6371	0.6395	0.6420	0.6445	0.6469
33	sin	0.5446	0.5461	0.5476	0.5490	0.5505	0.5519	0.5534	0.5548	0.5563	0.5577
	cos	0.8387	0.8377	0.8368	0.8358	0.8348	0.8339	0.8329	0.8320	0.8310	0.8300
	tan	0.6494	0.6519	0.6544	0.6569	0.6594	0.6619	0.6644	0.6669	0.6694	0.6720
34	sin	0.5592	0.5606	0.5621	0.5635	0.5650	0.5664	0.5678	0.5693	0.5707	0.5721
	cos	0.8290	0.8281	0.8271	0.8261	0.8251	0.8241	0.8231	0.8221	0.8211	0.8202
	tan	0.6745	0.6771	0.6796	0.6822	0.6847	0.6873	0.6899	0.6924	0.6950	0.6976
35	sin	0.5736	0.5750	0.5764	0.5779	0.5793	0.5807	0.5821	0.5835	0.5850	0.5864
	cos	0.8192	0.8181	0.8171	0.8161	0.8151	0.8141	0.8131	0.8121	0.8111	0.8100
	tan	0.7002	0.7028	0.7054	0.7080	0.7107	0.7133	0.7159	0.7186	0.7212	0.7239
36	sin	0.5878	0.5892	0.5906	0.5920	0.5934	0.5948	0.5962	0.5976	0.5990	0.6004
	cos	0.8090	0.8080	0.8070	0.8059	0.8049	0.8039	0.8028	0.8018	0.8007	0.7997
	tan	0.7265	0.7292	0.7319	0.7346	0.7373	0.7400	0.7427	0.7454	0.7481	0.7508
37	sin	0.6018	0.6032	0.6046	0.6060	0.6074	0.6088	0.6101	0.6115	0.6129	0.6143
	cos	0.7986	0.7976	0.7965	0.7955	0.7944	0.7934	0.7923	0.7912	0.7902	0.7891
	tan	0.7536	0.7563	0.7590	0.7618	0.7646	0.7673	0.7701	0.7729	0.7757	0.7785
38	sin	0.6157	0.6170	0.6184	0.6198	0.6211	0.6225	0.6239	0.6252	0.6266	0.6280
	cos	0.7880	0.7869	0.7859	0.7848	0.7837	0.7826	0.7815	0.7804	0.7793	0.7782
	tan	0.7813	0.7841	0.7869	0.7898	0.7926	0.7954	0.7983	0.8012	0.8040	0.8069
39	sin	0.6293	0.6307	0.6320	0.6334	0.6347	0.6361	0.6374	0.6388	0.6401	0.6414
	cos	0.7771	0.7760	0.7749	0.7738	0.7727	0.7716	0.7705	0.7694	0.7683	0.7672
	tan	0.8098	0.8127	0.8156	0.8185	0.8214	0.8243	0.8273	0.8302	0.8332	0.8361
40	sin	0.6428	0.6441	0.6455	0.6468	0.6481	0.6494	0.6508	0.6521	0.6534	0.6547
	cos	0.7660	0.7649	0.7638	0.7627	0.7615	0.7604	0.7593	0.7581	0.7570	0.7559
	tan	0.8391	0.8421	0.8451	0.8481	0.8511	0.8541	0.8571	0.8601	0.8632	0.8662
41	sin	0.6561	0.6574	0.6587	0.6600	0.6613	0.6626	0.6639	0.6652	0.6665	0.6678
	cos	0.7547	0.7536	0.7524	0.7513	0.7501	0.7490	0.7478	0.7466	0.7455	0.7443
	tan	0.8693	0.8724	0.8754	0.8785	0.8816	0.8847	0.8878	0.8910	0.8941	0.8972
42	sin	0.6691	0.6704	0.6717	0.6730	0.6743	0.6756	0.6769	0.6782	0.6794	0.6807
	cos	0.7431	0.7420	0.7408	0.7396	0.7385	0.7373	0.7361	0.7349	0.7337	0.7325
	tan	0.9004	0.9036	0.9067	0.9099	0.9131	0.9163	0.9195	0.9228	0.9260	0.9293
43	sin	0.6820	0.6833	0.6845	0.6858	0.6871	0.6884	0.6896	0.6909	0.6921	0.6934
	cos	0.7314	0.7302	0.7290	0.7278	0.7266	0.7254	0.7242	0.7230	0.7218	0.7206
	tan	0.9325	0.9358	0.9391	0.9424	0.9457	0.9490	0.9523	0.9556	0.9590	0.9623
44	sin	0.6947	0.6959	0.6972	0.6984	0.6997	0.7009	0.7022	0.7034	0.7046	0.7059
	cos	0.7193	0.7181	0.7169	0.7157	0.7145	0.7133	0.7120	0.7108	0.7096	0.7083
	tan	0.9657	0.9691	0.9725	0.9759	0.9793	0.9827	0.9861	0.9896	0.9930	0.9965
Degs.	Function	0′	6′	12′	18′	24′	30′	36′	42′	48′	54′

Degs.	Function	0.0°	0.1°	0.2°	0.3°	0.4°	0.5°	0.6°	0.7°	0.8°	0.9°
45	sin	0.7071	0.7083	0.7096	0.7108	0.7120	0.7133	0.7145	0.7157	0.7169	0.7181
	cos	0.7071	0.7059	0.7046	0.7034	0.7022	0.7009	0.6997	0.6984	0.6972	0.6959
	tan	1.0000	1.0035	1.0070	1.0105	1.0141	1.0176	1.0212	1.0247	1.0283	1.0319
46	sin	0.7193	0.7206	0.7218	0.7230	0.7242	0.7254	0.7266	0.7278	0.7290	0.7302
	cos	0.6947	0.6934	0.6921	0.6909	0.6896	0.6884	0.6871	0.6858	0.6845	0.6833
	tan	1.0355	1.0392	1.0428	1.0464	1.0501	1.0538	1.0575	1.0612	1.0649	1.0686
47	sin	0.7314	0.7325	0.7337	0.7349	0.7361	0.7373	0.7385	0.7396	0.7408	0.7420
	cos	0.6820	0.6807	0.6794	0.6782	0.6769	0.6756	0.6743	0.6730	0.6717	0.6704
	tan	1.0724	1.0761	1.0799	1.0837	1.0875	1.0913	1.0951	1.0990	1.1028	1.1067
48	sin	0.7431	0.7443	0.7455	0.7466	0.7478	0.7490	0.7501	0.7513	0.7524	0.7536
	cos	0.6691	0.6678	0.6665	0.6652	0.6639	0.6626	0.6613	0.6600	0.6587	0.6574
	tan	1.1106	1.1145	1.1184	1.1224	1.1263	1.1303	1.1343	1.1383	1.1423	1.1463
49	sin	0.7547	0.7559	0.7570	0.7581	0.7593	0.7604	0.7615	0.7627	0.7638	0.7649
	cos	0.6561	0.6547	0.6534	0.6521	0.6508	0.6494	0.6481	0.6468	0.6455	0.6441
	tan	1.1504	1.1544	1.1585	1.1626	1.1667	1.1708	1.1750	1.1792	1.1833	1.1875
50	sin	0.7660	0.7672	0.7683	0.7694	0.7705	0.7716	0.7727	0.7738	0.7749	0.7760
	cos	0.6428	0.6414	0.6401	0.6388	0.6374	0.6361	0.6347	0.6334	0.6320	0.6307
	tan	1.1918	1.1960	1.2002	1.2045	1.2088	1.2131	1.2174	1.2218	1.2261	1.2305
51	sin	0.7771	0.7782	0.7793	0.7804	0.7815	0.7826	0.7837	0.7848	0.7859	0.7869
	cos	0.6293	0.6280	0.6266	0.6252	0.6239	0.6225	0.6211	0.6198	0.6184	0.6170
	tan	1.2349	1.2393	1.2437	1.2482	1.2527	1.2572	1.2617	1.2662	1.2708	1.2753
52	sin	0.7880	0.7891	0.7902	0.7912	0.7923	0.7934	0.7944	0.7955	0.7965	0.7976
	cos	0.6157	0.6143	0.6129	0.6115	0.6101	0.6088	0.6074	0.6060	0.6046	0.6032
	tan	1.2799	1.2846	1.2892	1.2938	1.2985	1.3032	1.3079	1.3127	1.3175	1.3222
53	sin	0.7986	0.7997	0.8007	0.8018	0.8028	0.8039	0.8049	0.8059	0.8070	0.8080
	cos	0.6018	0.6004	0.5990	0.5976	0.5962	0.5948	0.5934	0.5920	0.5906	0.5892
	tan	1.3270	1.3319	1.3367	1.3416	1.3465	1.3514	1.3564	1.3613	1.3663	1.3713
54	sin	0.8090	0.8100	0.8111	0.8121	0.8131	0.8141	0.8151	0.8161	0.8171	0.8181
	cos	0.5878	0.5864	0.5850	0.5835	0.5821	0.5807	0.5793	0.5779	0.5764	0.5750
	tan	1.3764	1.3814	1.3865	1.3916	1.3968	1.4019	1.4071	1.4124	1.4176	1.4229
55	sin	0.8192	0.8202	0.8211	0.8221	0.8231	0.8241	0.8251	0.8261	0.8271	0.8281
	cos	0.5736	0.5721	0.5707	0.5693	0.5678	0.5664	0.5650	0.5635	0.5621	0.5606
	tan	1.4281	1.4335	1.4388	1.4442	1.4496	1.4550	1.4605	1.4659	1.4715	1.4770
56	sin	0.8290	0.8300	0.8310	0.8320	0.8329	0.8339	0.8348	0.8358	0.8368	0.8377
	cos	0.5592	0.5577	0.5563	0.5548	0.5534	0.5519	0.5505	0.5490	0.5476	0.5461
	tan	1.4826	1.4882	1.4938	1.4994	1.5051	1.5108	1.5166	1.5224	1.5282	1.5340
57	sin	0.8387	0.8396	0.8406	0.8415	0.8425	0.8434	0.8443	0.8453	0.8462	0.8471
	cos	0.5446	0.5432	0.5417	0.5402	0.5388	0.5373	0.5358	0.5344	0.5329	0.5314
	tan	1.5399	1.5458	1.5517	1.5577	1.5637	1.5697	1.5757	1.5818	1.5880	1.5941
58	sin	0.8480	0.8490	0.8499	0.8508	0.8517	0.8526	0.8536	0.8545	0.8554	0.8563
	cos	0.5299	0.5284	0.5270	0.5255	0.5240	0.5225	0.5210	0.5195	0.5180	0.5165
	tan	1.6003	1.6066	1.6128	1.6191	1.6255	1.6319	1.6383	1.6447	1.6512	1.6577
59	sin	0.8572	0.8581	0.8590	0.8599	0.8607	0.8616	0.8625	0.8634	0.8643	0.8652
	cos	0.5150	0.5135	0.5120	0.5105	0.5090	0.5075	0.5060	0.5045	0.5030	0.5015
	tan	1.6643	1.6709	1.6775	1.6842	1.6909	1.6977	1.7045	1.7113	1.7182	1.7251
Degs.	**Function**	0′	6′	12′	18′	24′	30′	36′	42′	48′	54′

60°–74.9°

Degs.	Function	0.0°	0.1°	0.2°	0.3°	0.4°	0.5°	0.6°	0.7°	0.8°	0.9°
60	sin	0.8660	0.8669	0.8678	0.8686	0.8695	0.8704	0.8712	0.8721	0.8729	0.8738
	cos	0.5000	0.4985	0.4970	0.4955	0.4939	0.4924	0.4909	0.4894	0.4879	0.4863
	tan	1.7321	1.7391	1.7461	1.7532	1.7603	1.7675	1.7747	1.7820	1.7893	1.7966
61	sin	0.8746	0.8755	0.8763	0.8771	0.8780	0.8788	0.8796	0.8805	0.8813	0.8821
	cos	0.4848	0.4833	0.4818	0.4802	0.4787	0.4772	0.4756	0.4741	0.4726	0.4710
	tan	1.8040	1.8115	1.8190	1.8265	1.8341	1.8418	1.8495	1.8572	1.8650	1.8728
62	sin	0.8829	0.8838	0.8846	0.8854	0.8862	0.8870	0.8878	0.8886	0.8894	0.8902
	cos	0.4695	0.4679	0.4664	0.4648	0.4633	0.4617	0.4602	0.4586	0.4571	0.4555
	tan	1.8807	1.8887	1.8967	1.9047	1.9128	1.9210	1.9292	1.9375	1.9458	1.9542
63	sin	0.8910	0.8918	0.8926	0.8934	0.8942	0.8949	0.8957	0.8965	0.8973	0.8980
	cos	0.4540	0.4524	0.4509	0.4493	0.4478	0.4462	0.4446	0.4431	0.4415	0.4399
	tan	1.9626	1.9711	1.9797	1.9883	1.9970	2.0057	2.0145	2.0233	2.0323	2.0413
64	sin	0.8988	0.8996	0.9003	0.9011	0.9018	0.9026	0.9033	0.9041	0.9048	0.9056
	cos	0.4384	0.4368	0.4352	0.4337	0.4321	0.4305	0.4289	0.4274	0.4258	0.4242
	tan	2.0503	2.0594	2.0686	2.0778	2.0872	2.0965	2.1060	2.1155	2.1251	2.1348
65	sin	0.9063	0.9070	0.9078	0.9085	0.9092	0.9100	0.9107	0.9114	0.9121	0.9128
	cos	0.4226	0.4210	0.4195	0.4179	0.4163	0.4147	0.4131	0.4115	0.4099	0.4083
	tan	2.1445	2.1543	2.1642	2.1742	2.1842	2.1943	2.2045	2.2148	2.2251	2.2355
66	sin	0.9135	0.9143	0.9150	0.9157	0.9164	0.9171	0.9178	0.9184	0.9191	0.9198
	cos	0.4067	0.4051	0.4035	0.4019	0.4003	0.3987	0.3971	0.3955	0.3939	0.3923
	tan	2.2460	2.2566	2.2673	2.2781	2.2889	2.2998	2.3109	2.3220	2.3332	2.3445
67	sin	0.9205	0.9212	0.9219	0.9225	0.9232	0.9239	0.9245	0.9252	0.9259	0.9265
	cos	0.3907	0.3891	0.3875	0.3859	0.3843	0.3827	0.3811	0.3795	0.3778	0.3762
	tan	2.3559	2.3673	2.3789	2.3906	2.4023	2.4142	2.4262	2.4383	2.4504	2.4627
68	sin	0.9272	0.9278	0.9285	0.9291	0.9298	0.9304	0.9311	0.9317	0.9323	0.9330
	cos	0.3746	0.3730	0.3714	0.3697	0.3681	0.3665	0.3649	0.3633	0.3616	0.3600
	tan	2.4751	2.4876	2.5002	2.5129	2.5257	2.5386	2.5517	2.5649	2.5782	2.5916
69	sin	0.9336	0.9342	0.9348	0.9354	0.9361	0.9367	0.9373	0.9379	0.9385	0.9391
	cos	0.3584	0.3567	0.3551	0.3535	0.3518	0.3502	0.3486	0.3469	0.3453	0.3437
	tan	2.6051	2.6187	2.6325	2.6464	2.6605	2.6746	2.6889	2.7034	2.7179	2.7326
70	sin	0.9397	0.9403	0.9409	0.9415	0.9421	0.9426	0.9432	0.9438	0.9444	0.9449
	cos	0.3420	0.3404	0.3387	0.3371	0.3355	0.3338	0.3322	0.3305	0.3289	0.3272
	tan	2.7475	2.7625	2.7776	2.7929	2.8083	2.8239	2.8397	2.8556	2.8716	2.8878
71	sin	0.9455	0.9461	0.9466	0.9472	0.9478	0.9483	0.9489	0.9494	0.9500	0.9505
	cos	0.3256	0.3239	0.3223	0.3206	0.3190	0.3173	0.3156	0.3140	0.3123	0.3107
	tan	2.9042	2.9208	2.9375	2.9544	2.9714	2.9887	3.0061	3.0237	3.0415	3.0595
72	sin	0.9511	0.9516	0.9521	0.9527	0.9532	0.9537	0.9542	0.9548	0.9553	0.9558
	cos	0.3090	0.3074	0.3057	0.3040	0.3024	0.3007	0.2990	0.2974	0.2957	0.2940
	tan	3.0777	3.0961	3.1146	3.1334	3.1524	3.1716	3.1910	3.2106	3.2305	3.2506
73	sin	0.9563	0.9568	0.9573	0.9578	0.9583	0.9588	0.9593	0.9598	0.9603	0.9608
	cos	0.2924	0.2907	0.2890	0.2874	0.2857	0.2840	0.2823	0.2807	0.2790	0.2773
	tan	3.2709	3.2914	3.3122	3.3332	3.3544	3.3759	3.3977	3.4197	3.4420	3.4646
74	sin	0.9613	0.9617	0.9622	0.9627	0.9632	0.9636	0.9641	0.9646	0.9650	0.9655
	cos	0.2756	0.2740	0.2723	0.2706	0.2689	0.2672	0.2656	0.2639	0.2622	0.2605
	tan	3.4874	3.5105	3.5339	3.5576	3.5816	3.6059	3.6305	3.6554	3.6806	3.7062
Degs.	Function	0′	6′	12′	18′	24′	30′	36′	42′	48′	54′

Degs.	Function	0.0°	0.1°	0.2°	0.3°	0.4°	0.5°	0.6°	0.7°	0.8°	0.9°
75	sin	0.9659	0.9664	0.9668	0.9673	0.9677	0.9681	0.9686	0.9690	0.9694	0.9699
	cos	0.2588	0.2571	0.2554	0.2538	0.2521	0.2504	0.2487	0.2470	0.2453	0.2436
	tan	3.7321	3.7583	3.7848	3.8118	3.8391	3.8667	3.8947	3.9232	3.9520	3.9812
76	sin	0.9703	0.9707	0.9711	0.9715	0.9720	0.9724	0.9728	0.9732	0.9736	0.9740
	cos	0.2419	0.2402	0.2385	0.2368	0.2351	0.2334	0.2317	0.2300	0.2284	0.2267
	tan	4.0108	4.0408	4.0713	4.1022	4.1335	4.1653	4.1976	4.2303	4.2635	4.2972
77	sin	0.9744	0.9748	0.9751	0.9755	0.9759	0.9763	0.9767	0.9770	0.9774	0.9778
	cos	0.2250	0.2232	0.2215	0.2198	0.2181	0.2164	0.2147	0.2130	0.2113	0.2096
	tan	4.3315	4.3662	4.4015	4.4374	4.4737	4.5107	4.5483	4.5864	4.6252	4.6646
78	sin	0.9781	0.9785	0.9789	0.9792	0.9796	0.9799	0.9803	0.9806	0.9810	0.9813
	cos	0.2079	0.2062	0.2045	0.2028	0.2011	0.1994	0.1977	0.1959	0.1942	0.1925
	tan	4.7046	4.7453	4.7867	4.8288	4.8716	4.9152	4.9594	5.0045	5.0504	5.0970
79	sin	0.9816	0.9820	0.9823	0.9826	0.9829	0.9833	0.9836	0.9839	0.9842	0.9845
	cos	0.1908	0.1891	0.1874	0.1857	0.1840	0.1822	0.1805	0.1788	0.1771	0.1754
	tan	5.1446	5.1929	5.2422	5.2924	5.3435	5.3955	5.4486	5.5026	5.5578	5.6140
80	sin	0.9848	0.9851	0.9854	0.9857	0.9860	0.9863	0.9866	0.9869	0.9871	0.9874
	cos	0.1736	0.1719	0.1702	0.1685	0.1668	0.1650	0.1633	0.1616	0.1599	0.1582
	tan	5.6713	5.7297	5.7894	5.8502	5.9124	5.9758	6.0405	6.1066	6.1742	6.2432
81	sin	0.9877	0.9880	0.9882	0.9885	0.9888	0.9890	0.9893	0.9895	0.9898	0.9900
	cos	0.1564	0.1547	0.1530	0.1513	0.1495	0.1478	0.1461	0.1444	0.1426	0.1409
	tan	6.3138	6.3859	6.4596	6.5350	6.6122	6.6912	6.7720	6.8548	6.9395	7.0264
82	sin	0.9903	0.9905	0.9907	0.9910	0.9912	0.9914	0.9917	0.9919	0.9921	0.9923
	cos	0.1392	0.1374	0.1357	0.1340	0.1323	0.1305	0.1288	0.1271	0.1253	0.1236
	tan	7.1154	7.2066	7.3002	7.3962	7.4947	7.5958	7.6996	7.8062	7.9158	8.0285
83	sin	0.9925	0.9928	0.9930	0.9932	0.9934	0.9936	0.9938	0.9940	0.9942	0.9943
	cos	0.1219	0.1201	0.1184	0.1167	0.1149	0.1132	0.1115	0.1097	0.1080	0.1063
	tan	8.1443	8.2636	8.3863	8.5126	8.6427	8.7769	8.9152	9.0579	9.2052	9.3572
84	sin	0.9945	0.9947	0.9949	0.9951	0.9952	0.9954	0.9956	0.9957	0.9959	0.9960
	cos	0.1045	0.1028	0.1011	0.0993	0.0976	0.0958	0.0941	0.0924	0.0906	0.0889
	tan	9.5144	9.6768	9.8448	10.02	10.20	10.39	10.58	10.78	10.99	11.20
85	sin	0.9962	0.9963	0.9965	0.9966	0.9968	0.9969	0.9971	0.9972	0.9973	0.9974
	cos	0.0872	0.0854	0.0837	0.0819	0.0802	0.0785	0.0767	0.0750	0.0732	0.0715
	tan	11.43	11.66	11.91	12.16	12.43	12.71	13.00	13.30	13.62	13.95
86	sin	0.9976	0.9977	0.9978	0.9979	0.9980	0.9981	0.9982	0.9983	0.9984	0.9985
	cos	0.0698	0.0680	0.0663	0.0645	0.0628	0.0610	0.0593	0.0576	0.0558	0.0541
	tan	14.30	14.67	15.06	15.46	15.89	16.35	16.83	17.34	17.89	18.46
87	sin	0.9986	0.9987	0.9988	0.9989	0.9990	0.9990	0.9991	0.9992	0.9993	0.9993
	cos	0.0523	0.0506	0.0488	0.0471	0.0454	0.0436	0.0419	0.0401	0.0384	0.0366
	tan	19.08	19.74	20.45	21.20	22.02	22.90	23.86	24.90	26.03	27.27
88	sin	0.9994	0.9995	0.9995	0.9996	0.9996	0.9997	0.9997	0.9997	0.9998	0.9998
	cos	0.0349	0.0332	0.0314	0.0297	0.0279	0.0262	0.0244	0.0227	0.0209	0.0192
	tan	28.64	30.14	31.82	33.69	35.80	38.19	40.92	44.07	47.74	52.08
89	sin	0.9998	0.9999	0.9999	0.9999	0.9999	1.000	1.000	1.000	1.000	1.000
	cos	0.0175	0.0157	0.0140	0.0122	0.0105	0.0087	0.0070	0.0052	0.0035	0.0017
	tan	57.29	63.66	71.62	81.85	95.49	114.6	143.2	191.0	286.5	573.0
Degs.	Function	0′	6′	12′	18′	24′	30′	36′	42′	48′	54′

Degs.	Function	0.0°	0.1°	0.2°	0.3°	0.4°	0.5°	0.6°	0.7°	0.8°	0.9°
	log sin	−∞	7.2419	7.5429	7.7190	7.8439	7.9408	8.0200	8.0870	8.1450	8.1961
0	log cos	0	0	0	0	0	0	0	0	0	9.9999
	log tan	−∞	7.2419	7.5429	7.7190	7.8439	7.9409	8.0200	8.0870	8.1450	8.1962
	log sin	8.2419	8.2832	8.3210	8.3558	8.3880	8.4179	8.4459	8.4723	8.4971	8.5206
1	log cos	9.9999	9.9999	9.9999	9.9999	9.9999	9.9999	9.9998	9.9998	9.9998	9.9998
	log tan	8.2419	8.2833	8.3211	8.3559	8.3881	8.4181	8.4461	8.4725	8.4973	8.5208
	log sin	8.5428	8.5640	8.5842	8.6035	8.6220	8.6397	8.6567	8.6731	8.6889	8.7041
2	log cos	9.9997	9.9997	9.9997	9.9996	9.9996	9.9996	9.9996	9.9995	9.9995	9.9994
	log tan	8.5431	8.5643	8.5845	8.6038	8.6223	8.6401	8.6571	8.6736	8.6894	8.7046
	log sin	8.7188	8.7330	8.7468	8.7602	8.7731	8.7857	8.7979	8.8098	8.8213	8.8326
3	log cos	9.9994	9.9994	9.9993	9.9993	9.9992	9.9992	9.9991	9.9991	9.9990	9.9990
	log tan	8.7194	8.7337	8.7475	8.7609	8.7739	8.7865	8.7988	8.8107	8.8223	8.8336
	log sin	8.8436	8.8543	8.8647	8.8749	8.8849	8.8946	8.9042	8.9135	8.9226	8.9315
4	log cos	9.9989	9.9989	9.9988	9.9988	9.9987	9.9987	9.9986	9.9985	9.9985	9.9984
	log tan	8.8446	8.8554	8.8659	8.8762	8.8862	8.8960	8.9056	8.9150	8.9241	8.9331
	log sin	8.9403	8.9489	8.9573	8.9655	8.9736	8.9816	8.9894	8.9970	9.0046	9.0120
5	log cos	9.9983	9.9983	9.9982	9.9981	9.9981	9.9980	9.9979	9.9978	9.9978	9.9977
	log tan	8.9420	8.9506	8.9591	8.9674	8.9756	8.9836	8.9915	8.9992	9.0068	9.0143
	log sin	9.0192	9.0264	9.0334	9.0403	9.0472	9.0539	9.0605	9.0670	9.0734	9.0797
6	log cos	9.9976	9.9975	9.9975	9.9974	9.9973	9.9972	9.9971	9.9970	9.9969	9.9968
	log tan	9.0216	9.0289	9.0360	9.0430	9.0499	9.0567	9.0633	9.0699	9.0764	9.0828
	log sin	9.0859	9.0920	9.0981	9.1040	9.1099	9.1157	9.1214	9.1271	9.1326	9.1381
7	log cos	9.9968	9.9967	9.9966	9.9965	9.9964	9.9963	9.9962	9.9961	9.9960	9.9959
	log tan	9.0891	9.0954	9.1015	9.1076	9.1135	9.1194	9.1252	9.1310	9.1367	9.1423
	log sin	9.1436	9.1489	9.1542	9.1594	9.1646	9.1697	9.1747	9.1797	9.1847	9.1895
8	log cos	9.9958	9.9956	9.9955	9.9954	9.9953	9.9952	9.9951	9.9950	9.9949	9.9947
	log tan	9.1478	9.1533	9.1587	9.1640	9.1693	9.1745	9.1797	9.1848	9.1898	9.1948
	log sin	9.1943	9.1991	9.2038	9.2085	9.2131	9.2176	9.2221	9.2266	9.2310	9.2353
9	log cos	9.9946	9.9945	9.9944	9.9943	9.9941	9.9940	9.9939	9.9937	9.9936	9.9935
	log tan	9.1997	9.2046	9.2094	9.2142	9.2189	9.2236	9.2282	9.2328	9.2374	9.2419
	log sin	9.2397	9.2439	9.2482	9.2524	9.2565	9.2606	9.2647	9.2687	9.2727	9.2767
10	log cos	9.9934	9.9932	9.9931	9.9929	9.9928	9.9927	9.9925	9.9924	9.9922	9.9921
	log tan	9.2463	9.2507	9.2551	9.2594	9.2637	9.2680	9.2722	9.2764	9.2805	9.2846
	log sin	9.2806	9.2845	9.2883	9.2921	9.2959	9.2997	9.3034	9.3070	9.3107	9.3143
11	log cos	9.9919	9.9918	9.9916	9.9915	9.9913	9.9912	9.9910	9.9909	9.9907	9.9906
	log tan	9.2887	9.2927	9.2967	9.3006	9.3046	9.3085	9.3123	9.3162	9.3200	9.3237
	log sin	9.3179	9.3214	9.3250	9.3284	9.3319	9.3353	9.3387	9.3421	9.3455	9.3488
12	log cos	9.9904	9.9902	9.9901	9.9899	9.9897	9.9896	9.9894	9.9892	9.9891	9.9889
	log tan	9.3275	9.3312	9.3349	9.3385	9.3422	9.3458	9.3493	9.3529	9.3564	9.3599
	log sin	9.3521	9.3554	9.3586	9.3618	9.3650	9.3682	9.3713	9.3745	9.3775	9.3806
13	log cos	9.9887	9.9885	9.9884	9.9882	9.9880	9.9878	9.9876	9.9875	9.9873	9.9871
	log tan	9.3634	9.3668	9.3702	9.3736	9.3770	9.3804	9.3837	9.3870	9.3903	9.3935
	log sin	9.3837	9.3867	9.3897	9.3927	9.3957	9.3986	9.4015	9.4044	9.4073	9.4102
14	log cos	9.9869	9.9867	9.9865	9.9863	9.9861	9.9859	9.9857	9.9855	9.9853	9.9851
	log tan	9.3968	9.4000	9.4032	9.4064	9.4095	9.4127	9.4158	9.4189	9.4220	9.4250
Degs.	Function	0′	6′	12′	18′	24′	30′	36′	42′	48′	54′

Degs.	Function	0.0°	0.1°	0.2°	0.3°	0.4°	0.5°	0.6°	0.7°	0.8°	0.9°
15	log sin	9.4130	9.4158	9.4186	9.4214	9.4242	9.4269	9.4296	9.4323	9.4350	9.4377
	log cos	9.9849	9.9847	9.9845	9.9843	9.9841	9.9839	9.9837	9.9835	9.9833	9.9831
	log tan	9.4281	9.4311	9.4341	9.4371	9.4400	9.4430	9.4459	9.4488	9.4517	9.4546
16	log sin	9.4403	9.4430	9.4456	9.4482	9.4508	9.4533	9.4559	9.4584	9.4609	9.4634
	log cos	9.9828	9.9826	9.9824	9.9822	9.9820	9.9817	9.9815	9.9813	9.9811	9.9808
	log tan	9.4575	9.4603	9.4632	9.4660	9.4688	9.4716	9.4744	9.4771	9.4799	9.4826
17	log sin	9.4659	9.4684	9.4709	9.4733	9.4757	9.4781	8.4805	9.4829	9.4853	9.4876
	log cos	9.9806	9.9804	9.9801	9.9799	9.9797	9.9794	9.9792	9.9789	9.9787	9.9785
	log tan	9.4853	9.4880	9.4907	9.4934	9.4961	9.4987	9.5014	9.5040	9.5066	9.5092
18	log sin	9.4900	9.4923	9.4946	9.4969	9.4992	9.5015	9.5037	9.5060	9.5082	9.5104
	log cos	9.9782	9.9780	9.9777	9.9775	9.9772	9.9770	9.9767	9.9764	9.9762	9.9759
	log tan	9.5118	9.5143	9.5169	9.5195	9.5220	9.5245	9.5270	9.5295	9.5320	9.5345
19	log sin	9.5126	9.5148	9.5170	9.5192	9.5213	9.5235	9.5256	9.5278	9.5299	9.5320
	log cos	9.9757	9.9754	9.9751	9.9749	9.9746	9.9743	9.9741	9.9738	9.9735	9.9733
	log tan	9.5370	9.5394	9.5419	9.5443	9.5467	9.5491	9.5516	9.5539	9.5563	9.5587
20	log sin	9.5341	9.5361	9.5382	9.5402	9.5423	9.5443	9.5463	9.5484	9.5504	9.5523
	log cos	9.9730	9.9727	9.9724	9.9722	9.9719	9.9716	9.9713	9.9710	9.9707	9.9704
	log tan	9.5611	9.5634	9.5658	9.5681	9.5704	9.5727	9.5750	9.5773	9.5796	9.5819
21	log sin	9.5543	9.5563	9.5583	9.5602	9.5621	9.5641	9.5660	9.5679	9.5698	9.5717
	log cos	9.9702	9.9699	9.9696	9.9693	9.9690	9.9687	9.9684	9.9681	9.9678	9.9675
	log tan	9.5842	9.5864	9.5887	9.5909	9.5932	9.5954	9.5976	9.5998	9.6020	9.6042
22	log sin	9.5736	9.5754	9.5773	9.5792	9.5810	9.5828	9.5847	9.5865	9.5883	9.5901
	log cos	9.9672	9.9669	9.9666	9.9662	9.9659	9.9656	9.9653	9.9650	9.9647	9.9643
	log tan	9.6064	9.6086	9.6108	9.6129	9.6151	9.6172	9.6194	9.6215	9.6236	9.6257
23	log sin	9.5919	9.5937	9.5954	9.5972	9.5990	9.6007	9.6024	9.6042	9.6059	9.6076
	log cos	9.9640	9.9637	9.9634	9.9631	9.9627	9.9624	9.9621	9.9617	9.9614	9.9611
	log tan	9.6279	9.6300	9.6321	9.6341	9.6362	9.6383	9.6404	9.6424	9.6445	9.6465
24	log sin	9.6093	9.6110	9.6127	9.6144	9.6161	9.6177	9.6194	9.6210	9.6227	9.6243
	log cos	9.9607	9.9604	9.9601	9.9597	9.9594	9.9590	9.9587	9.9583	9.9580	9.9576
	log tan	9.6486	9.6506	9.6527	9.6547	9.6567	9.6587	9.6607	9.6627	9.6647	9.6667
25	log sin	9.6259	9.6276	9.6292	9.6308	9.6324	9.6340	9.6356	9.6371	9.6387	9.6403
	log cos	9.9573	9.9569	9.9566	9.9562	9.9558	9.9555	9.9551	9.9548	9.9544	9.9540
	log tan	9.6687	9.6706	9.6726	9.6746	9.6765	9.6785	9.6804	9.6824	9.6843	9.6863
26	log sin	9.6418	9.6434	9.6449	9.6465	9.6480	9.6495	9.6510	9.6526	9.6541	9.6556
	log cos	9.9537	9.9533	9.9529	9.9525	9.9522	9.9518	9.9514	9.9510	9.9506	9.9503
	log tan	9.6882	9.6901	9.6920	9.6939	9.6958	9.6977	9.6996	9.7015	9.7034	9.7053
27	log sin	9.6570	9.6585	9.6600	9.6615	9.6629	9.6644	9.6659	9.6673	9.6687	9.6702
	log cos	9.9499	9.9495	9.9491	9.9487	9.9483	9.9479	9.9475	9.9471	9.9467	9.9463
	log tan	9.7072	9.7090	9.7109	9.7128	9.7146	9.7165	9.7183	9.7202	9.7220	9.7238
28	log sin	9.6716	9.6730	9.6744	9.6759	9.6773	9.6787	9.6801	9.6814	9.6828	9.6842
	log cos	9.9459	9.9455	9.9451	9.9447	9.9443	9.9439	9.9435	9.9431	9.9427	9.9422
	log tan	9.7257	9.7275	9.7293	9.7311	9.7330	9.7348	9.7366	9.7384	9.7402	9.7420
29	log sin	9.6856	9.6869	9.6883	9.6896	9.6910	9.6923	9.6937	9.6950	9.6963	9.6977
	log cos	9.9418	9.9414	9.9410	9.9406	9.9401	9.9397	9.9393	9.9388	9.9384	9.9380
	log tan	9.7438	9.7455	9.7473	9.7491	9.7509	9.7526	9.7544	9.7562	9.7579	9.7597
Degs.	Function	0′	6′	12′	18′	24′	30′	36′	42′	48′	54′

Degs.	Function	0.0°	0.1°	0.2°	0.3°	0.4°	0.5°	0.6°	0.7°	0.8°	0.9°
30	log sin	9.6990	9.7003	9.7016	9.7029	9.7042	9.7055	9.7068	9.7080	9.7093	9.7106
	log cos	9.9375	9.9371	9.9367	9.9362	9.9358	9.9353	9.9349	9.9344	9.9340	9.9335
	log tan	9.7614	9.7632	9.7649	9.7667	9.7684	9.7701	9.7719	9.7736	9.7753	9.7771
31	log sin	9.7118	9.7131	9.7144	9.7156	9.7168	9.7181	9.7193	9.7205	9.7218	9.7230
	log cos	9.9331	9.9326	9.9322	9.9317	9.9312	9.9308	9.9303	9.9298	9.9294	9.9289
	log tan	9.7788	9.7805	9.7822	9.7839	9.7856	9.7873	9.7890	9.7907	9.7924	9.7941
32	log sin	9.7242	9.7254	9.7266	9.7278	9.7290	9.7302	9.7314	9.7326	9.7338	9.7349
	log cos	9.9284	9.9279	9.9275	9.9270	9.9265	9.9260	9.9255	9.9251	9.9246	9.9241
	log tan	9.7958	9.7975	9.7992	9.8008	9.8025	9.8042	9.8059	9.8075	9.8092	9.8109
33	log sin	9.7361	9.7373	9.7384	9.7396	9.7407	9.7419	9.7430	9.7442	9.7453	9.7464
	log cos	9.9236	9.9231	9.9226	9.9221	9.9216	9.9211	9.9206	9.9201	9.9196	9.9191
	log tan	9.8125	9.8142	9.8158	9.8175	9.8191	9.8208	9.8224	9.8241	9.8257	9.8274
34	log sin	9.7476	9.7487	9.7498	9.7509	9.7520	9.7531	9.7542	9.7553	9.7564	9.7575
	log cos	9.9186	9.9181	9.9175	9.9170	9.9165	9.9160	9.9155	9.9149	9.9144	9.9139
	log tan	9.8290	9.8306	9.8323	9.8339	9.8355	9.8371	9.8388	9.8404	9.8420	9.8436
35	log sin	9.7586	9.7597	9.7607	9.7618	9.7629	9.7640	9.7650	9.7661	9.7671	9.7682
	log cos	9.9134	9.9128	9.9123	9.9118	9.9112	9.9107	9.9101	9.9096	9.9091	9.9085
	log tan	9.8452	9.8468	9.8484	9.8501	9.8517	9.8533	9.8549	9.8565	9.8581	9.8597
36	log sin	9.7692	9.7703	9.7713	9.7723	9.7734	9.7744	9.7754	9.7764	9.7774	9.7785
	log cos	9.9080	9.9074	9.9069	9.9063	9.9057	9.9052	9.9046	9.9041	9.9035	9.9029
	log tan	9.8613	9.8629	9.8644	9.8660	9.8676	9.8692	9.8708	9.8724	9.8740	9.8755
37	log sin	9.7795	9.7805	9.7815	9.7825	9.7835	9.7844	9.7854	9.7864	9.7874	9.7884
	log cos	9.9023	9.9018	9.9012	9.9006	9.9000	9.8995	9.8989	9.8983	9.8977	9.8971
	log tan	9.8771	9.8787	9.8803	9.8818	9.8834	9.8850	9.8865	9.8881	9.8897	9.8912
38	log sin	9.7893	9.7903	9.7913	9.7922	9.7932	9.7941	9.7951	9.7960	9.7970	9.7979
	log cos	9.8965	9.8959	9.8953	9.8947	9.8941	9.8935	9.8929	9.8923	9.8917	9.8911
	log tan	9.8928	9.8944	9.8959	9.8975	9.8990	9.9006	9.9022	9.9037	9.9053	9.9068
39	log sin	9.7989	9.7998	9.8007	9.8017	9.8026	9.8035	9.8044	9.8053	9.8063	9.8072
	log cos	9.8905	9.8899	9.8893	9.8887	9.8880	9.8874	9.8868	9.8862	9.8855	9.8849
	log tan	9.9084	9.9099	9.9115	9.9130	9.9146	9.9161	9.9176	9.9192	9.9207	9.9223
40	log sin	9.8081	9.8090	9.8099	9.8108	9.8117	9.8125	9.8134	9.8143	9.8152	9.8161
	log cos	9.8843	9.8836	9.8830	9.8823	9.8817	9.8810	9.8804	9.8797	9.8791	9.8784
	log tan	9.9238	9.9254	9.9269	9.9284	9.9300	9.9315	9.9330	9.9346	9.9361	9.9376
41	log sin	9.8169	9.8178	9.8187	9.8195	9.8204	9.8213	9.8221	9.8230	9.8238	9.8247
	log cos	9.8778	9.8771	9.8765	9.8758	9.8751	9.8745	9.8738	9.8731	9.8724	9.8718
	log tan	9.9392	9.9407	9.9422	9.9438	9.9453	9.9468	9.9483	9.9499	9.9514	9.9529
42	log sin	9.8255	9.8264	9.8272	9.8280	9.8289	9.8297	9.8305	9.8313	9.8322	9.8330
	log cos	9.8711	9.8704	9.8697	9.8690	9.8683	9.8676	9.8669	9.8662	9.8655	9.8648
	log tan	9.9544	9.9560	9.9575	9.9590	9.9605	9.9621	9.9636	9.9651	9.9666	9.9681
43	log sin	9.8338	9.8346	9.8354	9.8362	9.8370	9.8378	9.8386	9.8394	9.8402	9.8410
	log cos	9.8641	9.8634	9.8627	9.8620	9.8613	9.8606	9.8598	9.8591	9.8584	9.8577
	log tan	9.9697	9.9712	9.9727	9.9742	9.9757	9.9772	9.9788	9.9803	9.9818	9.9833
44	log sin	9.8418	9.8426	9.8433	9.8441	9.8449	9.8457	9.8464	9.8472	9.8480	9.8487
	log cos	9.8569	9.8562	9.8555	9.8547	9.8540	9.8532	9.8525	9.8517	9.8510	9.8502
	log tan	9.9848	9.9864	9.9879	9.9894	9.9909	9.9924	9.9939	9.9955	9.9970	9.9985
Degs.	Function	0′	6′	12′	18′	24′	30′	36′	42′	48′	54′

Degs.	Function	0.0°	0.1°	0.2°	0.3°	0.4°	0.5°	0.6°	0.7°	0.8°	0.9°
45	log sin	9.8495	9.8502	9.8510	9.8517	9.8525	9.8532	9.8540	9.8547	9.8555	9.8562
	log cos	9.8495	9.8487	9.8480	9.8472	9.8464	9.8457	9.8449	9.8441	9.8433	9.8426
	log tan	0.0000	0.0015	0.0030	0.0045	0.0061	0.0076	0.0091	0.0106	0.0121	0.0136
46	log sin	9.8569	9.8577	9.8584	9.8591	9.8598	9.8606	9.8613	9.8620	9.8627	9.8634
	log cos	9.8418	9.8410	9.8402	9.8394	9.8386	9.8378	9.8370	9.8362	9.8354	9.8346
	log tan	0.0152	0.0167	0.0182	0.0197	0.0212	0.0228	0.0243	0.0258	0.0273	0.0288
47	log sin	9.8641	9.8648	9.8655	9.8662	9.8669	9.8676	9.8683	9.8690	9.8697	9.8704
	log cos	9.8338	9.8330	9.8322	9.8313	9.8305	9.8297	9.8289	9.8280	9.8272	9.8264
	log tan	0.0303	0.0319	0.0334	0.0349	0.0364	0.0379	0.0395	0.0410	0.0425	0.0440
48	log sin	9.8711	9.8718	9.8724	9.8731	9.8738	9.8745	9.8751	9.8758	9.8765	9.8771
	log cos	9.8255	9.8247	9.8238	9.8230	9.8221	9.8213	9.8204	9.8195	9.8187	9.8178
	log tan	0.0456	0.0471	0.0486	0.0501	0.0517	0.0532	0.0547	0.0562	0.0578	0.0593
49	log sin	9.8778	9.8784	9.8791	9.8797	9.8804	9.8810	9.8817	9.8823	9.8830	9.8836
	log cos	9.8169	9.8161	9.8152	9.8143	9.8134	9.8125	9.8117	9.8108	9.8099	9.8090
	log tan	0.0608	0.0624	0.0639	0.0654	0.0670	0.0685	0.0700	0.0716	0.0731	0.0746
50	log sin	9.8843	9.8849	9.8855	9.8862	9.8868	9.8874	9.8880	9.8887	9.8893	9.8899
	log cos	9.8081	9.8072	9.8063	9.8053	9.8044	9.8035	9.8026	9.8017	9.8007	9.7998
	log tan	0.0762	0.0777	0.0793	0.0808	0.0824	0.0839	0.0854	0.0870	0.0885	0.0901
51	log sin	9.8905	9.8911	9.8917	9.8923	9.8929	9.8935	9.8941	9.8947	9.8953	9.8959
	log cos	9.7989	9.7979	9.7970	9.7960	9.7951	9.7941	9.7932	9.7922	9.7913	9.7903
	log tan	0.0916	0.0932	0.0947	0.0963	0.0978	0.0994	0.1010	0.1025	0.1041	0.1056
52	log sin	9.8965	9.8971	9.8977	9.8983	9.8989	9.8995	9.9000	9.9006	9.9012	9.9018
	log cos	9.7893	9.7884	9.7874	9.7864	9.7854	9.7844	9.7835	9.7825	9.7815	9.7805
	log tan	0.1072	0.1088	0.1103	0.1119	0.1135	0.1150	0.1166	0.1182	0.1197	0.1213
53	log sin	9.9023	9.9029	9.9035	9.9041	9.9046	9.9052	9.9057	9.9063	9.9069	9.9074
	log cos	9.7795	9.7785	9.7774	9.7764	9.7754	9.7744	9.7734	9.7723	9.7713	9.7703
	log tan	0.1229	0.1245	0.1260	0.1276	0.1292	0.1308	0.1324	0.1340	0.1356	0.1371
54	log sin	9.9080	9.9085	9.9091	9.9096	9.9101	9.9107	9.9112	9.9118	9.9123	9.9128
	log cos	9.7692	9.7682	9.7671	9.7661	9.7650	9.7640	9.7629	9.7618	9.7607	9.7597
	log tan	0.1387	0.1403	0.1419	0.1435	0.1451	0.1467	0.1483	0.1499	0.1516	0.1532
55	log sin	9.9134	9.9139	9.9144	9.9149	9.9155	9.9160	9.9165	9.9170	9.9175	9.9181
	log cos	9.7586	9.7575	9.7564	9.7553	9.7542	9.7531	9.7520	9.7509	9.7498	9.7487
	log tan	0.1548	0.1564	0.1580	0.1596	0.1612	0.1629	0.1645	0.1661	0.1677	0.1694
56	log sin	9.9186	9.9191	9.9196	9.9201	9.9206	9.9211	9.9216	9.9221	9.9226	9.9231
	log cos	9.7476	9.7464	9.7453	9.7442	9.7430	9.7419	9.7407	9.7396	9.7384	9.7373
	log tan	0.1710	0.1726	0.1743	0.1759	0.1776	0.1792	0.1809	0.1825	0.1842	0.1858
57	log sin	9.9236	9.9241	9.9246	9.9251	9.9255	9.9260	9.9265	9.9270	9.9275	9.9279
	log cos	9.7361	9.7349	9.7338	9.7326	9.7314	9.7302	9.7290	9.7278	9.7266	9.7254
	log tan	0.1875	0.1891	0.1908	0.1925	0.1941	0.1958	0.1975	0.1992	0.2008	0.2025
58	log sin	9.9284	9.9289	9.9294	9.9298	9.9303	9.9308	9.9312	9.9317	9.9322	9.9326
	log cos	9.7242	9.7230	9.7218	9.7205	9.7193	9.7181	9.7168	9.7156	9.7144	9.7131
	log tan	0.2042	0.2059	0.2076	0.2093	0.2110	0.2127	0.2144	0.2161	0.2178	0.2195
59	log sin	9.9331	9.9335	9.9340	9.9344	9.9349	9.9353	9.9358	9.9362	9.9367	9.9371
	log cos	9.7118	9.7106	9.7093	9.7080	9.7068	9.7055	9.7042	9.7029	9.7016	9.7003
	log tan	0.2212	0.2229	0.2247	0.2264	0.2281	0.2299	0.2316	0.2333	0.2351	0.2368
Degs.	Function	0′	6′	12′	18′	24′	30′	36′	42′	48′	54′

Common Logarithms of Sines, Cosines and Tangents

60°–74.9°

Degs.	Function	0.0°	0.1°	0.2°	0.3°	0.4°	0.5°	0.6°	0.7°	0.8°	0.9°
60	log sin	9.9375	9.9380	9.9384	9.9388	9.9393	9.9397	9.9401	9.9406	9.9410	9.9414
	log cos	9.6990	9.6977	9.6963	9.6950	9.6937	9.6923	9.6910	9.6896	9.6883	9.6869
	log tan	0.2386	0.2403	0.2421	0.2438	0.2456	0.2474	0.2491	0.2509	0.2527	0.2545
61	log sin	9.9418	9.9422	9.9427	9.9431	9.9435	9.9439	9.9443	9.9447	9.9451	9.9455
	log cos	9.6856	9.6842	9.6828	9.6814	9.6801	9.6787	9.6773	9.6759	9.6744	9.6730
	log tan	0.2562	0.2580	0.2598	0.2616	0.2634	0.2652	0.2670	0.2689	0.2707	0.2725
62	log sin	9.9459	9.9463	9.9467	9.9471	9.9475	9.9479	9.9483	9.9487	9.9491	9.9495
	log cos	9.6716	9.6702	9.6687	9.6673	9.6659	9.6644	9.6629	9.6615	9.6600	9.6585
	log tan	0.2743	0.2762	0.2780	0.2798	0.2817	0.2835	0.2854	0.2872	0.2891	0.2910
63	log sin	9.9499	9.9503	9.9506	9.9510	9.9514	9.9518	9.9522	9.9525	9.9529	9.9533
	log cos	9.6570	9.6556	9.6541	9.6526	9.6510	9.6495	9.6480	9.6465	9.6449	9.6434
	log tan	0.2928	0.2947	0.2966	0.2985	0.3004	0.3023	0.3042	0.3061	0.3080	0.3099
64	log sin	9.9537	9.9540	9.9544	9.9548	9.9551	9.9555	9.9558	9.9562	9.9566	9.9569
	log cos	9.6418	9.6403	9.6387	9.6371	9.6356	9.6340	9.6324	9.6308	9.6292	9.6276
	log tan	0.3118	0.3137	0.3157	0.3176	0.3196	0.3215	0.3235	0.3254	0.3274	0.3294
65	log sin	9.9573	9.9576	9.9580	9.9583	9.9587	9.9590	9.9594	9.9597	9.9601	9.9604
	log cos	9.6259	9.6243	9.6227	9.6210	9.6194	9.6177	9.6161	9.6144	9.6127	9.6110
	log tan	0.3313	0.3333	0.3353	0.3373	0.3393	0.3413	0.3433	0.3453	0.3473	0.3494
66	log sin	9.9607	9.9611	9.9614	9.9617	9.9621	9.9624	9.9627	9.9631	9.9634	9.9637
	log cos	9.6093	9.6076	9.6059	9.6042	9.6024	9.6007	9.5990	9.5972	9.5954	9.5937
	log tan	0.3514	0.3535	0.3555	0.3576	0.3596	0.3617	0.3638	0.3659	0.3679	9.3700
67	log sin	9.9640	9.9643	9.9647	9.9650	9.9653	9.9656	9.9659	9.9662	9.9666	9.9669
	log cos	9.5919	9.5901	9.5883	9.5865	9.5847	9.5828	9.5810	9.5792	9.5773	9.5754
	log tan	0.3721	0.3743	0.3764	0.3785	0.3806	0.3828	0.3849	0.3871	0.3892	0.3914
68	log sin	9.9672	9.9675	9.9678	9.9681	9.9684	9.9687	9.9690	9.9693	9.9696	9.9699
	log cos	9.5736	9.5717	9.5698	9.5679	9.5660	9.5641	9.5621	9.5602	9.5583	9.5563
	log tan	0.3936	0.3958	0.3980	0.4002	0.4024	0.4046	0.4068	0.4091	0.4113	0.4136
69	log sin	9.9702	9.9704	9.9707	9.9710	9.9713	9.9716	9.9719	9.9722	9.9724	9.9727
	log cos	9.5543	9.5523	9.5504	9.5484	9.5463	9.5443	9.5423	9.5402	9.5382	9.5361
	log tan	0.4158	0.4181	0.4204	0.4227	0.4250	0.4273	0.4296	0.4319	9.4342	0.4366
70	log sin	9.9730	9.9733	9.9735	9.9738	9.9741	9.9743	9.9746	9.9749	9.9751	9.9754
	log cos	9.5341	9.5320	9.5299	9.5278	9.5256	9.5235	9.5213	9.5192	9.5170	9.5148
	log tan	0.4389	0.4413	0.4437	0.4461	0.4484	0.4509	0.4533	0.4557	0.4581	0.4606
71	log sin	9.9757	9.9759	9.9762	9.9764	9.9767	9.9770	9.9772	9.9775	9.9777	9.9780
	log cos	9.5126	9.5104	9.5082	9.5060	9.5037	9.5015	9.4992	9.4969	9.4946	9.4923
	log tan	0.4630	0.4655	0.4680	0.4705	0.4730	0.4755	0.4780	0.4805	0.4831	0.4857
72	log sin	9.9782	9.9785	9.9787	9.9789	9.9792	9.9794	9.9797	9.9799	9.9801	9.9804
	log cos	9.4900	9.4876	9.4853	9.4829	9.4805	9.4781	9.4757	9.4733	9.4709	9.4684
	log tan	0.4882	0.4908	0.4934	0.4960	0.4986	0.5013	0.5039	0.5066	0.5093	0.5120
73	log sin	9.9806	9.9808	9.9811	9.9813	9.9815	9.9817	9.9820	9.9822	9.9824	9.9826
	log cos	9.4659	9.4634	9.4609	9.4584	9.4559	9.4533	9.4508	9.4482	9.4456	9.4430
	log tan	0.5147	0.5174	0.5201	0.5229	0.5256	0.5284	0.5312	0.5340	0.5368	0.5397
74	log sin	9.9828	9.9831	9.9833	9.9835	9.9837	9.9839	9.9841	9.9843	9.9845	9.9847
	log cos	9.4403	9.4377	9.4350	9.4323	9.4296	9.4269	9.4242	9.4214	9.4186	9.4158
	log tan	0.5425	0.5454	0.5483	0.5512	0.5541	0.5570	0.5600	0.5629	0.5659	0.5689
Degs.	Function	0′	6′	12′	18′	24′	30′	36′	42′	48′	54′

Degs.	Function	0.0°	0.1°	0.2°	0.3°	0.4°	0.5°	0.6°	0.7°	0.8°	0.9°
75	log sin	9.9849	9.9851	9.9853	9.9855	9.9857	9.9859	9.9861	9.9863	9.9865	9.9867
	log cos	9.4130	9.4102	9.4073	9.4044	9.4015	9.3986	9.3957	9.3927	9.3897	9.3867
	log tan	0.5719	0.5750	0.5780	0.5811	0.5842	0.5873	0.5905	0.5936	0.5968	0.6000
76	log sin	9.9869	9.9871	9.9873	9.9875	9.9876	9.9878	9.9880	9.9882	9.9884	9.9885
	log cos	9.3837	9.3806	9.3775	9.3745	9.3713	9.3682	9.3650	9.3618	9.3586	9.3554
	log tan	0.6032	0.6065	0.6097	0.6130	0.6163	0.6196	0.6230	0.6264	0.6298	0.6332
77	log sin	9.9887	9.9889	9.9891	9.9892	9.9894	9.9896	9.9897	9.9899	9.9901	9.9902
	log cos	9.3521	9.3488	9.3455	9.3421	9.3387	9.3353	9.3319	9.3284	9.3250	9.3214
	log tan	0.6366	0.6401	0.6436	0.6471	0.6507	0.6542	0.6578	0.6615	0.6651	0.6688
78	log sin	9.9904	9.9906	9.9907	9.9909	9.9910	9.9912	9.9913	9.9915	9.9916	9.9918
	log cos	9.3179	9.3143	9.3107	9.3070	9.3034	9.2997	9.2959	9.2921	9.2883	9.2845
	log tan	0.6725	0.6763	0.6800	0.6838	0.6877	0.6915	0.6954	0.6994	0.7033	0.7073
79	log sin	9.9919	9.9921	9.9922	9.9924	9.9925	9.9927	9.9928	9.9929	9.9931	9.9932
	log cos	9.2806	9.2767	9.2727	9.2687	9.2647	9.2606	9.2565	9.2524	9.2482	9.2439
	log tan	0.7113	0.7154	0.7195	0.7236	0.7278	0.7320	0.7363	0.7406	0.7449	0.7493
80	log sin	9.9934	9.9935	9.9936	9.9937	9.9939	9.9940	9.9941	9.9943	9.9944	9.9945
	log cos	9.2397	9.2353	9.2310	9.2266	9.2221	9.2176	9.2131	9.2085	9.2038	9.1991
	log tan	0.7537	0.7581	0.7626	9.7672	0.7718	0.7764	0.7811	0.7858	9.7906	0.7954
81	log sin	9.9946	9.9947	9.9949	9.9950	9.9951	9.9952	9.9953	9.9954	9.9955	9.9956
	log cos	9.1943	9.1895	9.1847	9.1797	9.1747	9.1697	9.1646	9.1594	9.1542	9.1489
	log tan	0.8003	0.8052	0.8102	0.8152	0.8203	0.8255	0.8307	0.8360	0.8413	0.8467
82	log sin	9.9958	9.9959	9.9960	9.9961	9.9962	9.9963	9.9964	9.9965	9.9966	9.9967
	log cos	9.1436	9.1381	9.1326	9.1271	9.1214	9.1157	9.1099	9.1040	9.0981	9.0920
	log tan	0.8522	0.8577	0.8633	0.8690	0.8748	0.8806	0.8865	0.8924	0.8985	0.9046
83	log sin	9.9968	9.9968	9.9969	9.9970	9.9971	9.9972	9.9973	9.9974	9.9975	9.9975
	log cos	9.0859	9.0797	9.0734	9.0670	9.0605	9.0539	9.0472	9.0403	9.0334	9.0264
	log tan	0.9109	0.9172	0.9236	0.9301	0.9367	0.9433	0.9501	0.9570	0.9640	0.9711
84	log sin	9.9976	9.9977	9.9978	9.9978	9.9979	9.9980	9.9981	9.9981	9.9982	9.9983
	log cos	9.0192	9.0120	9.0046	8.9970	8.9894	8.9816	8.9736	8.9655	8.9573	8.9489
	log tan	0.9784	0.9857	0.9932	1.0008	1.0085	1.0164	1.0244	1.0326	1.0409	1.0494
85	log sin	9.9983	9.9984	9.9985	9.9985	9.9986	9.9987	9.9987	9.9988	9.9988	9.9989
	log cos	8.9403	8.9315	8.9226	8.9135	8.9042	8.8946	8.8849	8.8749	8.8647	8.8543
	log tan	1.0580	1.0669	1.0759	1.0850	1.0944	1.1040	1.1138	1.1238	1.1341	1.1446
86	log sin	9.9989	9.9990	9.9990	9.9991	9.9991	9.9992	9.9992	9.9993	9.9993	9.9994
	log cos	8.8436	8.8326	8.8213	8.8098	8.7979	8.7857	8.7731	8.7602	8.7468	8.7330
	log tan	1.1554	1.1664	1.1777	1.1893	1.2012	1.2135	1.2261	1.2391	1.2525	1.2663
87	log sin	9.9994	9.9994	9.9995	9.9995	9.9996	9.9996	9.9996	9.9996	9.9997	9.9997
	log cos	8.7188	8.7041	8.6889	8.6731	8.6567	8.6397	8.6220	8.6035	8.5842	8.5640
	log tan	1.2806	1.2954	1.3106	1.3264	1.3429	1.3599	1.3777	1.3962	1.4155	1.4357
88	log sin	9.9997	9.9998	9.9998	9.9998	9.9998	9.9999	9.9999	9.9999	9.9999	9.9999
	log cos	8.5428	8.5206	8.4971	8.4723	8.4459	8.4179	8.3880	8.3558	8.3210	8.2832
	log tan	1.4569	1.4792	1.5027	1.5275	1.5539	1.5819	1.6119	1.6441	1.6789	1.7167
89	log sin	9.9999	9.9999	0	0	0	0	0	0	0	0
	log cos	8.2419	8.1961	8.1450	8.0870	8.0200	7.9408	7.8439	7.7190	7.5429	7.2419
	log tan	1.7581	1.8038	1.8550	1.9130	1.9800	2.0591	2.1561	2.2810	2.4571	2.7581
Degs.	Function	0′	6′	12′	18′	24′	30′	36′	42′	48′	54′

Angle	Function	0.00	0.01	0.02	0.03	0.04	0.05	0.06	0.07	0.08	0.09
0.0	sinh	0.0000	0.0100	0.0200	0.0300	0.0400	0.0500	0.0600	0.0701	0.0801	0.0901
	cosh	1.0000	1.0001	1.0002	1.0005	1.0008	1.0013	1.0018	1.0025	1.0032	1.0041
	tanh	0.0000	0.0100	0.0200	0.0300	0.0400	0.0500	0.0599	0.0699	0.0798	0.0898
0.1	sinh	0.1002	0.1102	0.1203	0.1304	0.1405	0.1506	0.1607	0.1708	0.1810	0.1911
	cosh	1.0050	1.0061	1.0072	1.0085	1.0098	1.0113	1.0128	1.0145	1.0162	1.0181
	tanh	0.0997	0.1096	0.1194	0.1293	0.1391	0.1489	0.1587	0.1684	0.1781	0.1878
0.2	sinh	0.2013	0.2115	0.2218	0.2320	0.2423	0.2526	0.2629	0.2733	0.2837	0.2941
	cosh	1.0201	1.0221	1.0243	1.0266	1.0289	1.0314	1.0340	1.0367	1.0395	1.0423
	tanh	0.1974	0.2070	0.2165	0.2260	0.2355	0.2449	0.2543	0.2636	0.2729	0.2821
0.3	sinh	0.3045	0.3150	0.3255	0.3360	0.3466	0.3572	0.3678	0.3785	0.3892	0.4000
	cosh	1.0453	1.0484	1.0516	1.0549	1.0584	1.0619	1.0655	1.0692	1.0731	1.0770
	tanh	0.2913	0.3004	0.3095	0.3185	0.3275	0.3364	0.3452	0.3540	0.3627	0.3714
0.4	sinh	0.4108	0.4216	0.4325	0.4434	0.4543	0.4653	0.4764	0.4875	0.4986	0.5098
	cosh	1.0811	1.0852	1.0895	1.0939	1.0984	1.1030	1.1077	1.1125	1.1174	1.1225
	tanh	0.3800	0.3885	0.3969	0.4053	0.4136	0.4219	0.4301	0.4382	0.4462	0.4542
0.5	sinh	0.5211	0.5324	0.5438	0.5552	0.5666	0.5782	0.5897	0.6014	0.6131	0.6248
	cosh	1.1276	1.1329	1.1383	1.1438	1.1494	1.1551	1.1609	1.1669	1.1730	1.1792
	tanh	0.4621	0.4700	0.4777	0.4854	0.4930	0.5005	0.5080	0.5154	0.5227	0.5299
0.6	sinh	0.6367	0.6485	0.6605	0.6725	0.6846	0.6967	0.7090	0.7213	0.7336	0.7461
	cosh	1.1855	1.1919	1.1984	1.2051	1.2119	1.2188	1.2258	1.2330	1.2402	1.2476
	tanh	0.5370	0.5441	0.5511	0.5581	0.5649	0.5717	0.5784	0.5850	0.5915	0.5980
0.7	sinh	0.7586	0.7712	0.7838	0.7966	0.8094	0.8223	0.8353	0.8484	0.8615	0.8748
	cosh	1.2552	1.2628	1.2706	1.2785	1.2865	1.2947	1.3030	1.3114	1.3199	1.3286
	tanh	0.6044	0.6107	0.6169	0.6231	0.6292	0.6352	0.6411	0.6469	0.6527	0.6584
0.8	sinh	0.8881	0.9015	0.9150	0.9286	0.9423	0.9561	0.9700	0.9840	0.9981	1.0122
	cosh	1.3374	1.3464	1.3555	1.3647	1.3740	1.3835	1.3932	1.4029	1.4128	1.4229
	tanh	0.6640	0.6696	0.6751	0.6805	0.6858	0.6911	0.6963	0.7014	0.7064	0.7114
0.9	sinh	1.0265	1.0409	1.0554	1.0700	1.0847	1.0995	1.1144	1.1294	1.1446	1 1598
	cosh	1.4331	1.4434	1.4539	1.4645	1.4753	1.4862	1.4973	1.5085	1.5199	1.5314
	tanh	0.7163	0.7211	0.7259	0.7306	0.7352	0.7398	0.7443	0.7487	0.7531	0.7574
1.0	sinh	1.1752	1.1907	1.2063	1.2220	1.2379	1.2539	1.2700	1.2862	1.3025	1.3190
	cosh	1.5431	1.5549	1.5669	1.5790	1.5913	1.6038	1.6164	1.6292	1.6421	1.6552
	tanh	0.7616	0.7658	0.7699	0.7739	0.7779	0.7818	0.7857	0.7895	0.7932	0.7969
1.1	sinh	1.3356	1.3524	1.3693	1.3863	1.4035	1.4208	1.4382	1.4558	1.4735	1.4914
	cosh	1.6685	1.6820	1.6956	1.7093	1.7233	1.7374	1.7517	1.7662	1.7808	1.7956
	tanh	0.8005	0.8041	0.8076	0.8110	0.8144	0.8178	0.8210	0.8243	0.8275	0.8306
1.2	sinh	1.5095	1.5276	1.5460	1.5645	1.5831	1.6019	1.6209	1.6400	1.6593	1.6788
	cosh	1.8107	1.8258	1.8412	1.8568	1.8725	1.8884	1.9045	1.9208	1.9373	1.9540
	tanh	0.8337	0.8367	0.8397	0.8426	0.8455	0.8483	0.8511	0.8538	0.8565	0.8591
1.3	sinh	1.6984	1.7182	1.7381	1.7583	1.7786	1.7991	1.8198	1.8406	1.8617	1.8829
	cosh	1.9709	1.9880	2.0053	2.0228	2.0404	2.0583	2.0764	2.0947	2.1132	2.1320
	tanh	0.8617	0.8643	0.8668	0.8693	0.8717	0.8741	0.8764	0.8787	0.8810	0.8832
1.4	sinh	1.9043	1.9259	1.9477	1.9697	1.9919	2.0143	2.0369	2.0597	2.0827	2.1059
	cosh	2.1509	2.1700	2.1894	2.2090	2.2288	2.2488	2.2691	2.2896	2.3103	2.3312
	tanh	0.8854	0.8875	0.8896	0.8917	0.8937	0.8957	0.8977	0.8996	0.9015	0.9033

Angle	Function	0.00	0.01	0.02	0.03	0.04	0.05	0.06	0.07	0.08	0.09
	sinh	2.1293	2.1529	2.1768	2.2008	2.2251	2.2496	2.2743	2.2993	2.3245	2.3499
1.5	cosh	2.3524	2.3738	2.3955	2.4174	2.4395	2.4619	2.4845	2.5074	2.5305	2.5538
	tanh	0.9052	0.9069	0.9087	0.9104	0.9121	0.9138	0.9154	0.9170	0.9186	0.9202
	sinh	2.3756	2.4015	2.4276	2.4540	2.4806	2.5075	2.5346	2.5620	2.5896	2.6175
1.6	cosh	2.5775	2.6013	2.6255	2.6499	2.6746	2.6995	2.7247	2.7502	2.7760	2.8020
	tanh	0.9217	0.9232	0.9246	0.9261	0.9275	0.9289	0.9302	0.9316	0.9329	0.9342
	sinh	2.6456	2.6740	2.7027	2.7317	2.7609	2.7904	2.8202	2.8503	2.8806	2.9112
1.7	cosh	2.8283	2.8549	2.8818	2.9090	2.9364	2.9642	2.9922	3.0206	3.0493	3.0782
	tanh	0.9354	0.9367	0.9379	0.9391	0.9402	0.9414	0.9425	0.9436	0.9447	0.9458
	sinh	2.9422	2.9734	3.0049	3.0367	3.0689	3.1013	3.1340	3.1671	3.2005	3.2341
1.8	cosh	3.1075	3.1371	3.1669	3.1972	3.2277	3.2585	3.2897	3.3212	3.3530	3.3852
	tanh	0.9468	0.9478	0.9488	0.9498	0.9508	0.9518	0.9527	0.9536	0.9545	0.9554
	sinh	3.2682	3.3025	3.3372	3.3722	3.4075	3.4432	3.4792	3.5156	3.5523	3.5894
1.9	cosh	3.4177	3.4506	3.4838	3.5173	3.5512	3.5855	3.6201	3.6551	3.6904	3.7261
	tanh	0.9562	0.9571	0.9579	0.9587	0.9595	0.9603	0.9611	0.9619	0.9626	0.9633
	sinh	3.6269	3.6647	3.7028	3.7414	3.7803	3.8196	3.8593	3.8993	3.9398	3.9806
2.0	cosh	3.7622	3.7987	3.8355	3.8727	3.9103	3.9483	3.9867	4.0255	4.0647	4.1043
	tanh	0.9640	0.9647	0.9654	0.9661	0.9668	0.9674	0.9680	0.9686	0.9693	0.9699
	sinh	4.0219	4.0635	4.1056	4.1480	4.1909	4.2342	4.2779	4.3221	4.3666	4.4117
2.1	cosh	4.1443	4.1847	4.2256	4.2668	4.3085	4.3507	4.3932	4.4362	4.4797	4.5236
	tanh	0.9705	0.9710	0.9716	0.9722	0.9727	0.9732	0.9738	0.9743	0.9748	0.9752
	sinh	4.4571	4.5030	4.5494	4.5962	4.6434	4.6912	4.7394	4.7880	4.8372	4.8868
2.2	cosh	4.5679	4.6127	4.6580	4.7037	4.7499	4.7966	4.8437	4.8914	4.9395	4.9881
	tanh	0.9757	0.9762	0.9767	0.9771	0.9776	0.9780	0.9785	0.9789	0.9793	0.9797
	sinh	4.9370	4.9876	5.0387	5.0903	5.1425	5.1951	5.2483	5.3020	5.3562	5.4109
2.3	cosh	5.0372	5.0868	5.1370	5.1876	5.2388	5.2905	5.3427	5.3954	5.4487	5.5026
	tanh	0.9801	0.9805	0.9809	0.9812	0.9816	0.9820	0.9823	0.9827	0.9830	0.9834
	sinh	5.4662	5.5221	5.5785	5.6354	5.6929	5.7510	5.8097	5.8689	5.9288	5.9892
2.4	cosh	5.5569	5.6119	5.6674	5.7235	5.7801	5.8373	5.8951	5.9535	6.0125	6.0721
	tanh	0.9837	0.9840	0.9843	0.9846	0.9849	0.9852	0.9855	0.9858	0.9861	0.9864
	sinh	6.0502	6.1118	6.1741	6.2369	6.3004	6.3645	6.4293	6.4946	6.5607	6.6274
2.5	cosh	6.1323	6.1931	6.2545	6.3166	6.3793	6.4426	6.5066	6.5712	6.6365	6.7024
	tanh	0.9866	0.9869	0.9871	0.9874	0.9876	0.9879	0.9881	0.9884	0.9886	0.9888
	sinh	6.6947	6.7628	6.8315	6.9009	6.9709	7.0417	7.1132	7.1854	7.2583	7.3319
2.6	cosh	6.7690	6.8363	6.9043	6.9729	7.0423	7.1123	7.1831	7.2546	7.3268	7.3998
	tanh	0.9890	0.9892	0.9895	0.9897	0.9899	0.9901	0.9903	0.9905	0.9906	0.9908
	sinh	7.4063	7.4814	7.5572	7.6338	7.7112	7.7894	7.8683	7.9480	8.0285	8.1098
2.7	cosh	7.4735	7.5479	7.6231	7.6991	7.7758	7.8533	7.9316	8.0106	8.0905	8.1712
	tanh	0.9910	0.9912	0.9914	0.9915	0.9917	0.9919	0.9920	0.9922	0.9923	0.9925
	sinh	8.1919	8.2749	8.3586	8.4432	8.5287	8.6150	8.7021	8.7902	8.8791	8.9689
2.8	cosh	8.2527	8.3351	8.4182	8.5022	8.5871	8.6728	8.7594	8.8469	8.9352	9.0244
	tanh	0.9926	0.9928	0.9929	0.9931	0.9932	0.9933	0.9935	0.9936	0.9937	0.9938
	sinh	9.0596	9.1512	9.2437	9.3371	9.4315	9.5268	9.6231	9.7203	9.8185	9.9177
2.9	cosh	9.1146	9.2056	9.2976	9.3905	9.4844	9.5792	9.6749	9.7716	9.8693	9.9680
	tanh	0.9940	0.9941	0.9942	0.9943	0.9944	0.9945	0.9946	0.9948	0.9949	0.9950

Angle	Function	0.00	0.01	0.02	0.03	0.04	0.05	0.06	0.07	0.08	0.09
	sinh	10.018	10.119	10.221	10.324	10.429	10.534	10.640	10.748	10.856	10.966
3.0	cosh	10.068	10.168	10.270	10.373	10.476	10.581	10.687	10.794	10.902	11.011
	tanh	0.9951	0.9952	0.9953	0.9953	0.9954	0.9955	0.9956	0.9957	0.9958	0.9959
	sinh	11.076	11.188	11.301	11.415	11.530	11.647	11.764	11.883	12.003	12.124
3.1	cosh	11.121	11.233	11.345	11.459	11.574	11.689	11.806	11.925	12.044	12.165
	tanh	0.9960	0.9960	0.9961	0.9962	0.9963	0.9963	0.9964	0.9965	0.9966	0.9966
	sinh	12.246	12.369	12.494	12.620	12.747	12.876	13.006	13.137	13.269	13.403
3.2	cosh	12.287	12.410	12.534	12.660	12.786	12.915	13.044	13.175	13.307	13.440
	tanh	0.9967	0.9968	0.9968	0.9969	0.9969	0.9970	0.9971	0.9971	0.9972	0.9972
	sinh	13.538	13.674	13.812	13.951	14.092	14.234	14.377	14.522	14.668	14.816
3.3	cosh	13.575	13.711	13.848	13.987	14.127	14.269	14.412	14.556	14.702	14.850
	tanh	0.9973	0.9973	0.9974	0.9974	0.9975	0.9975	0.9976	0.9976	0.9977	0.9977
	sinh	14.965	15.116	15.268	15.422	15.577	15.734	15.893	16.053	16.215	16.378
3.4	cosh	14.999	15.149	15.301	15.455	15.610	15.766	15.924	16.084	16.245	16.408
	tanh	0.9978	0.9978	0.9979	0.9979	0.9979	0.9980	0.9980	0.9981	0.9981	0.9981
	sinh	16.543	16.709	16.877	17.047	17.219	17.392	17.567	17.744	17.923	18.103
3.5	cosh	16.573	16.739	16.907	17.077	17.248	17.421	17.596	17.772	17.951	18.131
	tanh	0.9982	0.9982	0.9983	0.9983	0.9983	0.9984	0.9984	0.9984	0.9985	0.9985
	sinh	18.285	18.470	18.655	18.843	19.033	19.224	19.418	19.613	19.811	20.010
3.6	cosh	18.313	18.497	18.682	18.870	19.059	19.250	19.444	19.639	19.836	20.035
	tanh	0.9985	0.9985	0.9986	0.9986	0.9986	0.9987	0.9987	0.9987	0.9987	0.9988
	sinh	20.211	20.415	20.620	20.828	21.037	21.249	21.463	21.679	21.897	22.117
3.7	cosh	20.236	20.439	20.644	20.852	21.061	21.272	21.486	21.702	21.919	22.139
	tanh	0.9988	0.9988	0.9988	0.9989	0.9989	0.9989	0.9989	0.9989	0.9990	0.9990
	sinh	22.339	22.564	22.791	23.020	23.252	23.486	23.722	23.961	24.202	24.445
3.8	cosh	22.362	22.586	22.813	23.042	23.273	23.507	23.743	23.982	24.222	24.466
	tanh	0.9990	0.9990	0.9990	0.9991	0.9991	0.9991	0.9991	0.9991	0.9992	0.9992
	sinh	24.691	24.939	25.190	25.444	25.700	25.958	26.219	26.483	26.749	27.018
3.9	cosh	24.711	24.959	25.210	25.463	25.719	25.977	26.238	26.502	26.768	27.037
	tanh	0.9992	0.9992	0.9992	0.9992	0.9992	0.9993	0.9993	0.9993	0.9993	0.9993
	sinh	27.290	27.564	27.842	28.122	28.404	28.690	28.979	29.270	29.564	29.862
4.0	cosh	27.308	27.583	27.860	28.139	28.422	28.707	28.996	29.287	29.581	29.878
	tanh	0.9993	0.9993	0.9994	0.9994	0.9994	0.9994	0.9994	0.9994	0.9994	0.9994
	sinh	30.162	30.465	30.772	31.081	31.393	31.709	32.028	32.350	32.675	33.004
4.1	cosh	30.178	30.482	30.788	31.097	31.409	31.725	32.044	32.365	32.691	33.019
	tanh	0.9995	0.9995	0.9995	0.9995	0.9995	0.9995	0.9995	0.9995	0.9995	0.9995
	sinh	33.336	33.671	34.009	34.351	34.697	35.046	35.398	35.754	36.113	36.476
4.2	cosh	33.351	33.686	34.024	34.366	34.711	35.060	35.412	35.768	36.127	36.490
	tanh	0.9996	0.9996	0.9996	0.9996	0.9996	0.9996	0.9996	0.9996	0.9996	0.9996
	sinh	36.843	37.214	37.588	37.965	38.347	38.733	39.122	39.515	39.913	40.314
4.3	cosh	36.857	37.227	37.601	37.979	38.360	38.746	39.135	39.528	39.925	40.326
	tanh	0.9996	0.9996	0.9997	0.9997	0.9997	0.9997	0.9997	0.9997	0.9997	0.9997
	sinh	40.719	41.129	41.542	41.960	42.382	42.808	43.238	43.673	44.112	44.555
4.4	cosh	40.732	41.141	41.554	41.972	42.393	42.819	43.250	43.684	44.123	44.566
	tanh	0.9997	0.9997	0.9997	0.9997	0.9997	0.9997	0.9997	0.9997	0.9997	0.9998

Angle	Function	0.00	0.01	0.02	0.03	0.04	0.05	0.06	0.07	0.08	0.09
	sinh	45.003	45.455	45.912	46.374	46.840	47.311	47.787	48.267	48.752	49.242
4.5	cosh	45.014	45.466	45.923	46.385	46.851	47.321	47.797	48.277	48.762	49.252
	tanh	0.9998	0.9998	0.9998	0.9998	0.9998	0.9998	0.9998	0.9998	0.9998	0.9998
	sinh	49.737	50.237	50.742	51.252	51.767	52.288	52.813	53.344	53.880	54.422
4.6	cosh	49.747	50.247	50.752	51.262	51.777	52.297	52.823	53.354	53.890	54.431
	tanh	0.9998	0.9998	0.9998	0.9998	0.9998	0.9998	0.9998	0.9998	0.9998	0.9998
	sinh	54.969	55.522	56.080	56.643	57.213	57.788	58.369	58.955	59.548	60.147
4.7	cosh	54.978	55.531	56.089	56.652	57.221	57.796	58.377	58.964	59.556	60.155
	tanh	0.9998	0.9998	0.9998	0.9998	0.9999	0.9999	0.9999	0.9999	0.9999	0.9999
	sinh	60.751	61.362	61.979	62.601	63.231	63.866	64.508	65.157	65.812	66.473
4.8	cosh	60.759	61.370	61.987	62.609	63.239	63.874	64.516	65.164	65.819	66.481
	tanh	0.9999	0.9999	0.9999	0.9999	0.9999	0.9999	0.9999	0.9999	0.9999	0.9999
	sinh	67.141	67.816	68.498	69.186	69.882	70.584	71.293	72.010	72.734	73.465
4.9	cosh	67.149	67.823	68.505	69.193	69.889	70.591	71.300	72.017	72.741	73.472
	tanh	0.9999	0.9999	0.9999	0.9999	0.9999	0.9999	0.9999	0.9999	0.9999	0.9999
	sinh	74.203	74.949	75.702	76.463	77.232	78.008	78.792	79.584	80.384	81.192
5.0	cosh	74.210	74.956	75.709	76.470	77.238	78.014	78.798	79.590	80.390	81.198
	tanh	0.9999	0.9999	0.9999	0.9999	0.9999	0 9999	0.9999	0.9999	0.9999	0.9999
	sinh	82.008	82.832	83.665	84.506	85.355	86.213	87.079	87.955	88.839	89.732
5.1	cosh	82.014	82.838	83.671	84.512	85.361	86.219	87.085	87.960	88.844	89.737
	tanh	0.9999	0.9999	0.9999	0.9999	0.9999	0.9999	0.9999	0.9999	0.9999	0.9999
	sinh	90.633	91.544	92.464	93.394	94.332	95.281	96.238	97.205	98.182	99.169
5.2	cosh	90.639	91.550	92.470	93.399	94.338	95.286	96.243	97.211	98.188	99.174
	tanh	0.9999	0.9999	0.9999	0.9999	0.9999	0.9999	1.0000	1.0000	1.0000	1.0000
	sinh	100.17	101.17	102.19	103.22	104.25	105.30	106.36	107.43	108.51	109.60
5.3	cosh	100.17	101.18	102.19	103.22	104.26	105.31	106.37	107.43	108.51	109.60
	tanh	1.0000	1.0000	1.0000	1.0000	1.0000	1.0000	1.0000	1.0000	1.0000	1.0000
	sinh	110.70	111.81	112.94	114.07	115.22	116.38	117.55	118.73	119.92	121.13
5.4	cosh	110.71	111.82	112.94	114.08	115.22	116.38	117.55	118.73	119.93	121.13
	tanh	1.0000	1.0000	1.0000	1.0000	1.0000	1.0000	1.0000	1.0000	1.0000	1.0000
	sinh	122.34	123.57	124.82	126.07	127.34	128.62	129.91	131.22	132.53	133.87
5.5	cosh	122.35	123.58	124.82	126.07	127.34	128.62	129.91	131.22	132.54	133.87
	tanh	1.0000	1.0000	1.0000	1.0000	1.0000	1.0000	1.0000	1.0000	1.0000	1.0000
	sinh	135.21	136.57	137.94	139.33	140.73	142.14	143.57	145.02	146.47	147.95
5.6	cosh	135.22	136.57	137.95	139.33	140.73	142.15	143.58	145.02	146.48	147.95
	tanh	1.0000	1.0000	1.0000	1.0000	1.0000	1.0000	1.0000	1.0000	1.0000	1.0000
	sinh	149.43	150.93	152.45	153.98	155.53	157.09	158.67	160.27	161.88	163.51
5.7	cosh	149.44	150.94	152.45	153.99	155.53	157.10	158.68	160.27	161.88	163.51
	tanh	1.0000	1.0000	1.0000	1.0000	1.0000	1.0000	1.0000	1.0000	1.0000	1.0000
	sinh	165.15	166.81	168.48	170.18	171.89	173.62	175.36	177.12	178.90	180.70
5.8	cosh	165.15	166.81	168.49	170.18	171.89	173.62	175.36	177.13	178.91	180.70
	tanh	1.0000	1.0000	1.0000	1.0000	1.0000	1.0000	1.0000	1.0000	1.0000	1.0000
	sinh	182.52	184.35	186.20	188.08	189.97	191.88	193.80	195.75	197.72	199.71
5.9	cosh	182.52	184.35	186.21	188.08	189.97	191.88	193.81	195.75	197.72	199.71
	tanh	1.0000	1.0000	1.0000	1.0000	1.0000	1.0000	1.0000	1.0000	1.0000	1.0000

x	Function	0.00	0.01	0.02	0.03	0.04	0.05	0.06	0.07	0.08	0.09
0.0	ϵ^x	1.0000	1.0101	1.0202	1.0305	1.0408	1.0513	1.0618	1.0725	1.0833	1.0942
	ϵ^{-x}	1.0000	0.9900	0.9802	0.9704	0.9608	0.9512	0.9418	0.9324	0.9231	0.9139
0.1	ϵ^x	1.1052	1.1163	1.1275	1.1388	1.1503	1.1618	1.1735	1.1853	1.1972	1.2093
	ϵ^{-x}	0.9048	0.8958	0.8869	0.8781	0.8694	0.8607	0.8521	0.8437	0.8353	0.8270
0.2	ϵ^x	1.2214	1.2337	1.2461	1.2586	1.2712	1.2840	1.2969	1.3100	1.3231	1.3364
	ϵ^{-x}	0.8187	0.8106	0.8025	0.7945	0.7866	0.7788	0.7711	0.7634	0.7558	0.7483
0.3	ϵ^x	1.3499	1.3634	1.3771	1.3910	1.4049	1.4191	1.4333	1.4477	1.4623	1.4770
	ϵ^{-x}	0.7408	0.7334	0.7261	0.7189	0.7118	0.7047	0.6977	0.6907	0.6839	0.6771
0.4	ϵ^x	1.4918	1.5068	1.5220	1.5373	1.5527	1.5683	1.5841	1.6000	1.6161	1.6323
	ϵ^{-x}	0.6703	0.6637	0.6570	0.6505	0.6440	0.6376	0.6313	0.6250	0.6188	0.6126
0.5	ϵ^x	1.6487	1.6653	1.6820	1.6989	1.7160	1.7333	1.7507	1.7683	1.7860	1.8040
	ϵ^{-x}	0.6065	0.6005	0.5945	0.5886	0.5827	0.5769	0.5712	0.5655	0.5599	0.5543
0.6	ϵ^x	1.8221	1.8404	1.8589	1.8776	1.8965	1.9155	1.9348	1.9542	1.9739	1.9939
	ϵ^{-x}	0.5488	0.5434	0.5379	0.5326	0.5273	0.5220	0.5169	0.5117	0.5066	0.5017
0.7	ϵ^x	2.0138	2.0340	2.0544	2.0751	2.0959	2.1170	2.1383	2.1598	2.1815	2.2034
	ϵ^{-x}	0.4966	0.4916	0.4868	0.4819	0.4771	0.4724	0.4677	0.4630	0.4584	0.4538
0.8	ϵ^x	2.2255	2.2479	2.2705	2.2933	2.3164	2.3396	2.3632	2.3869	2.4109	2.4351
	ϵ^{-x}	0.4493	0.4449	0.4404	0.4360	0.4317	0.4274	0.4232	0.4190	0.4148	0.4107
0.9	ϵ^x	2.4596	2.4843	2.5093	2.5345	2.5600	2.5857	2.6117	2.6379	2.6645	2.6912
	ϵ^{-x}	0.4066	0.4025	0.3985	0.3946	0.3906	0.3867	0.3829	0.3791	0.3753	0.3716
1.0	ϵ^x	2.7183	2.7456	2.7732	2.8011	2.8292	2.8577	2.8864	2.9154	2.9447	2.9743
	ϵ^{-x}	0.3679	0.3642	0.3606	0.3570	0.3535	0.3499	0.3465	0.3430	0.3396	0.3362
1.1	ϵ^x	3.0042	3.0344	3.0649	3.0957	3.1268	3.1582	3.1899	3.2220	3.2544	3.2871
	ϵ^{-x}	0.3329	0.3296	0.3263	0.3230	0.3198	0.3166	0.3135	0.3104	0.3073	0.3042
1.2	ϵ^x	3.3201	3.3535	3.3872	3.4212	3.4556	3.4903	3.5254	3.5609	3.5966	3.6328
	ϵ^{-x}	0.3012	0.2982	0.2952	0.2923	0.2894	0.2865	0.2837	0.2808	0.2780	0.2753
1.3	ϵ^x	3.6693	3.7062	3.7434	3.7810	3.8190	3.8574	3.8962	3.9354	3.9749	4.0149
	ϵ^{-x}	0.2725	0.2698	0.2671	0.2645	0.2618	0.2592	0.2567	0.2541	0.2516	0.2491
1.4	ϵ^x	4.0552	4.0960	4.1371	4.1787	4.2207	4.2631	4.3060	4.3492	4.3929	4.4371
	ϵ^{-x}	0.2466	0.2441	0.2417	0.2393	0.2369	0.2346	0.2322	0.2299	0.2276	0.2254
1.5	ϵ^x	4.4817	4.5267	4.5722	4.6182	4.6646	4.7115	4.7588	4.8066	4.8550	4.9037
	ϵ^{-x}	0.2231	0.2209	0.2187	0.2165	0.2144	0.2122	0.2101	0.2080	0.2060	0.2039
1.6	ϵ^x	4.9530	5.0028	5.0531	5.1039	5.1552	5.2070	5.2593	5.3122	5.3656	5.4195
	ϵ^{-x}	0.2019	0.1999	0.1979	0.1959	0.1940	0.1920	0.1901	0.1882	0.1864	0.1845
1.7	ϵ^x	5.4739	5.5290	5.5845	5.6407	5.6973	5.7546	5.8124	5.8709	5.9299	5.9895
	ϵ^{-x}	0.1827	0.1809	0.1791	0.1773	0.1755	0.1738	0.1720	0.1703	0.1686	0.1670
1.8	ϵ^x	6.0496	6.1104	6.1719	6.2339	6.2965	6.3598	6.4237	6.4883	6.5535	6.6194
	ϵ^{-x}	0.1653	0.1637	0.1620	0.1604	0.1588	0.1572	0.1557	0.1541	0.1526	0.1511
1.9	ϵ^x	6.6859	6.7531	6.8210	6.8895	6.9588	7.0287	7.0993	7.1707	7.2427	7.3155
	ϵ^{-x}	0.1496	0.1481	0.1466	0.1451	0.1437	0.1423	0.1409	0.1395	0.1381	0.1367

x	Function	0.00	0.01	0.02	0.03	0.04	0.05	0.06	0.07	0.08	0.09
2.0	ϵ^x	7.3891	7.4633	7.5383	7.6141	7.6906	7.7679	7.8460	7.9248	8.0045	8.0849
	ϵ^{-x}	0.1353	0.1340	0.1327	0.1313	0.1300	0.1287	0.1275	0.1262	0.1249	0.1237
2.1	ϵ^x	8.1662	8.2482	8.3311	8.4149	8.4994	8.5849	8.6711	8.7583	8.8463	8.9352
	ϵ^{-x}	0.1225	0.1212	0.1200	0.1188	0.1177	0.1165	0.1153	0.1142	0.1130	0.1119
2.2	ϵ^x	9.0250	9.1157	9.2073	9.2999	9.3933	9.4877	9.5831	9.6794	9.7767	9.8749
	ϵ^{-x}	0.1108	0.1097	0.1086	0.1075	0.1065	0.1054	0.1044	0.1033	0.1023	0.1013
2.3	ϵ^x	9.9742	10.074	10.176	10.278	10.381	10.486	10.591	10.697	10.805	10.913
	ϵ^{-x}	0.1003	0.0993	0.0983	0.0973	0.0963	0.0954	0.0944	0.0935	0.0926	0.0916
2.4	ϵ^x	11.023	11.134	11.246	11.359	11.473	11.588	11.705	11.822	11.941	12.061
	ϵ^{-x}	0.0907	0.0898	0.0889	0.0880	0.0872	0.0863	0.0854	0.0846	0.0837	0.0829
2.5	ϵ^x	12.182	12.305	12.429	12.554	12.680	12.807	12.936	13.066	13.197	13.330
	ϵ^{-x}	0.0821	0.0813	0.0805	0.0797	0.0789	0.0781	0.0773	0.0765	0.0758	0.0750
2.6	ϵ^x	13.464	13.599	13.736	13.874	14.013	14.154	14.296	14.440	14.585	14.732
	ϵ^{-x}	0.0743	0.0735	0.0728	0.0721	0.0714	0.0707	0.0699	0.0693	0.0686	0.0679
2.7	ϵ^x	14.880	15.029	15.180	15.333	15.487	15.643	15.800	15.959	16.119	16.281
	ϵ^{-x}	0.0672	0.0665	0.0659	0.0652	0.0646	0.0639	0.0633	0.0627	0.0620	0.0614
2.8	ϵ^x	16.445	16.610	16.777	16.945	17.116	17.288	17.462	17.637	17.814	17.993
	ϵ^{-x}	0.0608	0.0602	0.0596	0.0590	0.0584	0.0578	0.0573	0.0567	0.0561	0.0556
2.9	ϵ^x	18.174	18.357	18.541	18.728	18.916	19.106	19.298	19.492	19.688	19.886
	ϵ^{-x}	0.0550	0.0545	0.0539	0.0534	0.0529	0.0523	0.0518	0.0513	0.0508	0.0503
3.0	ϵ^x	20.086	20.287	20.491	20.697	20.905	21.115	21.328	21.542	21.758	21.977
	ϵ^{-x}	0.0498	0.0493	0.0488	0.0483	0.0478	0.0474	0.0469	0.0464	0.0460	0.0455
3.1	ϵ^x	22.198	22.421	22.646	22.874	23.104	23.336	23.571	23.807	24.047	24.288
	ϵ^{-x}	0.0450	0.0446	0.0442	0.0437	0.0433	0.0429	0.0424	0.0420	0.0416	0.0412
3.2	ϵ^x	24.533	24.779	25.028	25.280	25.534	25.790	26.050	26.311	26.576	26.843
	ϵ^{-x}	0.0408	0.0404	0.0400	0.0396	0.0392	0.0388	0.0384	0.0380	0.0376	0.0373
3.3	ϵ^x	27.113	27.385	27.660	27.938	28.219	28.503	28.789	29.079	29.371	29.666
	ϵ^{-x}	0.0369	0.0365	0.0362	0.0358	0.0354	0.0351	0.0347	0.0344	0.0340	0.0337
3.4	ϵ^x	29.964	30.265	30.569	30.877	31.187	31.500	31.817	32.137	32.460	32.786
	ϵ^{-x}	0.0334	0.0330	0.0327	0.0324	0.0321	0.0317	0.0314	0.0311	0.0308	0.0305
3.5	ϵ^x	33.115	33.448	33.784	34.124	34.467	34.813	35.163	35.517	35.874	36.234
	ϵ^{-x}	0.0302	0.0299	0.0296	0.0293	0.0290	0.0287	0.0284	0.0282	0.0279	0.0276
3.6	ϵ^x	36.598	36.966	37.338	37.713	38.092	38.475	38.861	39.252	39.646	40.045
	ϵ^{-x}	0.0273	0.0271	0.0268	0.0265	0.0263	0.0260	0.0257	0.0255	0.0252	0.0250
3.7	ϵ^x	40.447	40.854	41.264	41.679	42.098	42.521	42.948	43.380	43.816	44.256
	ϵ^{-x}	0.0247	0.0245	0.0242	0.0240	0.0238	0.0235	0.0233	0.0231	0.0228	0.0226
3.8	ϵ^x	44.701	45.150	45.604	46.063	46.525	46.993	47.465	47.942	48.424	48.911
	ϵ^{-x}	0.0224	0.0221	0.0219	0.0217	0.0215	0.0213	0.0211	0.0209	0.0207	0.0204
3.9	ϵ^x	49.402	49.899	50.400	50.907	51.419	51.935	52.457	52.985	53.517	54.055
	ϵ^{-x}	0.0202	0.0200	0.0198	0.0196	0.0195	0.0193	0.0191	0.0189	0.0187	0.0185

x	Function	0.00	0.01	0.02	0.03	0.04	0.05	0.06	0.07	0.08	0.09
4.0	ϵ^x	54.598	55.147	55.701	56.261	56.826	57.397	57.974	58.557	59.145	59.740
	ϵ^{-x}	0.0183	0.0181	0.0180	0.0178	0.0176	0.0174	0.0172	0.0171	0.0169	0.0167
4.1	ϵ^x	60.340	60.947	61.559	62.178	62.803	63.434	64.072	64.715	65.366	66.023
	ϵ^{-x}	0.0166	0.0164	0.0162	0.0161	0.0159	0.0158	0.0156	0.0155	0.0153	0.0151
4.2	ϵ^x	66.686	67.357	68.033	68.717	69.408	70.105	70.810	71.522	72.240	72.966
	ϵ^{-x}	0.0150	0.0148	0.0147	0.0146	0.0144	0.0143	0.0141	0.0140	0.0138	0.0137
4.3	ϵ^x	73.700	74.440	75.189	75.944	76.708	77.478	78.257	79.044	79.838	80.640
	ϵ^{-x}	0.0136	0.0134	0.0133	0.0132	0.0130	0.0129	0.0128	0.0127	0.0125	0.0124
4.4	ϵ^x	81.451	82.269	83.096	83.931	84.775	85.627	86.488	87.357	88.235	89.121
	ϵ^{-x}	0.0123	0.0122	0.0120	0.0119	0.0118	0.0117	0.0116	0.0114	0.0113	0.0112
4.5	ϵ^x	90.017	90.922	91.836	92.759	93.691	94.632	95.583	96.544	97.514	98.494
	ϵ^{-x}	0.0111	0.0110	0.0109	0.0108	0.0107	0.0106	0.0105	0.0104	0.0103	0.0102
4.6	ϵ^x	99.484	100.48	101.49	102.51	103.54	104.58	105.64	106.70	107.77	108.85
	ϵ^{-x}	0.0101	0.0100	0.0099	0.0098	0.0097	0.0096	0.0095	0.0094	0.0093	0.0092
4.7	ϵ^x	109.95	111.05	112.17	113.30	114.43	115.58	116.75	117.92	119.10	120.30
	ϵ^{-x}	0.0091	0.0090	0.0089	0.0088	0.0087	0.0087	0.0086	0.0085	0.0084	0.0083
4.8	ϵ^x	121.51	122.73	123.97	125.21	126.47	127.74	129.02	130.32	131.63	132.95
	ϵ^{-x}	0.0082	0.0081	0.0081	0.0080	0.0079	0.0078	0.0078	0.0077	0.0076	0.0075
4.9	ϵ^x	134.29	135.64	137.00	138.38	139.77	141.17	142.59	144.03	145.47	146.94
	ϵ^{-x}	0.0074	0.0074	0.0073	0.0072	0.0072	0.0071	0.0070	0.0069	0.0069	0.0068
5.0	ϵ^x	148.41	149.90	151.41	152.93	154.47	156.02	157.59	159.17	160.77	162.39
	ϵ^{-x}	0.0067	0.0067	0.0066	0.0065	0.0065	0.0064	0.0063	0.0063	0.0062	0.0062
5.1	ϵ	164.02	165.67	167.34	169.02	170.72	172.43	174.16	175.91	177.68	179.47
	ϵ^{-x}	0.0061	0.0060	0.0060	0.0059	0.0059	0.0058	0.0057	0.0057	0.0056	0.0056
5.2	ϵ^x	181.27	183.09	184.93	186.79	188.67	190.57	192.48	194.42	196.37	198.34
	ϵ^{-x}	0.0055	0.0055	0.0054	0.0054	0.0053	0.0052	0.0052	0.0051	0.0051	0.0050
5.3	ϵ^x	200.34	202.35	204.38	206.44	208.51	210.61	212.72	214.86	217.02	219.20
	ϵ^{-x}	0.0050	0.0049	0.0049	0.0048	0.0048	0.0047	0.0047	0.0047	0.0046	0.0046
5.4	ϵ^x	221.41	223.63	225.88	228.15	230.44	232.76	235.10	237.46	239.85	242.26
	ϵ^{-x}	0.0045	0.0045	0.0044	0.0044	0.0043	0.0043	0.0043	0.0042	0.0042	0.0041
5.5	ϵ^x	244.69	247.15	249.64	252.14	254.68	257.24	259.82	262.43	265.07	267.74
	ϵ^{-x}	0.0041	0.0040	0.0040	0.0040	0.0039	0.0039	0.0038	0.0038	0.0038	0.0037
5.6	ϵ^x	270.43	273.14	275.89	278.66	281.46	284.29	287.15	290.03	292.95	295.89
	ϵ^{-x}	0.0037	0.0037	0.0036	0.0036	0.0036	0.0035	0.0035	0.0034	0.0034	0.0034
5.7	ϵ^x	298.87	301.87	304.90	307.97	311.06	314.19	317.35	320.54	323.76	327.01
	ϵ^{-x}	0.0033	0.0033	0.0033	0.0032	0.0032	0.0032	0.0032	0.0031	0.0031	0.0031
5.8	ϵ^x	330.30	333.62	336.97	340.36	343.78	347.23	350.72	354.25	357.81	361.41
	ϵ^{-x}	0.0030	0.0030	0.0030	0.0029	0.0029	0.0029	0.0029	0.0028	0.0028	0.0028
5.9	ϵ^x	365.04	368.71	372.41	376.15	379.93	383.75	387.61	391.51	395.44	399.41
	ϵ^{-x}	0.0027	0.0027	0.0027	0.0027	0.0026	0.0026	0.0026	0.0026	0.0025	0.0025

Fractions	Decimals	Fractions	Decimals	Fractions	Decimals	Fractions	Decimals
$\frac{1}{64}$	0.015625	$\frac{17}{64}$	0.265625	$\frac{33}{64}$	0.515625	$\frac{49}{64}$	0.765625
$\frac{1}{32}$	0.03125	$\frac{9}{32}$	0.28125	$\frac{17}{32}$	0.53125	$\frac{25}{32}$	0.78125
$\frac{3}{64}$	0.046875	$\frac{19}{64}$	0.296875	$\frac{35}{64}$	0.546875	$\frac{51}{64}$	0.796875
$\frac{1}{16}$	0.0625	$\frac{5}{16}$	0.3125	$\frac{9}{16}$	0.5625	$\frac{13}{16}$	0.8125
$\frac{5}{64}$	0.078125	$\frac{21}{64}$	0.328125	$\frac{37}{64}$	0.578125	$\frac{53}{64}$	0.828125
$\frac{3}{32}$	0.09375	$\frac{11}{32}$	0.34375	$\frac{19}{32}$	0.59375	$\frac{27}{32}$	0.84375
$\frac{7}{64}$	0.109375	$\frac{23}{64}$	0.359375	$\frac{39}{64}$	0.609375	$\frac{55}{64}$	0.859375
$\frac{1}{8}$	0.125	$\frac{3}{8}$	0.375	$\frac{5}{8}$	0.625	$\frac{7}{8}$	0.875
$\frac{9}{64}$	0.140625	$\frac{25}{64}$	0.390625	$\frac{41}{64}$	0.640625	$\frac{57}{64}$	0.890625
$\frac{5}{32}$	0.15625	$\frac{13}{32}$	0.40625	$\frac{21}{32}$	0.65625	$\frac{29}{32}$	0.90625
$\frac{11}{64}$	0.171875	$\frac{27}{64}$	0.421875	$\frac{43}{64}$	0.671875	$\frac{59}{64}$	0.921875
$\frac{3}{16}$	0.1875	$\frac{7}{16}$	0.4375	$\frac{11}{16}$	0.6875	$\frac{15}{16}$	0.9375
$\frac{13}{64}$	0.203125	$\frac{29}{64}$	0.453125	$\frac{45}{64}$	0.703125	$\frac{61}{64}$	0.953125
$\frac{7}{32}$	0.21875	$\frac{15}{32}$	0.46875	$\frac{23}{32}$	0.71875	$\frac{31}{32}$	0.96875
$\frac{15}{64}$	0.234375	$\frac{31}{64}$	0.484375	$\frac{47}{64}$	0.734375	$\frac{63}{64}$	0.984375
$\frac{1}{4}$	0.25	$\frac{1}{2}$	0.5	$\frac{3}{4}$	0.75	1	1

Factorials

n	n! = 1·2·3···n	1/n!	n	n! = 1·2·3····n	1/n!
1	1	1.	11	$399{,}168 \times 10^{2}$	0.250521×10^{-7}
2	2	0.5	12	$479{,}002 \times 10^{3}$	$.208768 \times 10^{-8}$
3	6	.166667	13	$622{,}702 \times 10^{4}$	$.160590 \times 10^{-9}$
4	24	$.416667 \times 10^{-1}$	14	$871{,}783 \times 10^{5}$	$.114707 \times 10^{-10}$
5	120	$.833333 \times 10^{-2}$	15	$130{,}767 \times 10^{7}$	$.764716 \times 10^{-12}$
6	720	$.138889 \times 10^{-2}$	16	$209{,}228 \times 10^{8}$	$.477948 \times 10^{-13}$
7	5,040	$.198413 \times 10^{-3}$	17	$355{,}687 \times 10^{9}$	$.281146 \times 10^{-14}$
8	40,320	$.248016 \times 10^{-4}$	18	$640{,}237 \times 10^{10}$	$.156192 \times 10^{-15}$
9	362,880	$.275573 \times 10^{-5}$	19	$121{,}645 \times 10^{12}$	$.822064 \times 10^{-17}$
10	3,628,800	$.275573 \times 10^{-6}$	20	$243{,}290 \times 10^{13}$	$.411032 \times 10^{-18}$

Greek Alphabet

A	α	Alpha	N ν	Nu
B	β	Beta	Ξ ξ	Xi
Γ	γ	Gamma	O ο	Omicron
Δ	δ	Delta	Π π	Pi
E	ϵ	Epsilon	P ρ	Rho
Z	ζ	Zeta	Σ σ	Sigma
H	η	Eta	T τ	Tau
Θ	θ	Theta	Υ υ	Upsilon
I	ι	Iota	Φ φ	Phi
K	κ	Kappa	X χ	Chi
Λ	λ	Lambda	Ψ ψ	Psi
M	μ	Mu	Ω ω	Omega

Length of arc (L), length of chord (C), height of segment (H) and area of segment (A) subtending an angle (θ) in a circle of radius (R)

θ	$\frac{L}{R}$	$\frac{C}{R}$	$\frac{H}{R}$	$\frac{A}{R^2}$	θ	$\frac{L}{R}$	$\frac{C}{R}$	$\frac{H}{R}$	$\frac{A}{R^2}$
1	0.017	0.017	0.0000	0.00000	46	0.803	0.781	0.0795	0.04176
2	0.035	0.035	0.0002	0.00000	47	0.820	0.797	0.0829	0.04448
3	0.052	0.052	0.0003	0.00001	48	0.838	0.813	0.0865	0.04731
4	0.070	0.070	0.0006	0.00003	49	0.855	0.829	0.0900	0.05025
5	0.087	0.087	0.0010	0.00006	50	0.873	0.845	0.0937	0.05331
6	0.105	0.105	0.0014	0.00010	51	0.890	0.861	0.0974	0.05649
7	0.122	0.122	0.0019	0.00015	52	0.908	0.877	0.1012	0.05978
8	0.140	0.140	0.0024	0.00023	53	0.925	0.892	0.1051	0.06319
9	0.157	0.157	0.0031	0.00032	54	0.942	0.908	0.1090	0.06673
10	0.175	0.174	0.0038	0.00044	55	0.960	0.923	0.1130	0.07039
11	0.192	0.192	0.0046	0.00059	56	0.977	0.939	0.1171	0.07417
12	0.209	0.209	0.0055	0.00076	57	0.995	0.954	0.1212	0.07808
13	0.227	0.226	0.0064	0.00097	58	1.012	0.970	0.1254	0.08212
14	0.244	0.244	0.0075	0.00121	59	1.030	0.985	0.1296	0.08629
15	0.262	0.261	0.0086	0.00149	60	1.047	1.000	0.1340	0.09059
16	0.279	0.278	0.0097	0.00181	61	1.065	1.015	0.1384	0.09502
17	0.297	0.296	0.0110	0.00217	62	1.082	1.030	0.1428	0.09958
18	0.314	0.313	0.0123	0.00257	63	1.100	1.045	0.1474	0.10428
19	0.332	0.330	0.0137	0.00302	64	1.117	1.060	0.1520	0.10911
20	0.349	0.347	0.0152	0.00352	65	1.134	1.075	0.1566	0.11408
21	0.367	0.364	0.0167	0.00408	66	1.152	1.089	0.1613	0.11919
22	0.384	0.382	0.0184	0.00468	67	1.169	1.104	0.1661	0.12443
23	0.401	0.399	0.0201	0.00535	68	1.187	1.118	0.1710	0.12982
24	0.419	0.416	0.0219	0.00607	69	1.204	1.133	0.1759	0.13535
25	0.436	0.433	0.0237	0.00686	70	1.222	1.147	0.1808	0.14102
26	0.454	0.450	0.0256	0.00771	71	1.239	1.161	0.1859	0.14683
27	0.471	0.467	0.0276	0.00862	72	1.257	1.176	0.1910	0.15279
28	0.489	0.484	0.0297	0.00961	73	1.274	1.190	0.1961	0.15889
29	0.506	0.501	0.0319	0.01067	74	1.292	1.204	0.2014	0.16514
30	0.524	0.518	0.0341	0.01180	75	1.309	1.218	0.2066	0.17154
31	0.541	0.534	0.0364	0.01301	76	1.326	1.231	0.2120	0.17808
32	0.559	0.551	0.0387	0.01429	77	1.344	1.245	0.2174	0.18477
33	0.576	0.568	0.0412	0.01566	78	1.361	1.259	0.2229	0.19160
34	0.593	0.585	0.0437	0.01711	79	1.379	1.272	0.2284	0.19859
35	0.611	0.601	0.0463	0.01864	80	1.396	1.286	0.2340	0.20573
36	0.628	0.618	0.0489	0.02027	81	1.414	1.299	0.2396	0.21301
37	0.646	0.635	0.0517	0.02198	82	1.431	1.312	0.2453	0.22045
38	0.663	0.651	0.0545	0.02378	83	1.449	1.325	0.2510	0.22804
39	0.681	0.668	0.0574	0.02568	84	1.466	1.338	0.2569	0.23578
40	0.698	0.684	0.0603	0.02767	85	1.484	1.351	0.2627	0.24367
41	0.716	0.700	0.0633	0.02976	86	1.501	1.364	0.2686	0.25171
42	0.733	0.717	0.0664	0.03195	87	1.518	1.377	0.2746	0.25990
43	0.750	0.733	0.0696	0.03425	88	1.536	1.389	0.2807	0.26825
44	0.768	0.749	0.0728	0.03664	89	1.553	1.402	0.2867	0.27677
45	0.785	0.765	0.0761	0.03915	90	1.571	1.414	0.2929	0.28540

θ	L/R	C/R	H/R	A/R²	θ	L/R	C/R	H/R	A/R²
91	1.588	1.427	0.2991	0.2942	136	2.374	1.854	0.6254	0.8395
92	1.606	1.439	0.3053	0.3032	137	2.391	1.861	0.6335	0.8545
93	1.623	1.451	0.3116	0.3123	138	2.409	1.867	0.6416	0.8697
94	1.641	1.463	0.3180	0.3215	139	2.426	1.873	0.6498	0.8850
95	1.658	1.475	0.3244	0.3309	140	2.443	1.879	0.6580	0.9003
96	1.676	1.486	0.3309	0.3405	141	2.461	1.885	0.6662	0.9158
97	1.693	1.498	0.3374	0.3502	142	2.478	1.891	0.6744	0.9313
98	1.710	1.509	0.3439	0.3601	143	2.496	1.897	0.6827	0.9470
99	1.728	1.521	0.3506	0.3701	144	2.513	1.902	0.6910	0.9627
100	1.745	1.532	0.3572	0.3803	145	2.531	1.907	0.6993	0.9786
101	1.763	1.543	0.3639	0.3906	146	2.548	1.913	0.7076	0.9945
102	1.780	1.554	0.3707	0.4010	147	2.566	1.918	0.7160	1.0105
103	1.798	1.565	0.3775	0.4117	148	2.583	1.923	0.7244	1.0266
104	1.815	1.576	0.3843	0.4224	149	2.601	1.927	0.7328	1.0427
105	1.833	1.587	0.3912	0.4333	150	2.618	1.932	0.7412	1.0590
106	1.850	1.597	0.3982	0.4444	151	2.635	1.936	0.7496	1.0753
107	1.868	1.608	0.4052	0.4556	152	2.653	1.941	0.7581	1.0917
108	1.885	1.618	0.4122	0.4669	153	2.670	1.945	0.7666	1.1082
109	1.902	1.628	0.4193	0.4784	154	2.688	1.949	0.7750	1.1247
110	1.920	1.638	0.4264	0.4901	155	2.705	1.953	0.7836	1.1413
111	1.937	1.648	0.4336	0.5019	156	2.723	1.956	0.7921	1.1580
112	1.955	1.658	0.4408	0.5138	157	2.740	1.960	0.8006	1.1747
113	1.972	1.668	0.4481	0.5259	158	2.758	1.963	0.8092	1.1915
114	1.990	1.677	0.4554	0.5381	159	2.775	1.967	0.8178	1.2083
115	2.007	1.687	0.4627	0.5504	160	2.793	1.970	0.8264	1.2252
116	2.025	1.696	0.4701	0.5629	161	2.810	1.973	0.8350	1.2422
117	2.042	1.705	0.4775	0.5755	162	2.827	1.975	0.8436	1.2592
118	2.059	1.714	0.4850	0.5883	163	2.845	1.978	0.8522	1.2763
119	2.077	1.723	0.4925	0.6012	164	2.862	1.981	0.8608	1.2933
120	2.094	1.732	0.5000	0.6142	165	2.880	1.983	0.8695	1.3105
121	2.112	1.741	0.5076	0.6273	166	2.897	1.985	0.8781	1.3277
122	2.129	1.749	0.5152	0.6406	167	2.915	1.987	0.8868	1.3449
123	2.147	1.758	0.5228	0.6540	168	2.932	1.989	0.8955	1.3621
124	2.164	1.766	0.5305	0.6676	169	2.950	1.991	0.9042	1.3794
125	2.182	1.774	0.5383	0.6812	170	2.967	1.992	0.9128	1.3967
126	2.199	1.782	0.5460	0.6950	171	2.985	1.994	0.9215	1.4140
127	2.217	1.790	0.5538	0.7090	172	3.002	1.995	0.9302	1.4314
128	2.234	1.798	0.5616	0.7230	173	3.019	1.996	0.9390	1.4488
129	2.251	1.805	0.5695	0.7372	174	3.037	1.997	0.9477	1.4662
130	2.269	1.813	0.5774	0.7514	175	3.054	1.998	0.9564	1.4836
131	2.286	1.820	0.5853	0.7658	176	3.072	1.999	0.9651	1.5010
132	2.304	1.827	0.5933	0.7803	177	3.089	1.999	0.9738	1.5185
133	2.321	1.834	0.6013	0.7950	178	3.107	2.000	0.9825	1.5359
134	2.339	1.841	0.6093	0.8097	179	3.124	2.000	0.9913	1.5533
135	2.356	1.848	0.6173	0.8245	180	3.142	2.000	1.0000	1.5708

Material	Lbs. per cu. ft.	Material	Lbs. per cu. ft.
Air *	0.0809	copper, pure	554
acetylene gas *	0.0733	" cast	549–558
alabaster	168	" wrought	552–558
alcohol	49–57	" wire	555–558
aluminum, pure	168	cork	15.6
" cast	160		
" wire	168	Erbium	297
amber	67	emery	250
ammonia *	0.0482		
antimony	414	Feldspar	158–162
argon *	0.113	flint	162
arsenic	357	fluorine *	0.0920
asbestos	125–175		
asphaltum	69–94	Germanium	341
		german silver	515–535
Barium	234	glass, common	150–175
basalt	180	" flint	180–280
bismuth	609	glucinum	122
boron	159	glycerine	78.6
brass	510–542	gold	1203
brick	100–150	granite	125–187
bromine	196	gravel	90–147
bronze	545–555	gum arabic	90
		gun metal	533
Cadmium	540	gutta percha	61
caesium	117	gypsum	144
calcium	98.6		
carbon	125–144	Hydrogen *	0.00562
" bisulphide	80.6		
" dioxide *	0.124	Ice	55–57
" monoxide *	0.0782	iodine	300
celluloid	90	iridium	1399
cement, loose	72–105	iron, pure	491
" set	168–187	" gray cast	439–445
cerium	437	" white cast	473–482
chalk	119–175	" wrought	487–492
charcoal	17–35	" steel	474–494
chlorine *	0.196	ivory	114
chromium	368		
clay, hard	129–133	Lead	710
" soft	118	leather, dry	54
coal, anthracite	81–106	" greased	64
" " loose	47–58	lime	53–75
" bituminous	78–88	limestone	156–162
" " loose	44–54	lithium	39
" lignite	52	loam	65–88
cobalt	530–563		
coke	62–105	Magnesium	107
" loose	23–32	" carbonate	150
columbium	452	manganese	462
concrete (1 : 2 : 4)	146	marble	157–177
" (1 : 1½ : 3)	139	masonry	100–165
" (1 : 3 : 6)	156	mercury *	849
		mica	165–200
		molybdenum	529

* At 0° Cent. and atmospheric pressure.

Material	Lbs. per cu. ft.	Material	Lbs. per cu. ft.
mortar, hard	103	steel	474–494
muck	40–74	strontium	158
mud	80–130	sulphur	120–130
Naptha	53	Talc	168
nickel	540–550	tantalum	1040
nitrogen *	0.0782	tar	62.4
nitrous oxide *	0.0838	tellurium	389
Oil, cotton-seed	60.2	thallium	739
" lard	57.4	thorium	686
" linseed	58.8	tile	113
" lubricating	56.2–57.7	" hollow	26–45
" petroleum	54.8	tin	455
" transformer	52.6–54.2	titanium	218
" turpentine	54.2	trap rock	187–190
" whale	57.3	tungsten	1174
osmium	1400	turf	20–30
oxygen *	0.0895	Uranium	1165
Palladium	711	Vanadium	343
paper	44–72	Water, max. dens.	62.4
paraffin	54–57	" sea	64.0–64.3
peat	20–30	wax, bees	60.5
phosphorus	146	wood, ash	45–47
pitch	67	" bamboo	22–25
plaster of Paris	144	" beech	43–56
platinum	1336	" birch	32–48
porcelain	143–156	" butternut	24–28
potassium	53.7	" cedar	37–38
pumice stone	23–56	" cherry	43–56
Quartz	165	" chestnut	38–40
Resin	67	" cypress	32–37
rhodium	773	" ebony	69–83
rubber, pure	58.0–60.5	" elm	35–36
" compound	106–124	" fir	34–35
" ebonite	74.9–78.0	" hemlock	25–29
rubidium	955	" hickory	53–58
ruthenium	767	" lig. vitæ	78–83
Salt	129–131	" mahogany	32–53
sand	90–120	" maple	49–50
sandstone	124–200	" oak	37–56
selenium	300	" pine	24–45
shale	162	" poplar	24–27
silicon	131	" red wood	30–32
silver	660	" spruce	25–32
slate	162–205	" walnut	38–45
snow, fresh fallen	5–12	" willow	24–37
" wet compact	15–50	Xenon *	0.284
soapstone	162–175	Zinc	448
sodium	60.5	zirconium	258
spermaceti	59		

* At 0° Cent. and atmospheric pressure.

Coefficients of discharge (c) for circular orifices, with full contractions *

Head from center of orifice in feet	Diameters in feet					
	0.02	0.05	0.1	0.2	0.6	1.0
0.5	....	0.627	0.615	0.600	0.592	......
0.8	0.648	0.620	0.610	0.601	0.594	0.591
1.0	0.644	0.617	0.608	0.600	0.595	0.591
1.5	0.637	0.613	0.605	0.600	0.596	0.593
2.0	0.632	0.610	0.604	0.599	0.597	0.595
2.5	0.629	0.608	0.603	0.599	0.598	0.596
3.0	0.627	0.606	0.603	0.599	0.598	0.597
3.5	0.625	0.606	0.602	0.599	0.598	0.596
4.0	0.623	0.605	0.602	0.599	0.597	0.596
6.0	0.618	0.604	0.600	0.598	0.597	0.596
8.0	0.614	0.603	0.600	0.598	0.596	0.596
10.0	0.611	0.601	0.598	0.597	0.596	0.595
20.0	0.601	0.598	0.596	0.596	0.596	0.594
50.0	0.596	0.595	0.594	0.594	0.594	0.593
100.0	0.593	0.592	0.592	0.592	0.592	0.592

Coefficients of discharge (c) for square orifices, with full contractions *

Head from center of orifice in feet	Length of side of square in feet					
	0.02	0.05	0.1	0.2	0.6	1.0
0.5	....	0.633	0.619	0.605	0.597	....
0.8	0.652	0.625	0.615	0.605	0.600	0.597
1.0	0.648	0.622	0.613	0.605	0.601	0.599
1.5	0.641	0.619	0.610	0.605	0.602	0.601
2.0	0.637	0.615	0.608	0.605	0.604	0.602
2.5	0.634	0.613	0.607	0.605	0.604	0.602
3.0	0.632	0.612	0.607	0.605	0.604	0.603
3.5	0.630	0.611	0.607	0.605	0.604	0.602
4.0	0.628	0.610	0.606	0.605	0.603	0.602
6.0	0.623	0.609	0.605	0.604	0.603	0.602
8.0	0.619	0.608	0.605	0.604	0.603	0.602
10.0	0.616	0.606	0.604	0.603	0.602	0.601
20.0	0.606	0.603	0.602	0.602	0.601	0.600
50.0	0.602	0.601	0.600	0.600	0.599	0.599
100.0	0.599	0.598	0.598	0.598	0.598	0.598

* From Hamilton Smith's Hydraulics.

Coefficients of discharge (c) for contracted weirs *
For use in the Hamilton Smith formula.

Effective head in feet	Length of weir in feet									
	0.66	1	2	3	4	5	7	10	15	19
0.1	0.632	0.639	0.646	0.652	0.653	0.653	0.654	0.655	0.655	0.656
0.2	0.611	0.618	0.626	0.630	0.631	0.631	0.632	0.633	0.634	0.634
0.25	0.605	0.612	0.621	0.624	0.625	0.626	0.627	0.628	0.628	0.629
0.3	0.601	0.608	0.616	0.619	0.621	0.621	0.623	0.624	0.624	0.625
0.4	0.595	0.601	0.609	0.613	0.614	0.615	0.617	0.618	0.619	0.620
0.5	0.590	0.596	0.605	0.608	0.610	0.611	0.613	0.615	0.616	0.617
0.6	0.587	0.593	0.601	0.605	0.607	0.608	0.611	0.613	0.614	0.615
0.8	......	......	0.595	0.600	0.602	0.604	0.607	0.611	0.612	0.613
1.0	......	......	0.590	0.595	0.598	0.601	0.604	0.608	0.610	0.611
1.2	......	......	0.585	0.591	0.594	0.597	0.601	0.605	0.608	0.610
1.4	......	......	0.580	0.587	0.590	0.594	0.598	0.602	0.606	0.609
1.6	......	......	......	0.582	0.587	0.591	0.595	0.600	0.604	0.607

Coefficients of discharge (c) for suppressed weirs *
For use in the Hamilton Smith formula.

Effective head in feet	Length of weir in feet								
	0.66	2	3	4	5	7	10	15	19
0.1	......	......	......	......	0.659	0.658	0.658	0.657	0.657
0.2	0.656	0.645	0.642	0.641	0.638	0.637	0.637	0.636	0.635
0.25	0.653	0.641	0.638	0.636	0.634	0.633	0.632	0.631	0.630
0.3	0.651	0.639	0.636	0.633	0.631	0.629	0.628	0.627	0.626
0.4	0.650	0.636	0.633	0.630	0.628	0.625	0.623	0.622	0.621
0.5	0.650	0.637	0.633	0.630	0.627	0.624	0.621	0.620	0.619
0.6	0.651	0.638	0.634	0.630	0.627	0.623	0.620	0.619	0.618
0.8	0.656	0.643	0.637	0.633	0.629	0.625	0.621	0.620	0.618
1.0	......	0.648	0.641	0.637	0.633	0.628	0.624	0.621	0.619
1.2	......	......	0.646	0.641	0.636	0.632	0.626	0.623	0.620
1.4	......	......	......	0.644	0.640	0.634	0.629	0.625	0.622
1.6	......	......	......	0.647	0.642	0.637	0.631	0.626	0.623

* From Hamilton Smith's Hydraulics.

Values of friction factor (f) for clean cast-iron pipes

Diameter in inches	Velocity in feet per second						
	0.5	1	2	3	6	10	20
1	0.0398	0.0353	0.0317	0.0299	0.0266	0.0244	0.0228
3	0.0354	0.0316	0.0288	0.0273	0.0248	0.0232	0.0218
6	0.0317	0.0289	0.0264	0.0252	0.0231	0.0219	0.0208
9	0.0290	0.0269	0.0247	0.0237	0.0220	0.0209	0.0200
12	0.0268	0.0251	0.0233	0.0224	0.0209	0.0201	0.0192
18	0.0238	0.0224	0.0211	0.0204	0.0193	0.0188	0.0181
24	0.0212	0.0194	0.0193	0.0187	0.0180	0.0176	0.0170
30	0.0194	0.0186	0.0179	0.0175	0.0170	0.0166	0.0161
36	0.0177	0.0172	0.0167	0.0164	0.0160	0.0156	0.0152
48	0.0153	0.0150	0.0147	0.0145	0.0143	0.0141	0.0138
60	0.0137	0.0135	0.0133	0.0132	0.0130	0.0128	0.0125
72	0.0125	0.0124	0.0122	0.0120	0.0118	0.0117	0.0117
96	0.0109	0.0107	0.0106	0.0106	0.0105	0.0104	0.0103

Values of friction factor (f) for old cast-iron pipes

Diameter in inches	Velocity in feet per second			
	1	3	6	10
3	0.0608	0.0556	0.0512	0.0488
6	0.0540	0.0468	0.0432	0.0412
9	0.0488	0.0420	0.0400	0.0368
12	0.0432	0.0384	0.0356	0.0336
15	0.0396	0.0348	0.0324	0.0312
18	0.0348	0.0312	0.0292	0.0276
24	0.0304	0.0268	0.0252	0.0240
30	0.0268	0.0244	0.0228	0.0220
36	0.0244	0.0224	0.0208	0.0200
42	0.0232	0.0208	0.0200	0.0192
48	0.0228	0.0204	0.0196	0.0184

Values of coefficient (c) in Chezy Formula

Radius in Feet	Velocity in feet per second						
	1	2	3	4	6	10	15
0.5	96	104	109	112	116	121	124
1.0	109	116	121	124	129	134	138
1.5	117	124	128	132	136	143	147
2.0	123	130	134	137	142	150	155
2.5	128	134	139	142	147	155	....
3.0	132	138	142	145	150	....	....
3.5	135	141	145	149	153	....	....
4.0	137	143	148	151	....	....	....

Values of coefficients (c) in Kutter's formula

Slope	n	Hydraulic radius r in feet										
		0.2	0.4	0.6	0.8	1.0	1.5	2.0	6.0	10.0	15.0	50.0
0.00005	0.010	87	109	123	133	140	154	164	199	213	220	245
	0.015	52	66	76	83	89	99	107	138	150	159	181
	0.020	35	45	53	59	64	72	80	105	116	125	148
	0.025	26	35	41	45	49	57	62	85	96	104	127
	0.030	22	28	33	37	40	47	51	72	83	90	112
	0.040	15	20	24	27	29	34	38	56	64	71	93
0.0001	0.010	98	118	131	140	147	158	167	196	206	212	227
	0.015	57	72	81	88	93	103	109	134	143	150	166
	0.020	38	50	57	63	67	75	81	102	111	118	134
	0.025	28	38	43	48	51	59	64	84	93	98	114
	0.030	23	30	35	39	42	48	52	72	78	85	100
	0.040	16	22	25	28	31	35	39	54	62	68	83
0.0002	0.010	105	125	137	145	150	162	169	193	202	206	220
	0.015	61	76	84	91	96	105	110	132	140	145	158
	0.020	42	53	60	65	68	76	82	100	108	113	126
	0.025	30	40	45	50	54	60	65	83	90	95	108
	0.030	25	32	37	40	43	49	53	69	77	82	94
	0.040	17	23	26	29	32	36	40	53	60	65	78
0.0004	0.010	110	128	140	148	153	164	171	192	198	203	215
	0.015	64	78	87	93	98	106	112	130	137	142	154
	0.020	43	55	61	67	70	77	83	99	106	110	123
	0.025	32	42	47	51	55	60	65	82	88	92	104
	0.030	26	33	38	41	44	50	54	68	75	80	91
	0.040	18	23	27	30	32	37	40	53	59	63	75
0.001	0.010	113	132	143	150	155	165	172	190	197	201	212
	0.015	66	80	88	94	98	107	112	130	135	141	151
	0.020	45	56	62	68	71	78	84	98	105	109	120
	0.025	33	43	48	52	55	61	65	81	87	91	101
	0.030	27	34	38	42	45	50	54	68	74	78	89
	0.040	18	24	27	30	33	37	40	53	58	61	72
0.01	0.010	114	133	143	151	156	165	172	190	196	200	210
	0.015	67	81	89	95	99	107	113	129	135	140	150
	0.020	46	57	63	68	72	78	84	98	105	108	119
	0.025	34	44	49	52	56	62	65	80	86	90	100
	0.030	27	35	39	43	45	51	55	67	73	77	87
	0.040	19	24	28	30	33	37	40	52	58	61	71

Values of coefficients (c) in Bazin's Formula *

Hydraulic radius in feet	Coefficient of roughness m					
	0.06	0.16	0.46	0.85	1.30	1.75
0.2	126	96	55	36	25	19
0.3	132	103	63	41	30	23
0.4	134	108	68	46	33	26
0.5	136	112	71	50	36	29
0.75	140	118	80	57	42	34
1.0	142	122	86	62	47	38
1.25	143	125	90	66	51	41
1.5	145	127	94	70	54	44
2.0	146	131	99	76	59	49
2.5	147	133	104	80	63	53
3.0	148	135	106	83	67	57
5.0	150	140	115	93	77	65
10.0	152	144	125	106	91	79
20.0	154	148	133	117	103	92

* From Russell's "Textbook on Hydraulics."

STEAM TABLES

Abridged from "Thermodynamic Properties of Steam" by Joseph H. Keenan and Frederick G. Keyes

Published by John Wiley & Sons, Inc., New York

Table 1. Saturation: Temperatures

Temp. Fahr. t	Abs. Press. Lb./Sq. In. p	Specific Volume Sat. Liquid v_f	Specific Volume Evap. v_{fg}	Specific Volume Sat. Vapor v_g	Enthalpy Sat. Liquid h_f	Enthalpy Evap. h_{fg}	Enthalpy Sat. Vapor h_g	Entropy Sat. Liquid s_f	Entropy Evap. s_{fg}	Entropy Sat. Vapor s_g	Temp. Fahr. t
32°	0.08854	0.01602	3306	3306	0.00	1075.8	1075.8	0.0000	2.1877	2.1877	32°
35	0.09995	0.01602	2947	2947	3.02	1074.1	1077.1	0.0061	2.1709	2.1770	35
40	0.12170	0.01602	2444	2444	8.05	1071.3	1079.3	0.0162	2.1435	2.1597	40
45	0.14752	0.01602	2036.4	2036.4	13.06	1068.4	1081.5	0.0262	2.1167	2.1429	45
50	0.17811	0.01603	1703.2	1703.2	18.07	1065.6	1083.7	0.0361	2.0903	2.1264	50
60°	0.2563	0.01604	1206.6	1206.7	28.06	1059.9	1088.0	0.0555	2.0393	2.0948	60°
70	0.3631	0.01606	867.8	867.9	38.04	1054.3	1092.3	0.0745	1.9902	2.0647	70
80	0.5069	0.01608	633.1	633.1	48.02	1048.6	1096.6	0.0932	1.9428	2.0360	80
90	0.6982	0.01610	468.0	468.0	57.99	1042.9	1100.9	0.1115	1.8972	2.0087	90
100	0.9492	0.01613	350.3	350.4	67.97	1037.2	1105.2	0.1295	1.8531	1.9826	100
110°	1.2748	0.01617	265.3	265.4	77.94	1031.6	1109.5	0.1471	1.8106	1.9577	110°
120	1.6924	0.01620	203.25	203.27	87.92	1025.8	1113.7	0.1645	1.7694	1.9339	120
130	2.2225	0.01625	157.32	157.34	97.90	1020.0	1117.9	0.1816	1.7296	1.9112	130
140	2.8886	0.01629	122.99	123.01	107.89	1014.1	1122.0	0.1984	1.6910	1.8894	140
150	3.718	0.01634	97.06	97.07	117.89	1008.2	1126.1	0.2149	1.6537	1.8685	150
160°	4.741	0.01639	77.27	77.29	127.89	1002.3	1130.2	0.2311	1.6174	1.8485	160°
170	5.992	0.01645	62.04	62.06	137.90	996.3	1134.2	0.2472	1.5822	1.8293	170
180	7.510	0.01651	50.21	50.23	147.92	990.2	1138.1	0.2630	1.5480	1.8109	180
190	9.339	0.01657	40.94	40.96	157.95	984.1	1142.0	0.2785	1.5147	1.7932	190
200	11.526	0.01663	33.62	33.64	167.99	977.9	1145.9	0.2938	1.4824	1.7762	200
210°	14.123	0.01670	27.80	27.82	178.05	971.6	1149.7	0.3090	1.4508	1.7598	210°
212	14.696	0.01672	26.78	26.80	180.07	970.3	1150.4	0.3120	1.4446	1.7566	212
220	17.186	0.01677	23.13	23.15	188.13	965.2	1153.4	0.3239	1.4201	1.7440	220
230	20.780	0.01684	19.365	19.382	198.23	958.8	1157.0	0.3387	1.3901	1.7288	230
240	24.969	0.01692	16.306	16.323	208.34	952.2	1160.5	0.3531	1.3609	1.7140	240

250°	29.825	0.01700	13.804	13.821	218.48	945.5	1164.0	0.3675	1.3323	1.6998	250°
260	35.429	0.01709	11.746	11.763	228.64	938.7	1167.3	0.3817	1.3043	1.6860	260
270	41.858	0.01717	10.044	10.061	238.84	931.8	1170.6	0.3958	1.2769	1.6727	270
280	49.203	0.01726	8.628	8.645	249.06	924.7	1173.8	0.4096	1.2501	1.6597	280
290	57.556	0.01735	7.444	7.461	259.31	917.5	1176.8	0.4234	1.2238	1.6472	290
300°	67.013	0.01745	6.449	6.466	269.59	910.1	1179.7	0.4369	1.1980	1.6350	300°
310	77.68	0.01755	5.609	5.626	279.92	902.6	1182.5	0.4504	1.1727	1.6231	310
320	89.66	0.01765	4.896	4.914	290.28	894.9	1185.2	0.4637	1.1478	1.6115	320
330	103.06	0.01776	4.289	4.307	300.68	887.0	1187.7	0.4769	1.1233	1.6002	330
340	118.01	0.01787	3.770	3.788	311.13	879.0	1190.1	0.4900	1.0992	1.5891	340
350°	134.63	0.01799	3.324	3.342	321.63	870.7	1192.3	0.5029	1.0754	1.5783	350°
360	153.04	0.01811	2.939	2.957	332.18	862.2	1194.4	0.5158	1.0519	1.5677	360
370	173.37	0.01823	2.606	2.625	342.79	853.5	1196.3	0.5286	1.0287	1.5573	370
380	195.77	0.01836	2.317	2.335	353.45	844.6	1198.1	0.5413	1.0059	1.5471	380
390	220.37	0.01850	2.0651	2.0836	364.17	835.4	1199.6	0.5539	0.9832	1.5371	390
400°	247.31	0.01864	1.8447	1.8633	374.97	826.0	1201.0	0.5664	0.9608	1.5272	400°
410	276.75	0.01878	1.6512	1.6700	385.83	816.3	1202.1	0.5788	0.9386	1.5174	410
420	308.83	0.01894	1.4811	1.5000	396.77	806.3	1203.1	0.5912	0.9166	1.5078	420
430	343.72	0.01910	1.3308	1.3499	407.79	796.0	1203.8	0.6035	0.8947	1.4982	430
440	381.59	0.01926	1.1979	1.2171	418.90	785.4	1204.3	0.6158	0.8730	1.4887	440
450°	422.6	0.0194	1.0799	1.0993	430.1	774.5	1204.6	0.6280	0.8513	1.4793	450°
460	466.9	0.0196	0.9748	0.9944	441.4	763.2	1204.6	0.6402	0.8298	1.4700	460
470	514.7	0.0198	0.8811	0.9009	452.8	751.5	1204.3	0.6523	0.8083	1.4606	470
480	566.1	0.0200	0.7972	0.8172	464.4	739.4	1203.7	0.6645	0.7868	1.4513	480
490	621.4	0.0202	0.7221	0.7423	476.0	726.8	1202.8	0.6766	0.7653	1.4419	490
500°	680.8	0.0204	0.6545	0.6749	487.8	713.9	1201.7	0.6887	0.7438	1.4325	500°
520	812.4	0.0209	0.5385	0.5594	511.9	686.4	1198.2	0.7130	0.7006	1.4136	520
540	962.5	0.0215	0.4434	0.4649	536.6	656.6	1193.2	0.7374	0.6568	1.3942	540
560	1133.1	0.0221	0.3647	0.3868	562.2	624.2	1186.4	0.7621	0.6121	1.3742	560
580	1325.8	0.0228	0.2989	0.3217	588.9	588.4	1177.3	0.7872	0.5659	1.3532	580
600°	1542.9	0.0236	0.2432	0.2668	617.0	548.5	1165.5	0.8131	0.5176	1.3307	600°
620	1786.6	0.0247	0.1955	0.2201	646.7	503.6	1150.3	0.8398	0.4664	1.3062	620
640	2059.7	0.0260	0.1538	0.1798	678.6	452.0	1130.5	0.8679	0.4110	1.2789	640
660	2365.4	0.0278	0.1165	0.1442	714.2	390.2	1104.4	0.8987	0.3485	1.2472	660
680	2708.1	0.0305	0.0810	0.1115	757.3	309.9	1067.2	0.9351	0.2719	1.2071	680
700°	3093.7	0.0369	0.0392	0.0761	823.3	172.1	995.4	0.9905	0.1484	1.1389	700°
705.4	3206.2	0.0503	0	0.0503	902.7	0	902.7	1.0580	0	1.0580	705.4

Table 2. Saturation: Pressures

Abs. Press. Lb./Sq. In. p	Temp. Fahr. t	Specific Volume Sat. Liquid v_f	Specific Volume Sat. Vapor v_g	Enthalpy Sat. Liquid h_f	Enthalpy Evap. h_{fg}	Enthalpy Sat. Vapor h_g	Entropy Sat. Liquid s_f	Entropy Evap. s_{fg}	Entropy Sat. Vapor s_g	Internal Energy Sat. Liquid u_f	Internal Energy Sat. Vapor u_g	Abs. Press Lb./Sq. In. p
1.0	101.74	0.01614	333.6	69.70	1036.3	1106.0	0.1326	1.8456	1.9782	69.70	1044.3	1.0
2.0	126.08	0.01623	173.73	93.99	1022.2	1116.2	0.1749	1.7451	1.9200	93.98	1051.9	2.0
3.0	141.48	0.01630	118.71	109.37	1013.2	1122.6	0.2008	1.6855	1.8863	109.36	1056.7	3.0
4.0	152.97	0.01636	90.63	120.86	1006.4	1127.3	0.2198	1.6427	1.8625	120.85	1060.2	4.0
5.0	162.24	0.01640	73.52	130.13	1001.0	1131.1	0.2347	1.6094	1.8441	130.12	1063.1	5.0
6.0	170.06	0.01645	61.98	137.96	996.2	1134.2	0.2472	1.5820	1.8292	137.94	1065.4	6.0
7.0	176.85	0.01649	53.64	144.76	992.1	1136.9	0.2581	1.5586	1.8167	144.74	1067.4	7.0
8.0	182.86	0.01653	47.34	150.79	988.5	1139.3	0.2674	1.5383	1.8057	150.77	1069.2	8.0
9.0	188.28	0.01656	42.40	156.22	985.2	1141.4	0.2759	1.5203	1.7962	156.19	1070.8	9.0
10	193.21	0.01659	38.42	161.17	982.1	1143.3	0.2835	1.5041	1.7876	161.14	1072.2	10
14.696	212.00	0.01672	26.80	180.07	970.3	1150.4	0.3120	1.4446	1.7566	180.02	1077.5	14.696
15	213.03	0.01672	26.29	181.11	969.7	1150.8	0.3135	1.4415	1.7549	181.06	1077.8	15
20	227.96	0.01683	20.089	196.16	960.1	1156.3	0.3356	1.3962	1.7319	196.10	1081.9	20
25	240.07	0.01692	16.303	208.42	952.1	1160.6	0.3533	1.3606	1.7139	208.34	1085.1	25
30	250.33	0.01701	13.746	218.82	945.3	1164.1	0.3680	1.3313	1.6993	218.73	1087.8	30
35	259.28	0.01708	11.898	227.91	939.2	1167.1	0.3807	1.3063	1.6870	227.80	1090.1	35
40	267.25	0.01715	10.498	236.03	933.7	1169.7	0.3919	1.2844	1.6763	235.90	1092.0	40
45	274.44	0.01721	9.401	243.36	928.6	1172.0	0.4019	1.2650	1.6669	243.22	1093.7	45
50	281.01	0.01727	8.515	250.09	924.0	1174.1	0.4110	1.2474	1.6585	249.93	1095.3	50
55	287.07	0.01732	7.787	256.30	919.6	1175.9	0.4193	1.2316	1.6509	256.12	1096.7	55
60	292.71	0.01738	7.175	262.09	915.5	1177.6	0.4270	1.2168	1.6438	261.90	1097.9	60
65	297.97	0.01743	6.655	267.50	911.6	1179.1	0.4342	1.2032	1.6374	267.29	1099.1	65
70	302.92	0.01748	6.206	272.61	907.9	1180.6	0.4409	1.1906	1.6315	272.38	1100.2	70
75	307.60	0.01753	5.816	277.43	904.5	1181.9	0.4472	1.1787	1.6259	277.19	1101.2	75
80	312.03	0.01757	5.472	282.02	901.1	1183.1	0.4531	1.1676	1.6207	281.76	1102.1	80
85	316.25	0.01761	5.168	286.39	897.8	1184.2	0.4587	1.1571	1.6158	286.11	1102.9	85
90	320.27	0.01766	4.896	290.56	894.7	1185.3	0.4641	1.1471	1.6112	290.27	1103.7	90
95	324.12	0.01770	4.652	294.56	891.7	1186.2	0.4692	1.1376	1.6068	294.25	1104.5	95
100	327.81	0.01774	4.432	298.40	888.8	1187.2	0.4740	1.1286	1.6026	298.08	1105.2	100
110	334.77	0.01782	4.049	305.66	883.2	1188.9	0.4832	1.1117	1.5948	305.30	1106.5	110

120	341.25	0.01789	3.728	312.44	877.9	1190.4	0.4916	1.0962	1.5878	312.05	1107.6	120
130	347.32	0.01796	3.455	318.81	872.9	1191.7	0.4995	1.0817	1.5812	318.38	1108.6	130
140	353.02	0.01802	3.220	324.82	868.2	1193.0	0.5069	1.0682	1.5751	324.35	1109.6	140
150	358.42	0.01809	3.015	330 51	863.6	1194.1	0.5138	1.0556	1.5694	330.01	1110.5	150
160	363.53	0.01815	2.834	335.93	859.2	1195.1	0.5204	1.0436	1.5640	335.39	1111.2	160
170	368.41	0.01822	2.675	341.09	854.9	1196.0	0.5266	1.0324	1.5590	340.52	1111.9	170
180	373.06	0.01827	2.532	346.03	850.8	1196.9	0.5325	1.0217	1.5542	345.42	1112.5	180
190	377.51	0.01833	2.404	350.79	846.8	1197.6	0.5381	1.0116	1.5497	350.15	1113.1	190
200	381.79	0.01839	2.288	355.36	843.0	1198.4	0.5435	1.0018	1.5453	354.68	1113.7	200
250	400.95	0.01865	1.8438	3760.0	825.1	1201.1	0.5675	0.9588	1.5263	375.14	1115.8	250
300	417.33	0.01890	1.5433	393.84	809.0	1202.8	0.5879	0.9225	1.5104	392.79	1117.1	300
350	431.72	0.01913	1.3260	409.69	794.2	1203.9	0.6056	0.8910	1.4966	408.45	1118.0	350
400	444.59	0.0193	1.1613	424.0	780.5	1204.5	0.6214	0.8630	1.4844	422.6	1118.5	400
450	456.28	0.0195	1.0320	437.2	767.4	1204.6	0 6356	0.8378	1.4734	435.5	1118.7	450
500	467.01	0.0197	0.9278	449.4	755.0	1204.4	0.6487	0.8147	1.4634	447.6	1118.6	500
550	476.94	0.0199	0.8424	460.8	743.1	1203.9	0.6608	0.7934	1.4542	458.8	1118.2	550
600	486.21	0.0201	0.7698	471.6	731.6	1203.2	0.6720	0.7734	1.4454	469.4	1117.7	600
650	494.90	0.0203	0.7083	481.8	720.5	1202.3	0.6826	0.7548	1.4374	479.4	1117.1	650
700	503.10	0.0205	0.6554	491.5	709.7	1201.2	0.6925	0.7371	1.4296	488.8	1116.3	700
750	510.86	0.0207	0.6092	500.8	699.2	1200.0	0.7019	0.7204	1.4223	498.0	1115.4	750
800	518.23	0.0209	0.5687	509 7	688.9	1198.6	0.7108	0.7045	1.4153	506.6	1114.4	800
850	525.26	0.0210	0.5327	518.3	678.8	1197.1	0.7194	0.6891	1.4085	515.0	1113.3	850
900	531.98	0.0212	0.5006	526.6	668.8	1195.4	0.7275	0.6744	1.4020	523.1	1112.1	900
950	538.43	0.0214	0.4717	534.6	659.1	1193.7	0.7355	0 6602	1.3957	530.9	1110.8	950
1000	544.61	0.0216	0.4456	542.4	649.4	1191.8	0.7430	0.6467	1.3897	538.4	1109.4	1000
1100	556.31	0.0220	0.4001	557.4	630.4	1187.8	0.7575	0.6205	1.3780	552.9	1106.4	1100
1200	567.22	0.0223	0.3619	571.7	611.7	1183.4	0.7711	0.5956	1.3667	566.7	1103.0	1200
1300	577.46	0.0227	0.3293	585.4	593.2	1178.6	0.7840	0.5719	1.3559	580.0	1099.4	1300
1400	587.10	0.0231	0.3012	598.7	574.7	1173.4	0.7963	0.5491	1.3454	592.7	1095.4	1400
1500	596.23	0.0235	0.2765	611.6	556.3	1167.9	0.8082	0.5269	1.3351	605.1	1091.2	1500
2000	635.82	0.0257	0.1878	671.7	463.4	1135.1	0.8619	0.4230	1.2849	662.2	1065.6	2000
2500	668.13	0.0287	0.1307	730.6	360.5	1091.1	0.9126	0.3197	1.2322	717.3	1030.6	2500
3000	695.36	0.0346	0.0858	802.5	217.8	1020.3	0.9731	0.1885	1.1615	783.4	972.7	3000
3206.2	705.40	0.0503	0.0503	902.7	0	920.7	1.0580	0	1.0580	872.9	872.9	3206.2

Table 3. Superheated Vapor

Abs. Press. Lb./Sq. In. (Sat. Temp.)		Temperature-Degrees Fahrenheit 200°	300°	400°	500°	600°	700°	800°	900°	1000°	1100°	1200°	1400°	1600°
1	v	392.6	452.3	512.0	571.6	631.2	690.8	750.4	809.9	869.5	929.1	988.7	1107.8	1227.0
(101.74)	h	1150.4	1195.8	1241.7	1288.3	1335.7	1383.8	1432.8	1482.7	1533.5	1585.2	1637.7	1745.7	1857.5
	s	2.0512	2.1153	2.1720	2.2233	2.2702	2.3137	2.3542	2.3923	2.4283	2.4625	2.4952	2.5566	2.6137
5	v	78.16	90.25	102.26	114.22	126.16	138.10	150.03	161.95	173.87	185.79	197.71	221.6	245.4
(162.24)	h	1148.8	1195.0	1241.2	1288.0	1335.4	1383.6	1432.7	1482.6	1533.4	1585.1	1637.7	1745.7	1857.4
	s	1.8718	1.9370	1.9942	2.0456	2.0927	2.1361	2.1767	2.2148	2.2509	2.2851	2.3178	2.3792	2.4363
10	v	38.85	45.00	51.04	57.05	63.03	69.01	74.98	80.95	86.92	92.88	98.84	110.77	122.69
(193.21)	h	1146.6	1193.9	1240.6	1287.5	1335.1	1383.4	1432.5	1482.4	1533.2	1585.0	1637.6	1745.6	1857.3
	s	1.7927	1.8595	1.9172	1.9689	2.0160	2.0596	2.1002	2.1383	2.1744	2.2086	2.2413	2.3028	2.3598
14.696	v	....	30.53	34.68	38.78	42.86	46.94	51.00	55.07	59.13	63.19	67.25	75.37	83.48
(212.00)	h	....	1192.8	1239.9	1287.1	1334.8	1383.2	1432.3	1482.3	1533.1	1584.8	1637.5	1745.5	1857.3
	s	....	1.8160	1.8743	1.9261	1.9734	2.0170	2.0576	2.0958	2.1319	2.1662	2.1989	2.2603	2.3174
20	v	....	22.36	25.43	28.46	31.47	34.47	37.46	40.45	43.44	46.42	49.41	55.37	61.34
(227.96)	h	....	1191.6	1239.2	1286.6	1334.4	1382.9	1432.1	1482.1	1533.0	1584.7	1637.4	1745.4	1857.2
	s	....	1.7808	1.8396	1.8918	1.9392	1.9829	2.0235	2.0618	2.0978	2.1321	2.1648	2.2263	2.2834
40	v	....	11.040	12.628	14.168	15.688	17.198	18.702	20.20	21.70	23.20	24.69	27.68	30.66
(267.25)	h	....	1186.8	1236.5	1284.8	1333.1	1381.9	1431.3	1481.4	1532.4	1584.3	1637.0	1745.1	1857.0
	s	....	1.6994	1.7608	1.8140	1.8619	1.9058	1.9467	1.9850	2.0212	2.0555	2.0883	2.1498	2.2069
60	v	....	7.259	8.357	9.403	10.427	11.441	12.449	13.452	14.454	15.453	16.451	18.446	20.44
(292.71)	h	....	1181.6	1233.6	1283.0	1331.8	1380.9	1430.5	1480.8	1531.9	1583.8	1636.6	1744.8	1856.7
	s	....	1.6492	1.7135	1.7678	1.8162	1.8605	1.9015	1.9400	1.9762	2.0106	2.0434	2.1049	2.1621
80	v	....	....	6.220	7.020	7.797	8.562	9.322	10.077	10.830	11.582	12.332	13.830	15.325
(312.03)	h	....	....	1230.7	1281.1	1330.5	1379.9	1429.7	1480.1	1531.3	1583.4	1636.2	1744.5	1856.5
	s	....	....	1.6791	1.7346	1.7836	1.8281	1.8694	1.9079	1.9442	1.9787	2.0115	2.0731	2.1303
100	v	....	....	4.937	5.589	6.218	6.835	7.446	8.052	8.656	9.259	9.860	11.060	12.258
(327.81)	h	....	....	1227.6	1279.1	1329.1	1378.9	1428.9	1479.5	1530.8	1582.9	1635.7	1744.2	1856.2
	s	....	....	1.6518	1.7085	1.7581	1.8029	1.8443	1.8829	1.9193	1.9538	1.9867	2.0484	2.1056

120	v	…	…	4.081	4.636	5.165	5.683	6.195	6.702	7.207	7.710	8.212	9.214	10.213
	h	…	…	1224.4	1277.2	1327.7	1377.8	1428.1	1478.8	1530.2	1582.4	1635.3	1743.9	1856.0
(341.25)	s	…	…	1.6287	1.6869	1.7370	1.7822	1.8237	1.8625	1.8990	1.9335	1.9664	2.0281	2.0854
140	v	…	…	3.468	3.954	4.413	4.861	5.301	5.738	6.172	6.604	7.035	7.895	8.752
	h	…	…	1221.1	1275.2	1326.4	1376.8	1427.3	1478.2	1529.7	1581.9	1634.9	1743.5	1855.7
(353.02)	s	…	…	1.6087	1.6683	1.7190	1.7645	1.8063	1.8451	1.8817	1.9163	1.9493	2.0110	2.0683
160	v	…	…	3.008	3.443	3.849	4.244	4.631	5.015	5.396	5.775	6.152	6.906	7.656
	h	…	…	1217.6	1273.1	1325.0	1375.7	1426.4	1477.5	1529.1	1581.4	1634.5	1743.2	1855.5
(363.53)	s	…	…	1.5908	1.6519	1.7033	1.7491	1.7911	1.8301	1.8667	1.9014	1.9344	1.9962	2.0535
180	v	…	…	2.649	3.044	3.411	3.764	4.110	4.452	4.792	5.129	5.466	6.136	6.804
	h	…	…	1214.0	1271.0	1323.5	1374.7	1425.6	1476.8	1528.6	1581.0	1634.1	1742.9	1855.2
(373.06)	s	…	…	1.5745	1.6373	1.6894	1.7355	1.7776	1.8167	1.8534	1.8882	1.9212	1.9831	2.0404
200	v	…	…	2.361	2.726	3.060	3.380	3.693	4.002	4.309	4.613	4.917	5.521	6.123
	h	…	…	1210.3	1268.9	1322.1	1373.6	1424.8	1476.2	1528.0	1580.5	1633.7	1742.6	1855.0
(381.79)	s	…	…	1.5594	1.6240	1.6767	1.7232	1.7655	1.8048	1.8415	1.8763	1.9094	1.9713	2.0287
220	v	…	…	2.125	2.465	2.772	3.066	3.352	3.634	3.913	4.191	4.467	5.017	5.565
	h	…	…	1206.5	1266.7	1320.7	1372.6	1424.0	1475.5	1527.5	1580.0	1633.3	1742.3	1854.7
(389.86)	s	…	…	1.5453	1.6117	1.6652	1.7120	1.7545	1.7939	1.8308	1.8656	1.8987	1.9607	2.0181
240	v	…	…	1.9276	2.247	2.533	2.804	3.068	3.327	3.584	3.839	4.093	4.597	5.100
	h	…	…	1202.5	1264.5	1319.2	1371.5	1423.2	1474.8	1526.9	1579.6	1632.9	1742.0	1854.5
(397.37)	s	…	…	1.5319	1.6003	1.6546	1.7017	1.7444	1.7839	1.8209	1.8558	1.8889	1.9510	2.0084
260	v	…	…	…	2.063	2.330	2.582	2.827	3.067	3.305	3.541	3.776	4.242	4.707
	h	…	…	…	1262.3	1317.7	1370.4	1422.3	1474.2	1526.3	1579.1	1632.5	1741.7	1854.2
(404.42)	s	…	…	…	1.5897	1.6447	1.6922	1.7352	1.7748	1.8118	1.8467	1.8799	1.9420	1.9995
280	v	…	…	…	1.9047	2.156	2.392	2.621	2.845	3.066	3.286	3.504	3.938	4.370
	h	…	…	…	1260.0	1316.2	1369.4	1421.5	1473.5	1525.8	1578.6	1632.1	1741.4	1854.0
(411.05)	s	…	…	…	1.5796	1.6354	1.6834	1.7265	1.7662	1.8033	1.8383	1.8716	1.9337	1.9912
300	v	…	…	…	1.7675	2.005	2.227	2.442	2.652	2.859	3.065	3.269	3.674	4.078
	h	…	…	…	1257.6	1314.7	1368.3	1420.6	1472.8	1525.2	1578.1	1631.7	1741.0	1853.7
(417.33)	s	…	…	…	1.5701	1.6268	1.6751	1.7184	1.7582	1.7954	1.8305	1.8638	1.9260	1.9835
350	v	…	…	…	1.4923	1.7036	1.8980	2.084	2.266	2.445	2.622	2.798	3.147	3.493
	h	…	…	…	1251.5	1310.9	1365.5	1418.5	1471.1	1523.8	1577.0	1630.7	1740.3	1853.1
(431.72)	s	…	…	…	1.5481	1.6070	1.6563	1.7002	1.7403	1.7777	1.8130	1.8463	1.9086	1.9663
400	v	…	…	…	1.2851	1.4770	1.6508	1.8161	1.9767	2.134	2.290	2.445	2.751	3.055
	h	…	…	…	1245.1	1306.9	1362.7	1416.4	1469.4	1522.4	1575.8	1629.6	1739.5	1852.5
(444.59)	s	…	…	…	1.5281	1.5984	1.6398	1.6842	1.7247	1.7623	1.7977	1.8311	1.8936	1.9513

Table 3. Superheated Vapor

Abs. Press. Lb./Sq. In. (Sat. Temp.)		Temperature-Degrees Fahrenheit 500°	550°	600°	620°	640°	660°	680°	700°	800°	900°	1000°	1200°	1400°	1600°
	v	1.1231	1.2155	1.3005	1.3332	1.3652	1.3967	1.4278	1.4584	1.6074	1.7516	1.8928	2.170	2.443	2.714
450	h	1238.4	1272.0	1302.8	1314.6	1326.2	1337.5	1348.8	1359.9	1414.3	1467.7	1521.0	1628.6	1738.7	1851.9
(456.28)	s	1.5095	1.5437	1.5735	1.5845	1.5951	1.6054	1.6153	1.6250	1.6699	1.7108	1.7486	1.8177	1.8803	1.9381
	v	0.9927	1.0800	1.1591	1.1893	1.2188	1.2478	1.2763	1.3044	1.4405	1.5715	1.6996	1.9504	2.197	2.442
500	h	1231.3	1266.8	1298.6	1310.7	1322.6	1334.2	1345.7	1357.0	1412.1	1466.0	1519.6	1627.6	1737.9	1851.3
(467.01)	s	1.4919	1.5280	1.5588	1.5701	1.5810	1.5915	1.6016	1.6115	1.6571	1.6982	1.7363	1.8056	1.8683	1.9262
	v	0.8852	0.9686	1.0431	1.0714	1.0989	1.1259	1.1523	1.1783	1.3038	1.4241	1.5414	1.7706	1.9957	2.219
550	h	1223.7	1261.2	1294.3	1306.8	1318.9	1330.8	1342.5	1354.0	1409.9	1464.3	1518.2	1626.6	1737.1	1850.6
(476.94)	s	1.4751	1.5131	1.5451	1.5568	1.5680	1.5787	1.5890	1.5991	1.6452	1.6868	1.7250	1.7946	1.8575	1.9155
	v	0.7947	0.8753	0.9463	0.9729	0.9988	1.0241	1.0489	1.0732	1.1899	1.3013	1.4096	1.6208	1.8279	2.033
600	h	1215.7	1255.5	1289.9	1302.7	1315.2	1327.4	1339.3	1351.1	1407.7	1462.5	1516.7	1625.5	1736.3	1850.0
(486.21)	s	1.4586	1.4990	1.5323	1.5443	1.5558	1.5667	1.5773	1.5875	1.6343	1.6762	1.7147	1.7846	1.8476	1.9056
	v	....	0.7277	0.7934	0.8177	0.8411	0.8639	0.8860	0.9077	1.0108	1.1082	1.2024	1.3853	1.5641	1.7405
700	h	....	1243.2	1280.6	1294.3	1307.5	1320.3	1332.8	1345.0	1403.2	1459.0	1513.9	1623.5	1734.8	1848.8
(503.10)	s	....	1.4722	1.5084	1.5212	1.5333	1.5449	1.5559	1.5665	1.6147	1.6573	1.6963	1.7666	1.8299	1.8881
	v	....	0.6154	0.6779	0.7006	0.7223	0.7433	0.7635	0.7833	0.8763	0.9633	1.0470	1.2088	1.3662	1.5214
800	h	....	1229.8	1270.7	1285.4	1299.4	1312.9	1325.9	1338.6	1398.6	1455.4	1511.0	1621.4	1733.2	1847.5
(518.23)	s	....	1.4467	1.4863	1.5000	1.5129	1.5250	1.5366	1.5476	1.5972	1.6407	1.6801	1.7510	1.8146	1.8729
	v	....	0.5264	0.5873	0.6089	0.6294	0.6491	0.6680	0.6863	0.7716	0.8506	0.9262	1.0714	1.2124	1.3509
900	h	....	1215.0	1260.1	1275.9	1290.9	1305.1	1318.8	1332.1	1393.9	1451.8	1508.1	1619.3	1731.6	1846.3
(531.98)	s	....	1.4216	1.4653	1.4800	1.4938	1.5066	1.5187	1.5303	1.5814	1.6257	1.6656	1.7371	1.8009	1.8595
	v	....	0.4533	0.5140	0.5350	0.5546	0.5733	0.5912	0.6084	0.6878	0.7604	0.8294	0.9615	1.0893	1.2146
1000	h	....	1198.3	1248.8	1265.9	1281.9	1297.0	1311.4	1325.3	1389.2	1448.2	1505.1	1617.3	1730.0	1845.0
(544.61)	s	....	1.3961	1.4450	1.4610	1.4757	1.4893	1.5021	1.5141	1.5670	1.6121	1.6525	1.7245	1.7886	1.8474
	v	....	....	0.4532	0.4738	0.4929	0.5110	0.5281	0.5445	0.6191	0.6866	0.7503	0.8716	0.9885	1.1031
1100	h	....	....	1236.7	1255.3	1272.4	1288.5	1303.7	1318.3	1384.3	1444.5	1502.2	1615.2	1728.4	1843.8
(556.31)	s	....	....	1.4251	1.4425	1.4583	1.4728	1.4862	1.4989	1.5535	1.5995	1.6405	1.7130	1.7775	1.8363
	v	....	....	0.4016	0.4222	0.4410	0.4586	0.4752	0.4909	0.5617	0.6250	0.6843	0.7967	0.9046	1.0101
1200	h	....	....	1223.5	1243.9	1262.4	1279.6	1295.7	1311.0	1379.3	1440.7	1499.2	1613.1	1726.9	1842.5
(567.22)	s	....	....	1.4052	1.4243	1.4413	1.4568	1.4710	1.4843	1.5409	1.5879	1.6293	1.7025	1.7672	1.8263

1400	v	....	....	0.3174	0.3390	0.3580	0.3753	0.3912	0.4062	0.4714	0.5281	0.5805	0.6789	0.7727	0.8640
	h	....	....	1193.0	1218.4	1240.4	1260.3	1278.5	1295.5	1369.1	1433.1	1493.2	1608.9	1723.7	1840.0
(587.10)	s	....	....	1.3639	1.3877	1.4079	1.4258	1.4419	1.4567	1.5177	1.5666	1.6093	1.6836	1.7489	1.8083
	v	....	....	....	0.2733	0.2936	0.3112	0.3271	0.3417	0.4034	0.4553	0.5027	0 5905	0.6738	0.7545
1600	h	....	....	....	1187.8	1215.2	1238.7	1259.6	1278.7	1358.4	1425.3	1487.0	1604.6	1720.5	1837.5
(604.90)	s	....	....	....	1.3489	1.3741	1.3952	1.4137	1.4303	1.4964	1.5476	1.5914	1.6669	1.7328	1.7926
		....	....	....	....	0.2407	0.2597	0.2760	0.2907	0.3502	0.3986	0.4421	0.5218	0.5968	0.6693
1800	h	....	....	....	....	1185.1	1214.0	1238.5	1260.3	1347.2	1417.4	1480.8	1600.4	1717.3	1835.0
(621.03)	v	....	....	....	....	1.3377	1.3638	1.3855	1.4044	1.4765	1.5301	1.5752	1.6520	1.7185	1.7786
	v	....	....	....	....	0.1936	0.2161	0.2337	0.2489	0.3074	0.3532	0.3935	0.4668	0.5352	0.6011
2000	h	....	....	....	....	1145.6	1184.9	1214.8	1240.0	1335.5	1409.2	1474.5	1596.1	1714.1	1832.5
(635.82)	s	....	....	....	....	1.2945	1.3300	1.3564	1.3783	1.4576	1.5139	1.5603	1.6384	1.7055	1.7660
	v	....	....	....	....	....	....	0.1484	0.1686	0.2294	0.2710	0.3061	0.3678	0.4244	0.4784
2500	h	....	....	....	....	....	....	1132.3	1176.8	1303.6	1387.8	1458.4	1585.3	1706.1	1826.2
(668.13)	s	....	....	....	....	....	....	1.2687	1.3073	1.4127	1.4772	1.5273	1.6088	1.6775	1.7389
	v	....	....	....	....	....	....	....	0.0984	0.1760	0.2159	0.2476	0.3018	0.3505	0.3966
3000	h	....	....	....	....	....	....	....	1060.7	1267.2	1365.0	1441.8	1574.3	1698.0	1819.9
(695.36)	s	....	....	....	....	....	....	....	1.1966	1.3690	1.4439	1.4984	1.5837	1.6540	1.7163
	v	....	....	....	....	....	....	....	....	0.1583	0.1981	0.2288	0.2806	0.3267	0.3703
3206.2	h	....	....	....	....	....	....	....	....	1250.5	1355.2	1434.7	1569.8	1694.6	1817.2
(705.40)	s	....	....	....	....	....	....	....	....	1.3508	1.4309	1.4874	1.5742	1.6452	1.7080
	v	....	....	....	....	....	....	....	0.0306	0.1364	0.1762	0.2058	0.2546	0.2977	0.3381
3500	h	....	....	....	....	....	....	....	780.5	1224.9	1340.7	1424.5	1563.3	1689.8	1813.6
	s	....	....	....	....	....	....	....	0.9515	1.3241	1.4127	1.4723	1.5615	1.6336	1.6968
	v	....	....	....	....	....	....	....	0.0287	0.1052	0.1462	0.1743	0.2192	0.2581	0.2943
4000	h	....	....	....	....	....	....	....	763.8	1174.8	1314.4	1406.8	1552.1	1681.7	1807.2
	s	....	....	....	....	....	....	....	0.9347	1.2757	1.3827	1.4482	1.5417	1.6154	1.6795
	v	....	....	....	....	....	....	....	0.0276	0.0798	0.1226	0.1500	0.1917	0.2273	0.2602
4500	h	....	....	....	....	....	....	....	753.5	1113.9	1286.5	1388.4	1540.8	1673.5	1800.9
	s	....	....	....	....	....	....	....	0.9235	1.2204	1.3529	1.4253	1.5235	1.5990	1.6640
	v	....	....	....	....	....	....	....	0.0268	0.0593	0.1036	0.1303	0.1696	0.2027	0.2329
5000	h	....	....	....	....	....	....	....	746.4	1047.1	1256.5	1369.5	1529.5	1665.3	1794.5
	s	....	....	....	....	....	....	....	0.9152	1.1622	1.3231	1.4034	1.5066	1.5839	1.6499
	v	....	....	....	....	....	....	....	0.0262	0.0463	0.0880	0.1143	0.1516	0.1825	0.2106
5500	h	....	....	....	....	....	....	....	741.3	985.0	1224.1	1349.3	1518.2	1657.0	1788.1
	s	....	....	....	....	....	....	....	0.9090	1.1093	1.2930	1.3821	1.4908	1.5699	1.6369

Specific Heat

Average values (0° to 100° C. unless otherwise stated) of **c** in the formula $Q = Mkc\ (t_2 - t_1)$, **c** being measured in gram-calories per gram per degree C. or British thermal units per pound per degree F. See page 156.

Acetylene * (15)	0.383	Ice (−20 to 0)	0.505
air * (−30 to +10)	0.238	iridium	0.0323
air † (−30 to +10)	0.169	iron, cast	0.119
alcohol, ethyl (30)	0.615	iron, wrought	0.115
aluminum	0.226		
ammonia (liq. 0)	1.098	Lead	0.0297
ammonia *	0.520	leather, dry	0.360
ammonia †	0.391		
antimony	0.0504	Marble	0.206
asbestos	0.195	mercury	0.0331
		mica	0.208
Beryllium	0.425		
bismuth	0.0297	Nickel	0.109
brass (60 Cu, 40 Zn)	0.0917	nitrogen *	0.244
bronze (80 Cu, 20 Sn)	0.0860	nitrogen †	0.173
		Oxygen *	0.224
Calcium	0.149	oxygen †	0.155
carbon, gas	0.315	osmium	0.0311
carbon, graphite	0.310		
carbon dioxide * (15 to 100)	0.202	Paraffin	0.589
carbon dioxide † (15 to 100)	0.168	petroleum	0.504
carbon monoxide *	0.243	platinum	0.0319
carbon monoxide †	0.173	porcelain (15 to 950)	0.260
cement, Portland	0.271		
chalk	0.220	Quartz (12 to 100)	0.188
chloroform (liq., 30)	0.235		
chloroform (gas, 100 to 200)	0.147	Rock salt (13 to 45)	0.219
chromium	0.111	rubber, hard	0.339
clay, dry (20 to 100)	0.220		
coal	0.201	Selenium (−188 to +18)	0.0680
cobalt	0.103	silicon	0.175
copper	0.0928	silver	0.0560
cork	0.485	steam (100 to 200)	0.480
cotton	0.362	steel	0.118
		sulphur (−188 to +18)	0.137
Gasoline	0.500		
german silver	0.0945	Tantalum (58)	0.0360
glass	0.180	tin	0.0556
glycerine (15 to 50)	0.576	tungsten	0.0340
gold	0.0312	turpentine (0)	0.411
granite (12 to 100)	0.192		
		Water (15)	1.000
		wood	0.420
Hydrogen *	3.41	wool	0.393
hydrogen †	2.42		
		Zinc	0.0950

* Constant pressure of one atmosphere. † Constant volume.

Average values (0° to 100° C. unless otherwise stated) of **a** in the formula, $l_t = l_0 (1 + at)$, **t** being measured in degrees C.

Substance	a × 10⁶	Substance	a × 10⁶
Aluminum (20 to 100)	23.8	Marble, Rutland blue (15 to 100)	15.0
antimony (15 to 101)	10.9	marble, Georgia gray (20 to 65)	1.00
Beryllium (20)	12.2	mercury (− 78 to − 38)	41.0
bismuth (19 to 101)	13.4	mica	7.60
brass	18.7	monel metal (25 to 100)	14.1
brick	9.50	Nickel (25 to 100)	12.9
bronze (80 Cu, 20 Sn) (0 to 800)	27.0	Osmium (40)	6.57
Cadmium	31.6	Paraffin (0 to 16)	107.
calcium (0 to 21)	25.0	paraffin (16 to 38)	130.
carbon, diamond (40)	1.18	phosphorous (6 to 44)	124.
" gas (40)	5.40	platinum (20)	8.93
" graphite (40)	7.86	porcelain, average	3.50
celluloid (20 to 70)	109.	Quartz, fused	0.500
cobalt (20)	12.3	Rubber, hard (20 to 60)	80.0
copper (25 to 100)	16.8	Selenium (40)	36.8
Duralumin, cast (20 to 100)	23.6	silicon (40)	7.63
duralumin, cold rolled (20 to 100)	23.7	silver (20)	18.8
German silver	18.4	slate (20)	8.00
glass, crown	8.97	solder	25.1
" flint (50 to 60)	7.88	sodium (−188 to +17)	62.2
gold (16 to 100)	14.3	steel, cast	13.6
granite	8.30	sulphur (40)	64.1
gutta percha	198.	Tin (18 to 100)	26.9
Ice (− 20 to − 1)	51.0	tungsten (0 to 500)	4.60
iridium (− 183 to + 19)	5.71	tungsten (1000 to 2000)	6.10
iron, pure	11.9	Wood, beech (2 to 34)	2.57
" cast (40)	10.6	wood, walnut (2 to 34)	6.58
" wrought (− 18 to + 100)	11.4	Zinc (10 to 100)	26.3
Lead (18 to 100)	29.4		

Melting and Boiling Points

(At atmospheric pressure)

Substance	Melts °C.	Boils °C.	Substance	Melts °C.	Boils °C.
Acetylene	−81.3	−72.2	Neodymium	840	.....
alcohol, ethyl	−115	78.3	neon	−248.7	−245.9
" methyl	−97.8	64.7	nickel	1455	2900
aluminum	659.7	1800	nitric oxide	−160.6	−153
ammonia	−75	−33.5	nitrogen	−209.9	−195.8
antimony	630.5	1380	Osmium	2700	>5300
argon	−189.2	−185.7	oxygen	−218.4	−183
Barium	850	1140	ozone	−251.4	−112
beryllium	1350	1500	Palladium	1553	2200
bismuth	271.3	1450	paraffin	52.4	.....
borax	561	.....	phosphorus	44.1	280
boron	2300	2550	platinum	1773.5	4300
brass	900±	.....	potassium	62.3	760
bromine	−7.2	58.8	praseodymium	940	.....
bronze	900±	.....	Radium	960	1140
Cadmium	320.9	767	radon	−110	.....
calcium	810	1170	rhenium	3000	.....
carbon	>3500	4200	rhodium	1985	>2500
" dioxide	−57	−80	rose's alloy	93.7	.....
" monoxide	−207	−191.5	rubber	100	.....
cerium	640	1400	rubidium	38.5	700
cesium	28.5	670	ruthenium	2450	>2700
chlorine	−101.6	−34.6	Samarium	>1300	.....
chromium	1615	2200	scandium	1200	2400
cobalt	1480	3000	selenium	220	688
columbium	1950	2900	silicon	1420	2600
copper	1083	2300	silver	960.5	1950
Fluorine	−223	−187	sodium	97.5	880
Gallium	29.75	>1600	" chloride	772	.....
german silver	1100±	.....	steel, Bessemer	1400	.....
germanium	958.5	2700	strontium	800	1150
glass, flint	1300	.....	sugar	160	.....
gold	1063	2600	sulphur	112.8	444.6
gutta percha	100	.....	Tantalum	2850	>4100
Hafnium	1700	>3200	tellurium	452	1390
helium	<−272.2	−268.9	thallium	303.5	1650
hydrogen	−259.1	−252.7	thorium	1845	>3000
Indium	155	1450	tin	231.9	2260
iodine	113.5	184.3	titanium	1800	>3000
iridium	2350	>4800	tungsten	3370	5900
iron, pure	1535	3000	turpentine	....	161
" gray pig	1200	.....	Uranium	<1850	.....
" white pig	1050	.....	Vanadium	1710	3000
Krypton	−169	−151.8	Wood's alloy	75.5	.....
Lanthanum	826	1800	Xenon	−140	−109
lead	327.4	1620	Ytterbium	1800	.....
lithium	186	>1220	yttrium	1490	2500
Magnesium	651	1110	Zinc	419.5	907
maganese	1260	1900	zirconium	1900	>2900
mercury	−38.87	356.9			
molybdenum	2620	3700			

Average values of **k** in the formula, $Q = \frac{ckS\theta t}{x}$. See page 181 for descriptions of units.

Substance	Temp. range	k × 10³
Air	0	0.0568
aluminum	18	480
antimony	0	44.2
argon	0	0.0389
asbestos, paper	....	0.6
Bismuth	0	17.7
blotting paper	....	0.15
brass	0	204
brick, alumina	0 to 700	2.0
" building	15 to 30	1.5
" carborundum	100 to 1000	23
" fire	0 to 1300	3.1
" graphite	100 to 1000	25
" magnesia	100 to 1000	7.1
" silica	100 to 1000	2.0
Cadmium	18	222
cambric, varn.	....	0.60
carbon, gas	100 to 942	130
" graphite	100 to 914	290
" dioxide	0	0.0307
" monox.	0	0.0499
carborundum	20 to 100	0.50
cardboard	....	0.50
cement, Port.	0 to 700	0.17
chalk	0 to 100	0.28
charcoal, powd'd	0 to 100	0.22
clinkers, small	0 to 700	1.1
coal	....	0.30
coke, powdered	0 to 100	0.44
concrete, cinder	....	0.81
" stone	....	2.2
copper	18	918
cotton wool	....	0.043
cotton batting, loose	....	0.11
cotton batting, packed	....	0.072
Earth, average	....	4.0
eiderdown, l'se	....	0.108
" packed	....	0.045
Feathers	20 to 155	0.16
felt	21 to 175	0.22
fiber, red	....	1.1
flannel	50	0.035
German silver	0 to 100	80
glass, crown	....	2.5
" flint	....	2.0
gold	18	700
granite	100	4.5
graphite	....	12
gutta percha	....	0.48
gypsum	....	3.1
Hair	20 to 155	0.15
" cloth, felt	....	0.042
helium	0	0.339
horn	....	0.087
hydrogen	0	0.327
Ice	....	3.9
iron, pure	18	161
" cast	18	109
" wrought	18	144
Lamp black	100	0.07
lead	18	83
leather, c'hide	....	0.42
" chamois	....	0.15
lime	....	0.29
linen	....	0.21
Magnesia	....	0.3
magnesium, carb.	100	0.23
marble	15 to 30	8.4
mercury	17	19.7
mica	....	0.86
Nickel	18	142
nitrogen	0	0.0524
Oxygen	0	0.0563
Paper	....	0.31
paraffin	....	0.62
pasteboard	....	0.45
plaster of Paris	20 to 155	0.42
plaster, mortar	....	1.3
platinum	18 to 100	170
plumbago	20 to 155	1.0
poplox (Na_2SiO_3)	200 to 500	0.13
porcelain	165 to 1055	4.3
petroleum	23	0.39
pumice stone	20 to 155	0.43
Quartz, pr. to ax.	....	30
" perp. to axis	....	160
Rubber, hard	....	0.43
" Para	....	0.38
Sand, dry	20 to 155	0.86
sandstone	....	5.5
sawdust	....	0.14
silica, fused	100	2.55
silk	50 to 100	0.13
silver	18	974
slate	94	4.8
snow	....	0.60
steel	18	115
Terra cotta	100 to 1000	2.3
tin	18	155
Water	0	1.4
"	30	1.6
wood, fir, with gr.	....	0.30
" fir, cross grain	....	0.09
wool, sheep's	20 to 100	0.14
" mineral	0 to 175	0.11
" steel	100	0.20
woolen, loose wadding	....	0.12
woolen, packed wadding	....	0.055
Zinc	18	265

Heat of Combustion

(British Thermal Units)

Substance*	Per Pound	Per gallon	Per cu. ft.†
Acetylene	21,500	....	1,480
alcohol, ethyl, denatured	11,600	78,900	....
" " pure (0.816)	12,400	84,300	....
" methyl (0.798)	9,540	63,700	....
Bagasse, dry	8,300	....	....
" 50% H_2O	3,000	....	....
benzine (0.879)	18,500	136,000	3,810
benzine (0.679)	17,900	102,000	....
Carbon, to CO	4,400	....	....
" to CO_2	14,500	....	....
carbon disulphide	5,820	62,700	....
carbon monoxide, to CO_2	4,370	....	323
charcoal, peat	11,600	....	....
" wood	13,500	....	....
coal, anthracite	11,500–14,000	....	....
" bituminous	11,000–15,300	....	....
" cannel	12,000–16,000	....	....
" lignite	5,500–11,000	....	....
" semi-bituminous	11,000–15,300	....	....
coke	12,000–14,400	....	....
Gas, blast furnace	....	....	90–110
" coal	....	....	630–680
" coke oven	....	....	430–600
" illuminating	....	....	550–600
" natural	....	....	700–2470
" oil	....	....	450–950
" producer	....	....	110–185
" water, blue	....	....	290–320
" " carburetted	....	....	400–680
gasoline (0.710)	21,200	126,000	....
" (0.770)	20,000	129,000	....
Hydrogen	62,000	....	326
Kerosene (0.783)	20,000	131,000	....
" (0.800)	20,160	136,000	....
Peat	3,500–10,000	....	....
petroleum (0.785)	20,000	131,000	....
" (1.000)	18,300	153,000	....
Straw	5,100–6,700	....	....
sulphur	4,020	....	....
Wood, air-dried	5,420–6,830	....	....

* Numbers indicate specific gravity. † At 60° F. and atmos. pressure.

Size	Area	Resistance	Weight	Current capacity in amperes		
American Wire Gage	Circular mils	Ohms per 1000 feet at 25° C.	Pounds per 1000 feet	Rubber Insulation	Varnished Cloth	Other Insulations
18	1,620	6.51	4.92	3		5
16	2,580	4.09	7.82	6		10
14	4,110	2.58	12.4	15	18	20
12	6,530	1.62	19.8	20	25	25
10	10,400	1.02	31.4	25	30	30
8	16,500	0.641	50.0	35	40	50
6	26,300	0.403	79.5	50	60	70
5	33,100	0.319	100	55	65	80
4	41,700	0.253	126	70	85	90
3	52,600	0.205	163	80	95	100
2	66,400	0.162	205	90	110	125
1	83,700	0.129	258	100	120	150
0	106,000	0.102	326	125	150	200
00	133,000	0.0811	411	150	180	225
000	168,000	0.0642	518	175	210	275
0000	212,000	0.0509	653	225	270	325
....	300,000	0.0360	926	275	330	400
....	400,000	0.0270	1240	325	390	500
....	500,000	0.0216	1540	400	480	600
....	600,000	0.0180	1850	450	540	680
....	700,000	0.0154	2160	500	600	760
....	800,000	0.0135	2470	550	660	840
....	900,000	0.0120	2780	600	720	920
....	1,000,000	0.0108	3090	650	780	1000
....	1,100,000	0.00981	3400	690	830	1080
....	1,200,000	0.00899	3710	730	880	1150
....	1,300,000	0.00830	4010	770	920	1220
....	1,400,000	0.00770	4320	810	970	1290
....	1,500,000	0.00719	4630	850	1020	1360
....	1,600,000	0.00674	4940	890	1070	1430
....	1,700,000	0.00634	5250	930	1120	1490
....	1,800,000	0.00599	5560	970	1160	1550
....	1,900,000	0.00568	5870	1010	1210	1610
....	2,000,000	0.00539	6180	1050	1260	1670

* For wires larger than No. 4 the values given are for stranded wires. The carrying capacity of insulated aluminum wires is 84 per cent of that given for copper.

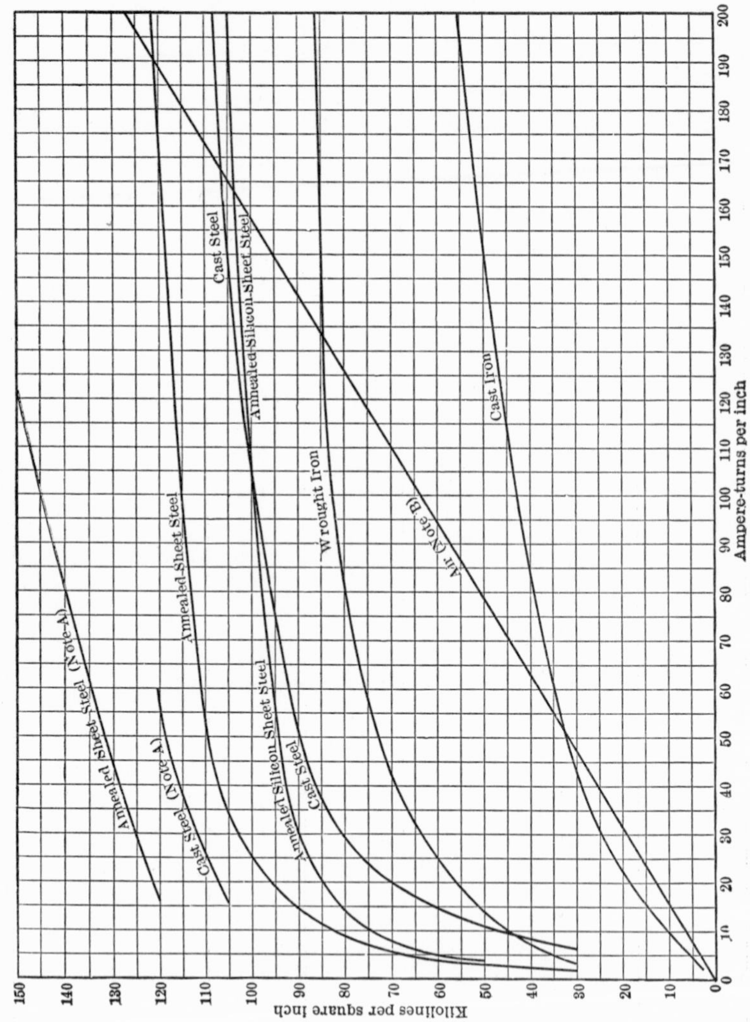

NOTE A. Multiply abscissa scale by 10.
NOTE B. Multiply abscissa scale by 200.

Resistivity (ρ) in microhms per cm. cube and Temperature Coefficient of Resistance (α) of Certain Conductors.

(Temperature is 20° C. unless otherwise specified.)

Material	ρ	α	Material	ρ	α
Aluminum.	2.688	0.00403	Mercury.....	95.8	0.00089
Antimony.	39.1 at 0	0.0036	Molybdenum	5.08 at 0	0.0047 (0–100)
Barium....	9.8	0.0033	Monel metal.	42	....
Beryllium.	10.1	....	Nickel.......	7.8	0.00537 (20–100)
Bismuth...	120	0.004	Osmium.....	9.5	0.0033
Carbon....	3500 at 0	−0.0009	Palladium...	11	....
Calcium...	4.59	0.00364 (0–600)	Platinum....	9.83 at 0	0.003
Cerium....	78	....	Potassium...	6.1 at 0	0.0055 at 0
Cesium....	19 at 0	....	Rhodium....	5.11 at 0	0.0043 at 0
Chromium.	2.6 at 0	....	Silver.......	1.629 at 18	0.0038
Cobalt....	9.7	0.00658 (0–100)	Sodium......	4.3 at 0	0.0054
Copper....	1.724	0.00393	Strontium...	24.8	....
Gold......	2.44	0.0034	Tantalum ...	15.5	0.0031
Graphite..	800 at 0	....	Tellurium...	2×10^5	....
Iron.......	9.8	0.0065 (0–100)	Thallium....	17.6 at 0	0.0040 at 0
" cast...	79–104	....	Thorium.....	18	0.0021 (20–1800)
Lead......	22.0	0.0039	Tin..........	11.5	0.0042
Lithium...	8.55 at 0	0.0047 at 0	Titanium....	3.0	....
Magnesium	4.46	0.0040	Tungsten....	5.5	0.0047 (0–100)
Manganese	5	....	Zinc.........	5.75 at 0	0.0037

Resistivity (ρ) in megohms per cm. cube and Dielectric Constant (k) of Certain Insulators at Room Temperature

Material	ρ	k	Material	ρ	k
Alcohol, ethyl	0.3	5.0–54.6	Oil, olive.....	5×10^6	3.11
" methyl	0.14	31.2–35.0	" paraffin ..	10^{10}	.....
Amber.......	5×10^{10}	....	" petroleum	2×10^{10}	2.13
Amylacetate..	....	4.81	Paper.......	10^4–10^9	1.7–3.8
Asbestos paper	1.6×10^5	2.7	Paraffin......	5×10^{10}–5×10^{12}	1.9–2.3
Asphalt.......	....	2.7	Porcelain.....	3×10^8	4.4
Bakelite......	10^5–10^{10}	4.5–5.5	Quartz.......	10^8–5×10^{12}	4.7–5.1
Beeswax......	6×10^8	....	Rosin........	7×10^9–5×10^{10}	2.5
Cellophane...	....	8	Rubber, hard.	3×10^{10}–10^{12}	2.0–3.5
Celluloid.....	2×10^4	13.3	Sealing wax...	10^9–8×10^9	.....
Cellulose acetate.....	....	5	Selenium.....	0.06	6.1–7.4
Glass.........	5×10^5–10^{10}	5.5–9.1	Shellac.......	10^{10}	3.0–3.7
Glycerine.....	....	56.2	Silica, fused..	10^8–10^{13}	3.5–3.6
Gutta percha.	3×10^4	2.9	Slate.........	10^2–10^4	6.6–7.4
Ice..........	720	86	Sulphur......	8×10^9–10^{11}	2.9–3.2
Ivory........	200	....	Turpentine...	....	2.23
Marble.......	10^3–10^5	8.3	Water, dist. ..	0.5	81
Mica.........	4×10^7–2×10^{11}	5–7	Wood, paraffined ..	3×10^4–4×10^7	4.1

The customary units of weight and mass are avoirdupois units unless designated otherwise. The symbol (δ) represents the density of a material expressed as a decimal fraction. *g* equals 980.7 cms. per sec. per sec.

Multiply	by	to obtain
Abamperes	10	amperes.
"	3×10^{10}	statamperes.
abamperes per square cm.	64.52	amperes per sq. inch.
abampere-turns	10	ampere-turns.
" "	12.57	gilberts.
abampere-turns per cm.	25.40	ampere-turns per inch.
abcoulombs	10	coulombs.
"	3×10^{10}	statcoulombs.
abcoulombs per square cm.	64.52	coulombs per sq. inch.
abfarads	10^{9}	farads.
"	10^{15}	microfarads.
"	9×10^{20}	statfarads.
abhenries	10^{-9}	henries.
"	10^{-6}	millihenries.
"	$1/9\times10^{-20}$	stathenries.
abmhos per cm. cube	$10^{5}/\delta$	mhos per meter-gram.
" " " "	1.662×10^{2}	mhos per mil foot.
" " " "	10^{3}	megmhos per cm. cube.
abohms	10^{-15}	megohms.
"	10^{-3}	microhms.
"	10^{-9}	ohms.
"	$1/9\times10^{-20}$	statohms.
abohms per cm. cube	10^{-3}	microhms per cm. cube.
" " " "	6.015×10^{-3}	ohms per mil foot.
" " " "	$10^{-5}\delta$	ohms per meter-gram.
abvolts	$1/3\times10^{-10}$	statvolts.
"	10^{-8}	volts.
acres	43,560	square feet.
"	6,272,640	square inches.
"	4047	square meters.
"	1.562×10^{-3}	square miles.
"	4840	square yards.
acre-feet	43,560	cubic-feet.
" "	3.259×10^{5}	gallons.
amperes	1/10	abamperes.
"	3×10^{9}	statamperes.
amperes per square cm.	6.452	amperes per sq. inch.
amperes per square inch	0.01550	abamperes per sq. cm.
" " " "	0.1550	amperes per sq. cm.
" " " "	4.650×10^{8}	statamperes per sq. cm.
ampere-turns	1/10	abampere-turns.
" "	1.257	gilberts.
ampere-turns per cm.	2.540	ampere-turns per inch.
ampere-turns per inch	0.03937	abampere-turns per cm.
" " " "	0.3937	ampere-turns per cm.
" " " "	0.4950	gilberts per cm.
ares	0.02471	acres.
"	100	square meters.
atmospheres	76	cms. of mercury.
"	29.92	inches of mercury.
"	33.90	feet of water.
"	10,332	kgs. per square meter.
"	14.70	pounds per sq. inch.
"	1.058	tons per sq. foot.

Multiply	by	to obtain
Bars	0.9869	atmospheres.
"	1	dynes per sq. cm.
"	1.020×10^{4}	kgs. per square meter.
"	2,089	pounds per square foot.
"	14.50	pounds per square inch.
board-feet	144 sq. in. × 1 in.	cubic inches.
British thermal units	778.2	foot-pounds.
" " "	3.930×10^{-4}	horse-power-hours.
" " "	1055	joules.
" " "	0.2520	kilogram-calories.
" " "	107.6	kilogram-meters.
" " "	2.930×10^{-4}	kilowatt hours.
B.t.u. per min.	12.97	foot-pounds per sec.
" " "	0.02358	horse-power.
" " "	0.01758	kilowatts.
" " "	17.58	watts.
B.t.u. per sq. ft. per min.	0.1221	watts per square inch.
bushels	1.244	cubic feet.
"	2150	cubic inches.
"	0.03524	cubic meters.
"	4	pecks.
"	64	pints (dry).
"	32	quarts (dry).
Centares.	1	square meters.
centigrams	0.01	grams.
centiliters	0.01	liters.
centimeters	3.281×10^{-2}	feet.
"	0.3937	inches.
"	0.01	meters.
"	6.214×10^{-6}	miles.
"	10	millimeters.
"	393.7	mils.
"	1.094×10^{-2}	yards.
centimeter-dynes	1.020×10^{-3}	centimeter-grams.
" "	1.020×10^{-8}	meter-kilograms.
" "	7.376×10^{-8}	pound-feet.
centimeter-grams	980.7	centimeter-dynes.
" "	10^{-5}	meter-kilograms.
" "	7.233×10^{-5}	pound-feet.
centimeters of mercury	0.01316	atmospheres.
" " "	0.4461	feet of water.
" " "	136.0	kgs. per square meter.
" " "	27.85	pounds per square foot.
" " "	0.1934	pounds per square inch.
centimeters per second	1.968	feet per minute.
" " "	0.03281	feet per second.
" " "	0.036	kilometers per hour.
" " "	0.6	meters per minute.
" " "	0.02237	miles per hour.
" " "	3.728×10^{-4}	miles per minute.
cms. per sec. per sec.	0.03281	feet per sec. per sec.
" " " " "	0.036	kms. per hour per sec.
" " " " "	0.02237	miles per hour per sec.
circular mils	5.067×10^{-6}	square centimeters.
" "	7.854×10^{-7}	square inches.

Multiply	by	to obtain
circular mils (*cont.*) . . .	0.7854	square mils.
cord-feet	4 ft.×4 ft.×1 ft.	cubic feet.
cords	8 ft.×4 ft.×4 ft.	cubic feet.
coulombs	1/10	abcoulombs.
"	3×10^{9}	statcoulombs.
coulombs per square inch .	0.01550	abcoulombs per sq. cm.
" " " " .	0.1550	coulombs per sq. cm.
" " " " .	4.650×10^{8}	statcouls. per sq. cm.
cubic centimeters	3.531×10^{-5}	cubic feet.
" "	6.102×10^{-2}	cubic inches.
" "	10^{-6}	cubic meters.
" "	1.308×10^{-6}	cubic yards.
" "	2.642×10^{-4}	gallons.
" "	10^{-3}	liters.
" "	2.113×10^{-3}	pints (liq.).
" "	1.057×10^{-3}	quarts (liq.).
cubic feet	2.832×10^{4}	cubic cms.
" "	1728	cubic inches.
" "	0.02832	cubic meters.
" "	0.03704	cubic yards.
" "	7.481	gallons.
" "	28.32	liters.
" "	59.84	pints (liq.).
" "	29.92	quarts (liq.).
cubic feet per minute . . .	472.0	cubic cms. per sec.
" " " " . . .	0.1247	gallons per sec.
" " " " . . .	0.4720	liters per second.
" " " " . . .	62.4	pounds of water per min.
cubic inches	16.39	cubic centimeters.
" "	5.787×10^{-4}	cubic feet.
" "	1.639×10^{-5}	cubic meters.
" "	2.143×10^{-5}	cubic yards.
" "	4.329×10^{-3}	gallons.
" "	1.639×10^{-2}	liters.
" "	1.061×10^{5}	mil-feet.
" "	0.03463	pints (liq.).
" "	0.01732	quarts (liq.).
cubic meters	10^{6}	cubic centimeters.
" "	35.31	cubic feet.
" "	61,023	cubic inches.
" "	1.308	cubic yards.
" "	264.2	gallons.
" "	10^{3}	liters.
" "	2113	pints (liq.).
" "	1057	quarts (liq.).
cubic yards	7.646×10^{5}	cubic centimeters.
" "	27	cubic feet.
" "	46,656	cubic inches.
" "	0.7646	cubic meters.
" "	202.0	gallons.
" "	764.6	liters.
" "	1616	pints (liq.).
" "	807.9	quarts (liq.).
cubic yards per minute . .	0.45	cubic feet per second.
" " " " . .	3.367	gallons per second.
" " " " . .	12.74	liters per second.

Multiply	by	to obtain
Days	24	hours.
"	1440	minutes.
"	86,400	seconds.
decigrams	0.1	grams.
deciliters	0.1	liters.
decimeters	0.1	meters.
degrees (angle)	60	minutes.
" "	0.01745	radians.
" "	3600	seconds.
degrees per second	0.01745	radians per second.
" " "	0.1667	revolutions per minute.
" " "	0.002778	revolutions per second.
dekagrams	10	grams.
dekaliters	10	liters.
dekameters	10	meters.
drams	1.772	grams.
"	0.0625	ounces.
dynes	1.020×10^{-3}	grams.
"	7.233×10^{-5}	poundals.
"	2.248×10^{-6}	pounds.
dynes per square cm.	1	bars.
Ergs	9.480×10^{-11}	British thermal units.
"	1	dyne-centimeters.
"	7.378×10^{-8}	foot-pounds.
"	1.020×10^{-3}	gram-centimeters.
"	10^{-7}	joules.
"	2.389×10^{-11}	kilogram-calories.
"	1.020×10^{-8}	kilogram-meters.
ergs per second	5.688×10^{-9}	B.t. units per minute.
" " "	4.427×10^{-6}	foot-pounds per minute.
" " "	7.378×10^{-8}	foot-pounds per second.
" " "	1.341×10^{-10}	horse-power.
" " "	1.433×10^{-9}	kg.-calories per minute.
" " "	10^{-10}	kilowatts.
Farads	10^{-9}	abfarads.
"	10^{6}	microfarads.
"	9×10^{11}	statfarads.
fathoms	6	feet.
feet	30.48	centimeters.
"	12	inches.
"	0.3048	meters.
"	1.894×10^{-4}	miles.
"	1/3	yards.
feet of water	0.02950	atmospheres.
" " "	0.8826	inches of mercury.
" " "	304.8	kgs. per square meter.
" " "	62.43	pounds per square foot.
" " "	0.4335	pounds per square inch.
feet per minute	0.5080	centimeters per second.
" " "	0.01667	feet per second.
" " "	0.01829	kilometers per hour.
" " "	0.3048	meters per minute.
" " "	0.01136	miles per hour.
feet per second	30.48	centimeters per second.

Multiply	by	to obtain
feet per second (*cont.*)	1.097	kilometers per hour.
" " "	0.5921	knots.
" " "	18.29	meters per minute.
" " "	0.6818	miles per hour.
" " "	0.01136	miles per minute.
feet per 100 feet	1	per cent grade.
feet per second per second	30.48	cms. per sec. per sec.
" " " " "	1.097	kms. per hour per sec.
" " " " "	0.3048	meters per sec. per sec.
" " " " "	0.6818	miles per hour per sec.
foot-pounds	1.285×10^{-3}	British thermal units.
" "	1.356×10^{7}	ergs.
" "	5.050×10^{-7}	horse-power-hours.
" "	1.356	joules.
" "	3.238×10^{-4}	kilogram-calories.
" "	0.1383	kilogram-meters.
" "	3.766×10^{-7}	kilowatt-hours.
foot-pounds per minute	1.285×10^{-3}	B.t. units per minute.
" " " "	0.01667	foot-pounds per second.
" " " "	3.030×10^{-5}	horse-power.
" " " "	3.238×10^{-4}	kg.-calories per min.
" " " "	2.260×10^{-5}	kilowatts.
foot-pounds per second	7.712×10^{-2}	B.t. units per minute.
" " " "	1.818×10^{-3}	horse-power.
" " " "	1.943×10^{-2}	kg.-calories per min.
" " " "	1.356×10^{-3}	kilowatts.
furlongs	40	rods.
Gallons	3785	cubic centimeters.
"	0.1337	cubic feet.
"	231	cubic inches.
"	3.785×10^{-3}	cubic meters.
"	4.951×10^{-3}	cubic yards.
"	3.785	liters.
"	8	pints (liq.).
"	4	quarts (liq.).
gallons per minute	2.228×10^{-3}	cubic feet per second.
" " "	0.06308	liters per second.
gausses	6.452	lines per square inch.
gilberts	0.07958	abampere-turns.
"	0.7958	ampere-turns.
gilberts per centimeter	2.021	ampere-turns per inch.
gills	0.1183	liters.
"	0.25	pints (liq.).
grains	1	grains (av.).
"	0.06480	grams.
"	0.04167	pennyweights (troy).
grams	980.7	dynes.
"	15.43	grains.
"	10^{-3}	kilograms.
"	10^{3}	milligrams.
"	0.03527	ounces.
"	0.03215	ounces (troy).
"	0.07093	poundals.
"	2.205×10^{-3}	pounds.
gram-calories (IT)	3.968×10^{-3}	British thermal units.

Multiply	by	to obtain
gram-centimeters	9.297×10^{-8}	British thermal units.
" "	980.7	ergs.
" "	7.235×10^{-5}	foot-pounds.
" "	9.807×10^{-5}	joules.
" "	2.343×10^{-8}	kilogram-calories.
" "	10^{-5}	kilogram-meters.
grams per cm.	5.600×10^{-3}	pounds per inch.
grams per cu. cm.	62.43	pounds per cubic foot.
" " " "	0.03613	pounds per cubic inch.
" " " "	3.405×10^{-7}	pounds per mil-foot.
Hectares	2.471	acres.
"	1.076×10^{5}	square feet.
hectograms	100	grams.
hectoliters	100	liters.
hectometers	100	meters.
hectowatts	100	watts.
hemispheres (solid angle)	0.5	sphere.
" " "	4	spherical right angles.
" " "	6.283	steradians.
henries	10^{9}	abhenries.
"	10^{3}	millihenries.
"	$1/9 \times 10^{-11}$	stathenries.
horse-power	42.40	B.t. units per min.
" "	33,000	foot-pounds per minute.
" "	550	foot-pounds per second.
" "	1.014	horse-power (metric).
" "	10.68	kg.-calories per minute.
" "	0.7457	kilowatts.
" "	745.7	watts.
horse-power (boiler)	33.520	B.t.u. per hour.
" " "	9.804	kilowatts.
horse-power-hours	2544	British thermal units.
" " "	1.98×10^{6}	foot-pounds.
" " "	2.684×10^{6}	joules.
" " "	641.1	kilogram-calories.
" " "	2.737×10^{5}	kilogram-meters.
" " "	0.7455	kilowatt-hours.
hours	4.167×10^{-2}	days.
"	60	minutes.
"	3600	seconds.
"	5.952×10^{-3}	weeks.
Inches	2.540	centimeters.
"	8.333×10^{-2}	feet.
"	1.578×10^{-5}	miles.
"	10^{3}	mils.
"	2.778×10^{-2}	yards.
inches of mercury	0.03342	atmospheres.
" " "	1.133	feet of water.
" " "	345.3	kgs. per square meter.
" " "	70.73	pounds per square foot.
" " "	0.4912	pounds per square inch.
inches of water	0.002458	atmospheres.
" " "	0.07355	inches of mercury.
" " "	25.40	kgs. per square meter.

Multiply	by	to obtain
inches of water (*cont.*)	0.5781	ounces per square inch.
" " "	5.204	pounds per square foot.
" " "	0.03613	pounds per square inch.
Joules (Int.)	9.480×10^{-4}	British thermal units.
" "	10^{7}	ergs.
" "	0.7378	foot-pounds.
" "	2.389×10^{-4}	kilogram-calories.
" "	0.1020	kilogram-meters.
" "	2.778×10^{-4}	watt-hours.
Kilograms	980,665	dynes.
"	10^{3}	grams.
"	70.93	poundals.
"	2.205	pounds.
"	1.102×10^{-3}	tons (short).
kilogram-calories	3.968	British thermal units.
" "	3088	foot-pounds.
" "	1.560×10^{-3}	horse-power-hours.
" "	4186	joules.
" "	427.0	kilogram-meters.
" "	1.163×10^{-3}	kilowatt-hours.
kilogram-calories per min.	51.47	foot-pounds per second.
" " " "	0.09358	horse-power.
" " " "	0.06977	kilowatts.
kgs.-cms. squared	2.373×10^{-3}	pounds-feet squared.
" " "	0.3417	pounds-inches squared.
kilogram-meters	9.294×10^{-3}	British thermal units.
" "	9.804×10^{7}	ergs.
" "	7.233	foot-pounds.
" "	9.804	joules.
" "	2.342×10^{-3}	kilogram-calories.
" "	2.723×10^{-6}	kilowatt-hours.
kilograms per cubic meter	10^{-3}	grams per cubic cm.
" " " "	0.06243	pounds per cubic foot.
" " " "	3.613×10^{-5}	pounds per cubic inch.
" " " "	3.405×10^{-10}	pounds per mil foot.
kgs. per meter	0.6720	pounds per foot.
kgs. per square meter	9.678×10^{-5}	atmospheres.
" " " "	98.07×10^{-6}	bars.
" " " "	3.281×10^{-3}	feet of water.
" " " "	2.896×10^{-3}	inches of mercury.
" " " "	0.2048	pounds per square foot.
" " " "	1.422×10^{-3}	pounds per square inch.
kgs. per square millimeter	10^{6}	kgs. per square meter.
kilolines	10^{3}	maxwells.
kiloliters	10^{3}	liters.
kilometers	10^{5}	centimeters.
"	3281	feet.
"	3.937×10^{4}	inches.
"	10^{3}	meters.
"	0.6214	miles.
"	1094	yards.
kilometers per hour	27.78	centimeters per second.
" " "	54.68	feet per minute.
" " "	0.9113	feet per second.
" " "	0.5396	knots.

Multiply	by	to obtain
kilometers per hour (*cont.*)	16.67	meters per minute.
" " "	0.6214	miles per hour.
kms. per hour per sec.	27.78	cms. per sec. per sec.
" " " " "	0.9113	ft. per sec. per sec.
" " " " "	0.2778	meters per sec. per sec.
" " " " "	0.6214	miles per hr. per sec.
kilometers per min.	60	kilometers per hour.
kilowatts	56.88	B.t. units per min.
"	4.427×10^4	foot-pounds per min.
"	737.8	foot-pounds per sec.
"	1.341	horse-power.
"	14.33	kg.-calories per min.
"	10^3	watts.
kilowatt-hours	3413	British thermal units.
" "	2.656×10^6	foot-pounds.
" "	1.341	horse-power-hours.
" "	3.6×10^6	joules.
" "	860	kilogram-calories.
" "	3.672×10^5	kilogram-meters.
knots (length)	6080	feet.
" "	1.853	kilometers.
" "	1.152	miles.
" "	2027	yards.
knots (speed)	51.48	centimeters per second.
" "	1.689	feet per second.
" "	1.853	kilometers per hour.
" "	1.152	miles per hour.
Lines per square cm.	1	gausses.
lines per square inch	0.1550	gausses.
links (engineer's)	12	inches.
links (surveyor's)	7.92	inches.
liters	10^3	cubic centimeters.
"	0.03531	cubic feet.
"	61.02	cubic inches.
"	10^{-3}	cubic meters.
"	1.308×10^{-3}	cubic yards.
"	0.2642	gallons.
"	2.113	pints (liq.).
"	1.057	quarts (liq.).
liters per minute	5.885×10^{-4}	cubic feet per second.
" " "	4.403×10^{-3}	gallons per second.
$\log_{10} N$	2.303	$\log_\epsilon N$ or $ln\ N$.
$\log_\epsilon N$ or $ln\ N$	0.4343	$\log_{10} N$.
lumens per sq. ft.	1	foot-candles.
Maxwells	10^{-3}	kilolines.
megalines	10^6	maxwells.
megmhos per cm. cube	10^{-3}	abmhos per cm. cube.
" " " "	2.540	megmhos per inch cube.
" " " "	$10^2/\delta$	mhos per meter-gram.
" " " "	0.1662	mhos per mil foot.
megmhos per inch cube	0.3937	megmhos per cm. cube.
megohms	10^6	ohms.
meters	100	centimeters.
"	3.281	feet.
"	39.37	inches.

Multiply	by	to obtain
meters (*cont.*)	10^{-3}	kilometers.
"	6.214×10^{-4}	miles.
"	10^{3}	millimeters.
"	1.094	yards.
meter-kilograms	9.807×10^{7}	centimeter-dynes.
" "	10^{5}	centimeter-grams.
" "	7.233	pound-feet.
meters per minute	1.667	centimeters per second.
" " "	3.281	feet per minute.
" " "	0.05468	feet per second.
" " "	0.06	kilometers per hour.
" " "	0.03728	miles per hour.
meters per second	196.8	feet per minute.
" " "	3.281	feet per second.
" " "	3.6	kilometers per hour.
" " "	0.06	kilometers per minute.
" " "	2.237	miles per hour.
" " "	0.03728	miles per minute.
meters per sec. per sec.	3.281	feet per sec. per sec.
" " " " "	3.6	kms. per hour per sec.
" " " " "	2.237	miles per hour per sec.
mhos per meter-gram	$10^{-5}\delta$	abmhos per cm. cube.
" " " "	$10^{-2}\delta$	megmhos per cm. cube.
" " " "	$2.540 \times 10^{-2}\delta$	megmhos per inch cube.
" " " "	$1.662 \times 10^{-3}\delta$	mhos per mil foot.
mhos per mil foot	6.015×10^{-3}	abmhos per cm. cube.
" " " "	6.015	megmhos per cm. cube.
" " " "	15.28	megmhos per in. cube.
" " " "	$601.5/\delta$	mhos per meter-gram.
microfarads	10^{-15}	abfarads.
"	10^{-6}	farads.
"	9×10^{5}	statfarads.
micrograms	10^{-6}	grams.
microliters	10^{-6}	liters.
microhms	10^{3}	abohms.
"	10^{-12}	megohms.
"	10^{-6}	ohms.
"	$1/9 \times 10^{-17}$	statohms.
microhms per cm. cube	10^{3}	abohms per cm. cube.
" " " "	0.3937	microhms per inch cube.
" " " "	$10^{-2}\delta$	ohms per meter-gram.
" " " "	6.015	ohms per mil foot.
microhms per inch cube	2.540	microhms per cm. cube.
microns	10^{-6}	meters.
miles	1.609×10^{5}	centimeters.
"	5280	feet.
"	6.336×10^{4}	inches.
"	1.609	kilometers.
"	1760	yards.
miles per hour	44.70	centimeters per sec.
" " "	88	feet per minute.
" " "	1.467	feet per second.
" " "	1.609	kilometers per hour.
" " "	0.8684	knots.
" " "	26.82	meters per minute.
miles per hour per second	44.70	cms. per sec. per sec.
" " " " "	1.467	feet per sec. per sec.

Multiply	by	to obtain
miles per hr. per sec. (*cont.*)	1.609	kms. per hour per sec.
" " " " "	0.4470	meters per sec. per sec.
miles per minute	2682	centimeters per second.
" " "	88	feet per second.
" " "	1.609	kilometers per min.
" " "	52.10	knots.
" " "	60	miles per hour.
mil-feet	9.425×10^{-6}	cubic inches.
milliers	10^{3}	kilograms.
milligrams	10^{-3}	grams.
millihenries	10^{6}	abhenries.
"	10^{-3}	henries.
"	$1/9 \times 10^{-14}$	stathenries.
milliliters	10^{-3}	liters.
millimeters	0.1	centimeters
"	3.281×10^{-3}	feet.
"	0.03937	inches.
"	6.214×10^{-7}	miles.
"	39.37	mils.
"	1.094×10^{-3}	yards.
mils	2.540×10^{-3}	centimeters.
"	8.333×10^{-5}	feet.
"	10^{-3}	inches.
"	2.540×10^{-8}	kilometers.
"	2.778×10^{-5}	yards.
miner's inches	1.5	cubic feet per min.
minutes	6.944×10^{-4}	days.
"	1.667×10^{-2}	hours.
"	9.921×10^{-5}	weeks.
minutes (angle)	2.909×10^{-4}	radians.
" "	60	seconds (angle).
months	30.42	days.
"	730	hours.
"	43.800	minutes.
"	2.628×10^{6}	seconds.
myriagrams	10	kilograms.
myriameters	10	kilometers.
myriawatts	10	kilowatts.
Ohms	10^{9}	abohms.
"	10^{-6}	megohms.
"	10^{6}	microhms.
"	$1/9 \times 10^{-11}$	statohms.
ohms per meter-gram . . .	$10^{5}/\delta$	abohms per cm. cube.
" " " " . . .	$10^{2}/\delta$	microhms per cm. cube.
" " " " . . .	$39.37/\delta$	microhms per in. cube.
" " " " . . .	$601.5/\delta$	ohms per mil foot.
ohms per mil foot	166.2	abohms per cm. cube.
" " " "	0.1662	microhms per cm. cube.
" " " "	0.06524	microhms per inch cube.
" " " "	$1.662 \times 10^{-3}\delta$	ohms per meter-gram.
ounces	16	drams.
"	437.5	grains.
"	28.35	grams.
"	0.0625	pounds.
ounces (fluid)	1.805	cubic inches.
" "	0.02957	liters.

Multiply	by	to obtain
ounces (troy)	480	grains.
" "	31.10	grams.
" "	20	pennyweights (troy).
" "	0.08333	pounds (troy).
ounces per square inch	0.0625	pounds per square inch.
Pennyweights (troy)	24	grains.
" "	1.555	grams.
" "	0.05	ounces (troy).
perches (masonry)	24.75	cubic feet.
pints (dry)	33.60	cubic inches.
pints (liq.)	473.2	cubic centimeters.
" "	1.671×10^{-2}	cubic feet.
" "	28.87	cubic inches.
" "	4.732×10^{-4}	cubic meters.
" "	6.189×10^{-4}	cubic yards.
" "	0.125	gallons.
" "	0.4732	liters.
poundals	13,826	dynes.
"	14.10	grams.
"	0.03108	pounds.
pounds	444,823	dynes.
"	7000	grains.
"	453.6	grams.
"	16	ounces.
"	32.17	poundals.
pounds (troy)	0.8229	pounds (av.).
pound-feet	1.356×10^{7}	centimeter-dynes.
" "	13,825	centimeter-grams.
" "	0.1383	meter-kilograms.
pounds-feet squared	421.3	kgs.-cms. squared.
" " "	144	pounds-inches squared.
pounds-inches squared	2.926	kgs.-cms. squared.
" " "	6.945×10^{-3}	pounds-feet squared.
pounds of water	0.01602	cubic feet.
" " "	27.68	cubic inches.
" " "	0.1198	gallons.
pounds of water per min.	2.669×10^{-4}	cubic feet per sec.
pounds per cubic foot	0.01602	grams per cubic cm.
" " " "	16.02	kgs. per cubic meter.
" " " "	5.787×10^{-4}	pounds per cubic inch.
" " " "	5.456×10^{-9}	pounds per mil foot.
pounds per cubic inch	27.68	grams per cubic cm.
" " " "	2.768×10^{4}	kgs. per cubic meter.
" " " "	1728	pounds per cubic foot.
" " " "	9.425×10^{-6}	pounds per mil foot.
pounds per foot	1.488	kgs. per meter.
pounds per inch	178.6	grams per cm.
pounds per mil foot	2.306×10^{6}	grams per cubic cm.
pounds per square foot	4.725×10^{-4}	atmospheres.
" " " "	0.01602	feet of water.
" " " "	1.414×10^{-2}	inches of mercury.
" " " "	4.882	kgs. per square meter.
" " " "	6.944×10^{-3}	pounds per square inch.
pounds per square inch	0.06804	atmospheres.
" " " "	2.307	feet of water.
" " " "	2.036	inches of mercury.

Multiply	by	to obtain
pounds per square in. (*cont.*)	703.1	kgs. per square meter.
“ “ “ “ . . .	144	pounds per square foot.
Quadrants (angle)	90	degrees.
“ “	5400	minutes.
“ “	1.571	radians.
quarts (dry)	67.20	cubic inches.
quarts (liq.)	946.4	cubic centimeters.
“ “	3.342×10^{-2}	cubic feet.
“ “	57.75	cubic inches.
“ “	9.464×10^{-4}	cubic meters.
“ “	1.238×10^{-3}	cubic yards.
“ “	0.25	gallons.
“ “	0.9463	liters.
quintals	100	pounds.
quires	25	sheets.
Radians	57.30	degrees.
“	3438	minutes.
“	0.6366	quadrants.
radians per second	57.30	degrees per second.
“ “ “	9.549	revolutions per minute.
“ “ “	0.1592	revolutions per second.
radians per sec. per sec. .	573.0	revs. per min. per min.
“ “ “ “ “ .	9.549	revs. per min. per sec.
“ “ “ “ “ .	0.1592	revs. per sec. per sec.
reams	500	sheets.
revolutions	360	degrees.
“	4	quadrants.
“	6.283	radians.
revolutions per minute . .	6	degrees per second.
“ “ “ . .	0.1047	radians per second.
“ “ “ . .	0.01667	revolutions per second.
revs. per min. per min. . .	1.745×10^{-3}	rads. per sec. per sec.
“ “ “ “ “ . .	0.01667	revs. per min. per sec.
“ “ “ “ “ . .	2.778×10^{-4}	revs. per sec. per sec
revolutions per second . .	360	degrees per second.
“ “ “ . .	6.283	radians per second.
“ “ “ . .	60	revs. per minute.
revs. per sec. per sec. . . .	6.283	radians per sec. per sec.
“ “ “ “ “ . . .	3600	revs. per min. per min.
“ “ “ “ “ . . .	60	revs. per min. per sec.
rods	16.5	feet.
Seconds	1.157×10^{-5}	days.
“	2.778×10^{-4}	hours.
“	1.667×10^{-2}	minutes.
“	1.654×10^{-6}	weeks.
seconds (angle)	4.848×10^{-6}	radians.
spheres (solid angle) . . .	12.57	steradians.
spherical right angles . . .	0.25	hemispheres.
“ “ “ . . .	0.125	spheres.
“ “ “ . . .	1.571	steradians.
square centimeters	1.973×10^{5}	circular mils.
“ “	1.076×10^{-3}	square feet.
“ “	0.1550	square inches.
“ “	10^{-4}	square meters.

Multiply	by	to obtain
square centimeters (*cont.*)	3.861×10^{-11}	square miles.
" "	100	square millimeters.
" "	1.196×10^{-4}	square yards.
sq. cms.-cms. sqd.	0.02402	sq. inches-inches sqd.
square feet	2.296×10^{-5}	acres.
" "	1.833×10^{8}	circular mils.
" "	929.0	square centimeters.
" "	144	square inches.
" "	0.09290	square meters.
" "	3.587×10^{-8}	square miles.
" "	1/9	square yards.
sq. feet-feet sqd.	2.074×10^{4}	sq. inches-inches sqd.
square inches	1.273×10^{6}	circular mils.
" "	6.452	square centimeters.
" "	6.944×10^{-3}	square feet.
" "	645.2	square millimeters.
" "	10^{6}	square mils.
" "	7.716×10^{-4}	square yards.
sq. inches-inches sqd. ..	41.62	sq. cms.-cms. sqd.
" " " " ..	4.823×10^{-5}	sq. feet-feet sqd.
square kilometers	247.1	acres.
" "	10.76×10^{6}	square feet.
" "	1.550×10^{9}	square inches.
" "	10^{6}	square meters.
" "	0.3861	square miles.
" "	1.196×10^{6}	square yards.
square meters	2.471×10^{-4}	acres.
" "	10.76	square feet.
" "	1550	square inches.
" "	3.861×10^{-7}	square miles.
" "	1.196	square yards.
square miles	640	acres.
" "	27.88×10^{6}	square feet.
" "	2.590	square kilometers.
" "	3.098×10^{6}	square yards.
square millimeters	1.973×10^{3}	circular mils.
" "	0.01	square centimeters.
" "	1.550×10^{-3}	square inches.
square mils	1.273	circular mils.
" "	6.452×10^{-6}	square centimeters.
" "	10^{-6}	square inches.
square yards	2.066×10^{-4}	acres.
" "	9	square feet.
" "	1296	square inches.
" "	0.8361	square meters.
" "	3.228×10^{-7}	square miles.
statamperes	$1/3 \times 10^{-10}$	abamperes.
"	$1/3 \times 10^{-9}$	amperes.
statcoulombs	$1/3 \times 10^{-10}$	abcoulombs.
"	$1/3 \times 10^{-9}$	coulombs.
statfarads	$1/9 \times 10^{-20}$	abfarads.
"	$1/9 \times 10^{-11}$	farads.
"	$1/9 \times 10^{-5}$	microfarads.
stathenries	9×10^{20}	abhenries.
"	9×10^{11}	henries.
"	9×10^{14}	millihenries.
statohms	9×10^{20}	abohms.

Multiply	by	to obtain
statohms (*cont.*)	9×10^5	megohms.
"	9×10^{17}	microhms.
"	9×10^{11}	ohms.
statvolts	3×10^{10}	abvolts.
"	300	volts.
steradians	0.1592	hemispheres.
"	0.07958	spheres.
"	0.6366	spherical right angles.
steres	10^3	liters.
Temp. (degs. Cent.)+273	1	abs. temp. (degs. Cent.).
" " " +17.8	1.8	temp. (degs. Fahr.).
temp. (degs. Fahr.)+460	1	abs. temp. (degs. Fahr.).
" " " −32	5/9	temp. (degs. Cent.).
tons (long)	1016	kilograms.
" "	2240	pounds.
tons (metric)	10^3	kilograms.
" "	2205	pounds.
tons (short)	907.2	kilograms.
" "	2000	pounds.
tons (short) per sq. ft.	9765	kgs. per square meter.
" " " " "	13.89	pounds per square inch.
tons (short) per sq. in.	1.406×10^6	kgs. per square meter.
" " " " "	2000	pounds per square inch.
Volts	10^8	abvolts.
"	1/300	statvolts.
volts per inch	3.937×10^7	abvolts per cm.
" " "	1.312×10^{-3}	statvolts per cm.
Watts	0.05688	B.t. units per min.
"	10^7	ergs per second.
"	44.27	foot-pounds per min.
"	0.7378	foot-pounds per second.
"	1.341×10^{-3}	horse-power.
"	0.01433	kg.-calories per minute.
"	10^{-3}	kilowatts.
watt-hours	3.413	British thermal units.
" "	2656	foot-pounds.
" "	1.341×10^{-3}	horse-power-hours.
" "	0.860	kilogram-calories.
" "	367.2	kilogram-meters.
" "	10^{-3}	kilowatt-hours.
webers	10^8	maxwells.
weeks	168	hours.
"	10,080	minutes.
"	604,800	seconds.
Yards	91.44	centimeters.
"	3	feet.
"	36	inches.
"	0.9144	meters.
"	5.682×10^{-4}	miles.
years (common)	365	days.
" "	8760	hours.
years (leap)	366	days.
" "	8784	hours.

INDEX

Note. — Numbers refer to pages.

Jesse J Mazie